State-Of-The-Art Materials Science in Belgium 2017

State-Of-The-Art Materials Science in Belgium 2017

Special Issue Editor

Dirk Poelman

MDPI • Basel • Beijing • Wuhan • Barcelona • Belgrade

MDPI

Special Issue Editor
Dirk Poelman
Ghent University
Belgium

Editorial Office
MDPI
St. Alban-Anlage 66
Basel, Switzerland

This is a reprint of articles from the Special Issue published online in the open access journal *Materials* (ISSN 1996-1944) in 2018 (available at: http://www.mdpi.com/journal/materials/special_issues/ Belgium_17)

For citation purposes, cite each article independently as indicated on the article page online and as indicated below:

LastName, A.A.; LastName, B.B.; LastName, C.C. Article Title. *Journal Name* **Year**, *Article Number*, Page Range.

ISBN 978-3-03897-242-6 (Pbk)
ISBN 978-3-03897-243-3 (PDF)

Cover image courtesy of Nicolas Eshraghi and Bénédicte Vertruyen.

Contents

About the Special Issue Editor

Dirk Poelman studied physics and obtained his Ph.D. on rare earth doped sulfide electroluminescent thin films at the Department of Solid State Sciences of Ghent University (Belgium). Since 2001, he has been a tenured member of staff at the same department, where he is leading the research group, Lumilab. As senior full professor, Dirk Poelman has experience in different fields of solid state physics research, including thin film deposition and optical characterization, photo-, electro- and cathodoluminescent materials, structural and electrical defects in semiconductors, photocatalysis for air purification, x-ray analytical techniques and human vision. His current research is focused on inorganic phosphors for white LEDs and displays, and persistent luminescent materials for safety illumination and medical imaging. Dirk Poelman is editor of the Journal of Luminescence and academic editor of *Materials*.

Preface to "State-Of-The-Art Materials Science in Belgium 2017"

This book is a collection of both review and original research papers which appeared in a Special Issue of the open access journal, Materials, "State-of-the-Art Materials Science in Belgium 2017". It covers a wide range of selected material topics, currently under investigation at universities throughout Belgium. As a country, Belgium has hardly any physical resources which can be exploited in materials research and development or industry. However, the resource drawback is compensated by focussing on highly technological fields of research, not requiring vast amounts of raw material. This 'high-tech' approach includes the development of new analytical methods for the characterization of materials, research into new advanced functional materials, as well as defining new industrial processes for existing materials. As such, the book presents a contemporary view on materials research activities in Belgium.

Dirk Poelman
Special Issue Editor

Review

The Use of Municipal Solid Waste Incineration Ash in Various Building Materials: A Belgian Point of View

Aneeta Mary Joseph [1,2], Ruben Snellings [3], Philip Van den Heede [1], Stijn Matthys [1] and Nele De Belie [1,*]

[1] Magnel Laboratory for Concrete Research, Department of Structural Engineering, Faculty of Engineering and Architecture, Ghent University, Tech Lane Ghent Science Park, Campus A, Technologiepark Zwijnaarde 904, B-9052 Ghent, Belgium; aneetamary.joseph@ugent.be (A.M.J.); philip.vandenheede@ugent.be (P.V.d.H.); stijn.matthys@ugent.be (S.M.)

[2] Strategic Initiative Materials (SIM vzw), Project ASHCEM within the Program "MARES", Tech Lane Ghent Science Park, Campus A, Technologiepark Zwijnaarde 935, B-9052 Ghent, Belgium

[3] Sustainable Materials Management, VITO, Boeretang 200, 2400 Mol, Belgium; ruben.snellings@vito.be

* Correspondence: nele.debelie@ugent.be; Tel.: +32-9-264-5522

Received: 19 November 2017; Accepted: 3 January 2018; Published: 16 January 2018

Abstract: Huge amounts of waste are being generated, and even though the incineration process reduces the mass and volume of waste to a large extent, massive amounts of residues still remain. On average, out of 1.3 billion tons of municipal solid wastes generated per year, around 130 and 2.1 million tons are incinerated in the world and in Belgium, respectively. Around 400 kT of bottom ash residues are generated in Flanders, out of which only 102 kT are utilized here, and the rest is exported or landfilled due to non-conformity to environmental regulations. Landfilling makes the valuable resources in the residues unavailable and results in more primary raw materials being used, increasing mining and related hazards. Identifying and employing the right pre-treatment technique for the highest value application is the key to attaining a circular economy. We reviewed the present pre-treatment and utilization scenarios in Belgium, and the advancements in research around the world for realization of maximum utilization are reported in this paper. Uses of the material in the cement industry as a binder and cement raw meal replacement are identified as possible effective utilization options for large quantities of bottom ash. Pre-treatment techniques that could facilitate this use are also discussed. With all the research evidence available, there is now a need for combined efforts from incineration and the cement industry for technical and economic optimization of the process flow.

Keywords: MSWI bottom ash; beneficiation; supplementary cementitious materials (SCMs); ceramics; clinker production; alternate fuel and raw materials (AFR)

1. Introduction

In the present consumer society, mass production and consumption of goods leads to the generation of large quantities of municipal waste. Initially, little attention was paid to the environmental burden of mass consumerism, and municipal waste was destined for disposal in dumps. Since the 1960s, increasing societal concern about health and ecological risks have invoked the implementation of advanced waste management systems in developed countries. The introduction of legislation and public waste management agencies aims to reduce the impact of waste disposal and control emissions of pollutants to the environment. Implementation of waste policies is one of the key policies of the European Commission. Waste reduction, collection, separation of recyclable or compostable materials and disposal of the residue to (sanitary) landfill or incineration facilities have now become common practice in solid waste management in many developed, densely-populated areas. In this respect, the hierarchy of waste management, also called Lansink's ladder of reduce-reuse-recycle adopted as the European Union (EU) waste hierarchy, can lead to better utilization of resources and reduction of waste. In addition, the circular economy policy adopted by the EU urges closing the material use loop by utilizing 'wastes' for suitable applications [1]. Urban waste generation has surged to 1.2 kg/cap/day and total generation to around 1.3 billion tonnes per year, out of which, around 15% is incinerated. The percentage incinerated is as high as 62% in industrialized countries [2–6].

The main types of wastes subjected to incineration are municipal wastes, non-hazardous industrial wastes, hazardous wastes, sewage sludge and clinical wastes. Mostly, all non-recyclable non-hazardous wastes (MSW, waste from paper and wood industry, etc.) are co-incinerated and constitute the majority of the volume. Further discussions are about the residues from this fraction. The incineration process generates heat, and this is recovered and utilized as such or converted to electrical energy. In some cases, incineration is used solely for disinfection and volume reduction. Waste gets reduced to 10–15% of its volume and 20–35% of its weight after incineration [7]. The fate of the residues depends mostly on their environmental quality. Bottom ash is usually classified as non-hazardous and can thus be considered for various applications such as in road construction as a subbase material. Finer fractions (mainly boiler and fly ashes) can be hazardous due to salts, heavy metals, biocides and organics. Fly ash accounts for about 1–3% of the total residues [6]. Until the 1980s, major concern went to the gaseous emissions from the incinerators. Emissions were monitored for particulate matter, acidic gases including HCl, HF, HBr, HI, SO_2, NOx, NH_3, greenhouse gases etc., heavy metals like Hg, Cd, Tl, As, Ni, Pb, etc., and organics like CO, VOCs, polychlorinated dibenzo-p-dioxins (PCDDs), polychlorinated dibenzofurans (PCDFs), etc. In response, improved flue gas treatment systems were implemented, and the focus shifted to the management of the solid incineration residues instead [8].

Landfilling is a 'go to' solution for wastes that do not have any other destination; however, it has many disadvantages. Maintaining an engineered sanitary landfill is a costly affair, and it is a dead end for all the valuable resources in it. Furthermore, its land requirement, as well as the secondary pollution possibilities like pollution of groundwater due to leaching and release of landfill gases like methane, which has a high global warming potential, make it the least favorable waste disposal method. In response, the European Union published a directive in 1999 that aims to ban landfilling for all but hazardous waste. Furthermore, countries such as the United States and New Zealand instituted a landfill tax to prevent the recyclable resources being landfilled [5].

A prime target for utilization of incineration residues is construction materials. Both in terms of market volumes and the required technical quality, there is clear compatibility. Currently, bottom ashes are mainly used as road subbase material and building sand and gravel alternatives. Additionally, in Belgium, this utilization is limited to 102 kT out of around 400 kT produced. The reasons for this limited utilization and options for improving this situation are discussed further. Substituting primary resources by secondary raw materials closes material cycles and reduces natural resource depletion, thus contributing to the realization of a circular economy. Furthermore, the embodied energy related to initial production of recycled materials is often lower than that of primary materials. In addition to the depletion of raw materials, the impact of resource mining needs to be taken into account [9]. Use of MSWI ash in building materials can be considered as a viable alternative to landfilling, given the past successes in recycling of different waste streams (e.g., fly ash, blast-furnace slag, phospho-gypsum, bauxite fines, etc., [10].

2. Solid Waste Management and Incineration Ash Recycling in Belgium

In Europe, the European commission has set waste legislation, mainly divided into five subcategories and organizes various action plans as guidance for sustainable waste management for its member states. It provided the revised waste framework directive in 2008 that focusses on recycling at least 70% construction waste by 2020. The circular economy is one of the main policies and action plans that focusses on reducing landfilling to a minimum [11]. Belgium is politically divided into three regions, Flanders, Wallonia and Brussels Capital Region (BCR), for administrative purposes, and there are significant differences between the waste management policies of these three regions. In Flanders, waste management is carried out and regulated by OVAM (de Openbare Vlaamse Afvalstoffenmaatschappij), the public waste agency of Flanders. Two agencies responsible for waste management in BCR are Bruxelles Environnement and Agence Bruxelles Propreté. Enforcing extended producer responsibility has been the main strategy employed to improve the recycling rates in BCR. In Wallonia, the waste management is regulated by Département du Sol et des Déchets-DSD (Soil and Waste Department) (Wallonia, Belgium). Out of five million tons of MSW generated in Belgium in 2010, 9% was contributed by the Brussels Capital Region, 60% by Flanders and 31% by Wallonia [12].

Overall, Belgium is a front runner in achieving sustainability in waste management. It can be seen from Figure 1a that MSW generation is lower compared to the EU average. Figure 1b represents the comparison of waste treatment techniques in various countries. This includes other waste streams in addition to municipal solid wastes. Furthermore, the recycling percentage is much higher than average, while that of landfilled and incinerated waste is much lower than the EU-28 average. The quantities of municipal solid wastes that underwent different treatments are depicted in Figure 2a. Policy measures have been the major driver to increase the recycling level, in part by enforcing incineration and landfilling taxes, but also by defining environmental quality requirements that need to be met.

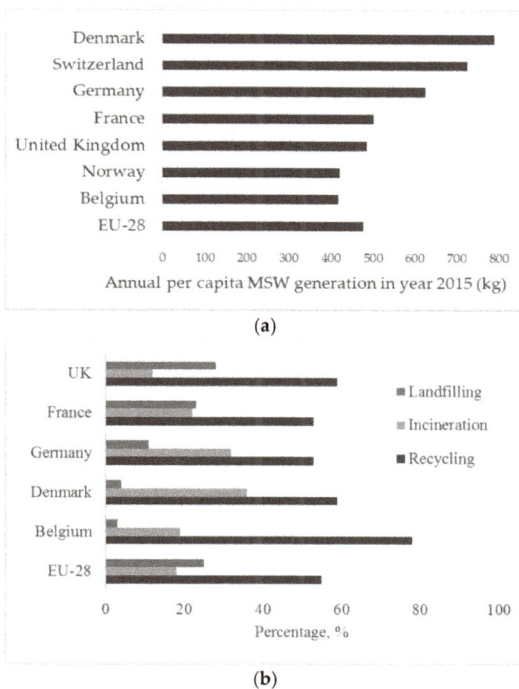

(a)

(b)

Figure 1. (**a**) Annual per capita generation of MSW in Belgium compared to the average value of EU nations and other prominent neighbors; (**b**) percentage of recycled, incinerated and landfilled waste cf. [13].

In Flanders, the residues generated by incineration need to be treated to facilitate safe recycling as set out by VLAREMA (Vlaams reglement betreffende het duurzaam beheer van materiaalkringlopen en afvalstoffen), which is the Flemish implementation order of the sustainable materials management decree. The VLAREMA order describes the legal requirements (so-called end-of waste criteria) for waste materials to lose their "waste" status and become resources (obtain an end-of-waste certificate).

VLAREMA specifically describes the environmental quality requirements for the use of recycled material as a construction material, by limiting the cumulative leached heavy metals at a liquid to solid ratio of 10, when tested according to the procedure in Compendium voor monsterneming en analyse CMA/II/A9.1. At present, these requirements are defined as contaminant leaching limits for metals and anions such as Cl and F and total concentrations for organic contaminants. In Wallonia, 'Arrêté du Gouvernement wallon favorisant la valorisation de certains déchets', published by the Walloon government, specifies the quality of bottom ash to be reused. It suggests to use the cleaned bottom ashes in shaped applications, and also, the environmental criteria limits with respect to leaching are more lenient compared to those of Flanders [14].

Two types of installations for bottom ash treatment are used in Europe, dry and wet installations. Dry installations give dry aggregates as end products with specific sizes. The processing consists of the following steps including cooling down the temperature of ash in air, primary ferrous metal separation, crushing, sieving, secondary ferrous separation, non-ferrous separation and ageing. Wet installations have an additional washing step to reduce the leachability of the final product, and the <2-mm fraction containing most contaminants is usually landfilled. However, the <2-mm portion constitutes around 50% of the total bottom ash generated. Incineration plants in Belgium adopt both wet and dry bottom ash treatment installations at different locations. At present, only waste incineration bottom ashes are treated in Belgium. The treatment processes adopted usually involve the recovery of ferrous metals by magnetic separation using over band magnets, which enables 55–60% ferrous metal recovery, mostly the coarser scrap. A ferrous metal content of 1.3–25.8% is typically left behind in the residue, mostly being fine or entangled metals. The non-ferrous metals (Al, Cu, Pb, etc.) can be recovered by eddy current separation, which also gives around a 50% recovery rate. The granulates are aged in open air to ensure immobilization of heavy metals and are then utilized as building materials, mainly as road subbase material [6,15].

According to Eurostat data, annual per capita generation of MSW in Belgium in 2015 was 475 kg (Figure 1), which translates to a total generation of 4.7 MT [16]. Thirty five percent of these wastes are incinerated in around 17 incineration installations [6]. In Flanders, 401 kT (Figure 2b) bottom ash are generated. A part of it is utilized in Flanders, a part outside Flanders and the rest outside Belgium; 174 kT bottom ash are processed, 24 kT as an alternative for gravel in Flanders; 72 kT are processed in Flanders and utilized in The Netherlands; 47 kT processed (in two treatment installations, one wet and one dry) and utilized in Germany; and 164 kT are landfilled [17]. Out of the processed fraction, 39 kT are used as alternative building sand in Flanders; 16 kT utilized in The Netherlands; and 40 kT are landfilled. To sum up, a total of 63 kT (15%) is utilized in Flanders; 134 Kt (34%) are utilized outside Flanders; and about 204 Kt (51%) are landfilled [18]. Utilization in Flanders is done as a road subbase material, landfill finishing material and elevation material for dike cores [17]. Furthermore, 134 kT are transported to the Netherlands and Germany due to stricter environmental regulations in Belgium, especially narrowed down by the allowable leaching levels of mainly copper (Cu) and at times zinc (Zn) and lead (Pb). The allowable content of Cu for shaped applications is 25 mg/m^2 in Belgium, while the value is as high as 98 mg/m^2 in Netherlands [19]. Possible beneficiation techniques to improve the utilization potential of these in specific applications are discussed further [13,18]. Treated bottom ashes also find application as aggregate for concrete blocks of dimension $150 \times 75 \times 40$ cm that are lighter than the conventional concrete products [20]. It is to be noted that the bottom ash quantities described here are not only from municipal waste, but also from certain other non-hazardous waste streams. In Flanders, 62 kT of fly ash are used as an alternative building material. In Wallonia, part of the bottom ash that conforms to environmental stipulations is used in road construction, and the rest (usually exceeding the limit molybdenum content) is used in the cement industry [17].

(a)

(b)

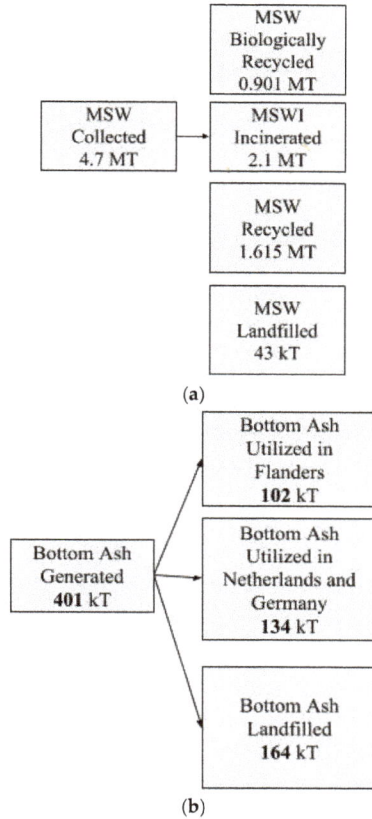

Figure 2. (a) Quantity of MSW generated and that composted, incinerated, recycled and landfilled in Belgium in 2015 [18]; (b) mass balance of bottom ash residue generated in Flanders in 2013 cf. [18].

3. Types of Incinerators and Ashes in Belgium

The quantity of wastes to be treated and the turbulence of waste or waste fuel mix undergoing incineration required determine the type of incinerator to be used. Types of incinerators used in MSW incineration plants are stationary hearth incinerators, rotary kilns, stationary and moving grates, modular/starved air systems, pyrolysis, gasification and fluidized bed systems [6]. A stationary hearth incinerator is used when the quantity of waste to be treated is very low, for instance in small-scale industrial, commercial and hospital incinerators. It has a refractory floor and air-inlet openings for complete combustion. Rotary kilns have been used in the past to burn MSW and are presently used to burn clinical wastes [6]. They consist of a steel cylinder lined with either refractory bricks or water tubes for cooling, inclined slightly towards the discharge end. Movement of the waste burned is controlled by the speed of rotation. The grate furnace is the most popular incinerator used now for mass burning of wastes. Types of grate systems include reciprocating grate, rocking grate, vibrating oscillating and impact grate, travelling grate and drum grate systems. Grates stay at a slope determined by the required residence time of the waste on the grates. The fluidized bed furnace consists of a homogeneous medium like sand, with air jets from below keeping the sand fluidized in high turbulence making the burn more homogeneous [21]. About 90% of the municipal solid wastes in Europe are incinerated in a grate incinerator [6]. All the incineration plants in Belgium have at least one grate furnace unit, and some have rotary kiln and a fluidized bed in addition to that [22]. There are 8 units with wet flue gas treatment systems, 9 with semidry and 2 with dry systems [6].

The composition of the residues depend on the input waste and the process parameters of the incinerator, even though it has little effect on the environmental quality of the material. The temperature profile in the incinerator has a large influence on the distribution of the volatile elements among different size fractions. Belevi and Langmeier found out that around 70% of zinc is transferred to the gaseous phase by 700 °C and about 80% by 800–900 °C. Eighty three percent of Sb is transferred to the gaseous phase at 500 °C and about 65% at 700 °C and 25% at 900 °C. About 67% Cd is volatilized at 700 °C, and about 86% is volatilized at 900 °C. The presence of chlorides enables the volatilization of heavy metals [23].

The ash or residue from the incinerator is named based on the point of collection of the ash from the incinerator. The International Ash Working Group (IAWG) recommended a nomenclature as given in Table 1. Large volumes of the total residues are composed of bottom ashes (more than 90%), and these contain least salts, heavy metals and POPs and are therefore best suited for recycling as construction material. Only 1–3% is fly ashes, and these are rich in chlorides, sulfates, PCDDs/PCDFs and other heavy metals, making them less suitable to be recycled [6]. Prior to recycling, all MSWI ashes require beneficiation treatments, the extent and intensity of which depend on the ash properties and the targeted use.

Table 1. Nomenclature of waste streams from MSWI ash [8,24].

Name	Point of Collection
Grate ash	Ash collected from the grate
Grate siftings	Material collected from the hoppers underneath the grate
Bottom ash	Combined grate ash and grate siftings and sometimes heat recovery ash; it is mainly composed of bottle glass, metals, ceramics and organic residues [25]
Heat recovery system ash (HRA)	Ash collected from boiler, economizer and super heater
Fly ash	Raw particulate matter entrained in the flue gas stream prior to addition of scrubbing reagents. It is a type of Air Pollution Control residue
Air pollution control (APC) residue	All particulate material captured downstream of any reagent injection and prior to discharge of gases to stack; its reuse will be more difficult due to the significant presence of heavy metals and toxic compounds like Polychlorinated dibenzo-p-dioxins (PCDDs) and polychlorinated dibenzofurans (PCDFs) [26]
Combined ash	Mixture of bottom ash, grate siftings and APC residues
Sintered ash	Bottom ash or fly ash is sometimes sintered and solidified, thus reducing the leaching and facilitating utilization

4. Characteristics of Incineration Ashes

The physical and chemical characteristics of ash depend on the type and operating parameters of the incinerator in which it is produced. Hyks and Astrup observed changes in bulk composition of bottom ashes with the change of input composition, but neither a change in input composition nor operational parameters had an effect on the leaching of heavy metals from bottom ashes [27].

4.1. Physical Properties

Bottom ash has dry densities of 0.95–1.75 g/cm^3 and a specific gravity of 1.1–2.7 [24,28–33]. The moisture content of the ash depends on post-combustion treatments and storage methods. Moisture is important for dust control and also for proper compaction. Loose bulk density of MSWI fly ashes decreases with decreasing ash separation temperature (temperature at which the fly ash is separated in the process flow) [34].

The morphology depends mainly on the type of incinerator and the temperature profile to which it was subjected. Bottom ash is usually very heterogeneous and irregularly shaped and sometimes contains vesicles formed as a result of melting and quenching, as can be seen from the scanning electron microscopy (SEM) images in Figure 3. From photo micrograph analysis, bottom ash particles were distinguished into two classes: a melt phase composed of amorphous glass phases and a loose fragmented phase from water quenching, which is mainly composed of hydrates and carbonates, which can be seen in Figure 4 [35]. Inkaew et al. studied the effect of quenching of bottom ash on the mineralogy and morphology of bottom ashes. They point out that the main quench products are amorphous and microcrystalline calcium-silicate-hydrate phases, portlandite, hydrocalumite and Friedel's salt, and they contain more chloride compared to other parts [36]. Image analysis indicates additional heterogeneities at the particle level with one particle often consisting of several phases with contrasting chemistries. Agglomeration, encrustation and rim formation add to the microstructural complexity and heterogeneity of the material [36]. In addition to phases formed during melting and quenching, bottom ashes show a large and heterogeneous residual fraction consisting of ceramic materials such as brick, stone, glass, ferrous and non-ferrous metals, non-combustible inorganic material and some residual organic matter [37]. For fly ashes, the morphology is affected by its separation temperature. The Blaine specific area of fly ash increases with decreasing separation temperature. At higher temperatures, the morphology of particles is more spherical. However, as temperature decreases, crystalline salts precipitate on silicate spheres, and irregular particles dominate [34,38].

Figure 3. SEM images of bottom ash. (**a**,**b**) shows the irregularly-shaped bottom ash particles, which are porous in nature; (**c**) shows crystallized anhydrite or gypsum on the surface of bottom ash particles; (**d**) shows rhombohedrally-shaped calcite crystals and other calcium-based minerals [35].

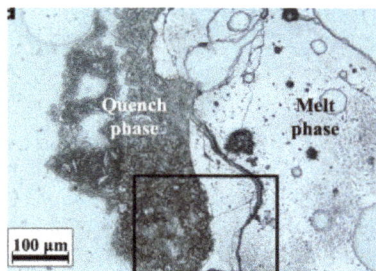

Figure 4. Photomicrographs of bottom ash particle differentiating the melt phase and quench phase [35].

4.2. Chemical Properties

4.2.1. Chemical Analysis by XRF

Obviously, the bulk composition of residues varies from place to place and over time in an incinerator facility. The major chemical components in ash are CaO, SiO_2 and Al_2O_3, which is illustrated in the Ca-Si-Al ternary diagram in Figure 5. In addition, minor components also have a significant effect on the technical and environmental quality of the material. The variation in minor components is due to the input waste characteristics and operational parameters [39]. Hyks and Astrup observed that the input of road salts and poly vinyl chloride (PVC) resulted in more chloride (Cl^-), impregnated wood resulted in higher content of arsenic (As), automotive shredder residues induced more barium (Ba), copper (Cu), molybdenum (Mo), nickel (Ni), antimony (Sb), tin (Sn) and zinc (Zn) and batteries increased the content of cadmium (Cd), cobalt (Co), sulfur (S) and strontium (Sr) [27]. The increase in combustion temperature results in the increase of pH due to conversion of calcite to portlandite [17]. The fly ash depicted in Figure 5 is that from thermal power plants and not from incinerators. This shows the bottom ash compositions superimposed on compositional regions of other residues that are commonly used as construction materials. Clearly, the bottom ashes cover a wide range of compositions that are closest to that of iron blast furnace slags and of coal combustion fly ashes.

The chemical composition of MSWI fly ashes depends on their separation temperature from the flue gas line. Keppert et al. studied the effect of separation temperatures on the chemical composition of three fly ashes with separation temperatures higher than 700 °C, between 500 and 700 °C and between 250 and 300 °C. It was observed that the content of soluble salts (Cl^- and SO_3^{2-}) and most heavy metals (As, Cd, Pb, Sb) increases with decreasing separation temperature [34].

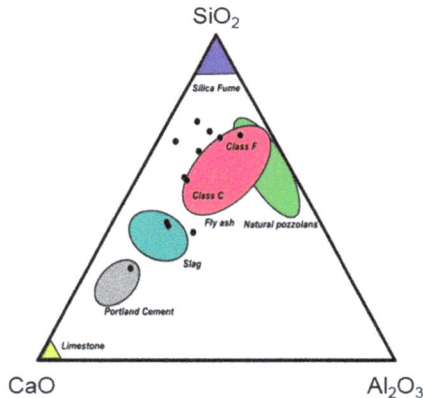

Figure 5. Composition of bottom ash from various incinerator facilities (black dots) superimposed on other materials commonly used in construction cf. [6,28,32,35,40–50].

4.2.2. Mineralogical Composition

The mineralogical composition often determines the technical properties and the end-use of a resource. Relative proportions of the constituent minerals are the underlying factors explaining why some materials are suitable for use as a reactive cement component and others are not. The major and minor mineral constituents reportedly found in bottom ashes, fly ashes and sintered ashes and their significance in various utilization scenarios are tabulated in Table 2. Ageing effects major changes in mineralogy, including the formation and increase of minerals such as calcite, quartz, sulfates and ettringite [48]. Ettringite and hydrocalumite are formed due to the quenching of bottom ash.

Table 2. Mineralogy of bottom ashes, fly ashes and sintered ashes from MSWI ash.

Minerals Identified	Reference	Occurrence and Potential Use
Quartz (SiO_2)	[26,33–35,49,51]	It acts as an inert filler when used in cement as SCM. Furthermore, it can have pozzolanic properties when very finely ground. It can be a source of silica when used as a cement raw material.
Calcite ($CaCO_3$)	[33–35]	It can contribute carbonate to the system, leading to stabilization of ettringite and mono-carbonate/hemi-carbonate when used as an SCM, depending on the content of C_3A. The rest of the calcite will act as a filler. Calcite is the commonly-used source of calcium, thus highly beneficial for clinker production.
Gehlenite ($Ca_2Al_2SiO_7$)	[26,33,35,51]	Inert constituent in calcium aluminate cements, carbonatable.
Hematite (Fe_2O_3)	[26,33,34]	Largely inert, formed during incineration.
Magnetite	[25,49]	High temperature phase/inert.
Ettringite	[49]	Mainly formed by quenching of bottom ash, from the reaction between sulfates and reactive aluminates.
Hydrocalumite	[49]	Mainly formed by quenching of bottom ash.
Diopside ($CaMgSi_2O_6$)	[33]	Principal crystalline phase of sintered ash.
Clinoenstatite ($MgSiO_3$)	[33]	Found in sintered ash/ceramics.
Wollastonite ($CaSiO_3$)	[33]	Found in sintered ash/ceramics.
Ingersonite (γ-Ca_2SiO_4)	[50]	Reactive towards CO_2.
Hedenbergite	[25]	Slag/ash component-inert.
Ferrohedenbergite	[25]	Slag/ash component-inert.
Feldspar	[25]	Common inert rock-forming mineral.
Melilite $(Ca,Na)_2(Al,Mg,Fe^{2+})[(Al,Si)SiO^7]$	[25,49]	Contains Mg; carbonatable.
Albite ($NaAlSi_3O_8$)	[33,34]	Found in sintered ash/ceramics.
Anorthite ($CaAl_2Si_2O_8$)	[35,47,51]	Common inert rock-forming mineral.
Anhydrite ($CaSO_4$)	[26,34,51]	Cement constituent, added to control setting.
Gypsum ($CaSO_4 \cdot 2H_2O$)	[36,49]	Cement constituent, added to control setting.
Gismondine ($CaAl_2Si_2O_8 \cdot 4H_2O$)	[50]	
Apatite($Ca5(PO_4)_3(OH,F,Cl)$)	[26]	Fly ash treated by washing, phosphation and calcination to 750 °C. Bone fragments can also be a source of apatite in ash.
Whitlockite (β-$Ca_3(PO_4)_2$)	[26]	Fly ash treated by washing, phosphation and calcination to 750 °C.
Titanite ($CaTiSiO_5$)	[26]	Fly ash treated by washing, phosphation and calcination to 750 °C.
Perovskite ($CaTiO_3$)	[26,52]	Inert.
Periclase	[25]	Carbonatable.

Apatite group minerals are formed in ashes as a result of phosphation. They are stable minerals, and during their formation, they encapsulate heavy metals. Gypsum and anhydrite are not inert; for instance, when used along with cement, they will affect the hydration reaction unlike most of the other crystalline minerals present in ash.

5. Beneficiation Needs for Use of MSWI Ash in Building Materials

5.1. Metallic Aluminum and Zinc

The various sources of aluminum in municipal waste include beverage cans, aluminum foils, nails, etc., and as a result of attrition within the waste, their size gets reduced greatly. The eddy current separation technique (cf. infra) can be used to recover the aluminum particles from the waste in incineration plants. However, small particles will not be recovered and will thus remain in the ash. When it is mixed in together with alkaline water, e.g., for utilization as aggregate in concrete or as a supplementary cementitious material, the aluminum in ash will oxidize and release hydrogen gas. This gas formation results in a high porosity of the matrix, and if continuous moisture is supplied, the reaction will continue after concrete hardening and result in spalling and cracking. The resulting aluminum hydroxide generated was reported to exist in different forms in the matrix: amorphous $Al(OH)_3$, bayerite, gibbsite and boehmite [46]. It is reported that the structure of the product formed depends on the rate of crystallization. Rapid crystallization leads to the formation of bayerite, and slow crystallization leads to the formation of gibbsite [53]. Elemental Zn present in ash also reacts similarly to generate hydrogen gas and the different stoichiometrical equations of the reactions are given below [41,54–56]

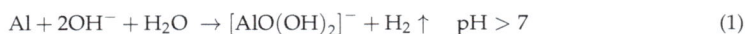

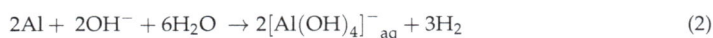

$$Al + 2OH^- + H_2O \rightarrow [AlO(OH)_2]^- + H_2 \uparrow \quad pH > 7 \tag{1}$$

$$2Al + 2OH^- + 6H_2O \rightarrow 2[Al(OH)_4]^-{}_{aq} + 3H_2 \tag{2}$$

$$Zn + 2OH^- \rightarrow ZnO_2{}^{2-} + H_2 \tag{3}$$

$$2Al + Ca(OH)_2 + 2H_2O \rightarrow Ca(AlO_2)_2 + 3H_2 \tag{4}$$

CUR (Civieltechnisch Centrum Uitvoering Research en Regelgeving) recommendation 116:2012 for use of MSWI bottom ash as aggregate in concrete regulates the content of elemental aluminum in the ash to a maximum of 1% by mass of bottom ash. It recommends a test set up to quantify the equivalent elemental aluminum content by measuring the volume of hydrogen released on reaction with 1 M sodium hydroxide solution [56]. It is called an equivalent aluminum because zinc even though present in smaller quantities undergoes a similar reaction, and the hydrogen collected is generated by aluminum and zinc jointly. However, it can be seen from Table 3, which is a theoretical calculation conducted by the authors, that the volume of hydrogen produced is significant even at a concentration of 0.1%. A certain part of this gets trapped inside concrete and reduces the strength by creating porosity.

Table 3. Estimation of hydrogen gas production.

Aluminum in Ash (%)	Theoretical Volume of Hydrogen Produced in 1 m^3 of Concrete (Assuming 25% Replacement of Cement by Ash and 450 kg/m^3 of Cement Content) in m^3; cf. Equation (2) at STP (Standard Temperature and Pressure)
0.1	0.150259
1	1.502592

5.2. Salts

Salts are mainly present in the form of chlorides and sulfates and, in combination with heavy metals, render MSWI fly ash hazardous, which prevents its utilization. It is rarely present in the bottom ash fraction, especially the coarser ones. The chloride content varies between 0.5% and 15% in fly ash and 0.2% and 5% in bottom ash. Most of the chlorides in the bottom ash are present in the finer size fractions [57]. This is due to metal chlorides and sulfides having a lower boiling point than their oxides and phosphates and thus becoming volatilized in the incinerator [58]. These mostly exist as soluble salts and will cause leaching problems when landfilled. Crystalline chlorides in ash exist as halite,

Friedel's salt and hydrocalumite [35]. When ash is used as a supplementary cementitious material, chloride ions released into pore solution accelerate the corrosion of steel and comprise a great threat to steel-reinforced concrete durability [59]. Free chlorides present in concrete react with $Ca(OH)_2$ to form calcium chloride and magnesium hydroxide, which eventually leaches out, exposes reinforcements and also leads to their corrosion [44]. Furthermore, when the chloride content in pore solution exceeds a critical value, the passive layer on the steel reinforcement disintegrates, and localized pitting corrosion is caused.

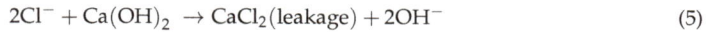

$$2Cl^- + Ca(OH)_2 \rightarrow CaCl_2(\text{leakage}) + 2OH^- \tag{5}$$

When high amounts of sulfates are present, the reaction of aluminate and calcium sulfates forms ettringite and monosulfate. Washing with water is the easiest solution to remove chlorides and sulfates, but it does not remove bound chlorides. To remove chlorides further, thermal treatments such as sintering, roasting and calcination have been tested. A single technique was not sufficient for maximum removal. A combination of thermal treatment and washing was able to remove about 90% of chlorides [35].

5.3. Heavy Metals

The presence of heavy metals in MSWI ash and its leaching into ground water have been a problem for a long time, since it was landfilled or used as a subbase material in road construction. Many countries are adopting source reduction techniques by limiting the toxic trace metals in different products. The heavy metals usually present in MSWI ash are arsenic (As), barium (Ba), cadmium (Cd), chromium (Cr), copper (Cu), molybdenum (Mo), nickel (Ni), lead (Pb), tin (Sn), antimony (Sb), selenium (Se) and zinc (Zn). The leaching of heavy metals is less affected by the change in bulk composition of ash or operating parameters of the plant [27]. Heavy metals are more cumulated in fly ashes than bottom ashes, except Cu, Cr and Pb due to their low volatilities [33,38]. In Flanders, the leaching of various heavy metals is regulated by a column leaching test done according to CMA/II/A.9.1 based on European Committee for Standardisation/Technical Specification, CEN/TS 14405:2004. Both procedures are different mainly in the maximum particle size (4 mm for CMA/II/A9.1 and 10 mm for CEN/TS 14405:2004) and column diameter (5 cm for CMA and 10 cm for CEN). The utilization of bottom ashes in Flanders is limited by the cumulative release of different species at a liquid to solid ratio of 10, and the limits are specified in VLAREMA (Vlaams reglement betreffende het duurzaam beheer van materiaalkringlopen en afvalstoffen). The round robin test reported by Geurts et al. verified the reproducibility of the test to confirm whether it complies with legal limits, even though the reproducibility of absolute leaching values of Cu and Zn was questionable. To achieve greater reproducibility, they recommend using an identical analysis technique for the eluates [60]. Utilization of bottom ash in Flanders is mainly limited by its Cu, Zn and Pb contents being past allowable limits as can be seen from Table 4. Some of these materials are transported to The Netherlands or Germany where more tolerance for these metals exists [61].

Table 4. Comparison of Flemish and Wallonia guidelines and typical concentrations in bottom ash cf. [62,63].

Parameter	Flanders Criteria (VLAREMA) Total Concentration Limit mg/kg Dry Matter	Wallonia Criteria Total Concentration Limit mg/kg Dry Matter	Bottom Ash Total Typical Concentration mg/kg Dry Matter
Arsenic	250	100	33 ± 17
Cadmium	10	8	
Chromium	1250	230	482 ± 73
Copper	375	210	4042 ± 888
Mercury	5	15	3 ± 2
Lead	1250	1150	1899 ± 396
Nickel	250	150	329 ± 69
Zinc	1250	680	5376 ± 782

The presence of metallic aluminum was found to have an effect on the leaching of chromium. The presence of Al in the (0) oxidation state appeared to prevent leaching that happens by reduction of Cr(VI) to Cr III, but the presence of manganese oxides accelerates leaching. This is a conflict of interest since the presence of Al is detrimental to the strength of concrete while advantageous in the case of leaching [64].

5.4. Persistent Organic Pollutants

Certain fly ashes from MSWI plants are classified as hazardous waste due to their dioxin and other POP content. Polychlorinated dibenzo-p-dioxins (PCDDs) and polychlorinated dibenzofurans (PCDFs) dioxins are polychlorinated biphenyls present in MSWI ashes. The most poisonous one is 2,3,7,8—tetra chloro benzo dioxin (TCDD) [65]. They can be either present in the feed, formed from chlorinated precursors such as PCBs, chlorophenols, chlorobenzene, etc., or formed spontaneously from the organics present in the fly ash, by chlorination and oxidation [66]. These are highly toxic substances. If inhaled, they are carcinogenic. Human exposure to these hazardous contaminants is possible by either inhalation of dust or ingestion of contaminated ground water [67]. Even though these compounds are formed by chlorination, the chlorine content in the input has little effect on the quantity of PCDD/Fs compared to the design and operating parameters of the plant [68]. As volatile compounds, they are removed from stack gas using activated carbon as the adsorbent and bag house filtration [69,70]. The addition of chemical suppressants after combustion is an effective method to reduce their formation. Various sulfur and nitrogen compounds have been identified as effective suppressants. In accordance with the Basel Convention on the Control of Transboundary Movements of Hazardous Wastes and Their Disposal of 1992, the limit for atmospheric emission of PCDDs is 0.14 ng total toxic equivalent (TEQ)/Nm^3 at a reference oxygen content of 11%. Total toxic equivalent (TEQ) corresponds to the sum of different cogeners multiplied by their toxic equivalency factor (TEF) values [65].

According to the same source, the limits for waste to be classified as low POP content are:

a. PCBs: 50 mg/kg
b. PCDDs and PCDFs: 15 µg TEQ/kg
c. Aldrin, chlordane, DDT, dieldrin, endrin, heptachlor, hexachlorobenzene (HCB), mirex and toxaphene: 50 mg/kg for each of these POPs.

5.5. Amorphous Silica Content

The amorphous silica content is an obstacle only when incineration ashes are used as aggregate in concrete. The glassy components in ash and alkali in cement react to form alkali silica gel, which has a very high volume and thus causes cracking. However, when metallic aluminum is present, it forms voids in the cement matrix as explained above; the gel formed as a result of alkali-silica reaction occupied these voids, and no cracks were generated, as reported [71].

6. Bottom Ash Pre-Treatment Techniques

MSWI ash is not an industrial product, but a by-product of the incineration of municipal solid wastes. Therefore, it is difficult to control its properties during production, and beneficiation of ash is mandatory for recycling. Metal recovery is the most important treatment step, as it is advantageous both for the economy of the plant and for enabling further use of the ash as a cementitious material. Berkhout et al. reviewed various techniques for metal recovery [72]. Other reasons for pre-treatment are the obstacles specified in Section 5. Some of the pre-treatment techniques are currently employed at an industrial scale, and others are still in the research stage.

6.1. Pre-Treatment Techniques Used at the Industrial Scale

6.1.1. Magnetic Separation

This is the most popular technique to separate materials based on their magnetic properties, ferromagnetic, paramagnetic or diamagnetic, and also their degree of magnetism. There are different types of magnetic separation techniques involving cross-belt magnetic separators, drum magnetic separators or magnetic pulley separators [73]. The magnetic density separation (MDS) technique can be used to separate the aluminum from other non-ferrous heavy metals based on their difference in apparent density in a gradient magnetic field [72]. It is presently used to separate larger fractions, but not employed in the treatment of other finer residues from the incinerator. It is reported that the efficiency of ferrous metal recovery by magnetic separation is as high as 57–83% [74]. De Boom et al. demonstrated the application of this technique to MSWI boiler fly ashes to recover magnetic particles enriched with Cr, Fe, Mn and Ni [75].

6.1.2. Eddy Current Separation

Eddy current separation (ECS) is the preferred method used in incineration plants and waste sorting facilities to separate the non-ferrous metals like aluminum from waste and ash. It has limited efficiency when applied to wet ash. Its efficiency depends on the size of the particles and ranges from up to 100% efficiency for particle sizes greater than 20 mm, down to 0% for those less than 5 mm. The efficiency can be improved by increasing the number of screening steps. On average, the efficiency of aluminum removal from MSWI ash by ECS is around 30% [46,74,76].

6.1.3. Washing

Washing is the simplest beneficiation technique to remove many of the deleterious compounds in MSWI ash preventing its utilization. Washing in alkaline conditions hydrates elemental aluminum and removes certain heavy metals and in acidic conditions removes certain other heavy metals and chlorides and sulfates. Since the washing process carries with itself the huge baggage of secondary pollution of water, some waste water streams have also been tested for this purpose, which resulted in some additional advantages including binding of heavy metals, which are discussed in further sections.

The presence of soluble leachate salts cuts the utilization potential of MSWI ashes. Water washing techniques have been tested for their effectiveness to remove excess chlorides and sulfates and also heavy metals present especially in the fine fractions. Many have reported that the chloride content in ashes can only be reduced to 0.5% by washing due to the presence of insoluble chlorides [35,77–79]. Chen et al. studied the effect of washing with water saturated with calcium hydroxide to control the ionic equilibrium of calcium ions and thus facilitate the dissolution of all chlorides except calcium chloride going into solution [78]. This prevents calcium compounds getting washed away, which would be a disadvantage when used as raw material for cement production. High waste water generation is the main disadvantage of this technique.

6.1.4. Shaking Table

Shaking tables/wet tables such as Wilfley tables are density separation technologies that separate heavy metal particles from other lightweight particles [80]. They have a sloping plank with ribs on the surface. Water/slurry flows perpendicular to the ribs, and the table oscillates parallel to the ribs. The technique is used at the industrial scale, and it successfully separates precious metal particles such as tin, copper, gold, lead, zinc, tungsten, etc., of a size of 50 μm–2 mm [81,82].

6.1.5. Jig Head Separation

These are devices to separate materials in wet condition that works on the principle for density separation. The particles are pushed upwards by pulsating motion, and when the pulse in water drops, the particles settle in order of their densities and thus can be separated. A tank filled with the slurry to be separated is pulsated up and down, and this results in settling of denser particles of gold, chromium, lead, etc., of a size of 75 μm–6 mm, which are collected and separated [81,82].

6.1.6. Ageing

Ageing is an effective and widely-used technique to reduce the leaching of heavy metals. It is a combined process of hydration, carbonation and oxidation. It is an exothermic process and raises the temperature of bottom ash from 70–80 °C [17]. During ageing, more minerals such as ettringite, hydrocalumite, C-S-H, carbonates, sulfates, etc., are formed, which bind heavy metals, thus reducing leaching. Sulfate minerals have lesser stability compared to carbonates and hydroxides [48,49].

6.2. Pre-Treatment Techniques in the Research Stage

6.2.1. Washing with Alkali

Higher alkalinity accelerates the oxidation of aluminum to form hydrogen and also increases the solubility of sulfates [83]. The disadvantages of alkaline washing include the dissolution of fine reactive particles, leading to some loss of cementitious properties during the washing process, and also the high cost of alkali.

6.2.2. Sulfide Rich Effluent

Effluent from waste water treatment contains sulfides generated by sulfate-reducing bacteria. When ash is treated with this sulfide-containing water, the heavy metals are precipitated as insoluble sulfides and are thus stabilized. The stabilization effect of anaerobic effluent also acts through carbonation, by CO_2 dissolved in the effluent. It is reported that the treatment can stabilize Ca, Cu, Pb, S and Zn, whereas it increased leaching of P and had no effect on As, Cr and Mo [84].

6.2.3. Wet Grinding

For utilization of ash as supplementary cementitious material, grinding of bottom ash is required to break it down to particle sizes comparable to that of cement. If the grinding process is done in wet condition, it provides enough water, turbulence and dissolution of alkaline phases like $Ca(OH)_2$, which gives enough pH for the aluminum reaction to occur. Bertolini et al. have reported that the 180-day compressive strength of concrete increased by four-fold compared to that of dry ground ash and was 25% higher than that of 100% ordinary Portland cement (OPC) mix. This was mainly because of the removal of expansive Al [7].

Further, wet milling is identified as an effective technique to stabilize certain heavy metals such as Pb. Wang et al. reported the effect of wet milling in stabilization of Cr, Cu, Zn and Pb for ash from grate furnaces and of Cr and Zn for ash from fluidized bed incinerators without a water extraction procedure [85]. Chen et al. reported stabilization of heavy metals except Cr by wet grinding in a liquid to solid ratio of nine at 93 rpm for 96 h. Wet grinding reduced the median diameter of particles (d50) of ashes from 72.5 μm down to 3.95 μm, as well as the crystallinity of the material. Concurrent fragmentation and agglomeration of the particles results in stabilization of heavy metals [86].

6.2.4. Carbonation

Carbonation is an effective technique to immobilize the heavy metals, if the ash needs to be disposed safely, or when used as an aggregate when it is not milled. Arickx et al. studied the effect of carbonation and different parameters related to it on leaching of heavy metals. Stack gas from the incinerator was also used as a source of CO_2. However, for applications that involve milling, carbonation does not prove to be very effective in immobilizing heavy metals [87]. Aguiar del Toro et al. report an increase in mobility of chloride ions as a result of carbonation [88]. Chlorides exist in MSWI ash, both as soluble salts such as NaCl, KCl, etc., and also in insoluble forms, such as Friedel's salt

($3CaO \cdot Al_2O_3 \cdot CaCl_2 \cdot 10H_2O$) and calcium chlorocalumite ($11CaO \cdot 7Al_2O_3 \cdot CaCl_2$) in heated ashes [35]. Chlorides existing as insoluble Ca-salts can be disintegrated by carbonation and then can be removed by washing.

6.2.5. Phosphation

Phosphation is the process of converting heavy metals into their insoluble phosphate form to prevent leaching into ground water. As a result of phosphation treatment, apatite group minerals are formed. They encapsulate the heavy metals present in the ash and thus stabilize them. Carbon dioxide gas evolves during phosphate reaction. The reaction also results in a reduction of pH, even though fly ash with high calcium hydroxide content is reported to maintain high pH [89,90].

6.2.6. Cement/Other Binder Stabilization

Cement-based stabilization is a widely-used technique to prevent toxic element leaching in landfills. It converts the liquid or loose waste into a monolithic form and also stabilizes chemically-reactive species into more stable forms. The high surface area of C-S-H also facilitates adsorption of species preventing leaching. It has been reported that Cr ion stabilization is most effective in binders with slag replacement. The stabilization is materialized by substitution in calcium aluminate hydrates. Ettringite stabilizes heavy metal oxyanions like chromate, arsenate and selenite. Even though cement stabilization is a technique presently used for disposal of ashes, the binding effect indicates the potential for reuse without the harmful effects of heavy metals [91–93]. Galiano et al. used coal fly ash-based geopolymers to stabilize fly ash and compared with other classical stabilizers like cement, lime, slag and metakaolin. Zn, Sb and Sn showed effective immobilization, whereas Mo, V and Cr were not immobilized much. The best immobilization was demonstrated by OPC, lime and a mix of blast furnace slag and potassium silicate [94]. Bosio et al. employed a mix of colloidal silica or rice husk ash with coal fly ash for stabilization of MSWI fly ash. Both showed considerable and comparable stabilization of heavy metals [95,96].

6.2.7. Hydrothermal Treatment

Hydrothermal treatment is the process of solidification by application of steam under high pressure. It results in the formation of new stable minerals such as tobermorite, katoite or C-S-H, and the composition should be optimized (Ca/Si ratio of 0.83) for maximum generation of tobermorite and thus maximum strength. Jing et al. reported a positive effect of the addition of NaOH in hydrothermal solvent on the strength [97]. The same researchers investigated the result of hydrothermal solidification of MSWI bottom ash mixed with quartz, slaked lime and water cooled blast furnace slag [98]. The hydrothermal solidification of MSWI bottom ash without any additives was studied, since it had an optimum Ca/Si ratio inherently [99]. Furthermore, they studied the effect of hydrothermal treatment in solvent with a mixture of ferric and ferrous sulfate on dioxin decomposition [100]. High temperature and ferric and ferrous sulfate promote decomposition, temperature having the highest effect. Bayuseno et al. studied the effect of hydrothermal solidification of MSWI fly ashes on heavy metal fixation. They washed the fly ash, subsequently mixed with alkaline solution and treated in an autoclave at 90–180 °C. This resulted in the formation of minerals such as tobermorite and katoite and a marked reduction in leaching of heavy metals [101].

An optimized wet extraction of heavy metals and chlorides in hydrothermal conditions was experimentally tested by [88] with acid extraction and carbonation. The liquid to solid ratio, carbonation and leaching time for maximum extraction were optimized. It was observed that hydrochloric acid was most effective in the extraction of metals [102].

6.2.8. Thermal Treatment

Calcination of the ashes at a temperature above 600–800 °C eliminates the presence of toxic organic compounds like poly-chlorinated dibenzo-p-dioxins (PCDDs) and polychlorinated dibenzo furan (PCDFs). Furthermore, calcination may result in oxidation of Cr(III) to Cr(VI), which is the soluble form of chromium and makes it more prone to leaching. Therefore, application of a calcination process should be done with enough care to avoid leaching of Cr(VI) [25]. Furthermore, heating above 600–800 °C disintegrates the quench phase and results in volatilization of chlorides [35]. Reaching a maximum temperature of above 900°C is necessary to avoid reformation of PCDD/Fs.

Washing of ground ash at comparatively high temperatures (65–80 °C) can accelerate the process of metallic aluminum hydration, thus mitigating its harmful effects if happening during utilization [103].

6.2.9. Electrodialytic Remediation

Electrodialytic remediation is a technique that combines the technique of electro-dialysis with the electro-migration of ions, explored for removal of heavy metals from different kinds of waste streams and affected materials. Pedersen studied the effect of different assisting agents (deionized water, 2.5% NH_3 solution, 0.25 M citric acid and 0.25 M ammonium citrate) in the extraction efficiency of five heavy metals (Cd, Pb, Zn, Cu and Cr) from MSWI fly ash. The optimized removal of all five heavy metals was obtained when ammonium nitrate was used as the assisting agent [104].

6.2.10. Revasol Process

This is an optimized process developed by Solvay Company (Neder-Over-Heembeek, Brussels, Belgium) in collaboration with Université Libre de Bruxelles (ULB) (Brussels, Belgium) in Belgium to increase the utilization of MSWI fly ash. The process consists of three steps [52]: water dissolution, in which the salts are dissolved in water without allowing heavy metals to get dissolved; phosphation with phosphoric acid, which stabilizes heavy metals; and calcination to remove organic impurities present.

Aubert et al. modified the process with the addition of Na_2CO_3 in the washing step to accelerate the oxidation process of aluminum and also to increase the solubility of sulfates [26].

Table 5 summarizes various obstacles and methods already employed and under research to resolve them.

Table 5. Summary table.

Obstacle	Pre-Treatment Technique	Advantages	Disadvantages
Metallic Aluminum and Zinc	Magnetic density separation	Versatile Simple	Initial cost, not for fines
	Eddy current separation	Can detect through several layers Can provide an accurate separation	Initial cost, not for fines Susceptible to magnetic permeability changes
	Wet grinding	Consumes lower power per ton of product. Enables the use of wet screening or classification for close product control. Eliminates the dust problem. Enables the use of simple handling and transport methods such as pumps, pipes and launders.	Storage of wet slurries
	Washing with alkali	Simple	Cost of alkali
Salts	Washing with water	Simple	Secondary pollution of water, unless the water in the slurry is used to make concrete from the material
	Carbonation	CO_2 from stack gas can be utilized and thus reduce the emission. Can be a method for carbon sequestration.	Not a very fast process, unless the CO_2 concentration is very high, which in turn will require air tight enclosures
	Thermal treatment	Simple technology	Energy Consumption Cost

Table 5. Cont.

Obstacle	Pre-Treatment Technique	Advantages	Disadvantages
Heavy metals	Washing with water	Simple	Secondary pollution of water
	Treatment with sulfide rich effluent	Simple and can utilize another waste stream	Applicable for specific heavy metals
	Wet grinding	Simple No addition of additional chemicals	Storage of wet slurries
	Phosphation	Stabilizes heavy metals	Applicable only for specific applications
	Cement stabilization	Stabilizes heavy metals	Cost of cement
	Hydrothermal treatment		Capital cost
	Electrodialytic remediation		Costly Energy Consumption

7. Research Regarding Ash Pre-Treatment in Belgium

Various research projects are carried out within the Strategic Initiative Materials under the MaRes program (Materials from solid and liquid industrial process Residues) that aims at creating a toolbox to separate useful fractions from waste streams and valorizing them in building materials. The ASHCEM project deals with developing novel cements and building materials from municipal solid waste incineration ashes. The SUPERMEX (sustainable transformation of new and landfilled solid industrial residues into metals and inorganic polymers using a SUPERMetalEXtractor) project investigates sustainable transformation of new and landfilled solid industrial residues using a SUPERMetalEXtractor; while the SMART (Sustainable Metal Extraction from Tailings) project generates a toolbox for the extraction of metals from various tailings [105].

Various other research groups in Belgium also investigated various pre-treatment techniques to improve the utilization potential of the residues. Van Gerven et al. studied the effect of carbonation in reducing heavy metal leaching [87,106]. Saikia et al. studied the effects of various treatments including heat treatment and treatment with sodium carbonate to curb the harmful effects of elemental aluminum and sulfates [41]. Arickx et al. studied the effect of accelerated carbonation on curbing leaching of heavy metals and identified that around four weeks of accelerated carbonation leads to the Cu leaching being under the limits. They also conducted accelerated carbonation using the stack gas from the incinerator, and condensate formation due to high humidity of stack gas was identified as the major hindrance towards its application [87]. The Revasol process was developed to render MSWI ashes non-hazardous and fit to be utilized by reducing the soluble content, reducing the leachable content of heavy metals and destroying POPs [52]. Van Den Heede et al. studied the effect of washing with NaOH and found it to be effective to reduce the metallic aluminum reaction when used as an aggregate to manufacture Lego bricks [71].

8. Utilization of MSWI Bottom Ash in Building Materials

Bottom ash has potential to be used in various applications, after specific treatments for each application, as depicted in Figure 6. It is presently used for certain applications such as in road construction, cement production as additive, concrete production as aggregate etc., which is discussed in detail further.

8.1. Present Areas of Utilization

At present, MSWI ash is mainly utilized as a road subbase material, landfill structure material, embankment fill, as cement raw material and concrete products, out of which, the use in road base applications tops the lot in terms of volumes [19].

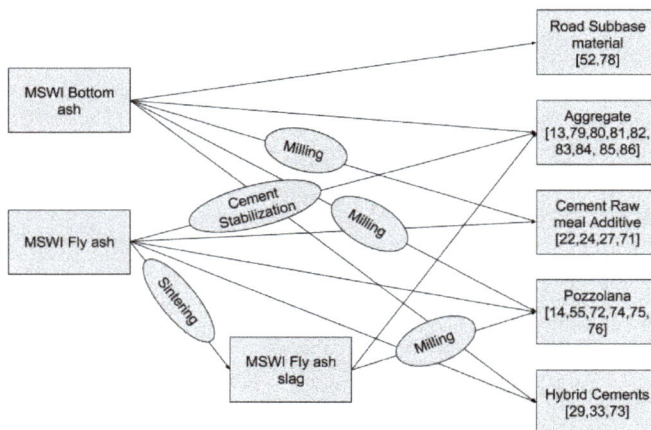

Figure 6. Routes of valorization for MSWI ash.

8.1.1. As a Road Construction Material and Landfill Stabilizer

Bottom ash can be used as aggregate in foundation layers, sub-base, embankment and as capping material in road construction [107]. When it is used in the hydraulically-bound form, washing of ash is done prior to utilization to prevent heavy metal leaching at the point of utilization. Since this does not involve milling of ash, the washing process can be more effective in curbing aluminum-induced expansion [77,108]. When used in bitumen bound applications, it resulted in higher bitumen demands due to the porosity of the material compared to conventional aggregates [107].

In landfills, it is used as a stabilizing layer to protect the geomembrane, intermediate layer and also as the leachate drainage layer [109,110].

8.1.2. As Cement Raw Meal Additive

The constituents of cement raw meal are calcium oxide, silica, alumina and iron oxide as the main components, along with other minor constituents, with specific limits to each of these. An appropriate composition is achieved by carefully proportioning different natural mineral sources like limestone, clay, etc. Presently, different alternate raw materials are added to the raw meal, usually as sources of silica. The replacement level is limited because of the low Ca content or the presence of other deleterious materials like alkalis and chlorine. Since cement is produced in huge quantities, utilization in even limited fractions can result in huge consumption of the MSWI ashes. A part of heavy metals, Cu, Zn and Pb, which is mostly affecting the utilization of bottom ashes, can be safely incorporated in mineral phases in clinker without affecting its properties. The proposed safe limits of Cu and Zn in clinker are 0.35% and 0.7%, respectively [111].

Eco-cement is a cement developed and being produced in bulk quantities in Japan with incineration ash mixed with limestone and other additives to adjust its composition being used as raw material for Portland cement production. Excess heavy metals in the ash are vaporized as chlorides and are collected in baghouses, and the heavy metals are regained from these by smelting, facilitating maximum recycling efficiency. Two types of eco-cements are being produced, normal Portland cement with alite, belite, aluminate, ferrite and calcium sulfate and a rapid hardening type with alite, belite, ferrite, $C_{11}A_7$ and $CaCl_2$ [112].

Replacement of cement raw meal by incineration bottom ash up to 15% was investigated by Shih et al. [42]. Five raw mixes were investigated: classic raw meal, raw meal with 3%, 10%, 15% replacement by bottom ash and with 15% magnetically-repelled MSWI ash after magnetic separation treatment. One mix was conditioned to required oxide proportions by the addition of pure calcium oxide. The lime saturation factor (LSF) of mixes were 1, 0.95, 0.84, 0.77 and 1.01 for 0%, 3%, 10%, 15% replacement and conditioned mix, respectively. Various literature works specify the required range of LSF as 0.92–1 [113]. A reduction in cement strength was observed when the lime saturation factor (LSF) was lower due to a lack of calcium for the formation of tri-calcium silicate. Higher strength was observed when additional calcium oxide was added to the system. A similar study with replacement of 5% and 10% of raw meal by dried and milled bottom ash without adjustment for the LSF was conducted by Krammart and Tangtermsirikul [32]. This replacement resulted in increased belite content at the expense of alite, resulting in a decrease in compressive strength and a delay in setting. Kikuchi produced Portland cement from raw meal with MSWI ash, sewage sludge, sewage dry powder, aluminum dross, limestone, clay and copper slag with up to 40% of MSWI fly ash. Significant fixing of chlorine in the form of calcium chloroaluminate was reported. Furthermore, the phosphate content in MSWI ash reduced the burnability of the mix and resulted in lower conversion of belite to alite [44]. Wang et al. manufactured clinker using washed fly ash as a component up to 15% with LSF, SM (silica modulus) and AM (alumina modulus) adjusted to prescribed limits. Washing of fly ash was optimized for maximum removal of chlorides and sulfates. Cement phases produced by sintering and the compressive strength of mortars were comparable to that of conventional raw meal [114].

8.1.3. MSWI Bottom Ash as Aggregate in Concrete

Research regarding use of bottom ash as aggregate in concrete started around two decades ago [55]. CUR Recommendation 116:2012 recommends utilization of MSWI bottom ash with less than 1% aluminum content to be used in concrete [56]. The main difficulties faced were expansion due to elemental Al, high water absorption, which increases water demand of concrete, and alkali-silica reaction. Washing with alkali, such as sodium hydroxide, is a possible solution for reducing expansion due to metallic aluminum.

Despite the mentioned practical challenges regarding the use of bottom ash as aggregate in concrete, research has been done on how a full replacement of traditional concrete aggregates by bottom ash-based aggregates would be attainable for certain concrete products, also in Belgium. For instance, Van den Heede et al. investigated whether non-steel reinforced prefabricated concrete Lego bricks of an acceptable quality could be produced with treated bottom ashes originating from a Belgian MSWI company as full aggregate replacement for limestone 2/6 and 6/20 [83]. It was found that the reduced concrete workability due to the high water absorption of the bottom ash can be dealt with by subjecting the fine bottom ash fraction to a crushing operation to eliminate the porous elements and by pre-wetting the fine and coarse bottom ash fractions in a controlled manner prior to concrete mixing. To overcome unacceptable expansion and longitudinal void formation related to the elemental aluminum present in the bottom ash, a reactive washing step with 1 M NaOH was found to be very effective. A modified Oberholster test indicated that the applicable 0.1% expansion limit was not surpassed. The required concrete strength class (C20/25) for prefabricated Lego brick production could be achieved without any problem [71]. Collivignarelli and Sorlini investigated the use of cement-lime stabilized MSWI fly ash as recycled aggregate in concrete. They reported to have achieved more than minimum compressive strength and good environmental compatibility with the tested mixes [115]. Results from the manufacture of lightweight aggregates with MSWI ash and reservoir sediment as additives are reported. The dosage of MSWI ash was 30%. The density of the aggregates manufactured was 0.99 g/cm^3 [116]. Ferraris et al. studied the effect of utilization of bottom ashes vitrified at 1450 °C without any agents as filler, sand and aggregate in concrete. Slump, alkali silica reactivity and mechanical properties were evaluated. When used as a filler, the 150-day compressive strength remained the same as that of a reference up to 20% replacement of cement and up to 75% effective replacement of natural aggregate [117]. Cioffi et al. studied the properties of artificial aggregates manufactured by binder stabilization of bottom ashes. These bottom ashes did not conform to the Italian environmental limitations for utilization. However, after stabilization, the cumulative leaching value of artificial aggregates conformed to environmental limitations. Moreover, the concrete blocks made with these aggregates conformed to the requirements for lightweight structural concrete [118]. Hwang et al. manufactured lightweight aggregates from reservoir sediment with MSWI fly ash as an additive. Concrete manufactured from these lightweight aggregates was tested for various properties such as compressive strength, electrical resistivity and ultrasonic pulse velocity, and it was concluded that the maximum addition of fly ash should be limited to 30% [119].

8.2. Potential Areas of Utilization

In addition to the low value applications except for as raw meal additive, for which bottom ashes are presently used, there are ongoing and past research works studying its application in higher value applications, which requires more pre-treatments to be done.

8.2.1. MSWI Ash as a Pozzolanic Addition in Cement

Supplementary cementitious materials are inorganic materials that, when used in conjunction with Portland cement, contribute to the properties of the hardened concrete through chemical reaction, e.g., hydraulic or pozzolanic activity. SCMs are mainly classified based on their origin as natural, thermally-activated and by-product SCM. The most used by-product SCMs include blast-furnace slag, fly ash from coal combustion and silica fume. These SCMs differ from each other in relative percentage of constituents, mineralogy, kinetics of reaction and also in the hydration products formed. ASTM 989 specifies the required fineness for use of slag as a pozzolana in terms of maximum percentage retained when wet screened on a 45-μm sieve to be a maximum of 20%. NBN 450-1 (Bureau voor Normalisatie, Belgium) sets the value to a maximum of 40% and 12% for two categories. Like slags, bottom ashes are generated as granulates and need a grinding step prior to their utilization as a supplementary cementitious material to reach this fineness. Like Portland cement, the main constituents of supplementary cementitious materials are CaO, SiO_2, Al_2O_3 and Fe_2O_3; however, most SCMs are enriched in silica compared to PC. NBN EN 450-1 and ASTM C 618 specify a minimum total content of 65% and 70% respectively for silica, alumina and iron oxide [120]. NBN EN 450-1 also limits the total organic carbon content to 5%, 7% and 9% for Category A, B and C fly ashes, respectively. For bottom ashes in Belgium, it is limited to 3% for ensuring burnout and is usually present in the range of 0.5–2.4% [17]. This limits free lime, alkali, chloride, sulfate and phosphate contents to 1.6%, 5.5% Na_2O_{eq} 0.1%, 3.5% and 5.5% by mass, respectively [121]. The washing and ageing in moist condition steps, which are already done in the field, converts the free lime present now into calcium hydroxide and eventually calcite. ASTM C 150 limits the maximum content of alkalis versus total binder to 0.6% expressed as sodium oxide equivalent (Na_2O_{eq}) when used in conjunction with reactive aggregates [122]. ASTM 989 limits the content of sulfur present as sulfide to a maximum of 2.5% and sulfate to a maximum of 4.0% in slag used in concrete [123]. MSWI ash is also a potential addition to the list of SCMs once the potential issues are resolved, which includes the presence of metallic aluminum, zinc, salts and heavy metals.

Remond et al. studied the effect of the replacement of cement by MSWI fly ashes on setting times, compressive strength, flow and microstructure of mortar and concrete. The pozzolanic phases were identified as CAS_2 (calcium aluminodisilicate) and AS (alumina silicate). The presence of zinc and lead resulted in the increase of setting times, and the presence of chloride had an accelerating effect, which increased the seven- and 28-day strengths for up to 15% ash replaced samples [124].

An optimized treatment procedure called the Revasol process and a modified process to eliminate metallic aluminum and sulfate were employed to get two treated fly ashes, which were further tested for reactivity and hydration as a pozzolanic addition by Aubert et al. They modified the previously developed Revasol process by including washing with Na_2CO_3 solution. Calcium aluminate hydrate, ettringite and calcium carbo-aluminate were generated as hydration products on reaction with lime. The modified treatment processes employed were successful in arresting expansion due to elemental aluminum, and mortars made with that ash gave a high activity index, for some more than 100%. However, leaching of antimony and chromium from the mortars was observed to be above limits [26,52,83].

To increase the activity of ash as a pozzolanic material, its reaction can be activated through the addition of chemical activators. After adding MSWI bottom ash as a pozzolanic material, alkaline activators can be added to activate the reaction. It was found that $CaCl_2$ is an effective activator, in contrast to Na-based activators, which did not improve the strength [28]. However, chloride in concrete can increase the susceptibility to reinforcement corrosion. Krammart and Tangtermsirikul used $Ca(OH)_2$ as an alkali activator, and a calcium aluminate monosulphate (AFm) type phase, calcium aluminum carbonate sulfate hydrate $(Ca_4Al_2O_6(CO_3)_{0.67}(SO_3)_{0.33} \cdot 11H_2O)$ and calcium silicate hydrate (CSH) with a low Ca-Si ratio were the hydration products formed. Since the material used was not treated to control the effect of aluminum, the compressive strength obtained was very low [32]. The use of bottom ash in a hybrid binder system consisting of OPC, MSWI bottom ash and fly ash, as well as a 5% mix of $CaSO_4$ and Na_2SO_4 as the activator was investigated [54]. It showed a 40% strength loss for this mix in comparison with a pure OPC mix. The cement was found to immobilize the elements Pb, Zr and Sn. Nevertheless, the leaching of Cl^- was very high, raising the question of its usage in reinforced concrete. Gao et al. studied the use of washed fly ash as pozzolanic addition. Washing reduced chloride content from 10.16% down to 1.28%, and the addition of 0.25% dithiocarbamic chelate reduced heavy metal leaching. Ten percent replacement resulted in the same compressive strength as that of reference samples [125].

Another route for utilization of MSWI fly ash as pozzolanic material is by sintering ash, producing fly ash slag, and milling the slag further to produce a pozzolanic material. Lin et al. studied the hydration properties of C_3S mixed with MSWI fly ash slag. A lower hydration degree of C_3S-slag pastes compared to pure C_3S pastes at all ages of hydration was observed. A pozzolanic reaction was observed after 28 days [126]. The effect of the addition of alumina prior to sintering was studied by Lin et al., who observed an increase in the degree of hydration up to 28 days for milled 10% alumina mixed slag blended cement pastes [127]. Similar research was carried out by Wang et al., which reported a reduction in early strength and an increase in 28-day strength due to 10–20% replacement of cement by milled fly ash slag [128]. Lin reported an increase in initial and final setting times for 10–40% cement replacement by municipal solid waste incineration fly ash and slag, possibly due to the presence of heavy metals like Zn, Pb and Cu [51]. It was also reported by Lee et al. that the early strength of mortars made with cement replaced with fly ash slag can be enhanced by increasing the basicity of slag by adding $CaCO_3$ prior to sintering. Mortar with 20% cement replacement with fly ash slag with 20% $CaCO_3$ additive had the same 14-day and 28-day compressive strength as that of reference OPC samples and roughly around an 8% increase compared to the reference at 90 days [129]. Lee et al. studied the effect of 20% cement replacement by vitrified slag made from fly ash and scrubber ash mix proportioned 1:3, with waste glass frit added to adjust the basicity of the mix. Leaching values of slags were well below the threshold limits, and the compressive strengths of mortars at 56 and 90 days were 5–10% more than those of the reference OPC mixes [130,131]. Lin et al. manufactured vitrified slag from a 50–50 mix of 1:3 fly ash scrubber ash mix and sludge from LED manufacturing. The ground slag, which proved to be safe in terms of leaching, was used to replace cement up to 30%, and the resulting mortars had a compressive strength 6–36% higher than that of the OPC reference [132].

8.2.2. Autoclaved Aerated Concrete

Autoclaved aerated concrete, also known as AAC, is industrially produced as aerated precast elements. Accelerated curing is done by autoclaving in high pressure steam, and this ensures higher strength development. Different technologies are prevalent today for aeration of concrete, and adding aluminum powder is one among them. MSWI ash inherently contains metallic aluminum, and thus, it can be an added advantage to be used as a raw material for AAC production. Furthermore, the aeration agent cost can be saved. Song et al. studied the effect of replacing fly ash or silica by incineration bottom ash. They reported an increase in compressive strength and a decrease in drying shrinkage with the increase in bottom ash replacement [46].

8.2.3. Manufacture of Ceramics

The high content of silica in the filter dust makes it suitable for the production of glass ceramics. There are already successful attempts in utilizing fly ash and slag for that purpose. High temperature sintering of the ash generates high density crystalline products that can be used as a building material, and at the same time, it will bind hazardous inorganic materials, making it environmentally safer. Glass ceramics were manufactured by heating at a rate of 10 °C/min up to 880 °C, maintaining this temperature for 4 h and then heating again at 5 °C/min to 950 °C and maintaining it for 10 h, then later cooled down [133]. Pyroxene group minerals, mainly diopside, were the main crystalline products formed by sintering. The density of glass ceramic formed was 2.89 g/cc. A fracture mirror study yielded a fracture strength of 240 MPa. The chemical composition and microstructure of ceramic produced by milling, compacting and sintering of bottom ash below 8 mm in size were also studied [33]. Quartz, calcite, gehlenite and hematite were the main crystalline minerals in ash before sintering. Diopside was the main mineral formed as a result of sintering to an optimum temperature of 1080 °C and controlled crystallization along with clinoenstatite, wollastonite and albite. The maximum density obtained was 2.6 g/cc.

9. Cement Industry and Utilization Potential of MSWI Ashes in Belgium

Utilization of bottom ash in the cement industry is a promising solution technically, as we have seen already. The cement industry already recycles many waste streams as alternate raw materials. Furthermore, cement being the most used building material is consumed in huge quantities. It can be seen from Tables 6 and 7 that around 2 MT of the cements CEM I, II and V were consumed in Belgium in 2015. Unutilized bottom ashes are available in 15% of this quantity by mass (~300 kT) [134].

Table 6. Cement consumption in Belgium cf. [134].

Categories	Cement Consumption in 2015 (MT)
Concrete products and fiber cement	1
Ready mix concrete	2.767
Delivery in construction site	0.743
Delivery in hardware shops	0.381
Total consumption	4.891
Import	1.513
Export	1.384

Table 7. Percentage consumption of various classes of cement in Belgium in 2015 cf. [134].

Cement Type	Strength Class	% Consumption
CEM I, II, V	32.5	12
	42.5	3
	52.5	26
CEM III	32.5	8
	42.5/52.5	51

At the same time, there is consumption of the CEM II type of cement that utilizes fly ashes from coal-fired thermal power plants. Since there are no longer thermal power plants in Belgium as of 2016, fly ashes for CEM II production need to be imported from neighboring countries. Transportation adds to the cost, energy consumption and CO_2 emissions considerably and thus affects the portfolio of cements produced. Incineration plants have good areal distribution and thus save transportation costs, and treated bottom ashes could act as a replacement for fly ash in CEM II, as can be seen from the map in Figure 7. Therefore, the use as pozzolanic material or as a raw meal for clinker production are promising options for utilization of more bottom ashes. The <2 mm fraction of bottom ash is mostly unutilized, due to more heavy metals. Another major obstacle to the utilization is the presence of metallic aluminum. There are a number of techniques available to stabilize heavy metals such as ageing, wet grinding, treatment with sulfide-rich effluent, binder stabilization, vitrification, etc. Ageing of samples is already done in the field. Vitrification/slagging processes are both capital and energy intensive. Both wet grinding and treatment with sulfide-rich effluent could be possible options. Wet grinding can solve both obstacles. Metallic aluminum can be hydrated to aluminum hydroxide during the process, and it can be accelerated by utilizing waste heat generated in the plant [103]. Sources of waste heat from the plant include heat from the stack gas and heat generated during ageing that raises the temperature of bottom ash to 70–80 °C. It would be a challenging job for an industrial engineer to harvest this waste heat, but if realized, it can make the process more economic and sustainable. The second option is to use the ashes as a cement raw material. A part of the heavy metals, especially Cu, Zn and Pb, can be incorporated in the phases in cement; however, if above certain limits, it can alter them. The presence of metallic aluminum is not known to affect the production process. Relatively high alkali content in bottom ashes can affect the clinkerization process, thus limiting the replacement ratios to a smaller value. Nevertheless, even small replacement ratios can result in huge utilization of material due to the enormous quantities of cement produced. The next step for more utilization is testing these options in pilot scale in cement/incineration plants and optimizing the process. This will allow full utilization of bottom ash in a more high value product, i.e., cement, than using it as aggregate or as road subbase material, making it more economical. It helps to reach the goal of the circular economy utilizing many waste streams. Various research works for realizing the utilization of fly ashes are available. Many studies reported pozzolanic activity of fly ash slag made by vitrification of fly ashes, thus reducing its leaching. Being a capital- and energy-intensive technique, it might be more challenging to optimize the process. More research to identify the advantages of materials and using them for the right application will be the key to achieving a circular economy for fly ashes.

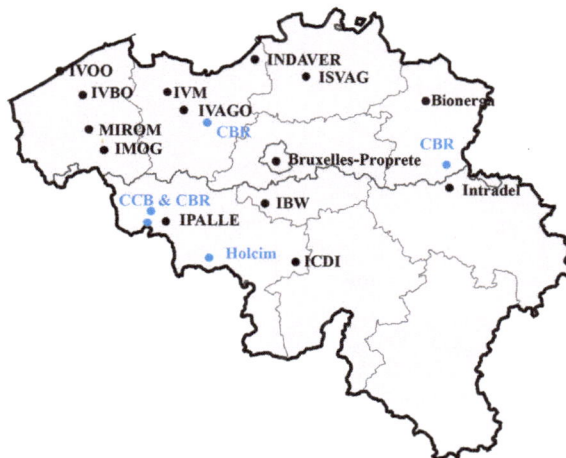

Figure 7. Areal distribution of major incineration plants and cement plants in Belgium (black, incineration plants; blue, cement plants) cf. [22,134].

10. Conclusions

This article discussed the utilization potential of MSWI ashes in a variety of fields, various obstacles preventing utilization and various options for treatment procedures that could overcome these obstacles. The composition of ash depends on the input waste, type of incinerator and process parameters, and this varies spatially and temporally; it does not hinder its utilization in one of the various potential fields of application. The most suitable area of utilization for each type of ash produced in around 2200 waste to energy plants around the world needs to be identified and put into use locally. Moreover, the cement industry, which is identified as the most suitable agent for valorizing bottom ashes as raw meal additive and supplementary cementitious material, is equipped for dealing with variability in raw material composition and still produces a final product with the required quality control. One of the sources of variation is the process parameters of incineration plants, which are optimized currently for maximum energy generation. However, the process parameters can have an effect on the chemical properties of final residues, which could in turn affect the technical quality of the final product after utilization. The process parameters of incineration plants need to be optimized for maximum utilization of residues. A comprehensive database of the effects of various characteristics of residues on the quality of final products after incineration needs to be developed after research.

In Belgium, we have already come a long way, and pretreatment installations solve many problems to an extent. Evidence of the successful application of pre-treatment techniques at the laboratory scale is available. The most suitable options for higher value applications are usage as a binder or raw meal substitute. Use as cement raw meal additive requires less pre-treatment and thus is another promising option for high value application. The present need is to optimize the whole process flow technically and economically by conducting pilot scale tests and doing plant-specific life cycle analysis to identify what is gained and what not. Use as a pozzolanic additive is a good option for Belgium considering its almost zero production of coal fly ashes. Sequencing and combining the right pre-treatment techniques will be the key to the most sustainable utilization. The Revasol technique modified by Aubert et al. is one such method addressing various obstacles for use as pozzolana. New research for localized solutions for treatment of ashes with minimal secondary pollutions needs to be done. In Belgium, Cu and Zn content is the major obstacle preventing utilization of 164 kT of bottom ashes. Another major problem is the metallic aluminum content preventing its effective use as a pozzolanic additive and aggregate in concrete. Wet grinding of the ash at a higher temperature can be a solution for both of these, and the process can be optimized by using local waste water preferably containing sulfides and thus reducing the impact of secondary pollution of water. Hydrogen gas generated by washing can be harvested and used for energy generation. Another solution can be carbonation of ash, for which the high CO_2 concentration of stack gas can be utilized or sequestrated. This can add to the economy of the plant, as well, due to the additional income from residue recycling. Capital-intensive techniques, like vitrification, which could be more effective for MSWI fly ash, can be employed after balancing the cost and benefit. Establishing the business model and development of standards and legislation are the key to a better recycling. Before that, comprehensive methodologies for utilization with emphasis on technical and environmental specifications need to be developed by the scientific community.

Acknowledgments: This research is a part of the ASHCEM project, which in itself is a part of the bigger program MaRes aimed at creating and demonstrating an operational, flexible toolbox to recover metals and valorize the residual matrix into building materials funded by SIM (Strategic Initiative Materials in Flanders) and VLAIO (Flanders Innovation & Entrepreneurship). The financial support from the foundations for this study is gratefully appreciated.

Author Contributions: Aneeta Mary Joseph did data collection and wrote the paper. Ruben Snellings and Philip Van den Heede helped in data collection, drafting and correcting the paper and was involved in the discussions throughout. Nele De Belie and Stijn Matthys have conceived the ASHCEM project, realized funding and gave critical analysis and final approval of the paper.

Conflicts of Interest: The authors declare no conflict of interest.

References

1. European Commision. Circular Economy. 2017. Available online: http://ec.europa.eu/environment/circular-economy/index_en.htm (accessed on 5 October 2017).
2. European Environment Agency. *Managing Municipal Solid Waste—A Review of Achievements in 32 European Countries*; European Environment Agency: Copenhagen, Denmark, 2013.
3. Hoornweg, D.; Bhada-Tata, P.; Joshi-Ghani, A. *What a Waste: A Global Review of Solid Waste Management*; Urban Development and Local Government Unit, World Bank: Washington, DC, USA, 2009.
4. Belevi, H.; Moench, H. Factors determining the element behavior in municipal solid waste incinerators. 1. Field studies. *Environ. Sci. Technol.* **2000**, *34*, 2501–2506. [CrossRef]
5. Cheng, H.; Hu, Y. Municipal solid waste (MSW) as a renewable source of energy: Current and future practices in China. *Bioresour. Technol.* **2010**, *101*, 3816–3824. [CrossRef] [PubMed]
6. European Commission. *Integrated Pollution Prevention and Control—Reference Document on the Best Available Techniques for Waste Incineration*; European Commission: Brussels, Belgium, 2006.
7. Bertolini, L.; Carsana, M.; Cassago, D.; Curzio, A.Q.; Collepardi, M. MSWI ashes as mineral additions in concrete. *Cem. Concr. Res.* **2004**, *34*, 1899–1906. [CrossRef]
8. Sawell, S.; Chandler, A.; Eighmy, T.; Hartlén, J.; Hjelmar, O.; Kosson, D.; van der Sloot, H.; Vehlow, J. An international perspective on the characterisation and management of residues from MSW incinerators. *Biomass Bioenergy* **1995**, *9*, 377–386. [CrossRef]
9. Csanyi, D.M.C. Environmental Hazards of Limestone Mining. Available online: http://education.seattlepi.com/environmental-hazards-limestone-mining-5608.html (accessed on 19 September 2016).
10. Beretka, J.; de Vito, B.; Santoro, L.; Sherman, N.; Valenti, G.L. Utilisation of industrial wastes and by-products for the synthesis of special cements. *Resour. Conserv. Recycl.* **1993**, *9*, 179–190. [CrossRef]
11. European Commision. EU Waste Legislation. 2016. Available online: http://ec.europa.eu/environment/waste/legislation/index.htm (accessed on 5 October 2017).
12. Gentil, E.C. *Municipal Waste Management in Belgium*; European Environment Agency: Copenhagen, Denmark, 2013.
13. Eurostat. Eurostat Data for Waste Management Indicators. Available online: http://ec.europa.eu/eurostat/statistics-explained/index.php/Waste_management_indicators (accessed on 5 October 2017).
14. Wallon, G. Arrêté du Gouvernement Wallon Favorisant la Valorisation de Certains Déchets. 2001. Available online: http://environnement.wallonie.be/legis/dechets/decat024.htm (accessed on 5 October 2017).
15. Vandecasteele, C.; Wauters, G.; Arickx, S.; Jaspers, M.; van Gerven, T. Integrated municipal solid waste treatment using a grate furnace incinerator: The Indaver case. *Waste Manag.* **2007**, *27*, 1366–1375. [CrossRef] [PubMed]
16. Eurostat. Municipal Waste Statistics. European Commission, 2017. Available online: http://ec.europa.eu/eurostat/statistics-explained/index.php/Municipal_waste_statistics (accessed on 11 November 2017).
17. Nielsen, P.; Kenis, C.; Vanassche, S.; Vrancken, K. *Beste Beschikbare Technieken (BBT) Voor Behandeling van Bodemas van Huisvuilverbranding*; Academia Press: New York, NY, USA, 2007.
18. Leefmilieu, N.E.; MOVA; OVIT. *Monitoringsysteem Duurzaam Oppervlaktedelfstoffenbeleid*; De Vlaamse Overheid, Flanders Government: Brussels, Belgium, 2013.
19. Born, J.P.; van Brecht, A. Recycling potentials of MSWI Bottom Ash. CEWEP 2014. Available online: http://www.cewep.eu/m_1318 (accessed on 5 October 2017).
20. Valomac. Valoblock®: Des Mâchefers au Matériau de Construction. 2013. Available online: https://suezbelgium.be/fr/economie-circulaire/creer-des-matieres-premieres/valoblock (accessed on 5 October 2017).
21. Niessen, W.R. *Combustion and Incineration Processes*; Marcel Dekker, Inc.: New York, NY, USA, 2002.
22. Belgian Waste to Energy BW2E. Available online: http://www.bw2e.be/nl/page/leden (accessed on 5 October 2017).
23. Belevi, H.; Langmeier, M. Factors determining the element behavior in municipal solid waste incinerators. 2. Laboratory experiments. *Environ. Sci. Technol.* **2000**, *34*, 2507–2512. [CrossRef]
24. Wiles, C.C. Municipal solid waste combustion ash: State of the knowledge. *J. Hazard. Mater.* **1995**, *3894*, 20. [CrossRef]
25. Müller, U.; Rübner, K. The microstructure of concrete made with municipal waste incinerator bottom ash as an aggregate component. *Cem. Concr. Res.* **2006**, *36*, 1434–1443. [CrossRef]

26. Aubert, J.E.; Husson, B.; Sarramone, N. Utilization of municipal solid waste incineration (MSWI) fly ash in blended cement. Part 1: Processing and characterization of MSWI fly ash. *J. Hazard. Mater.* **2006**, *136*, 624–631. [CrossRef] [PubMed]

27. Hyks, J.; Astrup, T. Influence of operational conditions, waste input and ageing on contaminant leaching from waste incinerator bottom ash: A full-scale study. *Chemosphere* **2009**, *76*, 1178–1184. [CrossRef] [PubMed]

28. Polettini, A.; Pomi, R.; Carcani, G. The effect of Na and Ca salts on MSWI bottom ash activation for reuse as a pozzolanic admixture. *Resour. Conserv. Recycl.* **2005**, *43*, 403–418. [CrossRef]

29. Alhassan, H.M.; Musa, A.; International, T.; Alhassan, H.M.; Tanko, A.M. Characterization of Solid Waste Incinerator Bottom Ash and the Potential for its Use. *Int. J. Eng. Res. Appl.* **2012**, *2*, 516–522.

30. Silva, R.V.; de Brito, J.; Lynn, C.J.; Dhir, R.K. Use of municipal solid waste incineration bottom ashes in alkali-activated materials, ceramics and granular applications: A review. *Waste Manag.* **2017**, *68*, 207–220. [CrossRef] [PubMed]

31. Izquierdo, M.; Querol, X.; Josa, A.; Vazquez, E.; López-Soler, A. Comparison between laboratory and field leachability of MSWI bottom ash as a road material. *Sci. Total Environ.* **2008**, *389*, 10–19. [CrossRef] [PubMed]

32. Krammart, P.; Tangtermsirikul, S. Properties of cement made by partially replacing cement raw materials with municipal solid waste ashes and calcium carbide waste. *Constr. Build. Mater.* **2004**, *18*, 579–583. [CrossRef]

33. Bethanis, S.; Sollars, C.J.; Cheeseman, C.R. Properties and Microstructure of Sintered Incinerator Bottom Ash. *Ceram. Int.* **2002**, *28*, 881–886. [CrossRef]

34. Keppert, M.; Pavlík, Z.; Tydlitát, V.; Volfová, P.; Švarcová, S.; Šyc, M.; Černý, R. Properties of municipal solid waste incineration ashes with respect to their separation temperature. *Waste Manag. Res.* **2012**, *30*, 1041–1048. [CrossRef] [PubMed]

35. Yang, S.; Saffarzadeh, A.; Shimaoka, T.; Kawano, T. Existence of Cl in municipal solid waste incineration bottom ash and dechlorination effect of thermal treatment. *J. Hazard. Mater.* **2014**, *267*, 214–220. [CrossRef] [PubMed]

36. Inkaew, K.; Saffarzadeh, A.; Shimaoka, T. Modeling the formation of the quench product in municipal solid waste incineration (MSWI) bottom ash. *Waste Manag.* **2015**, *52*, 159–168. [CrossRef] [PubMed]

37. Qiao, X.C.; Tyrer, M.; Poon, C.S.; Cheeseman, C.R. Characterization of alkali-activated thermally treated incinerator bottom ash. *Waste Manag.* **2008**, *28*, 1955–1962. [CrossRef] [PubMed]

38. Thipse, S.; Schoenitz, M.; Dreizin, E. Morphology and composition of the fly ash particles produced in incineration of municipal solid waste. *Fuel Process. Technol.* **2002**, *75*, 173–184. [CrossRef]

39. Song, G.J.; Kim, K.H.; Seo, Y.C.; Kim, S.C. Characteristics of ashes from different locations at the MSW incinerator equipped with various air pollution control devices. *Waste Manag.* **2004**, *24*, 99–106. [CrossRef]

40. Lothenbach, B.; Scrivener, K.; Hooton, R.D. Supplementary cementitious materials. *Cem. Concr. Res.* **2011**, *41*, 1244–1256. [CrossRef]

41. Saikia, N.; Mertens, G.; van Balen, K.; Elsen, J.; van Gerven, T.; Vandecasteele, C. Pre-treatment of municipal solid waste incineration (MSWI) bottom ash for utilisation in cement mortar. *Constr. Build. Mater.* **2015**, *96*, 76–85. [CrossRef]

42. Shih, P.H.; Chang, J.E.; Chiang, L.C. Replacement of raw mix in cement production by municipal solid waste incineration ash. *Cem. Concr. Res.* **2003**, *33*, 1831–1836. [CrossRef]

43. Tang, P.; Florea, M.V.A.; Spiesz, P.; Brouwers, H.J. The investigation of the MSWI bottom ash fines (0–2 mm) as binder substitute after combined treatments. In Proceedings of the Eurasia Waste Management Symposium, Istanbul, Turkey, 28–30 April 2014; pp. 126–130.

44. Kikuchi, R. Recycling of municipal solid waste for cement production: Pilot-scale test for transforming incineration ash of solid waste into cement clinker. *Resour. Conserv. Recycl.* **2001**, *31*, 137–147. [CrossRef]

45. Qiao, X.C.; Tyrer, M.; Poon, C.S.; Cheeseman, C.R. Novel cementitious materials produced from incinerator bottom ash. *Resour. Conserv. Recycl.* **2008**, *52*, 496–510. [CrossRef]

46. Song, Y.; Li, B.; Yang, E.; Liu, Y.; Ding, T. Feasibility study on utilization of municipal solid waste incineration bottom ash as aerating agent for the production of autoclaved aerated concrete. *Cem. Concr. Compos.* **2015**, *56*, 51–58. [CrossRef]

47. Rambaldi, E.; Esposito, L.; Andreola, F.; Barbieri, L.; Lancellotti, I.; Vassura, I. The recycling of MSWI bottom ash in silicate based ceramic. *Ceram. Int.* **2010**, *36*, 2469–2476. [CrossRef]

48. Piantone, P.; Bodénan, F.; Chatelet-Snidaro, L. Mineralogical study of secondary mineral phases from weathered MSWI bottom ash: Implications for the modelling and trapping of heavy metals. *Appl. Geochem.* **2004**, *19*, 1891–1904. [CrossRef]

49. Bayuseno, A.P.; Schmahl, W.W. Understanding the chemical and mineralogical properties of the inorganic portion of MSWI bottom ash. *Waste Manag.* **2010**, *30*, 1509–1520. [CrossRef] [PubMed]

50. Alba, N.; Gasso, S.; Lacorte, T.; Baldasano, J.M. Characterization of municipal solid waste incineration residues from facilities with different air pollution control systems. *J. Air Waste Manag. Assoc.* **1997**, *47*, 1170–1179. [CrossRef]

51. Lin, K. The influence of municipal solid waste incinerator fly ash slag blended in cement pastes. *Cem. Concr. Res.* **2005**, *35*, 979–986. [CrossRef]

52. Aubert, J.E.; Husson, B.; Vaquier, A. Use of municipal solid waste incineration fly ash in concrete. *Cem. Concr. Res.* **2004**, *34*, 957–963. [CrossRef]

53. Violante, A.; Huang, P.M. Formation mechanism of aluminum hydroxide polymorphs. *Clays Clay Miner.* **1993**, *41*, 590–597. [CrossRef]

54. Garcia-Lodeiro, I.; Carcelen-Taboada, V.; Fernández-Jiménez, A.; Palomo, A. Manufacture of hybrid cements with fly ash and bottom ash from a municipal solid waste incinerator. *Constr. Build. Mater.* **2016**, *105*, 218–226. [CrossRef]

55. Pera, J.; Courtaz, J.; Ambroise, J.; Chababbet, M. Use of Incinerator Bottom Ash in Concrete. *Cem. Concr. Res.* **1997**, *27*, 1–5. [CrossRef]

56. CUR. *Bouwen Met Kennis Materiaal Voor Beton*; CUR Bouw & Infra: Gouda, The Netherlands, 2012.

57. Zhou, J.; Wu, S.; Pan, Y.; Zhang, L.; Cao, Z.; Zhang, X.; Yonemochi, S.; Hosono, S.; Wang, Y.; Oh, K.; et al. Enrichment of heavy metals in fine particles of municipal solid waste incinerator (MSWI) fly ash and associated health risk. *Waste Manag.* **2015**, *43*, 239–246. [CrossRef] [PubMed]

58. Wu, S.; Xu, Y.; Sun, J.; Cao, Z.; Zhou, J.; Pan, Y.; Qian, G. Inhibiting evaporation of heavy metal by controlling its chemical speciation in MSWI fly ash. *Fuel* **2015**, *158*, 764–769. [CrossRef]

59. Bertolini, L.; Elsener, B.; Pedeferri, P.; Polder, R. *Corrosion of Steel in Concrete*; Wiley-VCH, Verlag GmbH & Co. KGaA: Weinheim, Germany, 2004.

60. Geurts, R.; Spooren, J.; Quaghebeur, M.; Broos, K.; Kenis, C.; Debaene, L. Round robin testing of a percolation column leaching procedure. *Waste Manag.* **2016**, *55*, 31–37. [CrossRef] [PubMed]

61. Van Gerven, T.; Geysen, D.; Stoffels, L.; Jaspers, M.; Wauters, G.; Vandecasteele, C. Management of incinerator residues in Flanders (Belgium) and in neighbouring countries. A comparison. *Waste Manag.* **2005**, *25*, 75–87. [CrossRef] [PubMed]

62. OVAM. VLAREMA Decree of the Flemish Government to Establish the Flemish Regulations on the Sustainable Management of Material Circuits and Waste. Government, Flemish. Available online: https://navigator.emis.vito.be/mijn-navigator?woId=44707 (accessed on 5 October 2017).

63. UGent; VITO; CRH; Indaver; Recmix. *ASHCEM Novel Cements & Building Materials from Incineration Ashes—Project Document*; SIM: Zwijnaarde, Belgium, 2014.

64. Astrup, T.; Rosenblad, C.; Trapp, S.; Christensen, T.H. Chromium release from waste incineration air-pollution-control residues. *Environ. Sci. Technol.* **2005**, *39*, 3321–3329. [CrossRef] [PubMed]

65. Van Caneghem, J.; Block, C.; Vandecasteele, C. Destruction and formation of dioxin-like PCBs in dedicated full scale waste incinerators. *Chemosphere* **2014**, *94*, 42–47. [CrossRef] [PubMed]

66. Hutzinger, O.; Blumich, M.J.; Berg, M.V.D.; Olie, K. Sources and fate of PCDDs and PCDFs: An overview. *Chemosphere* **1985**, *14*, 581–600. [CrossRef]

67. Zhan, M.; Chen, T.; Lin, X.; Fu, J.; Li, X.; Yan, J.; Buekens, A. Suppression of dioxins after the post-combustion zone of MSWIs. *Waste Manag.* **2016**, *54*, 153–161. [CrossRef] [PubMed]

68. HRigo, G.; Chandler, A.J. Is there a strong dioxin: Chlorine link in commercial scale systems? *Chemosphere* **1998**, *37*, 2031–2046.

69. Wu, S.; Zhou, J.; Pan, Y.; Zhang, J.; Zhang, L.; Ohtsuka, N.; Motegi, M.; Yonemochi, S.; Oh, K.; Hosono, S.; et al. Dioxin distribution characteristics and health risk assessment in different size particles of fly ash from MSWIs in China. *Waste Manag.* **2016**, *50*, 113–120. [CrossRef] [PubMed]

70. Wang, L.; Jin, Y.; Nie, Y. Investigation of accelerated and natural carbonation of MSWI fly ash with a high content of Ca. *J. Hazard. Mater.* **2010**, *174*, 334–343. [CrossRef] [PubMed]

71. Van den Heede, P.; Ringoot, N.; Beirnaert, A.; van Brecht, A.; van den Brande, E.; de Schutter, G.; de Belie, N. Sustainable high quality recycling of aggregates from waste-to-energy, treated in a wet bottom ash processing installation, for use in concrete products. *Materials* **2016**, *9*, 9. [CrossRef] [PubMed]

72. SBerkhout, P.M.; Oudenhoven, B.P.M.; Rem, P.C. Optimizing Non-Ferrous Metal Value from MSWI Bottom Ashes. *J. Environ. Prot. (Irvine Calif.)* **2011**, *2*, 564–570. [CrossRef]

73. Shen, H.; Forssberg, E. An overview of recovery of metals from slags. *Waste Manag.* **2003**, *23*, 933–949. [CrossRef]

74. Xia, Y.; He, P.; Shao, L.; Zhang, H. Metal distribution characteristic of MSWI bottom ash in view of metal recovery. *J. Environ. Sci.* **2016**, *52*, 1–12. [CrossRef] [PubMed]

75. De Boom, A.; Degrez, M.; Hubaux, P.; Lucion, C. MSWI boiler fly ashes: Magnetic separation for material recovery. *Waste Manag.* **2011**, *31*, 1505–1513. [CrossRef] [PubMed]

76. Biganzoli, L.; Ilyas, A.; van Praagh, M.; Persson, K.M.; Grosso, M. Aluminium recovery vs. hydrogen production as resource recovery options for fine MSWI bottom ash fraction. *Waste Manag.* **2013**, *33*, 1174–1181. [CrossRef] [PubMed]

77. Mulder, E. Pre-treatment of MSWI fly ash for useful application. *Waste Manag.* **1996**, *16*, 181–184. [CrossRef]

78. Chen, X.; Bi, Y.; Zhang, H.; Wang, J. Chlorides Removal and Control through Water-washing Process on MSWI Fly Ash. *Procedia Environ. Sci.* **2016**, *31*, 560–566. [CrossRef]

79. Chimenos, J.M.; Fernandez, A.I.; Cervantes, A.; Miralles, L.; Fernandez, M.A.; Espiell, F. Optimizing the APC residue washing process to minimize the release of chloride and heavy metals. *Waste Manag.* **2005**, *25*, 686–693. [CrossRef] [PubMed]

80. Hu, B.; Rem, P.C.; van de Winckel, T.P.M. Fine heavy non-ferrous and precious metals recovery in bottom ash treatment. *ISWA Knowl. Database* **2009**, *APESB2009*, 1–8.

81. Metso. *Basics in Minerals Processing*; Metso Corporation: Helsinki, Finland, 2015; p. 354.

82. Grewal, I. Mineral Processing Introduction. Available online: http://met-solvelabs.com/library/articles/mineral-processing-introduction (accessed on 10 December 2017).

83. Aubert, J.E.; Husson, B.; Sarramone, N. Utilization of municipal solid waste incineration (MSWI) fly ash in blended cement. Part 2. Mechanical strength of mortars and environmental impact. *J. Hazard. Mater.* **2007**, *146*, 12–19. [CrossRef] [PubMed]

84. Sivula, L.; Väisänen, A.; Rintala, J. Stabilisation of MSWI bottom ash with sulphide-rich anaerobic effluent. *Chemosphere* **2008**, *71*, 1–9. [CrossRef] [PubMed]

85. Wang, W.; Gao, X.; Li, T.; Cheng, S.; Yang, H.; Qiao, Y. Stabilization of heavy metals in fly ashes from municipal solid waste incineration via wet milling. *Fuel* **2018**, *216*, 153–159. [CrossRef]

86. Chen, C.; Sun, C.; Gau, S.; Wu, C.; Chen, Y. The effects of the mechanical—Chemical stabilization process for municipal solid waste incinerator fly ash on the chemical reactions in cement paste. *Waste Manag.* **2013**, *33*, 858–865. [CrossRef] [PubMed]

87. Arickx, S.; van Gerven, T.; Vandecasteele, C. Accelerated carbonation for treatment of MSWI bottom ash. *J. Hazard. Mater.* **2006**, *137*, 235–243. [CrossRef] [PubMed]

88. Del Toro, M.A.; Calmano, W.; Ecke, H. Wet extraction of heavy metals and chloride from MSWI and straw combustion fly ashes. *Waste Manag.* **2009**, *29*, 2494–2499. [CrossRef] [PubMed]

89. Bournonville, B.; Nzihou, A.; Sharrock, P.; Depelsenaire, G. Stabilisation of heavy metal containing dusts by reaction with phosphoric acid: Study of the reactivity of fly ash. *J. Hazard. Mater.* **2004**, *116*, 65–74. [CrossRef] [PubMed]

90. Nzihou, A.; Sharrock, P. Calcium phosphate stabilization of fly ash with chloride extraction. *Waste Manag.* **2002**, *22*, 235–239. [CrossRef]

91. Cornelis, G.; van Gerven, T.; Vandecasteele, C. Antimony leaching from uncarbonated and carbonated MSWI bottom ash. *J. Hazard. Mater.* **2006**, *137*, 1284–1292. [CrossRef] [PubMed]

92. Chen, Q.Y.; Tyrer, M.; Hills, C.D.; Yang, X.M.; Carey, P. Immobilisation of heavy metal in cement-based solidification/stabilisation: A review. *Waste Manag.* **2009**, *29*, 390–403. [CrossRef] [PubMed]

93. Shi, H.; Kan, L. Leaching behavior of heavy metals from municipal solid wastes incineration (MSWI) fly ash used in concrete. *J. Hazard. Mater.* **2009**, *164*, 750–754. [CrossRef] [PubMed]

94. Galiano, Y.L.; Pereira, C.F.; Vale, J. Stabilization/solidification of a municipal solid waste incineration residue using fly ash-based geopolymers. *J. Hazard. Mater.* **2011**, *185*, 373–381. [CrossRef] [PubMed]

95. Bosio, A.; Zacco, A.; Borgese, L.; Rodella, N.; Colombi, P.; Benassi, L.; Depero, L.E.; Bontempi, E. A sustainable technology for Pb and Zn stabilization based on the use of only waste materials: A green chemistry approach to avoid chemicals and promote CO_2 sequestration. *Chem. Eng. J.* **2014**, *253*, 377–384. [CrossRef]

96. Rodella, N.; Bosio, A.; Dalipi, R.; Zacco, A.; Borgese, L.; Depero, L.E.; Bontempi, E. Waste silica sources as heavy metal stabilizers for municipal solid waste incineration fly ash. *Arab. J. Chem.* **2017**, *10*, S3676–S3681. [CrossRef]

97. Jing, Z.; Matsuoka, N.; Jin, F.; Hashida, T.; Yamasaki, N. Municipal incineration bottom ash treatment using hydrothermal solidification. *Waste Manag.* **2007**, *27*, 287–293. [CrossRef] [PubMed]

98. Jing, Z.; Ran, X.; Jin, F.; Ishida, E.H. Hydrothermal solidification of municipal solid waste incineration bottom ash with slag addition. *Waste Manag.* **2010**, *30*, 1521–1527. [CrossRef] [PubMed]

99. Jing, Z.; Fan, X.; Zhou, L.; Fan, J.; Zhang, Y.; Pan, X.; Ishida, E.H. Hydrothermal solidification behavior of municipal solid waste incineration bottom ash without any additives. *Waste Manag.* **2003**, *33*, 1182–1189. [CrossRef] [PubMed]

100. Hu, Y.; Zhang, P.; Chen, D.; Zhou, B.; Li, J.; Li, X. Hydrothermal treatment of municipal solid waste incineration fly ash for dioxin decomposition. *J. Hazard. Mater.* **2012**, *207–208*, 79–85. [CrossRef] [PubMed]

101. Bayuseno, A.P.; Schmahl, W.W.; Müllejans, T. Hydrothermal processing of MSWI Fly Ash-towards new stable minerals and fixation of heavy metals. *J. Hazard. Mater.* **2009**, *167*, 250–259. [CrossRef] [PubMed]

102. Zhang, F.; Itoh, H. Extraction of metals from municipal solid waste incinerator fly ash by hydrothermal process. *J. Hazard. Mater.* **2006**, *136*, 663–670. [CrossRef] [PubMed]

103. Joseph, A.M.; van den Heede, P.; Matthys, S.; de Belie, N.; Snellings, R.; Van, A. Comparison of different beneficiation techniques to improve utilization potential of municipal solid waste incineration fly ash concrete. In Proceedings of the International Conference on Advances in Construction Materials and Systems, Chennai, India, 3–8 September 2017.

104. Pedersen, A.J. Evaluation of assisting agents for electrodialytic removal of Cd, Pb, Zn, Cu and Cr from MSWI fly ash. *J. Hazard. Mater.* **2002**, *95*, 185–198. [CrossRef]

105. Flanders, S. MARES, Strategic Initiative Materials in Flanders. 2017. Available online: http://www.sim-flanders.be/research-program/mares (accessed on 2 February 2017).

106. Van Gerven, T.; van Keer, E.; Arickx, S.; Jaspers, M.; Wauters, G.; Vandecasteele, C. Carbonation of MSWI-bottom ash to decrease heavy metal leaching, in view of recycling. *Waste Manag.* **2005**, *25*, 291–300. [CrossRef] [PubMed]

107. Lynn, C.J.; Ghataora, G.S.; OBE, R.K.D. Municipal incinerated bottom ash (MIBA) characteristics and potential for use in road pavements. *Int. J. Pavement Res. Technol.* **2017**, *10*, 185–201. [CrossRef]

108. Colangelo, F.; Cioffi, R.; Montagnaro, F.; Santoro, L. Soluble salt removal from MSWI fly ash and its stabilization for safer disposal and recovery as road basement material. *Waste Manag.* **2012**, *32*, 1179–1185. [CrossRef] [PubMed]

109. Kong, Q.; Yao, J.; Qiu, Z.; Shen, D. Effect of mass proportion of municipal solid waste incinerator bottom ash layer to municipal solid waste layer on the Cu and Zn discharge from landfill. *BioMed Res. Int.* **2016**, *2016*, 9687879. [CrossRef] [PubMed]

110. Li, W.B.; Yao, J.; Malik, Z.; di Zhou, G.; Dong, M.; Shen, D.S. Impact of MSWI bottom ash codisposed with MSW on landfill stabilization with different operational modes. *BioMed Res. Int.* **2014**, *2014*, 167197. [CrossRef] [PubMed]

111. Gineys, N.; Aouad, G.; Sorrentino, F.; Damidot, D. Effect of the clinker composition on the threshold limits for Cu, Sn or Zn. *Cem. Concr. Res.* **2012**, *42*, 1088–1093. [CrossRef]

112. Shimoda, T.; Yokoyama, S. Eco-cement: A new portland cement to solve municipal solid waste and industrial waste problems. In Proceedings of the Creating with Concrete, University of Dundee, Dundee, UK, 6–10 September 1999; ICE Publishing: London, UK, May 1999.

113. De Schepper, M. Completely Recyclable Concrete for a More Environment Friendly Construction. Ph.D. Dissertation, Ghent University, Ghent, Belgium, 1 July 2014.

114. Wang, L.; Jin, Y.; Nie, Y.; Li, R. Recycling of municipal solid waste incineration fly ash for ordinary Portland cement production: A real-scale test. *Resour. Conserv. Recycl.* **2010**, *54*, 1428–1435. [CrossRef]

115. Collivignarelli, C.; Sorlini, S. Reuse of municipal solid wastes incineration fly ashes in concrete mixtures. *Waste Manag.* **2002**, *22*, 909–912. [CrossRef]

116. Chen, H.; Wang, S.; Tang, C. Reuse of incineration fly ashes and reaction ashes for manufacturing lightweight aggregate. *Constr. Build. Mater.* **2010**, *24*, 46–55. [CrossRef]
117. Ferraris, M.; Salvo, M.; Ventrella, A.; Buzzi, L.; Veglia, M. Use of vitrified MSWI bottom ashes for concrete production. *Waste Manag.* **2009**, *29*, 1041–1047. [CrossRef] [PubMed]
118. Cioffi, R.; Colangelo, F.; Montagnaro, F.; Santoro, L. Manufacture of artificial aggregate using MSWI bottom ash. *Waste Manag.* **2011**, *31*, 281–288. [CrossRef] [PubMed]
119. Hwang, C.; Bui, L.A.; Lin, K.; Lo, C. Manufacture and performance of lightweight aggregate from municipal solid waste incinerator fly ash and reservoir sediment for self-consolidating lightweight concrete. *Cem. Concr. Compos.* **2012**, *34*, 1159–1166. [CrossRef]
120. ASTM International. *ASTM C 618—Standard Specification for Coal Fly Ash and Raw or Calcined Natural Pozzolan for Use*; ASTM International: West Conshohocken, PA, USA, 2010.
121. Bureau voor Normalisatie. *NBN EN 450-1 Fly Ash for Concrete- Part 1: Definitions, Specifications and Conformity Criteria*; Bureau voor Normalisatie: Brussels, Belgium, 2012.
122. ASTM International. *ASTM C 150—Standard Specification of Portland Cement.pdf*; ASTM International: West Conshohocken, PA, USA, 2007.
123. ASTM International. *ASTM C 989 Standard Specification for Slag Cement for Use in Concrete and Mortars*; ASTM International: West Conshohocken, PA, USA, 2004.
124. Remond, S.; Pimienta, P.; Bentz, D.P. Effects of the incorporation of Municipal Solid Waste Incineration fly ash in cement pastes and mortars I. Experimental study. *Cem. Concr. Res.* **2002**, *32*, 303–311. [CrossRef]
125. Gao, X.; Wang, W.; Ye, T.; Wang, F.; Lan, Y. Utilization of washed MSWI fly ash as partial cement substitute with the addition of dithiocarbamic chelate. *J. Environ. Manage.* **2008**, *88*, 293–299. [CrossRef] [PubMed]
126. Lin, K.L.; Wang, K.S.; Lee, T.Y.; Tzeng, B.Y. The hydration characteristics of MSWI fly ash slag present in C3S. *Cem. Concr. Res.* **2003**, *33*, 957–964. [CrossRef]
127. Lin, K.L.; Wang, K.S.; Tzeng, B.Y.; Wang, N.F.; Lin, C.Y. Effects of Al_2O_3 on the hydration activity of municipal solid waste incinerator fly ash slag. *Cem. Concr. Res.* **2004**, *34*, 587–592. [CrossRef]
128. Wang, K.; Lin, K.; Huang, Z. Hydraulic activity of municipal solid waste incinerator fly-ash-slag-blended eco-cement. *Cem. Concr. Res.* **2001**, *31*, 97–103. [CrossRef]
129. Lee, T.; Wang, W.; Shih, P.; Lin, K. Enhancement in early strengths of slag-cement mortars by adjusting basicity of the slag prepared from fl y-ash of MSWI. *Cem. Concr. Res.* **2009**, *39*, 651–658. [CrossRef]
130. Lee, T.; Chang, C.; Rao, M.; Su, X. Modified MSWI ash-mix slag for use in cement concrete. *Constr. Build. Mater.* **2011**, *25*, 1513–1520. [CrossRef]
131. Lee, T.; Rao, M. Recycling municipal incinerator fly- and scrubber-ash into fused slag for the substantial replacement of cement in cement-mortars. *Waste Manag.* **2009**, *29*, 1952–1959. [CrossRef] [PubMed]
132. Lin, K.; Lin, D.; Wang, W.; Chang, C.; Lee, T. Pozzolanic reaction of a mortar made with cement and slag vitrified from a MSWI ash-mix and LED sludge. *Constr. Build. Mater.* **2014**, *64*, 277–287. [CrossRef]
133. Boccaccini, A.R.; Kopf, M.; Stumpfe, W. Glass-ceramics from filter dusts from waste incinerators. *Ceram. Int.* **1995**, *21*, 231–235. [CrossRef]
134. Beauchemin, L. *Jaarverslag van de Belgische Cementnijverheid*; Federatie van de Belgische cementnijverheid vzw: Brussels, Belgium, 2015.

![materials logo] *materials*

MDPI

Review

Luminescent Lanthanide MOFs: A Unique Platform for Chemical Sensing

Shu-Na Zhao [1,2], **Guangbo Wang** [1], **Dirk Poelman** [2] and **Pascal Van Der Voort** [1,*]

[1] Department of Chemistry, Center for Ordered Materials, Organometallics and Catalysis (COMOC), Ghent University, Krijgslaan 281 (S3), 9000 Gent, Belgium; shuna.zhao@Ugent.be (S.-N.Z.); Guangbo.Wang@UGent.be (G.W.)

[2] LumiLab, Department of Solid State Sciences, Ghent University, Krijgslaan 281 (S1), 9000 Gent, Belgium; Dirk.Poelman@UGent.be

[*] Correspondence: Pascal.VanDerVoort@UGent.be; Tel.: +32-9-264-44-42

Received: 19 March 2018; Accepted: 5 April 2018; Published: 7 April 2018

Abstract: In recent years, lanthanide metal–organic frameworks (LnMOFs) have developed to be an interesting subclass of MOFs. The combination of the characteristic luminescent properties of Ln ions with the intriguing topological structures of MOFs opens up promising possibilities for the design of LnMOF-based chemical sensors. In this review, we present the most recent developments of LnMOFs as chemical sensors by briefly introducing the general luminescence features of LnMOFs, followed by a comprehensive investigation of the applications of LnMOF sensors for cations, anions, small molecules, nitroaromatic explosives, gases, vapors, pH, and temperature, as well as biomolecules.

Keywords: metal–organic frameworks; lanthanide codoping; chemical sensors; ratiometric luminescence sensing

1. Introduction

Metal–organic frameworks (MOFs) have attracted extensive attention over the past few decades. They are an emerging class of highly crystalline and porous materials formed by metal ions or metal clusters connected by multitopic organic linkers [1]. Their large surface areas, framework flexibility, and tunable pore surface properties, as well as "tailor-made" framework functionalities empower them to be promising candidates for a diverse range of applications, such as gas separation and sorption [2–4], luminescence [5,6], chemical catalysis [7,8], drug delivery [9], magnetism [10], chemical sensing [11–13], energy storage and conversion [14–16], proton conduction [17,18], and bio-imaging [19].

As a subclass of MOFs, luminescent MOFs possess potential for practical applications because of their explicit environments for luminophores in a crystalline state and characteristic optical performance [20]. Generally, the luminescent properties of MOFs generate from metal components and organic linkers with aromatic or conjugated π systems. The metal–ligand charge transfer (MLCT) related luminescence can extend their luminescence functionalities to another dimension. Moreover, some adsorbed guest molecules within MOFs are able to contribute to the luminescent properties. Until now, research on luminescent MOFs has mainly focused on the fundamental luminescent properties of MOFs, and the rational design of tunable luminescent MOFs for light emitting applications [21]. Recently, luminescent MOFs have been proven to be a unique platform for chemical sensing due to their special features, including (i) easily tunable luminescence that can be used as the appropriate sensing signal; (ii) specific functional groups (e.g., Lewis sites and open metal sites) that are able to promote preferred host–guest binding for selective sensing; and (iii) the permanent MOFs' porosity that could concentrate the guest molecules, thereby enhancing detective sensitivity. Numerous luminescent MOF sensors have been developed and reported in the literature for detecting cations [22,23], anions [24,25], small molecules [26–28], biological molecules [29,30], explosive chemicals [31–33], vapors [34,35], and pH [36], as well as temperature [37–39].

Lanthanide MOFs (LnMOFs) have drawn much attention among the luminescent MOFs because of the unique luminescent properties of lanthanide ions, such as long lifetime, characteristic sharp emissions, large Stokes shifts, and high color purity with high quantum yields in the near-infrared and visible regions [40–45]. Additionally, the luminescent properties of lanthanide ions highly depend on the structural details of their coordination environment, offering a unique platform as chemical sensors. The combination of these characteristic luminescent properties of lanthanide ions with the intriguing topological structures of MOFs opens up promising possibilities for developing luminescent materials with special applications.

In this review, we present the most recent developments of LnMOFs as chemical sensors. We begin by briefly introducing the general luminescence features of LnMOFs, followed by a comprehensive investigation of the applications of LnMOF sensors with single or multiple luminescent centers. More specifically, LnMOF sensors for cations, anions, small molecules, nitroaromatic explosives, gases, vapors, pH, and temperature, as well as biomolecules will be discussed in detail in this review.

2. Luminescent Properties of LnMOFs

Generally, lanthanide ions (Ln^{3+}) are characterized by successive filling of the 4f orbitals, with electronic configurations of $[Xe]4f^n$ ($n = 0$ to 14). These electronic configurations generate a rich variety of electronic levels with the number $14!/n!(14-n)!$, resulting in interesting optical properties [46–49]. All of the Ln^{3+}, except La^{3+} ($4f^0$) and Lu^{3+} ($4f^{14}$), exhibit luminescent f–f emissions, which almost cover the entire spectrum. Eu^{3+}, Tb^{3+}, Sm^{3+}, and Tm^{3+} emit in the visible region with the color red, green, orange, and blue, respectively. Pr^{3+}, Nd^{3+}, Sm^{3+}, Dy^{3+}, Ho^{3+}, Er^{3+}, Tm^{3+}, and Yb^{3+} show emissions the near-infrared region, while Ce^{3+} shows a broadband emission from 370 to 410 nm because of the 5d–4f transition [50].

Typically, the 4f–4f transitions of Ln^{3+} are Laporte forbidden due to the 4f orbitals that are well-shielded by the filled $5s^2 5p^6$ subshells [51]. Consequently, direct photoexcitation of Ln^{3+} ions rarely produces highly luminescent materials due to the low absorption efficiency of the 4f–4f transitions. This problem can be overcome by the "antenna effect" (Figure 1), which commonly uses a strong absorbing chromophore to sensitize Ln^{3+} [52,53]. The overall process of antenna sensitization involves the following characteristic steps: (i) the organic ligands can absorb light upon excitation; (ii) the excitation energy is then transferred into Ln^{3+} excited states through intramolecular energy transfer; and (iii) Ln^{3+} ions undergo a radiative process by characteristic luminescence. This process could effectively increase the luminescence quantum yield of Ln^{3+} in normal conditions at room temperature. Furthermore, the solvent quenching and self-quenching of Ln^{3+} ions are almost nullified in LnMOFs due to the separation of Ln^{3+} ions by organic ligands. Consequently, LnMOFs exhibit strong luminescence and can be utilized as chemical sensors.

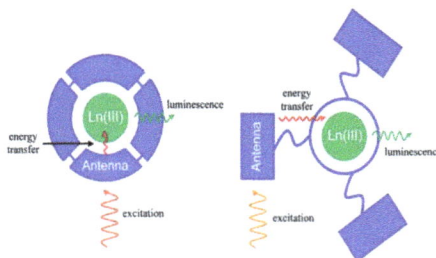

Figure 1. The antenna effect for lanthanide(III) (Ln(III)) sensitization, illustrated using the chromophoric chelate (right) and pendant chromophore (left) ligand designs. Reprinted with permission from [48]. Copyright 2009, American Chemical Society.

There are two other types of electronic transitions of Ln^{3+} ions: broad charge–transfer transitions (ligand–metal charge transfer (LMCT) and metal–ligand charge transfer (MLCT)) and broad 4f–5d transitions. They usually occur with high energies, resulting in rare observation in coordination compounds. However, the excitation energy of Sm^{3+}, Eu^{3+}, and Yb^{3+} can be transferred from an LMCT state to their 4f levels when the LMCT state lies at a high enough energy level. It is of great importance to investigate the numerous energy transfer processes for well-tuning the luminescent properties of LnMOFs.

The luminescence of Ln^{3+} ions is only possible from resonance levels, such as $^5D_0$ for Eu^{3+}, $^5D_4$ for Tb^{3+}, and $^2F_{5/2}$ for Yb^{3+}. The energies of resonance levels of Eu^{3+} ($^5D_0$), Tb^{3+} ($^5D_4$), and Yb^{3+} ($^2F_{5/2}$) lie at 17,250, 20,430, and 10,200 cm^{-1}, respectively [54]. If the Ln^{3+} ions are excited to a nonresonance level, the excitation energy is dissipated through a nonradiative process until a resonance level is reached. Therefore, the lowest triplet state of the organic ligands in LnMOFs must be located at an energy level nearly equal to or above the resonance level of the Ln^{3+} ions. If the energy difference between the organic linkers and Ln^{3+} ions is too small, a thermally activated energy back-transfer will occur. On the other hand, large energy differences may lead to slower energy transfer rates. The energy of the triplet state must be elaborately tuned to maximize the transfer and minimize the back-transfer. Thus, the rational design of suitable organic ligands with the appropriate energy level is of great significance for the synthesis of LnMOFs with the desired luminescent properties.

3. LnMOFs for Chemical Sensing

LnMOFs have been widely studied in various sensor applications owing to their inherent porosity and the particular luminescent properties of Ln^{3+} ions. Most of the LnMOF sensors show luminescence intensity changes, including luminescence enhancement (turn-on response) and quenching (turn-off response) upon recognition of the analytes. Eu^{3+} and Tb^{3+} are commonly used as luminescent centers in LnMOF sensors because of their strong, characteristic red emission at around 614 nm and green emission at around 541 nm, respectively [55]. LnMOFs succeed in sensing ionic species, small molecules, explosive chemicals, and pH, as well as temperature. In addition, the inherent structural and chemical features of LnMOFs make them considerably useful in biosensing and bioimaging applications [56]. In the remainder of this section, recent developments of LnMOFs for chemical sensing will be discussed in detail.

3.1. LnMOFs for Cation Sensing

Sensing and detecting metal ions is of great significance in environmental and ecological systems. Some transition-metal cations, such as Cu^{2+}, Fe^{2+}, Fe^{3+}, and Zn^{2+}, are essential in biological metabolism. The excess or deficiency of these metal cations can cause various diseases, such as Alzheimer's disease, Wilson's disease, anemia, mental decline, etc. [57–60]. Hg^{2+}, Pb^{2+}, and Cd^{2+} are well-known toxic metal ions that can give rise to serious damage to the human body and environment [61,62]. Therefore, the design and preparation of efficient and straightforward metal ion probes are urgently needed.

In 2009, Chen et al. reported a new LnMOF [Eu(PDC)$_{1.5}$(DMF)](DMF)$_{0.5}$(H$_2$O)$_{0.5}$ (PDC = pyridine-3,5-dicarboxylate, DMF = N′N-dimethylformamide) with Lewis basic pyridyl sites for sensing Cu^{2+} ions [63]. The desolated MOF Eu(PDC)$_{1.5}$ can selectively detect Co^{2+} and especially Cu^{2+} among other metal ions via a turn-off response. The authors hypothesized that the antenna efficiency of the PDC organic ligands was reduced by the binding of the pyridyl nitrogen atoms to Cu^{2+} or Co^{2+}, resulting in luminescence quenching. From then on, many LnMOF sensors with unsaturated Lewis basic sites have been synthesized based on this mechanism for detecting metal ions [64–68]. Recently, Yan and coworkers developed a FAM-ssDNA and Eu^{3+}@Bio-MOF-1 for sensing Cu^{2+} in aqueous solutions [69]. This luminescent hybrid material can simultaneously exhibit FAM and Eu^{3+} emissions by varying the ratio of Eu^{3+}@Bio-MOF-1 and FAM-ssDNA. Cu^{2+} can quench FAM emission, while enhancing the luminescence intensity of Eu^{3+} (Figure 2). The mechanism behind this is possibly based on the interaction of Cu^{2+} and ssDNA.

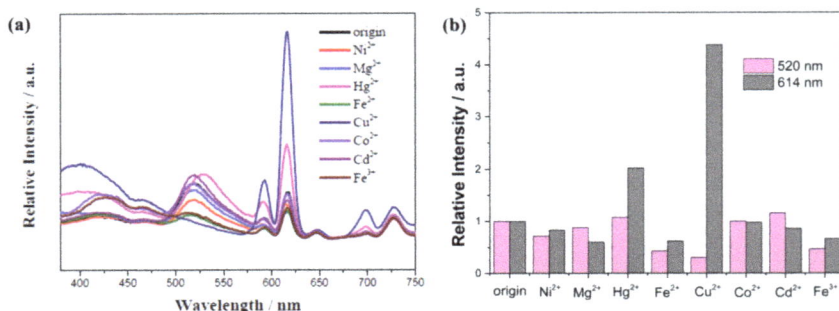

Figure 2. (**a**) PL spectra of FAM-ssDNA and Eu^{3+}@Bio-MOF-1 dispersed into aqueous solutions of various metal ions with the concentration of 10^{-5} mol/L when excited at 323 nm; (**b**) Relative luminescence intensity of FAM at 520 nm and Eu^{3+} at 614 nm. Reprinted with permission from [69]. Copyright 2017 Elsevier B.V., New York, NY, USA.

Additionally, Fe^{3+} detection was achieved by Zheng et al. with $[Eu(L_1)(BPDC)_{0.5}(NO_3)]\cdot H_3O$ (H_2L_1 = 2,5-di(pyridin-4-yl)terephthalic acid, BPDC = biphenyl-4,4'-dicarboxylic acid) based on an excellent luminescence turn-off response with a remarkable detection limit (5×10^{-7} mol/L) over various other metal cations, including Na^+, K^+, Cu^{2+}, Al^{3+}, Mg^{2+}, Cr^{3+}, Zn^{2+}, and Co^{2+} [70]. Sun and coworkers reported an anionic EuMOF, $[H_2N(CH_3)_2][Eu(H_2O)_2(BTMIPA)]\cdot 2H_2O$ (H_4BTMIPA = 5,5'-methylenebis(2,4,6-trimethylisophthalic acid)) with $[H_2N(CH_3)_2]^+$ cations in the tubular channels of the anionic frameworks, which exhibited luminescence quenching for Fe^{3+} and luminescence enhancement for Al^{3+} via ion-exchange between $[H_2N(CH_3)_2]^+$ cations and metal cations [71].

Tan et al. prepared adenine-based lanthanide coordination polymer nanoparticles (CPNPs), consisting of adenine (Ad), a Tb^{3+} ion, and dipicolinic acid (DPA). It showed a turn-on luminescence response for Hg^{2+} in aqueous solutions [72]. Due to the photoinduced electron transfer (PET) process, the Ad can transfer energy to the DPA and simultaneously prevent intramolecular energy transfer from DPA to Tb^{3+}, leading to the luminescence quenching of the CPNPs (Figure 3a). However, significantly enhanced luminescence (approximately fivefold) was observed in the CPNPs because of the suppression of the PET process from Ad to DPA by Hg^{2+}, which was further confirmed by Fourier-transform infrared spectroscopy (FTIR) and lifetime study (Figure 3b). This Hg^{2+} nanosensor also showed superior selectivity and exceptionally high sensitivity up to the detection limit of 0.2 nM and can be used in biosensing and imaging. Li and coworkers reported a EuMOF ($[Eu_2(FDC)_3DMA(H_2O)_3]\cdot DMA\cdot 4.5H_2O$, H_2FDC = 9,9-dimethyl-2,7-fluorenedicarboxylic acid,DMA = dimethylacetamide) for sensing Pb^{2+} in aqueous solutions through luminescence enhancement [73]. Another luminescence sensor for detecting Pb^{2+} based on a millimeter-sized TbMOF {$[Tb(L_2)(H_2O)_5]_n$·solvents $H_2L_2^-$ = 3, 5-dicarboxy-phenol anion ligand} was reported by Ji and coworkers [74]. It is the first high-efficiency MOF-based luminescence sensor for Pb^{2+} at a very low concentration and with the detection limit up to 10^{-7} M. A robust MOF, Sm-MIL-61(MIL-61 = Ga(OH)(btec)·0.5H_2O, H_4btec = Pyromellitic acid), was designed as an Ag^+ sensor in aqueous solutions with high efficiency and selectivity (Figure 3c,d). The luminescence enhancement was due to a more efficient energy transfer from organic linkers to Sm^{3+} evoked by Ag^+ (Figure 3e) [75].

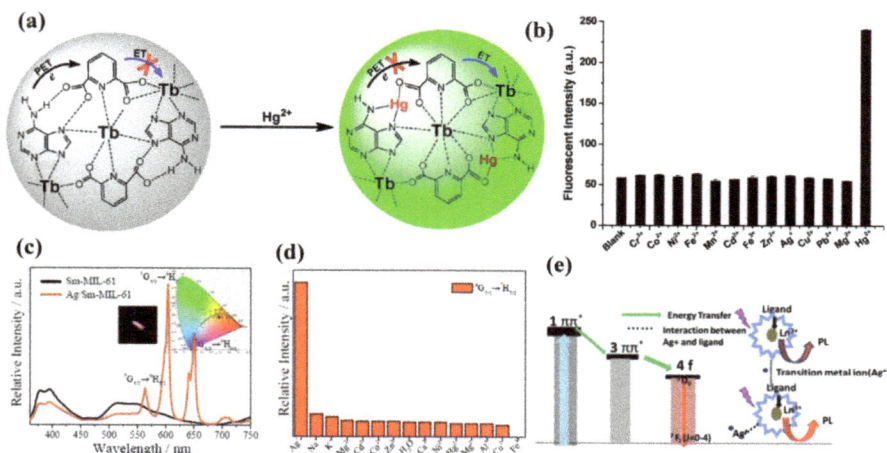

Figure 3. (**a**) Ad/Tb/DPA coordination polymer nanoparticles (CPNPs) for sensing of Hg^{2+} by photoinduced electron (PET) transfer; (**b**) Effect of various metal ions (1 μM) on the fluorescence intensity of Ad/Tb/DPA CPNPs at 545 nm. Reprinted with permission from [72]. Copyright 2012, American Chemical Society; (**c**) PL spectra of Sm-MIL-61 (black line) and Ag/Sm-MIL-61 (orange red line) when excited at 314 nm, the inset shows the CIE chromaticity diagram of Ag/Sm-MIL-61 (x: 0.4804 y: 0.3823) and corresponding photoluminescence color image under UV-] light irradiation at 314 nm; (**d**) The relative intensities of $^4G_{5/2} \rightarrow {}^6H_{7/2}$ at 603 nm upon Sm-MIL-61 in the presence of different metal ions; (**e**) Schematic illustration for the energy transfer from ligand to Ln^{3+} center. Reprinted with permission from [75]. Copyright 2017, The Royal Society of Chemistry.

3.2. LnMOFs for Anion Sensing

Various anions, such as halogen ions SO_4^{2-}, PO_4^{3-}, and CN^-, are fundamental in environmental and biological systems [76]. Therefore, the sensing of such anions is a remarkably interesting topic to investigate. In recent years, LnMOF-based sensors have been successfully utilized for sensing inorganic anions [77–80]. Chen and coworkers synthesized a TbMOF [Tb(BTC)·G] (BTC = benzene-1,3,5-tricarboxylate, G = guest solvent) with OH groups in the terminal solvents [78]. This TbMOF showed a fourfold luminescence enhancement in the presence of F^-, suggesting that this porous luminescent MOF is a promising candidate for sensing F^- (Figure 4a). The possible mechanism of luminescence enhancement by F^- ions lies in the stronger hydrogen bonding interactions between the F^- ion and the terminal methanol molecules that can restrict the stretching of the OH bond and thus reduce its quenching effect. The turn-off detection for F^- was achieved by Zhou and coworkers using an isostructural-doped LnMOF, $[Eu_{2x}Tb_{2(1-x)}(BPDC)(BDC)_2(H_2O)_2]_n$ (H_2BPDC = 2,2'-bipyridine-3,3'-dicarboxylic acid, H_2BDC = 1,4-benzenedicarboxylic acid) [79]. The emission intensity of this codoped LnMOF reduced to almost zero in the presence of F^- in aqueous solutions, while the emission intensities showed no change in the presence of Cl^-, Br^-, or I^-. The authors speculated that the F^- with smaller radii were trapped more easily in the MOF cavities than the other halogen anions, resulting in luminescence quenching. Shi et al. prepared two cationic hetero MOFs, $[Ln_2Zn(L_3)_3(H_2O)_4](NO_3)_2 \cdot 12H_2O_n$ (Ln = Eu and Tb, L_3 = 4,4'-dicarboxylate-2,2'-dipyridine anion) for selective and reversible I^- detection in aqueous solutions [80]. I^- ions can quench the luminescence of these two cationic MOFs with a fast response time (10 s) and low detection limit (0.001 ppm). It is believed that I_3^- ions are formed by the oxidation of I^- ions with the assistance of MOFs. They block the LMCT process by absorbing the excitation light, thus causing luminescence quenching.

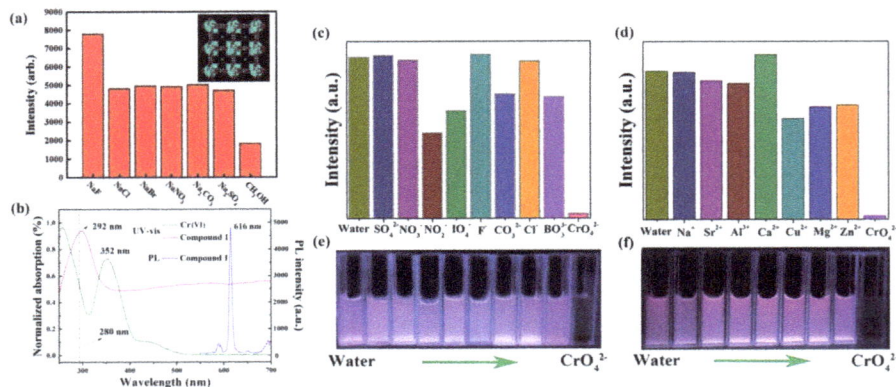

Figure 4. (**a**) $^5D_4 \rightarrow {}^7F_5$ transition intensities of [Tb(BTC)·G] activated in different types of 10^{-2} M NaX and Na_2X methanol solution (excited and monitored at 353 and 548 nm, respectively). The insert figure is the single crystal X-ray structure of [Tb(BTC)·G] activated in methanol containing NaF with the model of fluoride (green) at the center of the channel involving its hydrogen-bonding interaction with terminal methanol molecules (methanol oxygen, purple; the methyl group from methanol is omitted for clarity). Reprinted with permission from [78]. Copyright 2008, American Chemical Society. (**b**) Luminescence spectrum and UV−vis absorption spectra. (**c,d**) Luminescence intensity of $^5D_0 \rightarrow {}^7F_2$ of Eu^{3+} at 616 nm dispersed in different aqueous solutions of various anions and cations. (**e**) Luminescence in different anion solutions (excited at 365 nm), corresponding to figure (**c**). (**f**) Luminescence in different cation solutions (excited at 365 nm), corresponding to figure (**d**). Reprinted with permission from [81]. Copyright 2017, American Chemical Society.

Chromium is extensively used in various industrial processes causing Cr(VI) anions (CrO_4^{2-} and $Cr_2O_7^{2-}$) to often be present in all kinds of industrial wastewater It is one of the most prevalent, toxic heavy-metal ions of which excess intake can cause serious protein and DNA disruption, as well as damage the human enzyme system [82]. The detection of Cr(VI) anions was realized through a cationic EuMOF [$Eu_7(mtb)_5(H_2O)_{16}$]·NO_3·8DMA·18H_2O (H_4mtb = 4-[tris(4-carboxyphenyl)methyl]benzoic acid) with a luminescence turn-off response [81]. The Cr(VI) anions can absorb the excitation light and hinder the energy absorption of the EuMOF, resulting in luminescence quenching (Figure 4b–f). This highly stable EuMOF sensor with excellent sensitivity and selectivity can also be utilized in real environmental conditions, such as lake water and sea water, suggesting the possible application of MOF chemical sensors in environmental fields. In another study, Li et al. synthesized a EuMOF [$Eu(ipbp)_2(H_2O)_3$]·Br·6H_2O (H_2ipbpBr = 1-(3,5-dicarboxyphenyl)-4,4'-bipyridinium bromide) for the selective detection of $Cr_2O_7^{2-}$ and CrO_4^{2-} anions with K_{sv} of 8.98×10^3 M^{-1} and 7.08×10^3 M^{-1}, respectively [83].

The commonly used strong oxidant, MnO_4^-, causes serious damage to the environment and human health. A stable luminescence sensor for MnO_4^- was designed by Yan and coworkers using an In-MOF supporter encapsulated with Eu^{3+} ions [24]. Upon the addition of MnO_4^-, the luminescence of In-MOF-Eu was quenched to dark, corresponding to the competition of MnO_4^- with the organic linkers for absorption of excitation light. Moreover, the color of the MOF-based fluorescence test paper can be observed to go from red to black by the naked eye under UV light irradiation when immersed in different MnO_4^- concentrations. Li and coworkers utilized a heterometallic alkaline earth–lanthanide MOF {[$Ba_3La_{0.5}(\mu_3$-L)$_{2.5}(H_2O)_3$(DMF)]·(3DMF)}$_n$ (H_3L_4 = p-terphenyl-3,4'',5-tricarboxylic acid) to detect MnO_4^- with significant quenching over other anions, such as PO_4^{3-}, Cl^-, SiF_6^{2-}, CO_3^{2-}, HCO_3^-, BF_4^-, NO_3^-, Ac^-, SCN^-, SO_4^{2-}, Br^-, I^-, F^-, IO_3^-, BrO_3^-, and ClO_4^{2-} [84]. This probe exhibited high selectivity and sensitivity for MnO_4^- ions with high quench efficiency constants $K_{sv} = 7.73 \times 10^3$ M^{-1}, as well as a low fluorescence-detection limit (0.28 µM (S/N = 3)).

PO_4^{3-} ions are also a type of pollutant anion that can cause water eutrophication and serious pollution in aquatic ecosystems [85]. A luminescent TbMOF TbNTA·H_2O (NTA = nitrilotriacetate) for sensing PO_4^{3-} ions is provided by Qian and coworkers [86]. The luminescence intensity of TbNTA·H_2O quenched significantly in the presence of PO_4^{3-}, while it showed almost no change upon exposure to F^-, Cl^-, Br^-, I^-, NO_3^-, NO_2^-, HCO_3^-, CO_2^{2-}, or SO_4^{2-} (Figure 5a). They further discussed possible sensing mechanisms based on the matching degree of TbNTA·H_2O with different anions. The Tb–O bond may dilute the energy that transferred to Tb^{3+} via non-radioactive relaxation after incorporating PO_4^{3-} into TbNTA·H_2O (Figure 5b). Another PO_4^{3-} probe was achieved by Zhao and coworkers using a regenerable EuMOF {[$Eu_{1.5}(BTB)_{1.5}(H_2O)$]·$3DMF$}$_n$ (H_3BTB = 1,3,5-benzenetribenzoate) [87]. The recyclable performance of this EuMOF was investigated by fast and simple methods. Generally, this EuMOF was immersed in an PO_4^{3-} aqueous solution (10^{-3} M) for 20 s to completely form EuMOF-PO_4^{3-}, then EuMOF-PO_4^{3-} was washed in water several times to obtain the original EuMOF. The results demonstrate the promising practical applications of this recyclable PO_4^{3-} probe.

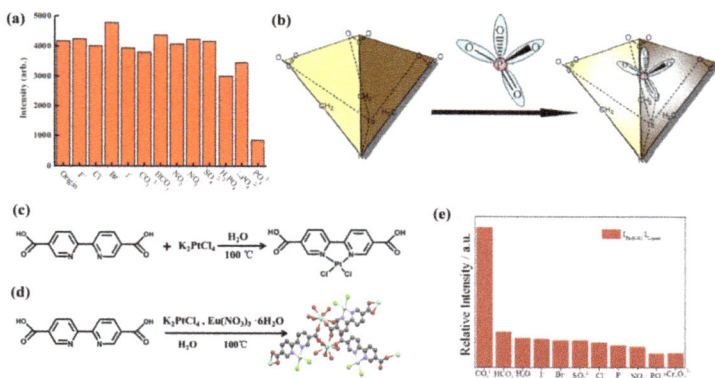

Figure 5. (**a**) Comparison of the $^5D_4 \rightarrow {^7}F_5$ transition intensities of TbNTA solid activated in 10^{-2} M NaX aqueous solution. (**b**) A schematic representation of the phosphate anion sensor. Reprinted with permission from [86]. Copyright 2010 Elsevier B.V. (**c,d**) Schematic illustration of the synthetic process of Eu/Pt-MOFs. The Eu, C, O, Pt, N, and Cl atoms are represented by blue, grey, red, white, purple and green, respectively. Hydrogen atoms and uncoordinated water molecules are excluded for clarity. (For interpretation of the references to color in this figure legend, the reader is referred to the web version of this article.) (**e**) A histogram demonstrating the value of $I_{Eu(614)}/I_{Ligand}$ according to the fluorescence spectrum. Reprinted with permission from [25]. Copyright 2017 Elsevier Ltd., Toronto, Canada.

Recently, Yan and coworkers reported a heterobimetallic Eu/Pt-MOF with dual emissions from both organic linkers and Eu^{3+} that exhibited facile, fast, and ratiometric detection of CO_3^{2-} (Figure 5c,d) [25]. The authors posited that the interaction with CO_3^{2-} suppressed the ligand-centered luminescence and enhanced the luminescence of Eu^{3+}, resulting in the maximum intensities ratio of Eu^{3+} (614 nm) to ligand (Figure 5e). The results indicate that the ratiometric sensing methodology could be an efficient platform for analytical monitoring of trace CO_3^{2-} in real samples due to the excellent orientation selectivity of CO_3^{2-}.

3.3. LnMOFs for Small Molecule Sensing

Formaldehyde (HCHO) is widespread in construction, furniture, and particle board, posing an impact on human health, such as watery eyes, asthma, and respiratory irritation [88]. Yu and coworkers developed a ratiometric luminescence HCHO probe through incorporation of Eu^{3+} ions into NH_2-UiO-66 under microwave irradiation conditions [89]. The dual-emitting luminescence originated from the characteristic red emission of Eu^{3+} ions (615 nm) and linker-to-cluster (Eu-oxo or Zr-oxo) charge transfer transition-related emission (465 nm). The interaction of the free amino groups with HCHO can drastically enhance emission around 465 nm due to the added electron transfer from the amino group with lone pair electrons to the positively charged HCHO. This is in contrast to the emission of Eu^{3+} at 615 nm that was only slightly enhanced. Then, a ratiometric luminescence HCHO probe was performed based on the intensity ratio of two emission bands at 465 nm and 615 nm. The results indicated that the fabrication of a ratiometric luminescence probe based on multiband luminescent MOFs can serve as a common sensing method for organic molecules. Another ratiometric luminescence sensor for HCHO was reported by Yang and coworkers [90]. This self-calibrating luminescent film was fabricated directly by growing Eu-NDC (H_2NDC = 2,6-naphthalenedicarboxylate) on hydrolyzed polyacrylonitrile (HPAN) via a layer-by-layer strategy (Figure 6a). The Eu-NDC@HPAN thin film can detect HCHO via a ratiometric luminescence approach with a 3.2-fold increase of the relative ratio of luminescence intensities at 453 nm and 616 nm. It has been proposed that the Eu-NDC frameworks will decompose after adding HCHO, while the NDC ligands regenerate, resulting in luminescence quenching and enhancing of Eu^{3+} ions and NDC, respectively (Figure 6b). The remarkable selectivity, sensitivity, and water stability of this film HCHO probe indicates its potential use in life sciences.

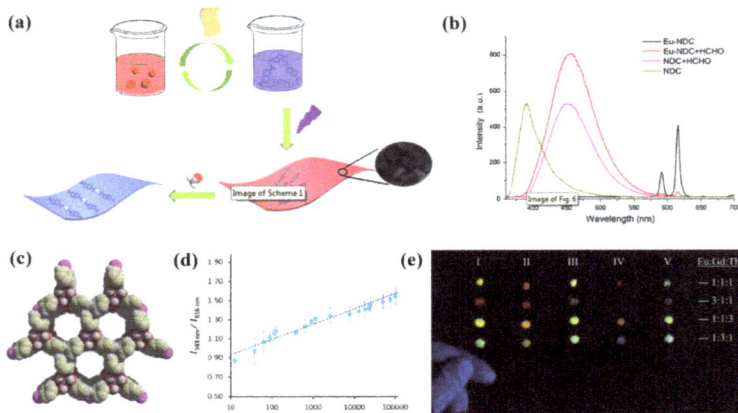

Figure 6. (**a**) Schematic representation of the synthesis process of Eu-NDC@HPAN and the luminescence quenching phenomenon of formaldehyde (HCHO) to Eu-NDC@HPAN; (**b**) Luminescence spectra of deprotonated NDC and Eu-NDC (0.25 mg/mL) (before and after treatment with formaldehyde) (λ_{ex} = 360 nm). Reprinted with permission from [90]. Copyright 2017 Elsevier B.V.; (**c**) Space-filling view along the c axis shows 1D hexagonal channels of PCM-22; (**d**) Relative photoemission response ratios obtained upon the addition of trace H_2O to Eu_1:Tb_5-PCM-22 presoaked in D_2O show a linear response. Error bars were obtained from three separate experiments; (**e**) Model dipstick detectors demonstrated for H_2O sensing: (I) as-synthesized, (II) desolvated in air with a heat gun, (III) after exposure to H_2O, (IV) reactivation using a heat gun, and (V) re-exposure to H_2O. Reprinted with permission from [91]. Copyright 2017 Elsevier Inc.

Recently, Humphrey and coworkers reported a rare example of a LnMOF probe for detecting trace H_2O in D_2O [91]. D_2O is an isotopically labeled version of H_2O and is widely used in chemical analysis and medicine [92]. High-purity D_2O is essential in various spectroscopic and synthetic applications. The codoped PCM-22 [Ln(tctp)(OH$_2$)$_3$]·3(1,4-dioxane) (Ln = Eu^{3+}, Tb^{3+} and Gd^{3+}, tctpH$_3$ = tris(p-carboxylato)triphenylphosphine (P(C_6H_4-*p*-CO_2H)$_3$)) has a 3D structure consisting of puckered

2D honeycomb sheets with large hexagonal channels and exhibits the characteristic luminescence of Eu^{3+} and Tb^{3+} (Figure 6c). This material allows for immediate solvent identification through color changes, which can easily be observed by the naked eye. Interestingly, the sensor can also be employed to quantitatively detect trace H_2O in D_2O (Figure 6d,e), as well as acetone, ethanol, and acetonitrile by uncomplicated spectrophotometry. To the best of our knowledge, this codoped LnMOF is the first material-based sensor for detecting H_2O in D_2O from 10 to 120,000 ppm. Buschbaum et al. proposed a new approach to obtain a ratiometric H_2O probe by using superparamagnetic microparticles Fe_3O_4/SiO_2 as a core and different LnMOFs as a shell [93]. $[Eu_2(BDC)_3] \cdot 2H_2O \cdot 2DMF$ (BDC^{2-} = benzene dicarboxylate) and $[Ln_2Cl_6(bipy)_3] \cdot 2bipy$ (Ln = Eu and Tb; bipy = 4,4'-bipyridine) were chosen to functionalize the Fe_3O_4/SiO_2 core, forming a color-tuned yellow-emitting Fe_3O_4/SiO_2@mixed-MOF composite system. The luminescence of two MOFs decreased unequally upon the presence of H_2O, allowing for a quantitative detection of H_2O content by the Tb^{3+}- and Eu^{3+}-based luminescence intensity ratio. In addition, the Fe_3O_4/SiO_2@mixed-MOF composite system can easily be removed from the liquid phase by means of a magnet.

A EuMOF $[Eu(FBPT)(H_2O)(DMF)]$ (FBPT = 2'-fluoro-biphenyl-3,4',5-tricarboxylate) for sensing acetone was reported by Zhang and coworkers [22]. The luminescence intensity of this EuMOF primarily depends on the organic solvents, particularly in the case of acetone, which exhibited the most significant quenching effect. It has been suggested that the competition of absorbing excited light energy between FBPT and acetone plays an important role in their luminescence diminishment. Guo et al. examined the capability of NIR luminescent YbMOF $Yb(BPT)(H_2O) \cdot (DMF)_{1.5}(H_2O)_{1.25}$ (BPT = biphenyl-3,4',5-tricarboxylate) for organic molecule sensing (Figure 7a–c) [94]. When excited at 326 nm, the active Yb(BPT) exhibits typical NIR emission of Yb^{3+} ions at 980 nm (Figure 7d). The NIR emission showed significant quenching and enhancement effects in the presence of acetone and DMF, respectively (Figure 7e). This study opens up a new approach for luminescent MOF-based sensors with NIR emission, demonstrating their potential applications in biological systems. Liu and coworkers synthesized a heterometallic MOF, $\{[Tb_2(ODA)_6Cd_3(H_2O)_6] \cdot 6H_2O\}_n$ (ODA = oxydiacetic acid), that can selectively detect ethanol and 2-propanol with luminescence turn-on and turn-off responses, respectively [95]. A series of MOFs composed of 4,4'-oxybis(benzoate) (OBA) ligands and suitable cations were reported by Hus and coworkers [96]. The MOF $Na[Tb(OBA)_2]_3 \cdot 0.4DMF_3 \cdot 1.5H_2O$ shows the strongest emission in the presence of BuOH and EtOH, whereas a much weaker emission was found in the presence of MeOH and H_2O. One possible mechanism explaining this is that MeOH and H_2O are trapped in the MOF cavities entering the coordination spheres of Tb^{3+}. This potentially causes the luminescence quenching effect. The EtOH and BuOH molecules then protect the Tb^{3+} from quenching by O–H oscillators because of the relatively sterical bulky alkyl groups. Wang et al. reported a codoped LnMOF, $[LnL_5(H_2O)_2] \cdot 2H_2O$ (Ln = Eu and Tb, H_3L_5 = 4-(2-carboxyphenoxy)benzene-1,3-dioic acid) showing good sensitivity to CH_3CN and nitrobenzene [97]. The emission can be enhanced remarkably in the presence of CH_3CN, while nitrobenzene can significantly quench the emission.

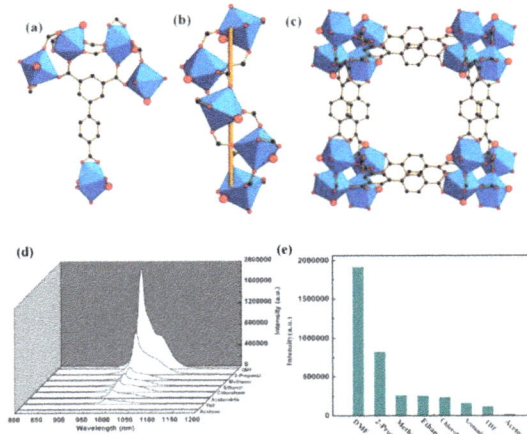

Figure 7. Crystal structure of YbMOF indicating (**a**) the BPT tricarboxylate linker and coordination environments of all Yb atoms related by symmetry; (**b**) 1D helical rod $[Yb(CO_2)_3]_n$ as the infinite SBU; and (**c**) 1D micropore of about 7.2×7.2 Å along c axis (Yb, blue polyhedra; C, black; O, red: terminal water molecules, large red sphere); (**d**) the PL spectra; and (**e**) the $^2F_{5/2}$–$^2F_{7/2}$ transition intensities of YbMOF introduced into various pure solvent emulsions when excited at 304 nm. Reprinted with permission from [94]. Copyright 2011 The Royal Society of Chemistry.

Benzene and its homologues, a prime type of toxic pollutant, bring great harm to both the environment and humans. It is therefore of significant importance to develop an efficient and easily processed approach to detect this kind of pollutant. Cheng and coworkers constructed a red luminescence sensor based on $\{[Eu_2(L_6)_3(DMF)_2] \cdot DMF \cdot MeOH\}_n(H_2L_6 = 5\text{-}(4H\text{-}1,2,4\text{-triazol-4-yl})$ benzene-1,3-dicarboxylic acid) to effectively detect polychlorinated benzenes [98]. This EuMOF represents a highly efficient quenching effect on detecting polychloriznated benzenes, including 1,2,4-trichlorobenzene, 1,2,3-tetrachlorobenzene, 1,2,4,5-tetrachlorobenzene, pentachlorobenzene, and hexachlorobenzene, which can be ascribed to the competition of the absorption of the excitation light between the analytes and ligands. Weng et al. fabricated a dual-emissive hybrid N-GQDs/Eu^{3+}@Mg-MOF (N-GQDs = N atom-doped graphene quantum dot, Mg-MOF = $\{[Mg_3(ndc)_{2.5}$ $(HCO_2)_2(H_2O)][NH_2Me_2] \cdot 2H_2O \cdot DMF\}$ 1,4-ndc = 1,4-naphthalenedicarboxylate) and employed it as a ratiometric luminescence sensor for decoding benzene homologues [53]. It exhibits dual-emission of N-GQDs and Eu^{3+} when excited at 394 nm, while the emission of the ligands and Eu^{3+} can be collected when excited at 349 nm. Thus, a 2D decoded map with I_L/I_{Eu} as abscissa and $I_{Eu}/I_{N\text{-}GQDs}$ as ordinate is established to identify benzene homologues. The results demonstrated that the decoded map can be used for the precise recognition of unknown compounds.

3.4. LnMOFs for Nitroaromatic Explosive Sensing

It is of great importance to selectively and rapidly detect nitroaromatic explosives in environmental monitoring, civilian safety, and homeland security [99]. The current methods for explosive detection are limited by their equipment demands and cost drawbacks [100]. However, luminescence sensing has proven to be an excellent detection technique for explosives owing to its speed and cost effectiveness, as well as to the fact that it is easily portable [101].

The detection of explosives using LnMOF-based luminescence sensors is usually performed in a turn-off manner. The luminescence quenching effect can be assigned to the photoinduced electron or energy transfer. The conduction band (CB) of the electron-rich MOF lies higher than the lowest unoccupied molecular orbitals (LUMOs) energy of the electron-deficient analytes. This allows for the electron transfer from the CB of the MOF sensors to the LUMOs of nitro analytes causing luminescence quenching [102]. Another possible mechanism for luminescence quenching is the competition of the absorption of the excitation light energy between the MOF ligands and nitro analytes. Based on these two possible sensing mechanisms, great success has been reported for sensing nitroaromatic explosives, such as nitrobenzene (NB), m-nitrotoluene (m-NT), o-nitrotoluene (o-NT), 3-nitrophenol (3-NP), 4-nitrophenol (4-NP), 2,4-dinitrophenol (2,4-NP), 2,4,6-trinitrophenol (TNP), and 2,4,6-trinitrotoluene (TNT) (Figure 8) [103–107].

The luminescence quenching efficiency of LnMOFs towards nitroaromatic explosives was analyzed using a quenching constant K_{sv} (M^{-1}) and detection limits. The quenching constant K_{sv} (M^{-1}) is calculated by using the Stern–Volmer (SV) equation, $(I_0/I) = K_{sv}[A] + 1$, where I_0 and I are the luminescence intensities before and after the addition of the analyte, respectively, and where [A] is the molar concentration of the analyte. The detection limit was calculated by K_{sv} values and the standard deviation (S_b), defined as nS_b/K_{sv} [108].

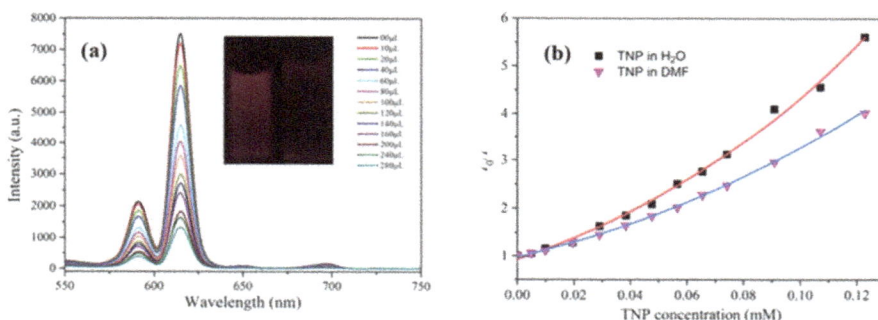

Figure 8. (**a**) Effect on the emission spectra of the activated EuMOF dispersed in H_2O upon incremental addition of a 2,4,6-trinitrophenol (TNP) aqueous solution (1 mM) (λ_{ex} = 362 nm). The legend indicates the overall concentration of TNP. Inset: A photograph showing the original fluorescence (**left**) and the fluorescence quenching (**right**) upon the addition of 280 µL TNP (UV light, 365 nm). (**b**) Stern–Volmer plots of I_0/I versus the TNP concentration in DMF and water. Reproduced with permission from [107]. Copyright 2014 Wiley-VCH.

3.5. LnMOFs for Gas and Vapor Sensing

The luminescent MOF films, CPM-5⊃∩Tb^{3+} and MIL-100(In)⊃∩Tb^{3+}, were designed by Qian and coworkers as a fast-response oxygen probe (Figure 9a,b) [109]. The luminescence intensities of the activated CPM-5⊃∩Tb^{3+} and MIL-100(In)⊃∩Tb^{3+} decreased gradually with increasing O_2 pressure. MIL-100(In)⊃∩Tb^{3+} showed higher quenching efficiencies (88%) than did CPM-5⊃∩Tb^{3+} (47%) at 1 atm of O_2 (Figure 9c,d). This is because the exposed carboxylate acids in MIL-100(In) can form Tb–O bonds with Tb^{3+} ions, leading to the intramolecular energy transfer, whilst Tb^{3+} merely balances cations in the pores of CPM-5, leading to intermolecular energy transfer. The high-oxygen sensitivity and short response/recovery time of MIL-100(In)⊃∩Tb^{3+} indicate their potential in sensing gases or vapors.

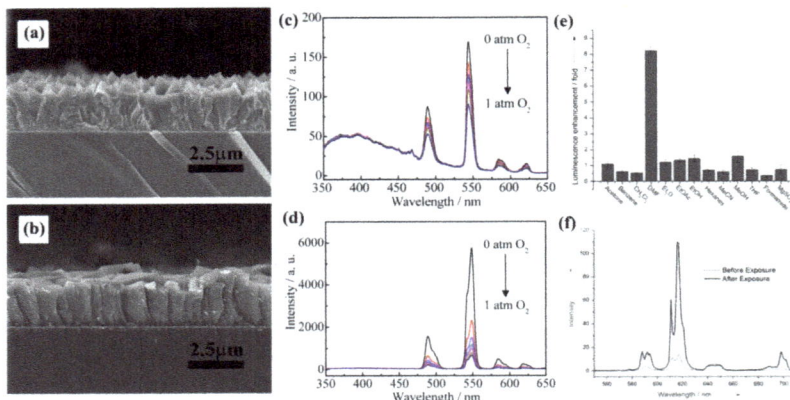

Figure 9. SEM images of CPM-5⊃ ∩Tb³⁺ (**a**) and MIL-100(In)⊃ ∩Tb³⁺ (**b**) films. Emission spectra of activated (**c**) CPM-5⊃ ∩Tb³⁺ and (**d**) MIL-100(In)⊃ ∩Tb³⁺ films under different oxygen partial pressure (Po₂). Reproduced with permission from [109]. Copyright 2014 American Chemical Society. (**e**) Increase in Eu emission intensity of [Eu₂L₃(H₂O)₄]·3DMF (I_{after}/I_{before}-1) after incubation for 24 h under various solvent vapors and with anhydrous MgSO₄. The intensity is measured at 616 nm. Error bars indicate the standard deviations of three or four parallel experiments. (**f**) Emission spectra of [Eu₂L₃(H₂O)₄]·3DMF before and after exposure to DMF vapor (excitation at 323 nm). The broad peak around 640 nm arises from scattering. Reproduced with permission from [110]. Copyright 2013 Wiley-VCH, Weinheim, Germany.

Song and coworkers exploited a EuMOF [Eu₂(L₇)₃(H₂O)₄]·3DMF (L₇ = 2′,5′-bis(methoxymethyl)-[1,1′:4′,1″-terphenyl]-4,4″-dicarboxylate) for sensing DMF vapor with a turn-on response [110]. A water-exchanged framework was formed by submerging the EuMOF in distilled water for 3 days and consequently showed much weaker Eu³⁺-based emission due to the quenching effect of the water molecules. The Eu³⁺ luminescence intensity exhibits a more than eightfold increase in the presence of DMF vapor. This is primarily due to partial replacement of the channel water by DMF molecules that reduce the quenching effect of the water molecules. This explanation was further confirmed by the fluorescence decay of deuteroxide- and water-exchanged samples. Moreover, DMF molecules within the channels of the compound can also modulate the energy levels of the ligands, thus promoting the LMCT process, all confirmed by NMR and XRD studies.

Besides the distinct rotten egg smell for which this toxic gas is commonly known, hydrogen sulfide (H₂S) is of great importance in biological systems, as well as the cause of acid rain and other environmental problems [111]. Tan and coauthors developed a ratiometric sensor for H₂S based on Cu²⁺-mediated fluorescence of LnCPs doped with carbon dots (CDs) (CDs@ZIF-8@GMP/Tb) [112]. GMP/Tb on the surface of ZIF-8 (zeolitic imidazolate framework-8) displays a typical ON-OFF-ON behavior upon the sequential addition of Cu²⁺ and H₂S, an observation that can be put to use in response signaling (Figure 10a,b). The fluorescence of the CDs of CDs@ZIF-8@GMP/Tb remains unchanged in the presence of Cu²⁺ or/and H₂S, empowering CDs to be of good reference. As a result, a ratiometric fluorescence sensor based on CDs@ZIF-8@GMP/Tb for sensing H₂S was fabricated (Figure 10c). The high selectivity towards H₂S against other anions (e.g., thiols and biological species) and the distinct feature of reversible sensing of this ratiometric sensor will promote the development of more sensitive ratiometric sensors based on LnMOFs. Another H₂S probe was reported by Yang and coworkers based on the postsynthetic modification of Tb³⁺@Cu-MOF [113]. The Tb³⁺@Cu-MOF (Cu-MOF: [Cu(HCPOC)₂]ₙ H₂CPOC = 5-(4′-carboxylphenoxy) nicotinic acid) exhibits a typically weak emission of Tb³⁺ yet a strong ligand-centered emission. The Tb³⁺-based emission can be strongly enhanced by H_2S due to its superior affinity towards Cu^{2+} ions. The detection performance of Tb^{3+}@Cu-MOF (1.20 μM) is capable of meeting that of biological systems indicating its potential in real-time organismal H_2S sensing.

Figure 10. (**a**) Schematic illustration of the preparation of CDs@ZIF-8@GMP/Tb and its working principle for ratiometric detection of H_2S; (**b**) SEM images of CDs@ZIF-8@GMP/Tb. Inset is the corresponding TEM image; (**c**) Emission spectra of CDs@ZIF-8@Tb/GMP in the presence of Cu^{2+} and Cu^{2+} + H_2S. Reproduced with permission from [112]. Copyright 2017 Elsevier B.V., New York, NY, USA.

Tanase and coworkers reported a dual-mode humidity sensor based on a EuMOF $[Eu(H_2O)_2(mpca)_2Eu(H_2O)_6W(CN)_8]\cdot nH_2O$ (mpca = 2-pyrazine-5-methyl-carboxylate) [114]. This EuMOF has a robust three-dimensional network with significant hydrophilic open channels filled with water molecules. The luminescence intensity of the EuMOF gradually decreases as the humidity increases. This effective and remarkably reliable humidity sensor also shows good linearity over a broad humidity range from 0% to 100% RH. Moreover, this sensing material was also examined for electrical detection methods. The recovery time of these methods was found to be similar to that in the photoluminescence measurement.

3.6. LnMOFs for pH Sensing

The need to explore fast pH sensing in industry, biomedicine, and many other environmental fields in order to monitor pH values and changes in biological systems and living cells has recently become of top priority [115]. The advantages of luminescence-based pH probes including quick response and high sensitivity, as well as easy operation, making them particularly desirable [116]. Chen and coworkers designed a pH-sensitive MOF nanoparticle using DMF and 1,10-phenanthroline (Phen) as ligands with Tb^{3+} ions based on the intramolecular-charge-transfer (ICT) effect [117]. A DMF molecule contains both an electron-donor and -acceptor part, allowing it to generate ICT [118]. It can furthermore change the Tb^{3+}-based luminescence through the antenna effect. Consequently, the protonation of H^+ could change the charge transfer of DMF and further change the antenna effect for Tb^{3+}, in turn resulting in a change of Tb^{3+}-based luminescence. The Phen molecule in the nanoparticle was used to improve such a change and reduce the luminescence quenching effect of Tb^{3+} by replacing the coordinated water molecules. The emission intensity of DMF–Tb was improved approximately 4 times, while the emission intensity of DMF–Tb–Phen was improved 10 times due to a decrease of the ICT effect and increase of the antenna effect on the Tb^{3+} ions upon adding H^+. This MOF nanoparticle pH sensor with high specificity and sensitivity could be used in strong acidic conditions, indicating its potential applications in biological systems. Qian and coworkers fabricated a fluorescence pH sensor by encapsulating Eu^{3+} ions into the pores of the nanoscale UiO-67-bpydc (bpydc = 2,2'-bipyridine-5,5'-dicarboxylic acid) [119]. The luminescence intensity of Eu^{3+}@UiO-67-bpydc shows a significant luminescence turn-off response in acidic solutions while exhibiting florescence enhancement in basic solutions. This is because protonation and deprotonation of the ligands first change the excited-state energy level of the ligands followed by a change in ligand-to-Eu energy transfer efficiency, explaining the different changes in the Eu^{3+}-based luminescence. This Eu^{3+}@UiO-67-bpydc pH sensor is stable within a wide pH range of 1.06 to 10.99 and can thus be used in physiological environments (pH = 6.80–7.60). The bio-compatibility of Eu^{3+}@UiO-67-bpydc was further confirmed by an MTT (MTT = 3-(4,5-dimethylthiazol-2-yl)-2,5-diphenyl-tetrazolium bromide) assay. Cell imaging results demonstrate that the Eu^{3+}@UiO-67-bpydc pH probe could be a promising candidate for monitoring pH both in vitro and in vivo. Very recently, the same group reported another luminescence pH sensor based on a nanoscale mixed LnMOF $Eu_{0.034}Tb_{0.966}(fum)_2(ox)(H_2O)_4$ (fum = fumarate, ox = oxalate) [120]. The $Eu_{0.034}Tb_{0.966}(fum)_2(ox)(H_2O)_4$ pH sensor shows high stability in aqueous solutions. Moreover, its morphology and size can easily be adjusted by changing the amount of CTAB surfactant. The mixed LnMOF exhibits both Tb^{3+} (545 nm) and Eu^{3+} (618 nm) emissions, which can be used for sensing pH values ranging from 3.00 to 7.00 in a ratiometric manner (Figure 11a–c). The MTT analysis and optical microscopy assay show that this mixed LnMOF sensor has low cytotoxicity and favorable biocompatibility (Figure 11e–f), indicating its potential to be applied as a pH sensor in physiological environments.

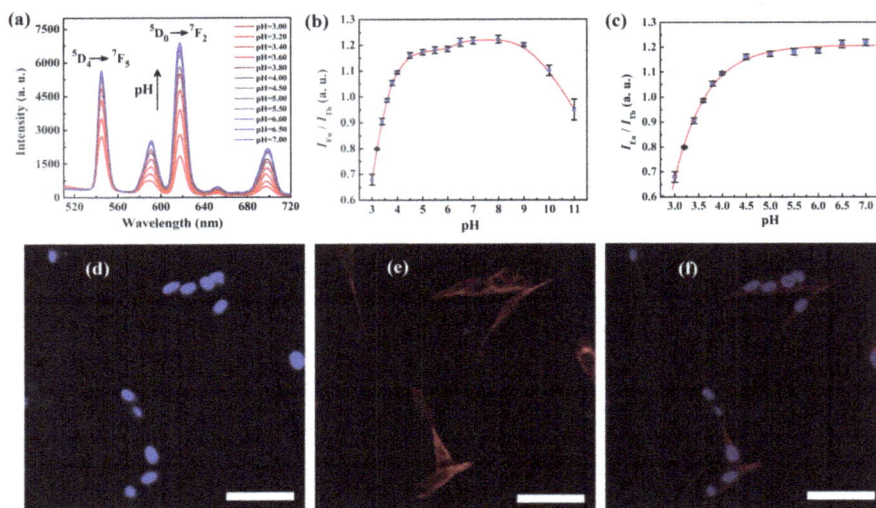

Figure 11. (**a**) pH-dependent emission spectra of $Eu_{0.034}Tb_{0.966}$-NMOF (W = 20, W is defined as the water-to-surfactant molar ratio) in the pH ranging from 3.00 to 7.00; (**b**) pH-dependent intensity ratio of Eu^{3+} (618 nm) to Tb^{3+} (545 nm) in the pH ranging from 3.00 to 11.00; (**c**) The fitted curve of $Eu_{0.034}Tb_{0.966}$-NMOF (W = 20) in the pH ranging from 3.00 to 7.00; (**d,e**) fluorescence and (**f**) overlapped confocal microscopy images of fixed PC12 cells incubated with 50 μg·mL^{-1} $Eu_{0.034}Tb_{0.966}$-NMOF for 24 h. Microtubular cytoskeleton (tubulin, red) and nuclei (blue) were fluorescently stained. The scale bar is 50 μm. Reproduced with permission from [120]. Copyright 2017 The Royal Society of Chemistry.

3.7. LnMOFs for Temperature Sensing

Temperature is an important thermodynamic parameter in human life and scientific investigations. Therefore, accurate temperature measurement is essential in both scientific and human development. Among the approaches for temperature determination, luminescence-based measurements have achieved tremendous attention with regards to their prominent advantages, including noninvasiveness, fast response, accuracy, high spatial resolution, and ability to work in strong electro or magnetic fields [121]. However, the most luminescent thermometers depend on a single emission susceptible to errors because of sample concentration changes and drifts of the optoelectronic system.

Qian and coworkers fabricated the first self-calibrated luminescent temperature sensor using a mixed LnMOF $Eu_{0.0069}Tb_{0.9931}$-DMBDC (DMBDC = 2,5-dimethoxy-1,4-benzenedicarboxylate) [37]. For Tb-DMBDC and Eu-DMBDC, the characteristic luminescence gradually decreases because of thermal activation of nonradiative decay pathways. However, $Eu_{0.0069}Tb_{0.9931}$-DMBDC exhibits a significant temperature-dependent luminescent behavior as the temperature increases from 10 to 300 K. The Tb^{3+}-based emission in $Eu_{0.0069}Tb_{0.9931}$-DMBDC decreases as the temperature increases, while that of the Eu^{3+} ions increases. This can be ascribed to the efficient energy transfer from Tb^{3+} to Eu^{3+} based on the phonon-assisted Förster transfer mechanism, an effect confirmed by luminescence lifetime measurements. The good linear relationship between the I_{Tb}/I_{Eu} ratio and temperature in the range of 50–200 K suggests that $Eu_{0.0069}Tb_{0.9931}$-DMBDC is an excellent temperature thermometer within this temperature range. These results suggest that mixed LnMOFs featuring temperature-dependent luminescence can be ideal candidates for self-referencing temperature sensing. Since then, many mixed LnMOFs have been fabricated for temperature measurement based on similar luminescent behavior [122].

In 2015, another mixed LnMOF $(Nd_{0.577}Yb_{0.423})_2(BDC-F_4)_3(DMF)(H_2O)\cdot DMF$ (H_2BDC-F_4 = 2,3,5,6-tetrafluoro-1,4-benzenedicarboxylate) with typical NIR emission for temperature sensing was designed by Qian and coworkers (Figure 12a) [123]. NIR emission can enter the biological system because of its relatively small adsorption and scattering. Thus, such NIR temperature thermometers have great potential for monitoring temperature in biological systems. The intensity ratio of Nd^{3+} at 1060 nm and Yb^{3+} at 980 nm is linearly related to temperatures in the physiological range (293–313 K) with a relative sensitivity of 0.816% K^{-1} at 313 K (Figure 12b,c), suitable for use in biomedical diagnosis.

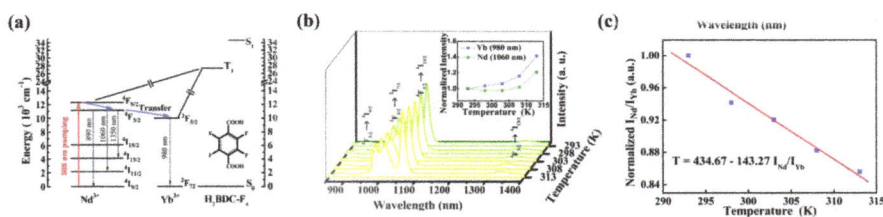

Figure 12. (a) Schematic representation of energy processes in $Nd_{0.577}Yb_{0.423}BDC-F_4$. (b) Emission spectra of $Nd_{0.577}Yb_{0.423}BDC-F_4$ in the range of 293–313 K excited at 808 nm; inset: temperature dependence of the normalized intensity of the corresponding transitions. (c) Temperature-dependent intensity ratio of Nd^{3+} (1060 nm) to Yb^{3+} (980 nm) and the fitted curve for $Nd_{0.577}Yb_{0.423}BDC-F_4$. Reproduced with permission from [123]. Copyright 2015 The Royal Society of Chemistry.

More recently, Qian and coworkers suggested that ratiometric temperature sensors can be achieved by the MOF⊃∩luminescent guest species composite method because of the energy transfer between luminescent guest species and Ln^{3+} ions [124]. The ZJU-88⊃∩perylene composite (ZJU⊃∩88 = $[Eu_2 (QPTCA)(NO_3)_2(DMF)_4]\cdot(CH_3CH_2OH)_3$, QPTCA = 1,1':4',1'':4'',1'''-quaterphenyl-3,3''',5,5'''-tetracarboxylic acid)) was designed as a dual-emitting thermometer with high sensitivity (1.28% °C^{-1} at 20 °C) (Figure 13a). Further results showed that the ZJU-88⊃∩perylene had good stability, an outstanding linear relationship, and excellent biocompatibility under simulated physiological conditions (Figure 13b,c), all indicating its potential use as a luminescent thermometer in biological applications.

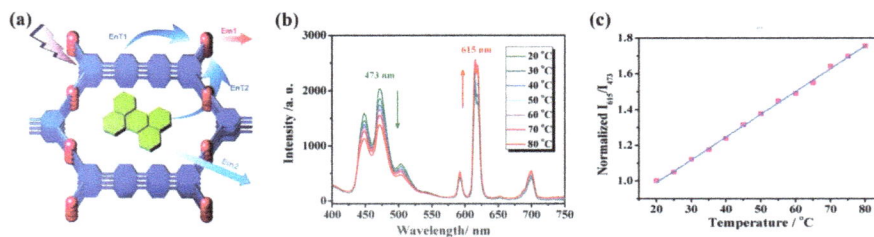

Figure 13. (**a**) Design of dual-emitting ZJU-88⊃ ∩perylene composite (EnT: energy transfer, Em: emission); (**b**) Emission spectra of ZJU-88⊃ ∩perylene recorded from 20 to 80 °C, excited at 388 nm; (**c**) Temperature-dependent intensity ratio of Eu^{3+} (615 nm) to perylene (473 nm) and the fitted curve for ZJU-88⊃ ∩perylene. Reproduced with permission from [124]. Copyright 2015 WILEY-VCH.

3.8. LnMOFs for Biosensing

Nitrofurans are a type of extensively used veterinary antibiotics effective for the treatment of protozoan and bacterial infections in human beings. It is, however, still urgently needed, as ell as very challenging to develop a rapid and effective approach to detect nitrofuran antibiotics (NFAs) [125]. Yang and coworkers fabricated a Eu-BCA ({[Eu_2(BCA)$_3$(H_2O)(DMF)$_3$]·0.5DMF·H_2O}$_n$, BCA = 2,2′-biquinoline-4,4′-dicarboxylate) thin-film sensor for NFAs by coating a cost-effective stainless-steel wire mesh using the Co_3O_4 nano-anchor fixation approach. The Eu-BCA thin-film sensor shows significant quenching effect for NFAs owing to the synergistic effect of electron-transfer and the inner-filter effect. It furthermore shows high selectivity and sensitivity to NFAs with detection limits of 0.21 and 0.16 mm for nitrofurantoin (NFT) and nitrofurazone (NFZ), respectively. NFAs were also successfully detected in real samples, indicating the potential of this Eu-BCA thin-film for biosensing [126].

Another pharmaceutical sensor was designed by Wang and coworkers based on a luminescent mixed-crystal LnMOF (MLMOF-3 = $Eu_{0.1}Tb_{0.9}$-BTC) thin film [127]. The uniform and continuous thin film was prepared by coating the monodisperse nanoscale MLMOF-3 on indium–tin–oxide (ITO) glass (Figure 14a,b). The luminescence intensity ratios of Eu^{3+} at 619 nm to Tb^{3+} at 547 nm of the MLMOF-3 film were used to calculate the intensity ratio change by (R–R_0)/R_0, where R_0 is the initial intensity ratio without the analyte, and R is the intensity ratio upon the addition of the analyte (Figure 14c). The luminescence intensity depended significantly on several pharmaceutical molecules (such as antipyrine, benzafibrate, caffeine, clofibrate, clotetracycline, coumarin, diclofenac, fluorouracil, nalidixic acid, naproxen, sulfachinoxalin, and tetracycline) Moreover, the MLMOF-3 thin film shows different guest-dependent colors that can intuitively be distinguished by the naked eye (Figure 14d,e). The authors presumed that the different functional groups and structures of these pharmaceutical molecules may not only modulate the antenna effect between organic linkers and Ln^{3+} ions but also affect energy transfer between Tb^{3+} and Eu^{3+}, causing the different luminescent changes in the MLMOF-3 thin film. These results demonstrate that the mixed LnMOF film can be used as luminescence sensors for different pharmaceutical molecules.

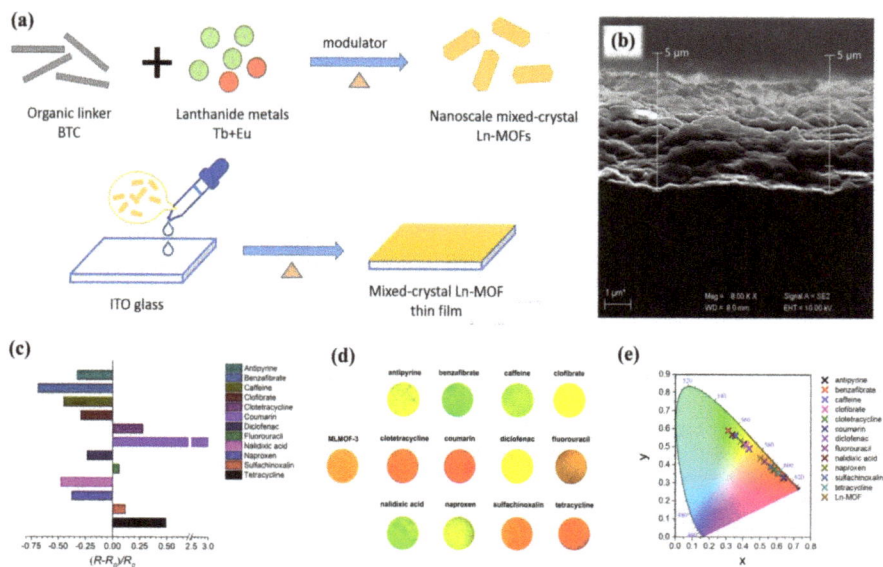

Figure 14. (a) Preparation process of mixed-crystal LnMOF thin film; (b) SEM image of cross-section of the film; (c) the emission intensity ratio changes; (d) the optical photographs; and (e) CIE chromaticity coordinates of the LnMOF thin film in the presence of different analytes. (20 mL, 10^{-4} M). Reproduced with permission from [127]. Copyright 2017 Elsevier B.V.

Yan and coworkers were the first to design a diagnosis platform for vinyl chloride carcinogen based on a 3d–4f–4d heterometallic MOF (Eu^{3+}/Cu^{2+}-$Zr_6O_4(OH)_4(O_2C$-C_6H_2-CO_2 $(CO_2H)_2)_6\cdot xH_2O$) [128]. The nanoprobe exhibits high selectivity to thiodiglycolic acid (TDGA) with a luminescence enhancement of about 27.5-fold, the main metabolite of vinyl chloride monomer (VCM) in human urine. It further shows a fast response to TDGA within 4 min and impressive sensitivity with a detection limit of 89 ng·mL^{-1} without interference of other coexisting species in urine. Such excellent sensing performance enables it to monitor TDGA levels in human urine. Furthermore, a portable urine dipstick based on the sensor has been developed to conveniently evaluate individual's intoxication degree of VCM.

4. Conclusions and Outlook

This comprehensive review covers the recent research progress on luminescent lanthanide MOFs and their applications in sensing cations, anions, small molecules, nitroaromatic explosives, gases, vapors, pH, temperature, and biomolecules. The sensing functionality of LnMOF probes is based on their luminescence changes in response to different analytes, all recognizable by means of spectrofluorometry or the naked eye. Most of the luminescent LnMOF sensors operate by a turn-off mechanism when detecting electron acceptors in which luminescence is quenched through both the electron and the energy transfer between LnMOF sensors and analytes. However, the turn-on detection mode with higher sensitivity and lower detection limits has also been implemented in luminescence-based LnMOF sensors resulting in luminescence enhancement or wavelength shifts. Furthermore, rational incorporation of the functional sites (e.g., Lewis acidic or basic sites and open metal sites) on the pores of the LnMOFs has made them very promising sensors to detect target compounds. Moreover, the ratiometric sensing approach has easily been achieved by embedding multi-luminescent motifs onto the frameworks, which can overcome the main drawbacks of the intensity-based measurements with only one transition.

Although the sensing behavior of LnMOFs has been studied comprehensively, some problems remain. While many investigations have shown excellent results for sensing hazardous materials, fast detection of nitroexplosives with a handheld device in public places, such as the airport and railway station, stays challenging. Furthermore, nanoscale luminescent MOFs with controllable size and morphology are very promising in applications for sensing in living cells. More efforts should be devoted to integrating different functionalities such as cellular sensing and imaging and molecular targeting, as well as drug delivery for practical applications in theranostic nanomedicine. Moreover, in-depth studies on the relationships between structure and luminescent behavior must be conducted using theoretical methods. In addition, the stabilities of recycling, material cost, and portability for practical applications need further improvement. With constant efforts being made to handle these challenges, we believe that the LnMOFs definitely hold a bright future in the field of luminescence sensing.

Acknowledgments: This research has been made possible by a postdoctoral scholarship of the Ghent University Special Research Fund (ref. number: BOF.PDO.2016.0030.01). G.W. would like to thank the Chinese Scholarship Council and UGENT-BOF grant for financial support.

Conflicts of Interest: The authors declare no conflict of interest.

Abbreviations

Ad	adenine
BCA	2,2′-biquinoline-4,4′-dicarboxylate
H_2BDC	1,4-benzenedicarboxylic acid
H_2BDC-F_4	2,3,5,6-tetrafluoro-1,4-benzenedicarboxylate
bipy	4,4′-bipyridine
H_2BPDC	2,2′-bipyridine-3,3′-dicarboxylic acid
BPT	biphenyl-3,4′,5-tricarboxylate
bpydc	2,2′-bipyridine-5,5′-dicarboxylic acid
H_3BTB	1,3,5-benzenetribenzoate
BTC	benzene-1,3,5- tricarboxylate
H_4btec	pyromellitic acid
$H_4BTMIPA$	5,5′-methylenebis(2,4,6-trimethylisophthalic acid)
CB	conduction band
CDs	carbon dots
HCHO	formaldehyde
CPNPs	coordination polymer nanoparticles
H_2CPOC	5-(4′-carboxylphenoxy) nicotinic acid
DMA	dimethylacetamide
DMBDC	2,5-dimethoxy-1,4-benzenedicarboxylate
DMF	N′N-dimethylformamide
DPA	dipicolinic acid
H_2FDC	9,9-dimethyl-2,7-fluorenedicarboxylic acid
FBPT	2′-fluoro-biphenyl-3,4′,5-tricarboxylate
FTIR	Fourier-transform infrared spectroscopy
fum	fumarate
ICT	intramolecular-charge-transfer
$H_2ipbpBr$	1-(3,5-dicarboxyphenyl)-4,4′-bipyridinium bromide
ITO	indium–tin–oxide
H_2L_1	2,5-di(pyridin-4-yl)terephthalic acid
$H_2L_2{}^-$	3, 5-dicarboxy-phenol anion ligand
L_3	4,4′-dicarboxylate-2,2′-dipyridine anion

H_3L_4	p-terphenyl-3,4'',5-tricarboxylic acid
H_3L_5	4-(2-carboxyphenoxy)benzene-1,3-dioic acid
H_2L_6	5-($4H$-1,2,4-triazol-4-yl)benzene-1,3-dicarboxylic acid
L_7	2',5'-bis(methoxymethyl)-[1,1':4',1''-terphenyl]-4,4''-dicarboxylate
LMCT	ligand-to-metal charge transfer
LnMOFs	lanthanide metal–organic frameworks
LUMOs	lowest unoccupied molecular orbitals
Mg-MOF	{[$Mg_3(ndc)_{2.5}(HCO_2)_2(H_2O)$][$NH_2Me_2$]·$2H_2O$·DMF}
MIL-61	$Ga(OH)(btec)$·$0.5H_2O$
MLCT	metal-to-ligand charge transfer
MOFs	Metal–organic frameworks
mpca	2-pyrazine-5-methyl-carboxylate
H_4mtb	4-[tris(4-carboxyphenyl)methyl]benzoic acid
MTT	3-(4,5-dimethylthiazol-2-yl)-2,5-diphenyl-tetrazolium bromide
H_2NDC	2,6-naphthalenedicarboxylate
NB	nitrobenzene
1,4-ndc	1,4-naphthalenedicarboxylate
NFAs	nitrofuran antibiotics
NFT	nitrofurantoin
NFZ	nitrofurazone
2,4-NP	2,4-dinitrophenol
3-NP	3-nitrophenol
4-NP	4-nitrophenol
N-GQDs	N atom-doped graphene quantum dots
m-NT	m-nitrotoluene
o-NT	o-nitrotoluene
NTA	nitrilotriacetate
OBA	4,4'-oxybis(benzoate)
ODA	oxydiacetic acid
ox	oxalate
HPAN	hydrolyzed polyacrylonitrile
PDC	pyridine-3,5-dicarboxylate
PET	photoinduced electron transfer
Phen	1,10-phenanthroline
QPTCA	1,1':4',1'':4'',1'''-quaterphenyl-3,3''',5,5'''-tetracarboxylic acid
H_2S	hydrogen sulfide
tctpH_3	tris(p-carboxylato)triphenylphosphine
TDGA	thiodiglycolic acid
TNP	2,4,6-trinitrophenol
TNT	2,4,6-trinitrotoluene
VCM	vinyl chloride monomer
ZIF-8	zeolitic imidazolate framework-8

References

1. Colon, Y.J.; Snurr, R.Q. High-throughput computational screening of metal-organic frameworks. *Chem. Soc. Rev.* **2014**, *43*, 5735–5749. [CrossRef] [PubMed]

2. Zhao, X.; Bu, X.; Nguyen, E.T.; Zhai, Q.G.; Mao, C.; Feng, P. Multivariable Modular Design of Pore Space Partition. *J. Am. Chem. Soc.* **2016**, *138*, 15102–15105. [CrossRef] [PubMed]

3. He, Y.; Zhou, W.; Qian, G.; Chen, B. Methane storage in metal-organic frameworks. *Chem. Soc. Rev.* **2014**, *43*, 5657–5678. [CrossRef] [PubMed]

4. Flaig, R.W.; Osborn Popp, T.M.; Fracaroli, A.M.; Kapustin, E.A.; Kalmutzki, M.J.; Altamimi, R.M.; Fathieh, F.; Reimer, J.A.; Yaghi, O.M. The Chemistry of CO_2 Capture in an Amine-Functionalized Metal-Organic Framework under Dry and Humid Conditions. *J. Am. Chem. Soc.* **2017**, *139*, 12125–12128. [CrossRef] [PubMed]

5. Medishetty, R.; Nalla, V.; Nemec, L.; Henke, S.; Mayer, D.; Sun, H.; Reuter, K.; Fischer, R.A. A New Class of Lasing Materials: Intrinsic Stimulated Emission from Nonlinear Optically Active Metal-Organic Frameworks. *Adv. Mater.* **2017**, *29*, 1605637. [CrossRef] [PubMed]

6. Zhu, M.; Hao, Z.M.; Song, X.Z.; Meng, X.; Zhao, S.N.; Song, S.Y.; Zhang, H.J. A new type of double-chain based 3D lanthanide(III) metal-organic framework demonstrating proton conduction and tunable emission. *Chem. Commun.* **2014**, *50*, 1912–1914. [CrossRef] [PubMed]

7. Nguyen, C.V.; Liao, Y.-T.; Kang, T.-C.; Chen, J.E.; Yoshikawa, T.; Nakasaka, Y.; Masuda, T.; Wu, K.C.W. A metal-free, high nitrogen-doped nanoporous graphitic carbon catalyst for an effective aerobic HMF-to-FDCA conversion. *Green Chem.* **2016**, *18*, 5957–5961. [CrossRef]

8. Fan, S.; Dong, W.; Huang, X.; Gao, H.; Wang, J.; Jin, Z.; Tang, J.; Wang, G. In Situ-Induced Synthesis of Magnetic Cu-CuFe$_2$O$_4$@HKUST-1 Heterostructures with Enhanced Catalytic Performance for Selective Aerobic Benzylic C–H Oxidation. *ACS Catal.* **2016**, *7*, 243–249. [CrossRef]

9. Du, P.-Y.; Gu, W.; Liu, X. A three-dimensional Nd(III)-based metal–organic framework as a smart drug carrier. *New J. Chem.* **2016**, *40*, 9017–9020. [CrossRef]

10. Vallejo, J.; Fortea-Pérez, F.R.; Pardo, E.; Benmansour, S.; Castro, I.; Krzystek, J.; Armentano, D.; Cano, J. Guest-dependent single-ion magnet behaviour in a cobalt(II) metal–organic framework. *Chem. Sci.* **2016**, *7*, 2286–2293. [CrossRef]

11. Zhao, S.N.; Song, X.Z.; Zhu, M.; Meng, X.; Wu, L.L.; Feng, J.; Song, S.Y.; Zhang, H.J. Encapsulation of Ln(III) Ions/Dyes within a Microporous Anionic MOF by Post-synthetic Ionic Exchange Serving as a Ln(III) Ion Probe and Two-Color Luminescent Sensors. *Chem. Eur. J.* **2015**, *21*, 9748–9752. [CrossRef] [PubMed]

12. Wales, D.J.; Grand, J.; Ting, V.P.; Burke, R.D.; Edler, K.J.; Bowen, C.R.; Mintova, S.; Burrows, A.D. Gas sensing using porous materials for automotive applications. *Chem. Soc. Rev.* **2015**, *44*, 4290–4321. [CrossRef] [PubMed]

13. Hao, Z.; Yang, G.; Song, X.; Zhu, M.; Meng, X.; Zhao, S.; Song, S.; Zhang, H. A europium(III) based metal–organic framework: Bifunctional properties related to sensing and electronic conductivity. *J. Mater. Chem. A* **2014**, *2*, 237–244. [CrossRef]

14. Niu, S.; Wang, Z.; Zhou, T.; Yu, M.; Yu, M.; Qiu, J. A Polymetallic Metal-Organic Framework-Derived Strategy toward Synergistically Multidoped Metal Oxide Electrodes with Ultralong Cycle Life and High Volumetric Capacity. *Adv. Funct. Mater.* **2017**, *27*, 1605332. [CrossRef]

15. Zhao, S.-N.; Song, X.-Z.; Song, S.-Y.; Zhang, H.-J. Highly efficient heterogeneous catalytic materials derived from metal-organic framework supports/precursors. *Coord. Chem. Rev.* **2017**, *337*, 80–96. [CrossRef]

16. Salunkhe, R.R.; Kaneti, Y.V.; Kim, J.; Kim, J.H.; Yamauchi, Y. Nanoarchitectures for Metal-Organic Framework-Derived Nanoporous Carbons toward Supercapacitor Applications. *Acc. Chem. Res.* **2016**, *49*, 2796–2806. [CrossRef] [PubMed]

17. Yan, H.; Hohman, J.N.; Li, F.H.; Jia, C.; Solis-Ibarra, D.; Wu, B.; Dahl, J.E.; Carlson, R.M.; Tkachenko, B.A.; Fokin, A.A.; et al. Hybrid metal-organic chalcogenide nanowires with electrically conductive inorganic core through diamondoid-directed assembly. *Nat. Mater.* **2017**, *16*, 349–355. [CrossRef] [PubMed]

18. Liu, S.; Yue, Z.; Liu, Y. Incorporation of imidazole within the metal-organic framework UiO-67 for enhanced anhydrous proton conductivity. *Dalton Trans.* **2015**, *44*, 12976–12980. [CrossRef] [PubMed]

19. Li, Y.; Tang, J.; He, L.; Liu, Y.; Liu, Y.; Chen, C.; Tang, Z. Core-Shell Upconversion Nanoparticle@Metal-Organic Framework Nanoprobes for Luminescent/Magnetic Dual-Mode Targeted Imaging. *Adv. Mater.* **2015**, *27*, 4075–4080. [CrossRef] [PubMed]

20. Cui, Y.; Yue, Y.; Qian, G.; Chen, B. Luminescent functional metal-organic frameworks. *Chem. Rev.* **2012**, *112*, 1126–1162. [CrossRef] [PubMed]

21. Wang, C.; Zhang, T.; Lin, W. Rational synthesis of noncentrosymmetric metal-organic frameworks for second-order nonlinear optics. *Chem. Rev.* **2012**, *112*, 1084–1104. [CrossRef] [PubMed]

22. Hao, Z.; Song, X.; Zhu, M.; Meng, X.; Zhao, S.; Su, S.; Yang, W.; Song, S.; Zhang, H. One-dimensional channel-structured Eu-MOF for sensing small organic molecules and Cu^{2+} ion. *J. Mater. Chem. A* **2013**, *1*, 11043–11050. [CrossRef]

23. Liu, W.; Huang, X.; Xu, C.; Chen, C.; Yang, L.; Dou, W.; Chen, W.; Yang, H.; Liu, W. A Multi-responsive Regenerable Europium-Organic Framework Luminescent Sensor for Fe^{3+}, Cr^{VI} Anions, and Picric Acid. *Chem. Eur. J.* **2016**, *22*, 18769–18776. [CrossRef] [PubMed]

24. Wu, J.X.; Yan, B. Eu(III)-functionalized In-MOF (In(OH)bpydc) as fluorescent probe for highly selectively sensing organic small molecules and anions especially for CHCl$_3$ and MnO$_4$. *J. Colloid Interface Sci.* **2017**, *504*, 197–205. [CrossRef] [PubMed]

25. Sun, N.-N.; Yan, B. Rapid and facile ratiometric detection of CO$_3$$^{2-}$ based on heterobimetallic metal-organic frameworks (Eu/Pt-MOFs). *Dyes Pigment.* **2017**, *142*, 1–7. [CrossRef]

26. Kreno, L.E.; Leong, K.; Farha, O.K.; Allendorf, M.; Van Duyne, R.P.; Hupp, J.T. Metal-organic framework materials as chemical sensors. *Chem. Rev.* **2012**, *112*, 1105–1125. [CrossRef] [PubMed]

27. Kang, X.M.; Cheng, R.R.; Xu, H.; Wang, W.M.; Zhao, B. A Sensitive Luminescent Acetylacetone Probe Based on Zn-MOF with Six-Fold Interpenetration. *Chem. Eur. J.* **2017**, *23*, 13289–13293. [CrossRef] [PubMed]

28. Cui, Y.; Chen, B.; Qian, G. Lanthanide metal-organic frameworks for luminescent sensing and light-emitting applications. *Coord. Chem. Rev.* **2014**, *273–274*, 76–86. [CrossRef]

29. Zhao, S.-N.; Wu, L.-L.; Feng, J.; Song, S.-Y.; Zhang, H.-J. An ideal detector composed of a 3D Gd-based coordination polymer for DNA and Hg^{2+} ion. *Inorg. Chem. Front.* **2016**, *3*, 376–380. [CrossRef]

30. Doonan, C.; Ricco, R.; Liang, K.; Bradshaw, D.; Falcaro, P. Metal-Organic Frameworks at the Biointerface: Synthetic Strategies and Applications. *Acc. Chem. Res.* **2017**, *50*, 1423–1432. [CrossRef] [PubMed]

31. Roales, J.; Moscoso, F.G.; Gamez, F.; Lopes-Costa, T.; Sousaraei, A.; Casado, S.; Castro-Smirnov, J.R.; Cabanillas-Gonzalez, J.; Almeida, J.; Queiros, C.; et al. Preparation of Luminescent Metal-Organic Framework Films by Soft-Imprinting for 2,4-Dinitrotoluene Sensing. *Materials* **2017**, *10*, 992. [CrossRef] [PubMed]

32. Lustig, W.P.; Mukherjee, S.; Rudd, N.D.; Desai, A.V.; Li, J.; Ghosh, S.K. Metal-organic frameworks: Functional luminescent and photonic materials for sensing applications. *Chem. Soc. Rev.* **2017**, *46*, 3242–3285. [CrossRef] [PubMed]

33. Yang, X.-L.; Chen, X.; Hou, G.-H.; Guan, R.-F.; Shao, R.; Xie, M.-H. A Multiresponsive Metal-Organic Framework: Direct Chemiluminescence, Photoluminescence, and Dual Tunable Sensing Applications. *Adv. Funct. Mater.* **2016**, *26*, 393–398. [CrossRef]

34. Chen, Y.; Wang, B.; Wang, X.; Xie, L.H.; Li, J.; Xie, Y.; Li, J.R. A Copper(II)-Paddlewheel Metal-Organic Framework with Exceptional Hydrolytic Stability and Selective Adsorption and Detection Ability of Aniline in Water. *ACS Appl. Mater. Interfaces* **2017**, *9*, 27027–27035. [CrossRef] [PubMed]

35. Wang, D.; Liu, J.; Liu, Z. A chemically stable europium metal-organic framework for bifunctional chemical sensor and recyclable on–off–on vapor response. *J. Solid State Chem.* **2017**, *251*, 243–247. [CrossRef]

36. Wang, J.; Li, Y.; Jiang, M.; Liu, Y.; Zhang, L.; Wu, P. A Highly Chemically Stable Metal-Organic Framework as a Luminescent Probe for the Regenerable Ratiometric Sensing of pH. *Chem. Eur. J.* **2016**, *22*, 13023–13027. [CrossRef] [PubMed]

37. Cui, Y.; Xu, H.; Yue, Y.; Guo, Z.; Yu, J.; Chen, Z.; Gao, J.; Yang, Y.; Qian, G.; Chen, B. A luminescent mixed-lanthanide metal-organic framework thermometer. *J. Am. Chem. Soc.* **2012**, *134*, 3979–3982. [CrossRef] [PubMed]

38. Wang, H.; Zhao, D.; Cui, Y.; Yang, Y.; Qian, G. A Eu/Tb-mixed MOF for luminescent high-temperature sensing. *J. Solid State Chem.* **2017**, *246*, 341–345. [CrossRef]

39. Rao, X.; Song, T.; Gao, J.; Cui, Y.; Yang, Y.; Wu, C.; Chen, B.; Qian, G. A highly sensitive mixed lanthanide metal-organic framework self-calibrated luminescent thermometer. *J. Am. Chem. Soc.* **2013**, *135*, 15559–15564. [CrossRef] [PubMed]

40. Zhang, Y.; Li, G.; Geng, D.; Shang, M.; Peng, C.; Lin, J. Color-tunable emission and energy transfer in Ca$_3$Gd$_7$(PO$_4$)(SiO$_4$)$_5$O$_2$: Ce^{3+}/Tb^{3+}/Mn^{2+} phosphors. *Inorg. Chem.* **2012**, *51*, 11655–11664. [CrossRef] [PubMed]

41. Zhang, Y.; Geng, D.; Kang, X.; Shang, M.; Wu, Y.; Li, X.; Lian, H.; Cheng, Z.; Lin, J. Rapid, large-scale, morphology-controllable synthesis of YOF:Ln^{3+} (Ln = Tb, Eu, Tm, Dy, Ho, Sm) nano-/microstructures with multicolor-tunable emission properties. *Inorg. Chem.* **2013**, *52*, 12986–12994. [CrossRef] [PubMed]

42. Zhang, Y.; Li, X.; Li, K.; Lian, H.; Shang, M.; Lin, J. Crystal-site engineering control for the reduction of Eu^{3+} to Eu^{2+} in CaYAlO$_4$: Structure refinement and tunable emission properties. *ACS Appl. Mater. Interfaces* **2015**, *7*, 2715–2725. [CrossRef] [PubMed]

43. Mao, J. Structures and luminescent properties of lanthanide phosphonates. *Coord. Chem. Rev.* **2007**, *251*, 1493–1520. [CrossRef]

44. Song, J.L.; Mao, J.G. New types of blue, red or near IR luminescent phosphonate-decorated lanthanide oxalates. *Chem. Eur. J.* **2005**, *11*, 1417–1424. [CrossRef] [PubMed]

45. Parmentier, A.B.; Smet, P.F.; Poelman, D. Broadband Luminescence in Rare Earth Doped Sr_2SiS_4: Relating Energy Levels of Ce^{3+} and Eu^{2+}. *Materials* **2013**, *6*, 3663–3675. [CrossRef] [PubMed]

46. Binnemans, K. Lanthanide-based luminescent hybrid materials. *Chem. Rev.* **2009**, *109*, 4283–4374. [CrossRef] [PubMed]

47. Choppin, G.R.; Peterman, D.R. Applications of lanthanide luminescence spectroscopy to solution studies of coordination chemistry. *Coord. Chem. Rev.* **1998**, *174*, 283–299. [CrossRef]

48. Moore, E.G.; Samuel, A.P.S.; Raymond, K.N. From Antenna to Assay: Lessons Learned in Lanthanide Luminescence. *Acc. Chem. Res.* **2009**, *42*, 542–552. [CrossRef] [PubMed]

49. Bünzli, J.-C.G. Lanthanide Luminescence for Biomedical Analyses and Imaging. *Chem. Rev.* **2010**, *110*, 2729–2755. [CrossRef] [PubMed]

50. Eliseeva, S.V.; Bunzli, J.C. Lanthanide luminescence for functional materials and bio-sciences. *Chem. Soc. Rev.* **2010**, *39*, 189–227. [CrossRef] [PubMed]

51. Bunzli, J.C.; Piguet, C. Taking advantage of luminescent lanthanide ions. *Chem. Soc. Rev.* **2005**, *34*, 1048–1077. [CrossRef] [PubMed]

52. Xu, L.-J.; Xu, G.-T.; Chen, Z.-N. Recent advances in lanthanide luminescence with metal-organic chromophores as sensitizers. *Coord. Chem. Rev.* **2014**, *273–274*, 47–62. [CrossRef]

53. Weng, H.; Yan, B. N-GQDs and Eu^{3+} co-encapsulated anionic MOFs: Two-dimensional luminescent platform for decoding benzene homologues. *Dalton Trans.* **2016**, *45*, 8795–8801. [CrossRef] [PubMed]

54. Parker, D. Luminescent lanthanide sensors for pH, pO_2 and selected anions. *Coord. Chem. Rev.* **2000**, *205*, 109–130. [CrossRef]

55. Sun, Y.-G.; Gu, X.-F.; Ding, F.; Smet, P.F.; Gao, E.-J.; Poelman, D.; Verpoort, F. Synthesis, Crystal Structures, and Properties of Novel Heterometallic La/Pr−Cu−K and Sm/Eu/Tb−Cu Coordination Polymers. *Cryst. Growth Des.* **2010**, *10*, 1059–1067. [CrossRef]

56. Liu, D.; Lu, K.; Poon, C.; Lin, W. Metal-organic frameworks as sensory materials and imaging agents. *Inorg. Chem.* **2014**, *53*, 1916–1924. [CrossRef] [PubMed]

57. Yi, F.-Y.; Chen, D.; Wu, M.-K.; Han, L.; Jiang, H.-L. Chemical Sensors Based on Metal-Organic Frameworks. *ChemPlusChem* **2016**, *81*, 675–690. [CrossRef]

58. Gaetke, L.M.; Chow-Johnson, H.S.; Chow, C.K. Copper: Toxicological relevance and mechanisms. *Arch. Toxicol.* **2014**, *88*, 1929–1938. [CrossRef] [PubMed]

59. Song, Y.; Qu, K.; Xu, C.; Ren, J.; Qu, X. Visual and quantitative detection of copper ions using magnetic silica nanoparticles clicked on multiwalled carbon nanotubes. *Chem. Commun.* **2010**, *46*, 6572–6574. [CrossRef] [PubMed]

60. Wen, R.M.; Han, S.D.; Ren, G.J.; Chang, Z.; Li, Y.W.; Bu, X.H. A flexible zwitterion ligand based lanthanide metal-organic framework for luminescence sensing of metal ions and small molecules. *Dalton Trans.* **2015**, *44*, 10914–10917. [CrossRef] [PubMed]

61. Song, C.; Yang, W.; Zhou, N.; Qian, R.; Zhang, Y.; Lou, K.; Wang, R.; Wang, W. Fluorescent theranostic agents for Hg^{2+} detection and detoxification treatment. *Chem. Commun.* **2015**, *51*, 4443–4446. [CrossRef] [PubMed]

62. Liu, T.; Che, J.X.; Hu, Y.Z.; Dong, X.W.; Liu, X.Y.; Che, C.M. Alkenyl/thiol-derived metal-organic frameworks (MOFs) by means of postsynthetic modification for effective mercury adsorption. *Chem. Eur. J.* **2014**, *20*, 14090–14095. [CrossRef] [PubMed]

63. Chen, B.; Wang, L.; Xiao, Y.; Fronczek, F.R.; Xue, M.; Cui, Y.; Qian, G. A luminescent metal-organic framework with Lewis basic pyridyl sites for the sensing of metal ions. *Angew Chem. Int. Ed.* **2009**, *48*, 500–503. [CrossRef] [PubMed]

64. Liu, B.; Hou, L.; Wu, W.P.; Dou, A.N.; Wang, Y.Y. Highly selective luminescence sensing for Cu^{2+} ions and selective CO_2 capture in a doubly interpenetrated MOF with Lewis basic pyridyl sites. *Dalton Trans.* **2015**, *44*, 4423–4427. [CrossRef] [PubMed]

65. Tang, Q.; Liu, S.; Liu, Y.; Miao, J.; Li, S.; Zhang, L.; Shi, Z.; Zheng, Z. Cation sensing by a luminescent metal-organic framework with multiple Lewis basic sites. *Inorg. Chem.* **2013**, *52*, 2799–2801. [CrossRef] [PubMed]

66. Hou, S.; Liu, Q.K.; Ma, J.P.; Dong, Y.B. Cd(II)-coordination framework: Synthesis, anion-induced structural transformation, anion-responsive luminescence, and anion separation. *Inorg. Chem.* **2013**, *52*, 3225–3235. [CrossRef] [PubMed]

67. Wu, Y.; Yang, G.P.; Zhao, Y.; Wu, W.P.; Liu, B.; Wang, Y.Y. Three new solvent-directed Cd(II)-based MOFs with unique luminescent properties and highly selective sensors for Cu^{2+} cations and nitrobenzene. *Dalton Trans.* **2015**, *44*, 3271–3277. [CrossRef] [PubMed]

68. Wen, G.X.; Wu, Y.P.; Dong, W.W.; Zhao, J.; Li, D.S.; Zhang, J. An Ultrastable Europium(III)-Organic Framework with the Capacity of Discriminating Fe^{2+}/Fe^{3+} Ions in Various Solutions. *Inorg. Chem.* **2016**, *55*, 10114–10117. [CrossRef] [PubMed]

69. Weng, H.; Yan, B. A Eu(III) doped metal-organic framework conjugated with fluorescein-labeled single-stranded DNA for detection of Cu(II) and sulfide. *Anal. Chim. Acta* **2017**, *988*, 89–95. [CrossRef] [PubMed]

70. Yan, W.; Zhang, C.; Chen, S.; Han, L.; Zheng, H. Two Lanthanide Metal-Organic Frameworks as Remarkably Selective and Sensitive Bifunctional Luminescence Sensor for Metal Ions and Small Organic Molecules. *ACS Appl. Mater. Interfaces* **2017**, *9*, 1629–1634. [CrossRef] [PubMed]

71. Chen, Z.; Sun, Y.; Zhang, L.; Sun, D.; Liu, F.; Meng, Q.; Wang, R.; Sun, D. A tubular europium-organic framework exhibiting selective sensing of Fe^{3+} and Al^{3+} over mixed metal ions. *Chem. Commun.* **2013**, *49*, 11557–11559. [CrossRef] [PubMed]

72. Tan, H.; Liu, B.; Chen, Y. Lanthanide Coordination Polymer Nanoparticles for Sensing of Mercury(II) by Photoinduced Electron Transfer. *ACS Nano* **2012**, *6*, 10505–10511. [CrossRef] [PubMed]

73. Li, L.; Chen, Q.; Niu, Z.; Zhou, X.; Yang, T.; Huang, W. Lanthanide metal–organic frameworks assembled from a fluorene-based ligand: Selective sensing of Pb^{2+} and Fe^{3+} ions. *J. Mater. Chem. C* **2016**, *4*, 1900–1905. [CrossRef]

74. Ji, G.; Liu, J.; Gao, X.; Sun, W.; Wang, J.; Zhao, S.; Liu, Z. A luminescent lanthanide MOF for selectively and ultra-high sensitively detecting Pb^{2+} ions in aqueous solution. *J. Mater. Chem. A* **2017**, *5*, 10200–10205. [CrossRef]

75. Sun, N.; Yan, B. Ag^+-induced photoluminescence enhancement in lanthanide post-functionalized MOFs and Ag^+ sensing. *Phys. Chem. Chem. Phys.* **2017**, *19*, 9174–9180. [CrossRef] [PubMed]

76. Mahata, P.; Mondal, S.K.; Singha, D.K.; Majee, P. Luminescent rare-earth-based MOFs as optical sensors. *Dalton Trans.* **2017**, *46*, 301–328. [CrossRef] [PubMed]

77. Wang, D.; Fan, J.; Shang, M.; Li, K.; Zhang, Y.; Lian, H.; Lin, J. Pechini-type sol–gel synthesis and multicolor-tunable emission properties of $GdY(MoO^4)_3:RE^{3+}$ (RE = Eu, Dy, Sm, Tb) phosphors. *Opt. Mater.* **2016**, *51*, 162–170. [CrossRef]

78. Chen, B.; Wang, L.; Zapata, F.; Qian, G.; Lobkovsky, E.B. A Luminescent Microporous Metal-Organic Framework for the Recognition and Sensing of Anions. *J. Am. Chem. Soc.* **2008**, *130*, 6718–6719. [CrossRef] [PubMed]

79. Zhou, J.M.; Shi, W.; Xu, N.; Cheng, P. Highly selective luminescent sensing of fluoride and organic small-molecule pollutants based on novel lanthanide metal-organic frameworks. *Inorg. Chem.* **2013**, *52*, 8082–8090. [CrossRef] [PubMed]

80. Shi, P.F.; Hu, H.C.; Zhang, Z.Y.; Xiong, G.; Zhao, B. Heterometal-organic frameworks as highly sensitive and highly selective luminescent probes to detect I(−) ions in aqueous solutions. *Chem. Commun.* **2015**, *51*, 3985–3988. [CrossRef] [PubMed]

81. Liu, W.; Wang, Y.; Bai, Z.; Li, Y.; Wang, Y.; Chen, L.; Xu, L.; Diwu, J.; Chai, Z.; Wang, S. Hydrolytically Stable Luminescent Cationic Metal Organic Framework for Highly Sensitive and Selective Sensing of Chromate Anions in Natural Water Systems. *ACS Appl. Mater. Interfaces* **2017**, *9*, 16448–16457. [CrossRef] [PubMed]

82. Thompson, C.M.; Kirman, C.R.; Proctor, D.M.; Haws, L.C.; Suh, M.; Hays, S.M.; Hixon, J.G.; Harris, M.A. A chronic oral reference dose for hexavalent chromium-induced intestinal cancer. *J. Appl. Toxicol.* **2014**, *34*, 525–536. [CrossRef] [PubMed]

83. Zhang, C.; Sun, L.; Yan, Y.; Shi, H.; Wang, B.; Liang, Z.; Li, J. A novel photo- and hydrochromic europium metal–organic framework with good anion sensing properties. *J. Mater. Chem. C* **2017**, *5*, 8999–9004. [CrossRef]

84. Ding, B.; Liu, S.X.; Cheng, Y.; Guo, C.; Wu, X.X.; Guo, J.H.; Liu, Y.Y.; Li, Y. Heterometallic Alkaline Earth-Lanthanide Ba^{II}-La^{III} Microporous Metal-Organic Framework as Bifunctional Luminescent Probes of Al^{3+} and $MnO_4{}^-$. *Inorg. Chem.* **2016**, *55*, 4391–4402. [CrossRef] [PubMed]

85. Zhao, D.; Wan, X.; Song, H.; Hao, L.; Su, Y.; Lv, Y. Metal–organic frameworks (MOFs) combined with ZnO quantum dots as a fluorescent sensing platform for phosphate. *Sens. Actuators B* **2014**, *197*, 50–57. [CrossRef]

86. Xu, H.; Xiao, Y.; Rao, X.; Dou, Z.; Li, W.; Cui, Y.; Wang, Z.; Qian, G. A metal–organic framework for selectively sensing of PO_4^{3-} anion in aqueous solution. *J. Alloys Compd.* **2011**, *509*, 2552–2554. [CrossRef]

87. Xu, H.; Cao, C.S.; Zhao, B. A water-stable lanthanide-organic framework as a recyclable luminescent probe for detecting pollutant phosphorus anions. *Chem. Commun.* **2015**, *51*, 10280–10283. [CrossRef] [PubMed]

88. El Sayed, S.; Pascual, L.; Licchelli, M.; Martinez-Manez, R.; Gil, S.; Costero, A.M.; Sancenon, F. Chromogenic Detection of Aqueous Formaldehyde Using Functionalized Silica Nanoparticles. *ACS Appl. Mater. Interfaces* **2016**, *8*, 14318–14322. [CrossRef] [PubMed]

89. Li, C.; Huang, J.; Zhu, H.; Liu, L.; Feng, Y.; Hu, G.; Yu, X. Dual-emitting fluorescence of Eu/Zr-MOF for ratiometric sensing formaldehyde. *Sens. Actuators B* **2017**, *253*, 275–282. [CrossRef]

90. Wang, Y.; Zhang, G.; Zhang, F.; Chu, T.; Yang, Y. A novel lanthanide MOF thin film: The highly performance self-calibrating luminescent sensor for detecting formaldehyde as an illegal preservative in aquatic product. *Sens. Actuators B* **2017**, *251*, 667–673. [CrossRef]

91. Dunning, S.G.; Nuñez, A.J.; Moore, M.D.; Steiner, A.; Lynch, V.M.; Sessler, J.L.; Holliday, B.J.; Humphrey, S.M. A Sensor for Trace H_2O Detection in D_2O. *Chem* **2017**, *2*, 579–589. [CrossRef]

92. Wiberg, K.B. The deuterium isotope effect. *Chem. Rev.* **1995**, *55*, 713–743.

93. Wehner, T.; Seuffert, M.T.; Sorg, J.R.; Schneider, M.; Mandel, K.; Sextl, G.; Müller-Buschbaum, K. Composite materials combining multiple luminescent MOFs and superparamagnetic microparticles for ratiometric water detection. *J. Mater. Chem. C* **2017**, *5*, 10133–10142. [CrossRef]

94. Guo, Z.; Xu, H.; Su, S.; Cai, J.; Dang, S.; Xiang, S.; Qian, G.; Zhang, H.; O'Keeffe, M.; Chen, B. A robust near infrared luminescent ytterbium metal-organic framework for sensing of small molecules. *Chem. Commun.* **2011**, *47*, 5551–5553. [CrossRef] [PubMed]

95. Ma, J.X.; Huang, X.F.; Song, X.Q.; Liu, W.S. Assembly of framework-isomeric 4 d–4 f heterometallic metal-organic frameworks with neutral/anionic micropores and guest-tuned luminescence properties. *Chem. Eur. J.* **2013**, *19*, 3590–3595. [CrossRef] [PubMed]

96. Lin, Y.W.; Jian, B.R.; Huang, S.C.; Huang, C.H.; Hsu, K.F. Synthesis and characterization of three ytterbium coordination polymers featuring various cationic species and a luminescence study of a terbium analogue with open channels. *Inorg. Chem.* **2010**, *49*, 2316–2324. [CrossRef] [PubMed]

97. Wang, S.; Shan, L.; Fan, Y.; Jia, J.; Xu, J.; Wang, L. Fabrication of Ln-MOFs with color-tunable photoluminescence and sensing for small molecules. *J. Solid State Chem.* **2017**, *245*, 132–137. [CrossRef]

98. Wang, L.; Fan, G.; Xu, X.; Chen, D.; Wang, L.; Shi, W.; Cheng, P. Detection of polychlorinated benzenes (persistent organic pollutants) by a luminescent sensor based on a lanthanide metal–organic framework. *J. Mater. Chem. A* **2017**, *5*, 5541–5549. [CrossRef]

99. Hu, Z.; Deibert, B.J.; Li, J. Luminescent metal-organic frameworks for chemical sensing and explosive detection. *Chem. Soc. Rev.* **2014**, *43*, 5815–5840. [CrossRef] [PubMed]

100. Salinas, Y.; Martinez-Manez, R.; Marcos, M.D.; Sancenon, F.; Costero, A.M.; Parra, M.; Gil, S. Optical chemosensors and reagents to detect explosives. *Chem. Soc. Rev.* **2012**, *41*, 1261–1296. [CrossRef] [PubMed]

101. Zhang, S.R.; Du, D.Y.; Qin, J.S.; Bao, S.J.; Li, S.L.; He, W.W.; Lan, Y.Q.; Shen, P.; Su, Z.M. A fluorescent sensor for highly selective detection of nitroaromatic explosives based on a 2D, extremely stable, metal-organic framework. *Chem. Eur. J.* **2014**, *20*, 3589–3594. [CrossRef] [PubMed]

102. Pramanik, S.; Zheng, C.; Zhang, X.; Emge, T.J.; Li, J. New microporous metal-organic framework demonstrating unique selectivity for detection of high explosives and aromatic compounds. *J. Am. Chem. Soc.* **2011**, *133*, 4153–4155. [CrossRef] [PubMed]

103. Sun, Z.; Li, Y.; Ma, Y.; Li, L. Dual-functional recyclable luminescent sensors based on 2D lanthanide-based metal-organic frameworks for highly sensitive detection of Fe^{3+} and 2,4-dinitrophenol. *Dyes Pigment.* **2017**, *146*, 263–271. [CrossRef]

104. Zhao, S.-N.; Song, X.-Z.; Zhu, M.; Meng, X.; Wu, L.-L.; Song, S.-Y.; Wang, C.; Zhang, H.-J. Highly thermostable lanthanide metal–organic frameworks exhibiting unique selectivity for nitro explosives. *RSC Adv.* **2015**, *5*, 93–98. [CrossRef]

105. Zhu, M.; Song, X.Z.; Song, S.Y.; Zhao, S.N.; Meng, X.; Wu, L.L.; Wang, C.; Zhang, H.J. A Temperature-Responsive Smart Europium Metal-Organic Framework Switch for Reversible Capture and Release of Intrinsic Eu^{3+} Ions. *Adv. Sci.* **2015**, *2*, 1500012. [CrossRef] [PubMed]

106. Zhao, S.; Hao, X.-M.; Liu, J.-L.; Wu, L.-W.; Wang, H.; Wu, Y.-B.; Yang, D.; Guo, W.-L. Construction of Eu(III)- and Tb(III)-MOFs with photoluminescence for sensing small molecules based on furan-2,5-dicarboxylic acid. *J. Solid State Chem.* **2017**, *255*, 76–81. [CrossRef]
107. Song, X.-Z.; Song, S.-Y.; Zhao, S.-N.; Hao, Z.-M.; Zhu, M.; Meng, X.; Wu, L.-L.; Zhang, H.-J. Single-Crystal-to-Single-Crystal Transformation of a Europium(III) Metal-Organic Framework Producing a Multi-responsive Luminescent Sensor. *Adv. Funct. Mater.* **2014**, *24*, 4034–4041. [CrossRef]
108. Zhang, Y.; Yuan, S.; Day, G.; Wang, X.; Yang, X.; Zhou, H.-C. Luminescent sensors based on metal-organic frameworks. *Coord. Chem. Rev.* **2018**, *354*, 28–45. [CrossRef]
109. Dou, Z.; Yu, J.; Cui, Y.; Yang, Y.; Wang, Z.; Yang, D.; Qian, G. Luminescent metal-organic framework films as highly sensitive and fast-response oxygen sensors. *J. Am. Chem. Soc.* **2014**, *136*, 5527–5530. [CrossRef] [PubMed]
110. Li, Y.; Zhang, S.; Song, D. A luminescent metal-organic framework as a turn-on sensor for DMF vapor. *Angew. Chem. Int. Ed.* **2013**, *52*, 710–713. [CrossRef] [PubMed]
111. Wu, P.; Zhang, J.; Wang, S.; Zhu, A.; Hou, X. Sensing during in situ growth of Mn-doped ZnS QDs: A phosphorescent sensor for detection of H_2S in biological samples. *Chem. Eur. J.* **2014**, *20*, 952–956. [CrossRef] [PubMed]
112. Gao, J.; Li, Q.; Wang, C.; Tan, H. Copper (II)-mediated fluorescence of lanthanide coordination polymers doped with carbon dots for ratiometric detection of hydrogen sulfide. *Sens. Actuators B* **2017**, *253*, 27–33. [CrossRef]
113. Zheng, X.; Fan, R.; Song, Y.; Wang, A.; Xing, K.; Du, X.; Wang, P.; Yang, Y. A highly sensitive turn-on ratiometric luminescent probe based on postsynthetic modification of Tb^{3+}@Cu-MOF for H_2S detection. *J. Mater. Chem. C* **2017**, *5*, 9943–9951. [CrossRef]
114. Gao, Y.; Jing, P.; Yan, N.; Hilbers, M.; Zhang, H.; Rothenberg, G.; Tanase, S. Dual-mode humidity detection using a lanthanide-based metal-organic framework: Towards multifunctional humidity sensors. *Chem. Commun.* **2017**, *53*, 4465–4468. [CrossRef] [PubMed]
115. Han, J.; Burgess, K. Fluorescent Indicators for Intracellular pH. *Chem. Rev.* **2010**, *110*, 2709–2728. [CrossRef] [PubMed]
116. Chen, X.; Pradhan, T.; Wang, F.; Kim, J.S.; Yoon, J. Fluorescent chemosensors based on spiroring-opening of xanthenes and related derivatives. *Chem. Rev.* **2012**, *112*, 1910–1956. [CrossRef] [PubMed]
117. Qi, Z.; Chen, Y. Charge-transfer-based terbium MOF nanoparticles as fluorescent pH sensor for extreme acidity. *Biosens. Bioelectron.* **2017**, *87*, 236–241. [CrossRef] [PubMed]
118. Dimitriev, O.P.; Kislyuk, V.V. Processes of molecular association and excimeric emission in protonated *N,N*-dimethylformamide. *Spectrochim. Acta Part A* **2007**, *68*, 29–35. [CrossRef] [PubMed]
119. Zhang, X.; Jiang, K.; He, H.; Yue, D.; Zhao, D.; Cui, Y.; Yang, Y.; Qian, G. A stable lanthanide-functionalized nanoscale metal-organic framework as a fluorescent probe for pH. *Sens. Actuators B* **2018**, *254*, 1069–1077. [CrossRef]
120. Xia, T.; Zhu, F.; Jiang, K.; Cui, Y.; Yang, Y.; Qian, G. A luminescent ratiometric pH sensor based on a nanoscale and biocompatible Eu/Tb-mixed MOF. *Dalton Trans.* **2017**, *46*, 7549–7555. [CrossRef] [PubMed]
121. Zhao, S.-N.; Li, L.-J.; Song, X.-Z.; Zhu, M.; Hao, Z.-M.; Meng, X.; Wu, L.-L.; Feng, J.; Song, S.-Y.; Wang, C.; et al. Lanthanide Ion Codoped Emitters for Tailoring Emission Trajectory and Temperature Sensing. *Adv. Funct. Mater.* **2015**, *25*, 1463–1469. [CrossRef]
122. Cui, Y.; Zhu, F.; Chen, B.; Qian, G. Metal-organic frameworks for luminescence thermometry. *Chem. Commun.* **2015**, *51*, 7420–7431. [CrossRef] [PubMed]
123. Lian, X.; Zhao, D.; Cui, Y.; Yang, Y.; Qian, G. A near infrared luminescent metal-organic framework for temperature sensing in the physiological range. *Chem. Commun.* **2015**, *51*, 17676–17679. [CrossRef] [PubMed]
124. Cui, Y.; Song, R.; Yu, J.; Liu, M.; Wang, Z.; Wu, C.; Yang, Y.; Wang, Z.; Chen, B.; Qian, G. Dual-emitting MOF supersetdye composite for ratiometric temperature sensing. *Adv. Mater.* **2015**, *27*, 1420–1425. [CrossRef] [PubMed]
125. Taokaenchan, N.; Tangkuaram, T.; Pookmanee, P.; Phaisansuthichol, S.; Kuimalee, S.; Satienperakul, S. Enhanced electrogenerated chemiluminescence of tris(2,2′-bipyridyl)ruthenium(II) system by L-cysteine-capped CdTe quantum dots and its application for the determination of nitrofuran antibiotics. *Biosens. Bioelectron.* **2015**, *66*, 231–237. [CrossRef] [PubMed]

Materials **2018**, *11*, 572

126. Zhang, F.; Yao, H.; Chu, T.; Zhang, G.; Wang, Y.; Yang, Y. A Lanthanide MOF Thin-Film Fixed with Co$_3$O$_4$ Nano-Anchors as a Highly Efficient Luminescent Sensor for Nitrofuran Antibiotics. *Chem. Eur. J.* **2017**, *23*, 10293–10300. [CrossRef] [PubMed]

127. Gao, Y.; Yu, G.; Liu, K.; Wang, B. Luminescent mixed-crystal Ln-MOF thin film for the recognition and detection of pharmaceuticals. *Sens. Actuators B* **2018**, *257*, 931–935. [CrossRef]

128. Hao, J.N.; Xu, X.Y.; Lian, X.; Zhang, C.; Yan, B. A Luminescent 3d-4f-4d MOF Nanoprobe as a Diagnosis Platform for Human Occupational Exposure to Vinyl Chloride Carcinogen. *Inorg. Chem.* **2017**, *56*, 11176–11183. [CrossRef] [PubMed]

materials

MDPI

Review

Spray-Drying of Electrode Materials for Lithium- and Sodium-Ion Batteries

Benedicte Vertruyen *, Nicolas Eshraghi, Caroline Piffet, Jerome Bodart, Abdelfattah Mahmoud and Frederic Boschini

GREENMAT, CESAM Research Unit, University of Liege, Chemistry Institute B6, Quartier Agora, Allée du 6 août, 13, B-4000 Liege, Belgium; neshraghi@uliege.be (N.E.); caroline.piffet@uliege.be (C.P.); jerome.bodart@uliege.be (J.B.); abdelfattah.mahmoud@uliege.be (A.M.); frederic.boschini@uliege.be (F.B.)
* Correspondence: b.vertruyen@uliege.be; Tel.: +32-4-366-3452

Received: 31 May 2018; Accepted: 21 June 2018; Published: 25 June 2018

Abstract: The performance of electrode materials in lithium-ion (Li-ion), sodium-ion (Na-ion) and related batteries depends not only on their chemical composition but also on their microstructure. The choice of a synthesis method is therefore of paramount importance. Amongst the wide variety of synthesis or shaping routes reported for an ever-increasing panel of compositions, spray-drying stands out as a versatile tool offering demonstrated potential for up-scaling to industrial quantities. In this review, we provide an overview of the rapidly increasing literature including both spray-drying of solutions and spray-drying of suspensions. We focus, in particular, on the chemical aspects of the formulation of the solution/suspension to be spray-dried. We also consider the post-processing of the spray-dried precursors and the resulting morphologies of granules. The review references more than 300 publications in tables where entries are listed based on final compound composition, starting materials, sources of carbon etc.

Keywords: spray-drying; batteries; lithium ion batteries; sodium ion batteries; electrode materials; solution synthesis; suspensions

1. Introduction

Secondary batteries such as Li-ion, Na-ion, or related batteries are complex electrochemical devices [1,2]. Their optimal performance relies on the harmonious operation of all parts, which depends not only on the individual characteristics of the positive electrode (cathode), the negative electrode (anode) and the electrolyte, but also on the interfaces between them. It is well known that microstructure effects have a strong impact on properties as can be illustrated by the case of the electrodes. On the one hand, the composition of the active electrode material determines electrode voltage and theoretical capacity. On the other hand, the microstructure (both of the active material component and of the composite electrode as a whole) strongly influences the actual electrochemical performance at high charge-discharge rates (rate capability). The microstructure also determines the specific surface area in contact with the electrolyte, with effects on kinetics and cycling stability. Finally, the microstructure has an influence on the packing efficiency and therefore on the energy density (=energy per unit of volume) of the battery.

This key role of the microstructure means that the selection of a synthesis and/or shaping method can have a decisive impact on practical performance indicators. As a result, the literature on the synthesis of electrode materials has been increasing at a tremendous rate, with reports of a wide variety of routes for each active electrode material candidate. Searching for the most appropriate preparation procedure(s) in each particular case is a legitimate and sound objective. However, the possibility to transfer results from the laboratory scale of typically a few grams to industrially relevant production conditions should be taken into account from an early stage. This is especially important in the case of electrode materials, since the microstructure is often one of the most impacted characteristics in case of upscaling, due to heat-transfer issues when going from small volumes to larger batches or continuous production. Comparatively easy upscaling is one of the strengths of spray-drying [3], a versatile and robust technique whose classical fields of applications (in the food and pharmaceutical industries) have recently been expanding to include the synthesis/shaping of electrode materials (Figure 1a).

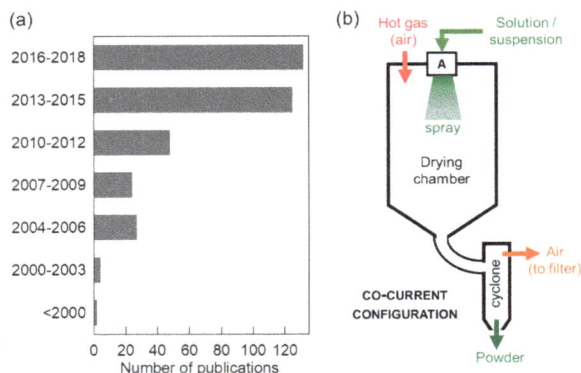

Figure 1. (**a**) Number of publications related to spray-drying of electrode materials for Li-ion, Na-ion and related batteries; (**b**) Schematic of a spray-dryer, showing the case of a co-current configuration and bi-fluid nozzle atomization.

In a spray-dryer (Figure 1b), a solution or suspension is sprayed into droplets and the solvent or liquid in each droplet is evaporated by a hot gas flow (usually air), resulting in a dry powder (see Figure 2 for a few examples of granule morphologies). Larger quantities can be obtained simply by spraying a larger volume over a longer time, without modification of the conditions experienced by each individual droplet. Several experimental configurations exist, as briefly discussed in Section 2.

Figure 2. Examples of morphology of as-sprayed granules: (**a**) precursor of $Na_3V_2(PO_4)_2F_3$, spray-drying of aqueous solution, bi-fluid nozzle atomization; (**b**) same as (**a**) with addition of carbon nanotubes in the solution; (**c**) silicon, spray-drying of suspension in alcohol, fountain mode. All three micrographs are unpublished scanning electron microscope (SEM) micrographs from the authors' own work.

Spray-drying can be applied to suspensions (Figure 3a) or solutions (Figure 3c) but also to the intermediate case of suspensions in solutions (Figure 3b). In all of these cases it can be used as a shaping technique, typically to obtain spherical granules. This application of spray-drying is commonly encountered in the food and pharmaceutical industries, and to granulate nanopowders into re-dispersible micrometric granules for safer handling and transport. In the context of electrode materials, this version of spray-drying (i.e., without post-processing heat treatment) is usually applied to suspensions containing both small particles of active material and some form of solid conducting carbon. The objective is then to achieve a good mixing of active material and carbon and to obtain granules with good flowability and packing properties for efficient electrode formulation.

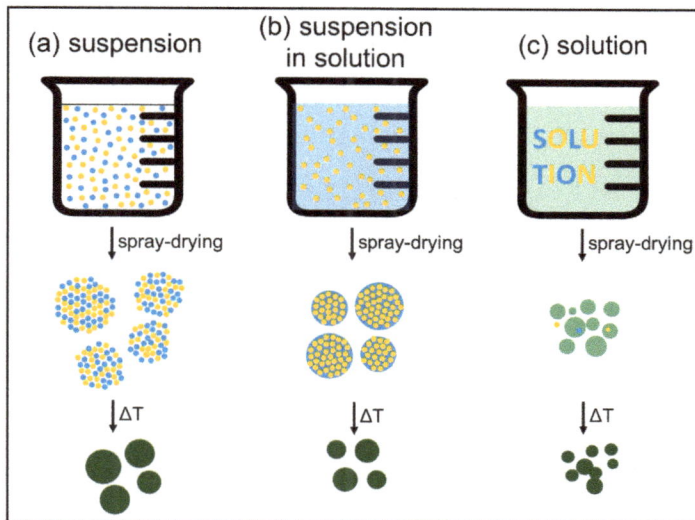

Figure 3. Spray-drying of (**a**) a suspension of solid particles (blue and yellow) dispersed in a non-solvent (transparent); (**b**) a suspension of solid particles (yellow) in a solution (light blue); (**c**) a solution (light green) of soluble precursors. All schematics consider the case where the spray-dried precursor is further transformed into the final phase (dark green) by heat treatment.

As depicted in Figure 3, spray-drying can also be used to intimately mix reactants in view of ulterior transformation into the final product by heat treatment. This version of spray-drying is the most common in the field of electrode materials, as will be seen in this review. Mixing of the reactants can occur at the atomic scale when starting from a solution, whereas homogeneity is determined by the (nano)particle size when starting from a suspension or a suspension in a solution. In spray-drying, the objective is the evaporation of the droplet liquid, and decomposition of the solid is not supposed to happen at this stage (especially in the case of heat sensitive pharmaceuticals or food). If further heat treatment is needed to form the final compound, partial decomposition during spray-drying is obviously not a problem. The technique of spray pyrolysis for powder synthesis targets decomposition and requires much higher temperatures, which are reached by spraying into a tubular furnace setup or in a flame. Spray pyrolysis will not be discussed here (see [4–8] for a few examples).

The present review is focused on spray-drying for electrode materials (see Table 1) and is to our knowledge the first of its kind. Readers interested in a more general overview of the technique and its broad-ranging scope of applications can refer to reviews such as those by Nandiyanto and Okuyama [9] (on particle sizes and morphologies), Mezghericher et al. [10] (on models of droplet drying), Zbicinski [11] (on modeling of industrial spray-dryers), Stunda-Zujeva et al. [3] (on spray-drying for ceramics), Deshmukh et al. [12] and Singh et al. [13] (on spray-drying for drug delivery), Gharsallaoui et al. [14] (on microencapsulation of food ingredients), Schuck et al. [15] (on spray-drying for the dairy industry) and references therein.

This review deals primarily with chemistry- and microstructure-related topics such as the formulation of solutions and suspensions, the impact of spray-drying parameter selection, or strategies to create composites with conducting carbon. It should be seen as a complement to available reviews that focus on the discussion and benchmarking of electrochemical performance in materials based on the same (family of) compound(s) or intended for one type of battery/electrode (see for examples [2,16–19]), where much less attention is paid to the details of the synthesis procedures.

Table 1. Bibliographical overview.

Compound Types, Formulas and References
Borates $LiMnBO_3$ [20], $LiFeBO_3$ [21], $Li(Fe,Ni)BO_3$ [22]
Elements   C [23–35], P [36], S [37–39], Sb [40], Si [41–81], Sn [82], Se [83]
Fluorides Li_2TiF_6 [84], Li_2NiF_4 [85], Li_3FeF_6 [86]
Fluorophosphates Na_2FePO_4F [87–89], Na_2MnPO_4F [90,91], Na_2CoPO_4F [92], $Na_3V_2(PO_4)_2F_3$ [93], $Na_3V_2O_2(PO_4)_2F$ [94]
Organic salts   Dilithium terephtalate $Li_2C_8H_4O_4$ [95], Disodium terephtalate $Na_2C_8H_4O_4$ [96],   Disodium 2,5-dihydroxy-1,4-benzoquinone $Na_2C_6H_2O_4$ [97],
Oxides M_xO_y CeO_2 [98], CoO_x [99], CoO [100], Co_3O_4 [100–102], Cr_2O_3 [103], CuO [104–106], Fe_2O_3 [107–110], GeO_x [111], GeO_2 [112], La_2O_3 [113], MnO [114], MoO_3 [115], Nb_2O_5 [116], NiO [117], SiO [118,119], SiO_2 [120,121], SnO_2 [122–124], TiO_2 [125–133], V_2O_5 [134]
Oxides $M_xM'_yO_z$ $ZnFe_2O_4$ [135,136], $Mn_{0.5}Co_{0.5}Fe_2O_4$ [137], $NiCo_2O_4$ [138], $(Ni,Co)O_x$ [139], $Cu_{1.5}Mn_{1.5}O_4$ [140], $NiMoO_4$ [141], $TiNb_2O_7$ [142]

Oxides $Li_xM_yO_z$
 Layered oxides $Li_xM_yO_2$ (M = Li, Ni, Co, Mn, Al, …) [143–189]—see Table 3 for compositions
 $LiMn_2O_4$ [190–199],
 Co-doped $LiMn_2O_4$ [200,201], Cr-doped $LiMn_2O_4$ [202,203], Ni-doped $LiMn_2O_4$ [204,205]
 $LiNi_{0.5}Mn_{1.5}O_4$ [206–210],
     Ti-doped $LiNi_{0.5}Mn_{1.5}O_4$ [211], Fe,Ti-doped $LiNi_{0.5}Mn_{1.5}O_4$ [212],
     Ru,Ti-doped $LiNi_{0.5}Mn_{1.5}O_4$ [212], Co-doped $LiNi_{0.5}Mn_{1.5}O_4$ [213–215]
 $Li_4Ti_5O_{12}$ [216–249], $Li_{3.98}Al_{0.06}Ti_{4.96}O_{12}$ [250], $Li_{4+x}Ti_{4.95-x}Nb_{0.05}O_{12-d}$ [251]
 $Li_xV_3O_8$ [252–257], Li_3VO_4 [258–261], $Li_4Mn_5O_{12}$ [262]

Oxides $Na_xM_yO_z$ $Na_{2/3}Ni_{1/3}Mn_{2/3}O_2$ [263,264], $Na_2Ti_3O_7$ [265], $Na_4Mn_9O_{18}$ [266,267]
Phosphates $FePO_4$ [268–270] $LiFePO_4$ [271–309], $Li(Fe,Mn)PO_4$ [310–320], $LiMnPO_4$ [321,322], $Li(Mn_{0.85}Fe_{0.15})_{0.92}Ti_{0.08}PO_4$ [323] $LiVOPO_4$ [324] $Li_3V_2(PO_4)_3$ [325–336], $Li_3(V,Al/Fe)_2(PO_4)_3$ [337], electrolyte $Li_{1.3}Al_{0.3}Ti_{1.7}(PO_4)_3$ [338] $NaTi_2(PO_4)_3$ [339,340], $Na_3V_2(PO_4)_3$ [341–345], $Na_3V_{2-x}Cu_x(PO_4)_3$ [346],
Pyrophosphates $Na_2FeP_2O_7$ [347], SnP_2O_7 [348]
Silicates Li_2FeSiO_4 [349,350], $Li_{1.95}Na_{0.05}FeSiO_4$ [351], $Li_2Fe_{0.5}V_{0.5}SiO_4$ [352]
Sulfides and selenides   MnS [114], MoS_2 [353], $FeSe_2$ [354]
Composites (not with carbon) $Sn–Sn_2Co_3@CoSnO_3–Co_3O_4$ [355], Fe_2O_3-SnO_2 [356], $LiNi_{0.5}Mn_{1.5}O_4$-$Li_7La_3Zr_2O_{12}$ [357], $3Li_4Ti_5O_{12}.NiO$ [358], $LiFePO_4$-$Li_3V_2(PO_4)_3$ [359–363], $LiMnPO_4$-$Li_3V_2(PO_4)_3$ [364,365], Si-$FeSi_2$-$Cu_{3.17}Si$ [366], $MoS_2–Ni_9S_8$ [367], $MoSe_2–NiSe(–NiSe_2)$ [367]

2. Experimental Parameters in Spray-drying

Spray-dryers exist in all sizes, from table-top systems to industrial production units. In the primary scientific literature, the most common systems are home-made equipment, commercial table-top systems [368,369] and commercial (small) pilot-scale systems. As an example, our group started working on spray-drying at the beginning of the 2000s with a table-top Buchi Mini Spray-dryer B-190 (Büchi Labortechnik AG, Switzerland) and now owns two Mobile Minor™ units, which can evaporate up to 5.5 kg H_2O per hour and correspond to the smallest-but-one R&D systems on the catalogue of a provider of industrial spray-drying technology (GEA). One of the largest-scale tests for electrode materials (in scientific publications) was reported by Han et al. [221] for the synthesis of 15 kg $Li_4Ti_5O_{12}$.

Basically, all spray-dryers include (i) an atomizer (most often a bi-fluid nozzle or a rotating wheel) where the liquid feedstock is sprayed into droplets; (ii) a drying chamber where a hot gas flow (injected in co-current or counter-current configuration) evaporates the liquid and (iii) a final section where the dry powder is separated from the gas flow and collected, sometimes at several collection points depending on particle size. A typical configuration is schematized in Figure 1b. Ancillary equipment can be added to filter the outgoing gas, to carry out spray-drying using an inert gas instead of air or to condense vapors of organic solvents when non-aqueous solutions/suspensions are used. In this latter case, systems specially designed to prevent explosion/fire risks should be used.

When reporting on spray-drying experiments, good practice would be to provide information not only about the composition of the liquid feedstock but also about the spray-drying setup and experimental parameters such as inlet temperature, outlet temperature, and feedstock flow rate. When commercial equipment is used, additional parameters (such as air/gas pressure of the bifluid nozzle or rotating atomizer, etc.) should also be reported. A recent review by Arpagaus et al. [368] includes a section about electrode materials, focusing on a few publications where detailed spray-drying parameters are provided together with data on particle morphology and electrochemical performance. In most papers, however, information on the spray-drying parameters is missing or incomplete as illustrated by Table A1 in Appendix A for the case of layered oxide compounds.

Some of these parameters (for example the inlet temperature or the flow rate) can be selected independently but others, notably the outlet temperature, are the consequence of the selected parameters. Typically, increasing the inlet temperature or decreasing the feedstock flow rate results in an increase of the outlet temperature. In practice, the 'selectable' parameters are often adjusted to target a specific outlet temperature. Indeed, due to the wet-bulb effect [370], the outlet temperature is often the highest temperature experienced by the material in the spray-dryer (at least in the most common co-current configuration). The outlet temperature, therefore, determines to a large extent how dry the final powder will be and must be carefully controlled especially when spray-drying heat sensitive compounds.

3. Formulation of Solutions/Suspensions: Inorganic Components

As shown in Table 1, electrode materials prepared by spray-drying span a broad range of compositions, from elements to oxides, phosphates, sulfides, fluorides, and others. In most cases, the spray-drying step results in the formation of a precursor, which will be transformed into the final phase through ulterior treatment (most commonly through heat treatment). This section describes and discusses the formulation of solutions or suspensions used as feedstock for spray-drying. Some specific cases are taken as illustrative examples. More systematic information is provided in Table A2 in Appendix A, which consists of an inventory of the starting materials used in the publications referenced in this review.

3.1. Solvent/Liquid Phase

The most common solvent (for solutions) or liquid medium (for suspensions) is water. This is easily explained by considering that water is cheap, safe and non-toxic. As shown in Table 2, alcohols are also used, either pure or mixed with water. Other liquids are much less common (see Table 2). From the physico-chemical point of view, the two most important selection criteria are the vaporization temperature of the liquid (which must be in the adequate range for the spray-drying equipment) and its solvent/non-solvent character with respect to the reactants. However, safety, recycling, and prevention of release in the atmosphere must be addressed when using organic liquids, typically through appropriate equipment (fire/explosion-proof equipment, condensing of solvent vapors, etc.).

Table 2. Spray-drying synthesis of active materials involving organic or partially organic suspensions.

Liquid	Active Material
Ethanol	S [38], Si [42,45,47,52,57,58,69,76], SiO [118], SiO_x [371], TiO_2 [129,130], $Li_xMn_{1/3}Co_{1/3}Ni_{1/3}O_2$ [150], $Li_4Ti_5O_{12}$ [218], $LiFePO_4$ [273,277,289], $Li_3V_2(PO_4)_3$ [326], $LiFePO_4$-$Li_3V_2(PO_4)_3$ [359], $Li_2Fe_{0.5}V_{0.5}SiO_4$ [352]
Alcohol (unspecified)	$Li_4Ti_5O_{12}$ [231,232]
Ethanol-water	C [23], Si [54,60,65,72], SiO_2 [120], SnO_2 [123], TiO_2 [132], $LiMn_2O_4$ [199], $Li_4Ti_5O_{12}$ [229,238,241,243,250,251], $Na_2Ti_3O_7$ [265], $LiFePO_4$ [292]
Alcohol-water	Si [73,78]
Other liquid(s)	DMF for Sb/C [40], EG for Si/C [43], Ethylene glycol—cyclohexane for $ZnFe_2O_4$ [135], THF for Si/C [44,58], water-THF for Li_3PO_4-coated $Li_4Ti_5O_{12}$ [219]

3.2. Solutions

The discussion in this section and the next is illustrated with the case of the AMO_2 layered oxide compounds (A = Li^+, Na^+; M = one or several of Li, Ni, Mn, Co, Al, ...). The references in Table 3 are sorted into categories labeled A to H according to the type of solution/suspension.

An essential point to consider when preparing a solution for spray-drying is that, except volatiles, all components will be present in the spray-dried powder. This restricts the choice of counter-ions and of all additives to compounds that will be decomposed during ulterior heat treatment, or do not interfere with functionality. With this in mind, aqueous solutions can be prepared (1) by adding soluble salts in water or (2) by dissolving less soluble but cheaper precursors.

In the first case, nitrates and acetates (for cations) or ammonium salts (for anions) are common choices due to their low decomposition temperatures. This is illustrated by Category A in Table 3 where acetates and/or nitrates were selected as water-soluble salts of Ni, Co, Mn or Al. Regarding ammonium salts as a source of anions, the most common example is probably $(NH_4)H_2PO_4$ which is a popular precursor in the synthesis of phosphates (see Table A2 in Appendix A).

In the second case, dissolution in (aqueous) acid is the most frequent. Since hard acids (such as HNO_3) usually drive the pH to very low values which might damage the spray-drying equipment, dissolution in milder acids such as citric or acetic acids (or, more imaginatively, polyacrylic acid [159]) is often preferred when possible (see Categories B and D in Table 3). The pH can also be brought back to less acidic values by addition of bases that do not introduce foreign cations, such as ammonia solution. Incidentally, the possibility of auto-combustion occurring during the early stages of the heat treatment of the spray-dried material should be kept in mind when nitrates and organics are simultaneously present. The probability is enhanced if ammonium nitrate has been formed by neutralizing an excess of nitric acid by ammonia solution.

Table 3. Spray-drying for layered oxides AMO_2 (A = Li^+, Na^+; M = one/several of Li, Ni, Mn, Co, Al, ...).

	Li	Co	Ni	Mn	other	Comments
SPRAY-DRYING OF SOLUTIONS						
A. Spray-drying of aqueous solution of nitrates and/or acetates						
Duvigneaud et al. [145]	1	$0.18 - y$	0.82	-	Al	+ polyvinyl alcohol
He et al. [146]	1	0.105	0.35	0.545	Cr	0 to 6% Cr
He et al. [148]	$\checkmark$	$\checkmark$	$\checkmark$	$\checkmark$	-	-
Kim et al. [151]	$1 + x$	1/3	1/3	1/3	-	-
Kim et al. [152]	$1 + x$	$1 - 2z$	z	z	-	$x = 0$–0.1; $z = 0.1$–0.4
Kim et al. [187]	$1 + x$	0.4	0.3	0.3	-	-
Konstantinov et al. [153]	1	1	-	-	-	-
Li et al. [154,156]	1	1/3	1/3	1/3	-	-
Li et al. [157]	1	1/3	1/3	1/3	-	+ LiF
Li et al. [160]	1	1	-	-	-	+ polyethylene glycol
Liu et al. [166]	1	1/3	1/3	1/3	-	+ PVA
Wang et al. [263]	$Na_{2/3}$	-	1/3	2/3	-	-
Wang et al. [172]	1.57	1/6	1/6	2/3	-	-
Wang et al. [173]	$1 + x$	$1 - x$	-	x	-	-
Wu et al. [175]	1	0.2	0.8	-	-	-
Yue et al. [179,180]	$1 + x$	0.2	0.6	0.2	-	$x = 0$; 0.04
Zhang et al. [183]	$1 + x$	-	$0.5 - x/2$	$0.5 - x/2$	-	$x = 0$–0.2
Zhang et al. [186]	1	1/3	1/3	1/3	-	-
Zhao et al. [264]	$Na_{2/3}$	-	1/3	2/3	-	Followed by Li^+/Na^+ ion exchange
B. Spray-drying of aqueous solution of salts dissolved in aqueous citric acid						
Li et al. [158]	$\checkmark$	-	$\checkmark$	$\checkmark$	Fe	nitrates
Sun et al. [171]	$\checkmark$	$\checkmark$	$\checkmark$	$\checkmark$	-	acetates
Watanabe et al. [174]	1.2	0.03	0.18	0.58	-	acetates
Zhang et al. [184,185]	$\checkmark$	-	$\checkmark$	-	Ti	LiOH, Ni acetate and $[NH_4]_2[Ti(C_2O_4)_3]$
C. Spray-drying of aqueous solution of citrates						
Li et al. [155]	1	$2x$	$0.5 - x$	$0.5 - x$	-	$x = 0$–0.1
Qiao et al. [169]	1.17	-	0.25	$0.58 - x$	Sn	$x = 0$–0.05
Yuan et al. [178]	1.17	0.05	0.2	0.58	-	-
D. Spray-drying of aqueous solution (others)						
Li et al. [159]	1	1	-	-	-	hydroxides dissolved in polyacrylic acid solution
Oh et al. [167]	1	0.2	0.8	-	-	hydroxides and carbonate dissolved in acrylic acid solution
SPRAY-DRYING OF SUSPENSIONS						
E. Spray-drying of an aqueous suspension to mix reactants						
Hou et al. [149]	1.2	0.13	0.13	0.54	-	Li_2CO_3 and hydroxide co-precipitate
Lin et al. [164]	1.2	-	0.2	0.6	-	carbonates and oxides
Liu et al. [165]	1	1/3	1/3	1/3	-	in situ polymerized Li polyacrylate and hydroxide co-precipitate
Wang et al. [189]	1.2	0.13	0.13	0.54	-	carbonates and oxides
Yue et al. [181]	1	0.2	0.6	0.2	-	Li_2CO_3 and hydroxide co-precipitate
F. Spray-drying of an ethanol suspension to mix reactants						
Hu et al. [150]	1	1/3	1/3	1/3	-	LiOH and hydroxide co-precipitate
Lin et al. [161,162]	1	1/3	1/3	$1/3 - x$	Zr	$x = 0$–0.02-carbonates and oxides
G. Mixing of AMO_2 active material with conductive carbon or conductive carbon precursor						
Cheng et al. [144]	1.2	0.13	0.13	0.54	-	graphene oxide
Xia et al. [176]	1	1	-	-	-	P3DT (in CH_2Cl_2)
Yang et al. [177]	1.2	0.13	0.13	0.54	-	CNT
Yue et al. [182]	1	0.2	0.6	0.2	-	graphene oxide
H. Shaping of AMO_2 as spheres						
Chen et al. [143]	1	0.15	0.8	-	Al	0.05% Al-starch binder

In many cases, complexation of the metallic cations may be needed, either to prevent precipitation of a less soluble salt when different soluble salts are mixed in solution or if the solubility product of a metal hydroxide (less commonly a carbonate) is exceeded when adjusting the pH to more basic values.

Formation of stable complexes (such as citrates, see Category C in Table 3) is also a strategy to favor a homogeneous distribution of the chemical species in the spray-dried particles. This is relevant for cases of complex compositions where sequential precipitation might occur during the drying of the droplet, i.e., precipitation of several phases starting with the least soluble and going on to the most soluble. This raises the more general question of the extent to which the homogeneity of a solution can be retained in a spray-dried precursor. On the one hand, the actual impact of this issue is limited since, by comparison with other synthesis techniques, the degree of inhomogeneity is restricted by the small size of the droplets. On the other hand, maximum homogeneity remains desirable for ulterior formation of the target phase. This is a case-by-case issue since it depends on solubilities of specific compounds, however helpful guidelines could be achieved if more authors reported relevant data in their publications. Even if a detailed characterization of the homogeneity in the as-sprayed material is difficult to obtain, valuable insight might be gained by simpler procedures. One such procedure is to collect X-ray diffractograms on samples taken out of the furnace at lower temperatures during the heating ramp, in order to identify which phases form first.

The above discussion focused on electrode materials (such as oxides or (fluoro)phosphates) for which soluble precursors are available. In the case of titanate or silicate electrode materials, the preparation of solutions is more difficult because few precursors are soluble in aqueous solutions of less-than-extreme pH. Chloride and/or alkoxide precursors (such as TEOS $Si(OC_2H_5)_4$ [350,352], titanium isopropoxide $Ti(OC_3H_7)_4$ [128,132,212,227,235,243,248,251] or tert-butoxide $Ti(OC_4H_9)_4$ [211,222,229,238–241,249,250,265], ...) can be solubilized in alcohol but hydrolysis takes place when mixing with water, leading to the precipitation of SiO_2 or TiO_2 unless special care is taken as in the strategies summarized in Figure 4.

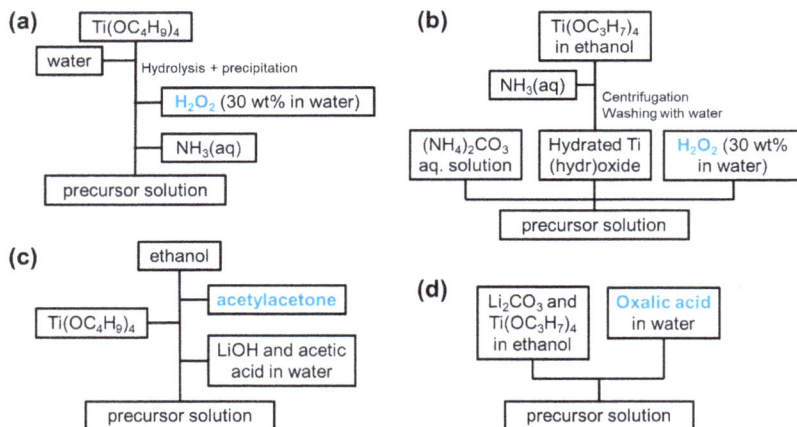

Figure 4. Procedures to prepare an aqueous solution starting from titanium alkoxide, as proposed by (**a**) [222,249]; (**b**) [127]; (**c**) [229,238]; (**d**) [243,248,251].

When the composition for the solution has been settled, other parameters still need to be decided on. One of them is the concentration of the solution. A naïve view is that it should be as high as possible, in order to minimize the amount of solvent to be evaporated. However, too high concentrations can lead to gel formation or precipitation in the atomization nozzle. Besides, the solution concentration influences the morphology of spray-dried particles. All other parameters being equal (esp. the inlet temperature and the droplet size), a higher concentration means that the solubility limit is exceeded sooner during vaporization of the solvent and crust formation, therefore, occurs at a larger droplet diameter (collapse or cracking may take place later if the mechanical strength of the crust is too low). The concentration of the solution is, therefore, best adjusted in conjunction with other parameters (inlet temperature, flow rate, atomization parameters) in order to optimize the temperature profile of the drying process as a function of the priorities. Examples of such priorities can be a specific type of morphology but also the avoidance of partial decomposition or the minimization of residual humidity. This last point is important in relation to the possible post-spray-drying aging of the spray-dried material. Generally speaking, it is not recommended to store as-sprayed materials if they are made up of hygroscopic compounds (such as nitrates or, to a lesser extent, acetates) or if inorganic condensation mechanisms can take place (typically for Ti- or Si-based precursors of oxides). In such cases, the as-sprayed materials should be heat-treated to a temperature selected to obtain a stable (and, therefore, reproducible) intermediate state of the material.

3.3. Suspensions

Here we consider as a suspension all cases where at least one component is insoluble or only partially soluble in the liquid medium (=cases a and b in Figure 3). Spray-drying of suspensions can be used to mix reactants before heat treatment (Categories E and F in Table 3), to mix an active material with conductive carbon (Category G in Table 3), or both. It can also be used as a "shaping-only" method to prepare spherical granules of an active material (Category H in Table 3).

The first point made when discussing solutions is also valid for suspensions: except volatiles, everything that is added to the suspension will be present in the spray-dried powder. Therefore, oxides, carbonates, oxalates or hydroxides are common choices since they decompose during the heat treatment in air without leaving residues. In theory, this also applies to the selection of additives such as cationic dispersing agents, where ammonium counter-ions should be preferred to sodium counter-ions, although quantities remain low.

When considering spray-drying of suspensions, the stability of the suspension is obviously an important requirement. What is called a "stable" suspension in this context may however vary. At one end of the spectrum, the criterion may be that there is no visible sign of sedimentation when the suspension is under stirring and when it is pumped through tubes to the atomization head. At the other end, a stable suspension can be characterized by long-term stability and low aggregation thanks to efficient repulsion between individual particles. Whatever the case, a particle size of about 1 μm or below is always preferable. If the good mixing of small particles is retained in the spray-dried material, the small particle size is also favorable for the formation of the final phase since diffusion distances during heat treatment will be correspondingly short. Minimizing diffusion distances is also the reason why some suspensions involve pre-synthesized co-precipitates of several cations (e.g., $(Co,Ni,Mn)(OH)_x$ [149,165,188], $(Co,Ni,Mn)O_x$ [150], (Ni,Mn) oxalate [210], $(Fe,Mn)_3(PO_4)_2 \cdot xH_2O$ [312–314,319]). If the coprecipitate is isolated by filtration or centrifugation before being redispersed into the suspension, its stoichiometry should be checked and the possibility of partial redissolution in suspension should be kept in mind.

Decreasing the particle size can be achieved by ball-milling a suspension of the larger particles in a liquid medium. An advantage of spray-drying is that the ball-milled suspension can often be used directly as feedstock for spray-drying if the liquid medium is suitable (see [143,216,275,283] for a few examples). Another possibility is to use commercial nanopowders to prepare the suspension. It should be noted that the viscosity of suspensions of very small particles (typically below about 100 nm) increases rapidly with solid loading. Also, depending on the fabrication process and/or aging on storage, the surface of the nanoparticles may be chemically different from the core (e.g., hydroxyl-rich or carbonated surface of some oxide particles), which might strongly affect their dispersion behavior and should also be taken into account when calculating stoichiometric proportions in multi-component suspensions. Finally, the high surface area of nanopowders means that they will be particularly affected if surface reaction or partial dissolution of the particle can occur in the liquid medium. These effects are rarely spectacular but should be considered when adjusting pH or when an unexpected behavior needs explaining.

As briefly mentioned above, the formulation of a suspension may involve the addition of a dispersing agent, which may be cationic, neutral or anionic and acts through electrostatic and/or steric effects. The formulation of suspensions of several powders (multicomponent suspensions) with long-term stability often becomes a formidable task, made even more complicated if the solid powders are suspended in a solution instead of in a simple liquid medium. At least in the context of electrode material synthesis, the formulation of multicomponent suspensions usually targets only practical stability, where "practical" means long enough for the spray-drying procedure.

In the case of suspensions prepared for spray-drying, other additives such as polyethylene glycol or polyvinyl alcohol may be added as binder to increase the cohesion and mechanical strength of the spray-dried granules. These binders usually tend to increase the viscosity of the suspension, which brings us back to the selection of the solid loading. Besides this practical limit associated with the maximum viscosity acceptable for the spray-drying equipment, the criteria for selecting solid loading are similar to those discussed for deciding the concentration of solutions: the solid loading in a suspension should be adjusted in conjunction with the primary spray-drying parameters (injection mode, inlet temperature, feed rate, atomization parameters) depending on the targeted size and morphology of granules. In the case of multicomponent suspensions, additional complexity is created if the different components have different particle sizes or in the case of suspension-in-solutions (Figure 3b), which may lead to distribution gradients in the dried granules. This phenomenon has not yet been studied in detail in the case of active electrode materials but other (simpler) systems have been investigated [372–374].

4. Formulation of Solutions/Suspensions: Organic/Carbon Components

This section focuses on organic (macro)molecules (listed in Table 4, with references) or carbon compounds (listed in Table 5, with references) which may be added to the solution/suspension for several reasons.

Table 4. Organic (macro)molecules used for the formulation of solutions/suspensions in view of spray-drying preparation of electrode materials.

Organic Compound Types, Compound and References
Carboxylic Acids
Acetic acid [87–89,211,212,229,238,265], Acrylic acid [165,167],
Citric acid [21,22,43,44,52,58,76,78,81,87–92,98,100–102,105,106,113,117,118,139,140,155,158,168,169,171,174,178,184,185,207,213, 214,235,241,278,284,286,295,296,301,302,310,311,325,327,329,331,332,334,337,342,345,346,349,351,360,362,364,365],
Ascorbic acid [93], Formic acid [268], Lactic acid [235], Malic acid [235], Malonic acid [235],
Oxalic acid [135,227,243,248,251,278,293,311,324,335,344,365], Tartaric acid [300,303]
Saccharides
Monosaccharides: Glucose [53,56,71,77,258–260,272,274,275,277,283,285,287–289,298,299,306,308,312,314,318,319,335,359,364]
Disaccharides: Sucrose
[33,46,63,64,75,80,107,108,110,120,242,279–281,286,293,294,296,297,302,315–317,320,323,326,336,343,347,348,363], Sugar [217,247]
Polysaccharides: Cellulose [234], Starch [143,202,203,276,288,290–292,313],
Dextrin [114,115,122,136,141,354,367], Cyclodextrin [142,299], Maltodextrin [128]
Synthetic Polymers
Melamine-formaldehyde resin [210]
Phenol-formaldehyde resin [31,42,45,47,65,273,292,371]
Polyacrylic acid [159,227]
Polyacrylonitrile [40]
Poly(3-decaylthiophene) [176] (for thermal protection via shut-down action at 110 °C)
Polyethylene glycol [160,230,280,286,299,302,304,315,316,326,359]
Polystyrene-acrylonitrile [25,43]
Polyvinylalcohol [59,62,67,79,121,145,166,227,269,270,274,275,286,296,307,330]
Polyvinylbutyral [161–163,231,232]
Polyvinylpyrrolidone [58,71,74,76,80,103,118,124,208,230,349,351]
Triblock copolymer PEO-PPO-PEO F127 [82]
Others
$C_2H_4N_4$ [241], Acetylacetone [229,238], Chitosan [23,61], Diethylene glycol [237], EDTA [82],
Ethylene glycol [101,139], Formamide [188], Pitch [43,44,58,73,328], Urea [269]

Table 5. Spray-drying synthesis of active material/carbon composites: references to publications where solid conducting carbon or graphene oxide is added to the spray-drying solution/suspension.

Carbon	Active Material
CNT	C [24], S [38], Si [48,49,55,69,73,77,79,80], SiO$_x$ [371], Na$_2$FePO$_4$F [87,89], Na$_3$V$_2$(PO$_4$)$_2$F$_3$ [93], Disodium terephtalate Na$_2$C$_8$H$_4$O$_4$ [96], Disodium 2,5-dihydroxy-1,4-benzoquinone Na$_2$C$_6$H$_2$O$_4$ [97], GeO$_x$ [111], V$_2$O$_5$ [134], Li$_x$M$_y$O$_2$ (M = Ni, Co, Mn, Al, ...) [177], Li$_4$Ti$_5$O$_{12}$ [218,248], Li$_3$VO$_4$ [258], Na$_4$Mn$_9$O$_{18}$ [266], LiFePO$_4$ [289,305], Li(Mn,Fe)PO$_4$ [313,316], NaTi$_2$(PO$_4$)$_3$ [339], Na$_3$V$_2$(PO$_4$)$_3$ [342], Li$_3$V$_2$(PO$_4$)$_3$ [329], Li$_2$FeSiO$_4$ [350]
Graphene oxide GO (reduced to RGO)	C [28,30–32], P [36], S [37,39], Se [83], Si [50,51,54,60,63,66,67,80], Na$_3$V$_2$O$_2$(PO$_4$)$_2$F [94], Cr$_2$O$_3$ [103], CuO [105], Fe$_2$O$_3$ [109], GeO$_2$ [112], MoO$_3$ [115], SiO$_2$ [120], SnO$_2$ [123], TiO$_2$ [127,133], NiCo$_2$O$_4$ [138], Li$_x$M$_y$O$_2$ (M = Li, Ni, Co, Mn, Al, ...) [144,147,182], Li$_4$Ti$_5$O$_{12}$ [244,245], Li$_3$VO$_4$ [260], Na$_4$Mn$_9$O$_{18}$ [267], LiFePO$_4$ [282,292,296,304], LiMnPO$_4$ [321], NaTi$_2$(PO$_4$)$_3$ [340], Na$_3$V$_2$(PO$_4$)$_3$ [344], Li$_3$V$_2$(PO$_4$)$_3$ [325,327], NiS [375], MoS$_2$ [353]
Carbon black (CB)	C [33], S [38], LiMnBO$_3$ [20], Na$_2$FePO$_4$F [89], Mn$_{0.5}$Co$_{0.5}$Fe$_2$O$_4$ [137], Li$_4$Ti$_5$O$_{12}$ [220,246], LiFePO$_4$ [298,302]
Graphite	C [25–27,29], Si [43,44,50,52,53,56,58,61,65,66,68,70,71,73,78,79,118], SiO [119]
Others	Carbon (nano)fibers: Si [52], Li$_4$Ti$_5$O$_{12}$ [234]; Graphitized needle coke: Si [64]; Graphitized carbon black: Si [75]

As already mentioned above, soluble organic (macro)molecules may function as complexing agents, dispersing agents, binders, etc. For example, carboxylic acids can be used as acids, as reducing agents or as complexing agents (especially when transformed into carboxylate ions by pH adjustment). Citric acid is an extremely popular choice, as can be seen in Table 4.

Another example is that of synthetic polymers which are used as dispersing agents, thickeners and/or binders. Their exact role is not always defined and depends in part on the molecular mass. Common choices are polyethylene glycol (PEG), polyvinylalcohol (PVA) and polyvinylpyrrolidone (PVP) (see Table 4). PEG, PVA and PVP are of the non-ionic (steric) type but cationic additives are also reported (ammonium polycarboxylate [216,221], sodium carboxymethylcellulose [68,70,119,227,333], sodium dodecyl benzene sulfonate (SDBS) [53,65]).

All these (macro)molecules and a whole range of other organic compounds (see Table 4) can also be used as precursors transforming into carbon during heat treatment in inert/reducing atmosphere. Indeed, a frequent concern when synthesizing electrode materials is that the (relatively) low intrinsic electronic conductivity of many active materials is a limit to the kinetics of the electrochemical reactions. In order to improve electron transport to the active material, common approaches are the formation of a coating and/or a composite with some form of conducting carbon.

Another reason for using composites with carbon is that some active materials (such as Si) undergo very large expansions/contractions on electrochemical cycling; in such cases carbon can be used as a buffer to limit the volume variations and the degradation of performance that results from loss of connectivity inside the electrode.

Since spray-drying usually yields relatively large particles (a few microns to a few tens of microns), surface coating of the spray-dried particles is not good enough for compounds that require intimate mixing with carbon. One possibility is to grind the spray-dried particles and mix them with carbon. Another approach is to include carbon or a carbon precursor in the spray-drying feedstock solution/suspension. Citric acid and saccharides such as glucose or sucrose are amongst the most common soluble carbon precursors (see Table 4), transforming into more or less graphitic carbon during the heat treatment. Interestingly, Choi and Kang [122] reported that dextrin might be preferable to glucose and sucrose to reduce the hygroscopicity of spray-dried powders (Figure 5).

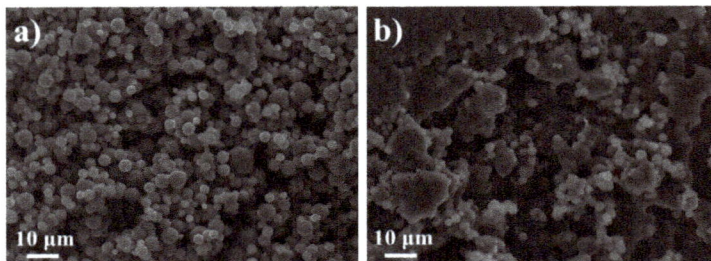

Figure 5. SEM images of as-sprayed powders after 6-h exposure to atmosphere: (**a**) the tin oxalate-dextrin composite is stable; (**b**) the tin oxalate-sucrose composite is hygroscopic. (Adapted from [122] with permission—© 2014 Wiley-VCH Verlag).

As can be seen in Table 5, carbon nanotubes (CNT) are a possible choice amongst conducting carbons that can be added to a solution/suspension before spray-drying. Most often, CNTs are added as a (commercial) dispersion. Sometimes there is little or no information about the characteristics of the CNTs (size distribution, residues of synthesis, dispersing agents, etc.) and even where reference and provider are reported it often turns out that the corresponding commercial datasheets are less than detailed. To some extent, the same comments apply to carbon blacks, although they are usually bought in powder form and easier to characterize. Also, they can be selected amongst the relatively well-known references commonly used for electrode formulation. Since pristine graphene does not disperse in water-based solution/suspensions, graphene oxide (GO) nanosheets suspensions (about which even less is usually known than in the case of CNT) are used and reduction to graphene (reduced graphene oxide—RGO) is achieved by heat treatment or, much less often, by chemical reduction with hydrazine vapor [37,60,83].

Similar principles apply to electrode materials that are made up of carbon only, typically as negative electrodes for Li-ion or Na-ion batteries [25–31,33,35] or as hosting material in Li-O_2 or Li-S batteries [23,24,32,34].

One of the electrode materials for which the broadest variety of carbon sources has been investigated is silicon, because the formation of Si/C composites is one of the most common strategies to buffer the expansions/contractions of Si during electrochemical cycling vs. Li.

Table 6 provides brief descriptions of the suspension compositions and post-spray-drying (post-SD) treatments. The last column reports the percentage of Si in the final Si/C composite materials. The references are sorted into categories depending on the role of spray-drying in the experimental procedure.

Table 6. Spray-drying in the preparation of Si-carbon composites, starting from Si. For synthesis of Si/C composites starting from SiO_2, see [48,49]. Unless otherwise stated, Si is "nano" (either purchased as such or ground by ball-milling). CNT = carbon nanotubes; GO = graphene oxide; n.a. = not available.

Reference	Suspension Composition	Post-SD Treatment	%Si
A. Spray-drying of suspension			
Li et al. [55]	Hydroxylated Si and carboxylic-functionalized CNT in water	-	70
Wang et al. [69]	Functionalized Si and functionalized CNT in ethanol	-	56 (EDX)
Yang et al. [72]	Si, lithium acetate and ammonium fluoride in ethanol-water	-	94
B. Spray-drying of suspension followed by heat treatment in inert/reducing atmosphere			
Bie et al. [42]	Si, CNT and phenol-formaldehyde resin in ethanol	900 °C in Ar	69
Gan et al. [50]	Si and graphite dispersed in GO suspension	600 °C in Ar	10
He et al. [51]	Si in GO suspension	700 °C in Ar/H_2	81
Lai et al. [53]	Si, graphite, glucose and sodium dodecyl benzene sulfonate in water	800 °C in Ar	25
Lee et al. [54]	Si and GO in aqueous ethanol	700 °C in Ar	63
Liu et al. [61]	Si, graphite and chitosan in water	700 °C in Ar	15
Pan et al. [63]	Si, GO and sucrose	800 °C in Ar/H_2	72
Su et al. [65]	Si, graphite, phenolic resin and sodium dodecyl benzene sulfonate in water-ethanol	700 °C in Ar	n.a.
Su et al. [66]	Si, graphite and GO in water with 5% alcohol	450 °C in Ar	16
Tao et al. [67]	Si, GO and polyvinyl alcohol in water	700 °C in Ar/H_2	49
Wang et al. [68]	Si/poly (acrylonitrile-co-divinylbenzene) hybrid microspheres, graphite and sodium carboxymethyl cellulose in water	900 °C in Ar	10
Wang et al. [81]	Micron-sized Si (with SiO_x surface layer) and citric acid in water (SiO_x not reduced by heat treatment)	600 °C in Ar	85–94
Wang et al. [70]	Microspheres of Si with in situ polymerized styrene-acrylonitrile copolymer, added to a dispersion of graphite and sodium carboxymethyl cellulose in water	900 °C in Ar	6.7
Yang et al. [73]	Si, pitch, CNT and graphite in alcohol-water	850 °C in Ar	30–35
Zhang et al. [75]	Si, graphitized carbon black and sucrose in water	900 °C in N_2	5-10
Zhang et al. [77]	Si, CNT and glucose in water	800 °C in Ar	n.a.
C. Two consecutive spray-dryings of suspension with intermediate and final heat treatment in inert/reducing atmosphere			
Chen et al. [43]	(Step 1) Si, polystyrene-acrylonitrile, citric acid and graphite in ethylene-glycol ; (Step 2) Powder from step 1 mixed with pitch in tetrahydrofuran	(1) 380 °C in N_2 (2) 500 °C and 900 °C in N_2	25
Chen et al. [44]	(Step 1) Si, graphite and citric acid in water; (Step 2) Powder from step 1 mixed with pitch in tetrahydrofuran	(1) 380 °C in N_2 (2) 500 °C and 900 °C in N_2	6
Chen et al. [45]	(Step 1) Si, graphite and phenol-formaldehyde in ethanol; (Step 2) Powder from step 1 mixed in phenol-formaldehyde solution	(1) and (2) 1000 °C in Ar/H_2	20
Li et al. [58]	(Step 1) Si, graphite, citric acid, polyvinylpyrrolidone in ethanol; (Step 2) Powder from step 1 mixed with pitch in tetrahydrofuran	(1) 380 °C in N_2 (2) 500 °C and 900 °C in N_2	8

Table 6. *Cont.*

Reference	Suspension Composition	Post-SD Treatment	%Si
D. Spray-drying of suspension followed by more complex post-processing			
Li et al. [56]	Si, graphite and glucose in water	Dispersion in pitch solution; drying at 80 °C in vacuum; 1050 °C in Ar; crushing	15
Li et al. [57]	Ball-milled Si in ethanol	HF etching of amorphous SiO_x surface layer	100
Li et al. [59]	Si and polyvinyl alcohol in water	Coating with poly-acrylonitrile; 800 °C in Ar	70
Lin et al. [60]	Si and GO in water-ethanol	Reduction and N-doping of GO by hydrazine hydrate vapor	89
Paireau et al. [62]	Si and polyvinyl alcohol in water	PVA crosslinking; 1050 °C in N_2	40–98
Ren et al. [64]	Si, graphitized needle coke and sucrose in water	900 °C in N_2; carbon coating by CVD	17
Zhang et al. [74]	Si, NaCl and polyvinyl pyrrolidone in water	900 °C in N_2; washing of NaCl in water	30
Zhang et al. [76]	Si, polyvinyl pyrrolidone, nickel acetate and citric acid in ethanol (spray-drying in N_2 atmosphere)	380 °C in N_2; growth of carbon nanotubes and nanofibers in C_2H_2/H_2 at 700 °C (NiO catalyst)	70
Zhou et al. [78]	Si, graphite and citric acid in alcohol-water	400 °C in Ar; coating in dopamine solution; treatment in Ar at temperatures from 600 to 900 °C	n.a.

It can be seen that in many cases, the suspension formulation includes a combination of several carbons or carbon precursors. In some cases (Category C in Table 6), Si is mixed with carbon and carbon precursors in a first spray-drying step, then, the heat-treated composites are again mixed with carbon in a second spray-drying step.

5. Post-Processing of the Spray-Dried Precursors

Spray-drying can be used as a shaping-only method to prepare microspheres and/or as a mixing method for components that do not require further transformation. However, the spray-dried powder is often an intermediate in the synthesis procedure. The very common case of a heat treatment is considered in Section 5.1 while more complex post-spray-drying procedures are described in Section 5.2.

5.1. Heat Treatment

Spray-dried powders often require a heat treatment to transform into the final phase. Depending on the composition of the as-sprayed material, this heat treatment involves thermal decomposition of precursors and/or solid state diffusion and/or crystallization. Thermal analysis (TGA/TDA) and X-ray diffraction are standard characterization techniques helping to optimize the temperature and duration of the heat treatment. Regarding the inorganic active material, heat treatment usually aims at a homogeneous, single-phase composition. Occasionally (see Composites at the end of Table 1), the precursor obtained by spray-drying of a solution is deliberately meant to crystallize into a mixture of two active phases, for example $LiFePO_4$-$Li_3V_2(PO_4)_3$ [359–363].

In the case of electrode compounds in which elements are not at their maximum oxidation state, the solution, suspension or spray-dried precursor may contain species susceptible to oxidation. If necessary, oxidation in solution can be suppressed by reducing additives, complexation and/or removal of dissolved oxygen by degassing. During spray-drying in air, oxygen might lead to some oxidation but most authors do not pay much attention to this effect, due to the short residence time in the spray-dryer. On the contrary, the atmosphere during the heat treatment step is a parameter of major importance to prevent oxidation or even promote reduction (typically in Ar/H_2 with 2 to 10 vol % H_2). This is illustrated by Categories B and C in Table 6 for the case of the synthesis of Si/C composites: oxidation of Si and existing carbon (such as CNT, carbon black, etc.) must be prevented and carbon precursors should transform into more or less graphitized carbon. An overview of the heat treatments reported in Table 6 (B&C) reveals a rather broad range of temperatures and atmospheres.

5.2. More Complex Post-Processing

In some cases, the spray-dried material is only an intermediate and is used as one of the reactants in an ulterior synthesis step. An unlithiated spray-dried (hydr)oxide of several transition metals can be mixed with a lithium salt to provide the electrode material by solid state reaction (see for example [214,376]). In a work by Wang et al. [377], a spray-dried composite of graphene-polyacrylonitrile was reacted with elemental sulfur in a nitrogen atmosphere at 300 °C. Similarly, Liu et al. [378] used mesoporous carbon microspheres prepared by spray-drying as a host for selenium. Oxides in spray-dried metal oxide/carbon composites can be transformed into sulfides or selenides by reaction with appropriate gaseous atmospheres (thiourea in Ar/H_2 [114,375,379] or Se in Ar/H_2 [367]). Wang et al. [380] reported the impregnation of molten lithium in CNT spray-dried spheres. Some authors [48,49] proposed the reduction of SiO_2 in spray-dried SiO_2/CNT composites by reaction with magnesium metal followed by dissolution of MgO in HCl.

The powders obtained in the spray-drying step can also be dispersed in a solution/suspension that is expected to form a coating of a different phase by sol-gel process (ZrO_2, TiO_2 or Al_2O_3 on $LiNi_{1/3}Co_{1/3}Mn_{1/3}O_2$ [150,156]; $Li_4Ti_5O_{12}$ on $LiMn_2O_4$ [190,309] or $LiFePO_4$ [298]), by evaporation of the solvent ($LiFePO_4$ on $Li_3V_2(PO_4)_3$ [336], $LiMnPO_4$ [168] or CeO_2 [178] on $Li_{1.17}Ni_{0.25}Mn_{0.58}O_2$), or by another spray-drying step ($LiCoO_2$ on $LiMn_2O_4$ [381]; Li_3PO_4 on $Li_4Ti_5O_{12}$ [219]; LiF on Si [72]).

Chemical vapor deposition (CVD) is sometimes used to create an additional carbon layer [49,64,71,95,352] or to grow carbon nanotubes/nanofibers if the necessary catalyst was included in the spray-drying step [76,118]. In a work by Shi et al. [382], sacrificial spray-dried layered double oxide (LDO) microspheres act as a template and a catalyst for the CVD growth of graphene; chemical etching of LDO yields a 3D graphene host for sulfur in Li-S batteries. Zhang et al. [383] reported CVD growth of a Si/C layer on graphitized spray-dried carbon black porous microspheres.

The variety of post-spray-drying processing can be further illustrated by the examples in Category D of Table 6, focusing on spray-dried Si.

6. Microstructure

This section is devoted to the microstructural aspects of spray-dried materials. As already mentioned in the introduction, these aspects are extremely important in the case of electrode materials. Basically, anything that favors (i) the penetration of the liquid electrolyte in the electrode material; (ii) short solid state diffusion paths of Li^+/Na^+ ions or (iii) fast transport of electrons is expected to improve the cycling performance. However, it should be kept in mind that high porosity or high content of compounds that do not store charge (e.g., carbon added to facilitate electron transport) will be paid for in terms of energy density (per volume or per mass, respectively).

Here the discussion focuses on the morphology of the individual granules (as-sprayed or after heat treatment) and on possibilities to influence it by various deliberate strategies. It is well-known that spray-drying tends to produce microspheres (Figure 2a) as the result of droplet drying. However, fast drying can also result in the precipitation/solidification of thin crusts leading to hollow or collapsed spheres (Figure 2b,c), depending on the mechanical strength of the crust. Hydrodynamic and/or visco-elastic effects are believed to be at the origin of more exotic shapes such as the "doughnut" particles [384]. The reader is referred to the review by Nandiyanto and Okuyama [9] for a catalogue and discussion of possible morphologies.

The concentration/solid loading of the solution/suspension (see for example [236]) and the spray-drying experimental parameters (equipment, inlet/outlet temperature, atomization parameters) all influence the average size, size distribution, and shape of spray-dried granules. Spray-drying of a solution often yields hollow, thin-shell spheres; the inside volume can be considered as lost space from the point of view of energy density. Breaking these spheres by grinding/milling and shaping the broken pieces into denser—but still porous—spheres by spray-drying of a suspension allows for a large gain in volumic efficiency (see Figure 6 adapted from [100]).

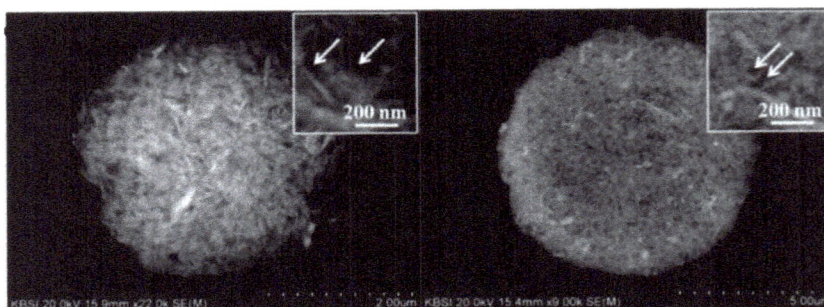

Figure 6. SEM images of cross-sections in (**left**) Co$_3$O$_4$ and (**right**) CoO–carbon composite powders. Both were obtained by a sequence of solution spray-drying—heat treatment in N$_2$—milling—suspension spray-drying—heat treatment (in air for Co$_3$O$_4$, in N$_2$ for CoO/C). (Adapted from [100] by permission of The Royal Society of Chemistry).

Spray-drying of suspensions is indeed recognized as a technique favoring packing efficiency, as illustrated in Figure 7 (adapted from [55]), showing a comparison of the volume occupied by equivalent masses of Si/CNT spray-dried composite spheres and of original Si nanoparticles.

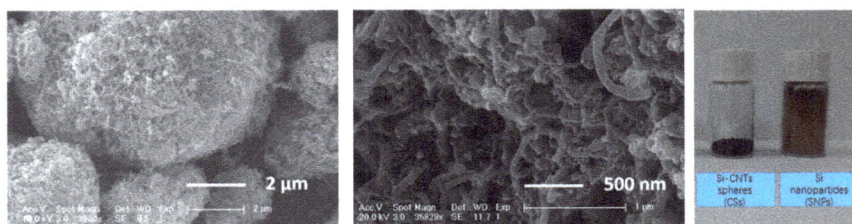

Figure 7. (**left** and **middle**) SEM images of Si/carbon nanotubes (CNT) composite microspheres; (**right**) Comparison of the volume occupied by equivalent masses of Si/CNT spray-dried composite spheres and of original Si nanoparticles. (Adapted from [55]—Published by The Royal Society of Chemistry under CC BY 3.0—https://creativecommons.org/licenses/by/3.0/).

The microstructure and porosity of as-sprayed granules can further evolve during heat treatment due to decomposition/graphitization of organics, crystallization, crystal growth or sintering. The porosity created by the decomposition of organics during a heat treatment in air is expected to help penetration of the electrolyte in the electrode material. Some authors have proposed a hard templating strategy based on polystyrene beads [234,260,318] to introduce controlled macroporosity. For example Nowack et al. [234] investigated the combined effects of nanoporosity (created by thermal decomposition of cellulose) and macroporosity (created by thermal decomposition of polystyrene spheres or carbon fibers) in $Li_4Ti_5O_{12}$ spray-dried granules (Figure 8 reproduced from [234]).

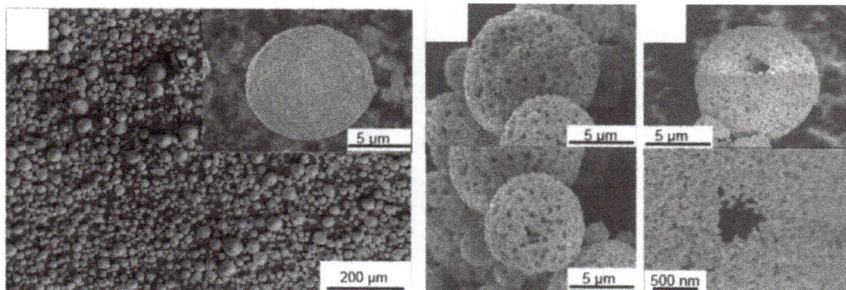

Figure 8. $Li_4Ti_5O_{12}$ spray-dried granules after heat treatment in air to decompose the organic templates: (**left**) nanoporous microspheres obtained from spray-drying with 3 wt % cellulose; (**middle**) macroporous spheres obtained from spray-drying with polystyrene beads as template and (**right**) microspheres with channel structures obtained from spray-drying with carbon fiber templates. (Reproduced from [234] under CC BY 4.0—https://creativecommons.org/licenses/by/4.0/).

Similar strategies rely on other sacrificial phases, such as SiO_2 spheres [32,34,385], in situ formed metal [128] or NaCl [46,74,80,82] particles, all of which are removed at a later stage by chemical etching (SiO_2, metals) or washing (NaCl).

As already explained in Sections 4 and 5, spray-dried electrode materials are frequently designed as composites with carbon in order to improve electron transport and/or buffer volume variations. Figure 9 shows an example of Sb nanoparticles embedded in a carbon matrix formed by carbonization of the organic precursor during heat treatment of the spray-dried precursor in inert atmosphere.

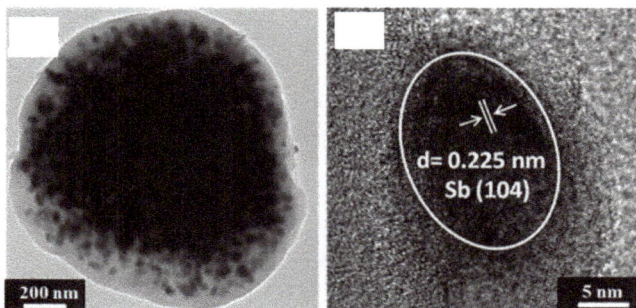

Figure 9. Sb nanoparticles embedded in carbon matrix: (**left**) transmission electron microsopy (TEM) image; (**right**) high resolution TEM (HRTEM) image. (Adapted from [40] with permission from The Royal Society of Chemistry).

When carbon is added as CNT, carbon black, graphite or graphene oxide in the solution/suspension before spray-drying, there is an (often implicit) assumption that the distribution of carbon in the granules will be of sufficient homogeneity. In the case of composites with reduced graphene oxide, some authors have been able to supplement the usual SEM and TEM images (see

Figure 7 for a CNT example) by cross-sectional TEM (Figure 10—adapted from [310]) or imaging of the graphene network after chemical etching of the inorganic phase (Figure 11—adapted from [344]).

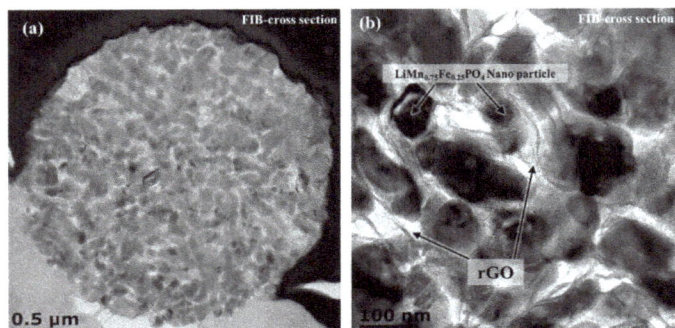

Figure 10. (**a**,**b**) Cross-sectional TEM images of LiMn$_{0.75}$Fe$_{0.25}$PO$_4$/reduced graphene oxide composite microsphere. (Adapted from [310] under CC BY 4.0—https://creativecommons.org/licenses/by/4.0/).

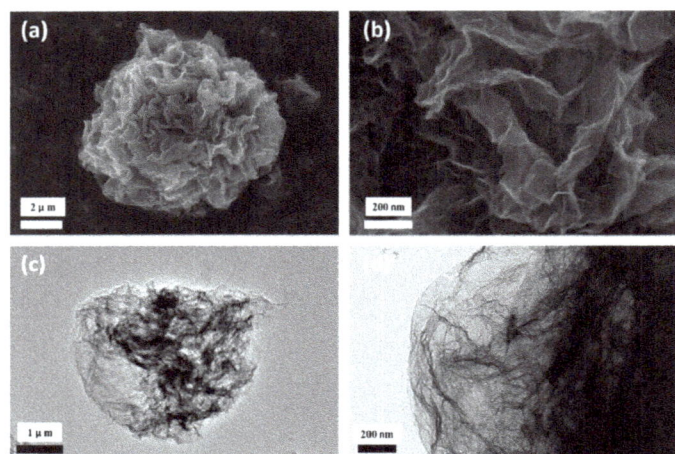

Figure 11. Graphene network after chemical etching of the Na$_3$V$_2$(PO$_4$)$_3$ phase: (**a**,**b**) SEM images; (**c**,**d**) TEM images. (Reproduced with permission from [344]. Copyright (2017) American Chemical Society.).

This overview of morphologies cannot be exhaustive. The examples shown in Figures 6–11 correspond to morphologies that retain a (roughly) spherical appearance, but Figure 2b,c should remind the reader that crumpled morphologies are also common. As a final illustration of the microstructural variety, Figure 12 displays a more unexpected, multi-shelled morphology which has been reported and studied by several groups [101,107,139]. Yolk-shell granules [103,122,136] are a less extreme case of a similar phenomenon.

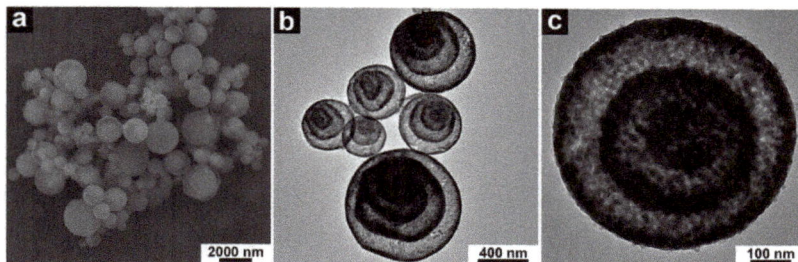

Figure 12. Hematite Fe_2O_3 multi-shelled hollow spheres obtained by heat treatment of precursors spray-dried from an iron(III) citrate and sucrose solution: (**a**) SEM image; (**b,c**) TEM images. (Adapted from [107] with permission of The Royal Society of Chemistry).

7. Electrochemical Properties

The overwhelming majority of spray-dried materials reported in the literature for Li-ion and Na-ion batteries are used as electrode materials. Amongst the few exceptions are (i) $Li_{1.3}Al_{0.3}Ti_{1.7}(PO_4)_3$ [338] which is used as a solid state electrolyte and (ii) La_2O_3 [113] or CeO_2 [98] hollow spheres which are coated on the separator of Li-sulfur batteries and are supposed to block lithium polysulfides and act as a catalyst for the sulfur redox reaction.

Literature on spray-dried materials for positive or negative electrodes follows the general trend: the largest number of publications concerns materials for Li-ion batteries but research on compounds for Na-ion batteries is increasing strongly in recent years. Regarding emerging technologies, spray-drying is receiving interest as a tool to prepare porous carbon hosts for sulfur/selenium in Li-sulfur [23,24,34,37–39,116,120,128,377,385] or Li-selenium [83,378] batteries. Similarly, reduced graphene oxide microspheres with high surface area were tested in Li-air batteries [32]. In the field of "beyond Li/Na" technologies, $Na_3V_2(PO_4)_3$/C [343] and Li_3VO_4/C [261] obtained by spray-drying have recently been mentioned in research on Mg-ion batteries.

As explained at the end of the introduction, the main focus of this review is on guidelines for the formulation of spray-drying feedstock solutions/suspensions and how it can affect microstructure. In the following of this section, a few examples are selected to illustrate the link between formulation, microstructure and electrochemical properties. As a complement, Table 3 in Appendix A lists values of experimental discharge capacities after 50 cycles.

The first examples concern layered oxides, including Li-rich compositions sometimes written as xLi_2MnO_3-$(1-x)LiMO_2$ (M = Ni, Co, Mn, ...), which are studied because of their high theoretical reversible capacity (above 250 mAh/g). Hou et al. [149] reported the synthesis of $0.5Li_2MnO_3$-$0.5LiMn_{1/3}Ni_{1/3}Co_{1/3}O_2$ (=$Li_{1.2}Mn_{0.54}Ni_{0.13}Co_{0.13}O_2$) by heat treatment of a precursor obtained by spray-drying of an aqueous suspension of Li_2CO_3 and a coprecipitated metal hydroxide (SD-LLO sample). For comparison, another sample was prepared by heat treatment of a dry mixture of Li_2CO_3 and coprecipitated metal hydroxide (CP-LLO sample). The authors found that the spray-drying procedure was more efficient to promote the homogeneity of the distribution of metal cations in the final oxide and resulted in better electrochemical performance (see Figure 13 reproduced from [149]). In particular, the decrease in average cell voltage was much less marked (Figure 13d), which was considered as an indication of the better stability of the layered structure against transformation into spinel structure on cycling [149].

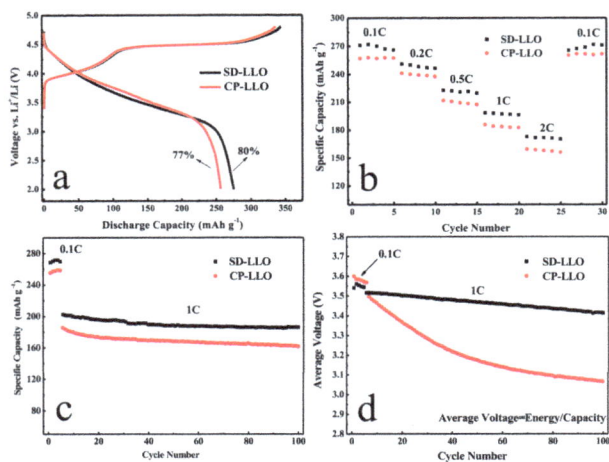

Figure 13. Comparison of two samples of Li-rich oxide $0.5Li_2MnO_3$-$0.5LiMn_{1/3}Ni_{1/3}Co_{1/3}O_2$ obtained by a spray-drying procedure (SD-LLO) or by a dry mixing procedure (CP-LLO)—see text for details. (**a**) First cycle charge/discharge profiles; (**b**) Rate performance; (**c**) Cycling performance between 2 and 4.8 V; (**d**) Average discharge voltage as a function of cycle number during cycling. (Reproduced from [149]. Copyright (2015), with permission from Elsevier).

The work by Hou et al. [149] described above can be considered as a demonstration of the superiority of wet mixing over dry mixing. In a study of Chen et al. on $LiNi_{0.8}Co_{0.15}Al_{0.05}O_2$ [143], a suspension of a ball-milled precursor was dried either by spray-drying (SD-NCA sample) or by common drying (CD-NCA sample). The mixing by ball milling is the same in the two samples so that the much better electrode performance of the SD-NCA sample (e.g., a capacity retention of 75% after 500 cycles at 2 C, against only 12% for the CD-NCA sample) can be attributed to a more favorable microstructure induced by spray-drying.

These two examples highlight positive features of the spray-drying of suspensions. This should not mask the fact that spray-drying of suspensions is a variant of solid state synthesis and is, therefore, subject to the usual limitations associated to diffusion lengths in the solid state. This was recently illustrated in a work by Wang et al. [189] where the formation of $Li[Li_{0.2}Mn_{0.54}Ni_{0.13}Co_{0.13}]O_2$ was followed by in-situ high-energy X-ray diffraction during the heat treatment. Irregularities in the temperature dependence of the crystallographic cell parameters and the presence of secondary phases were observed in the case of a precursor obtained by spray-drying a ball-milled suspension of the individual oxides and carbonates (Li_2CO_3, $MnCO_3$, Co_3O_4 and NiO). As could be expected, these irregularities and the content in secondary phases decreased when the suspension was prepared by ball-milling a precalcined mixture. Minimizing diffusion lengths is the usual reason to turn from solid state synthesis to solution routes. In the case of spray-drying, this means going from suspensions to solutions. For example, Watanabe et al. [174] could obtain a discharge specific capacity of 275 mAh/g for $Li_{1.2}Mn_{0.58}Ni_{0.18}Co_{0.03}O_2$ obtained by spray-drying of a solution of acetates in aqueous citric acid.

In the case of compounds with relatively low intrinsic electronic conductivity, the microspheres obtained by spray-drying are often too large for good performance. One of the works demonstrating this effect was published by Nakahara et al. [233] in 2003, where the authors compare as-obtained (LT-2 sample) and ball-milled (LT-FP sample) $Li_4Ti_5O_{12}$ prepared by spray-drying and heat treatment of an aqueous suspension of LiOH and TiO_2. The 5–10 μm sintered granules were broken by ball-milling into sub-micron particles; electrodes were prepared by mixing with acetylene black and PVDF and tested in half-cells against lithium metal. The rate capability test showed that the discharge capacity of the ball-milled LT-FP sample decreased by less than 15% when going from 0.15 C to 10 C, whereas the discharge capacity of the LT-2 sample had already decreased by more than 40% at 5 C.

As already mentioned in the previous sections, another way to deal with the issue of electronic conductivity is to form/include conductive carbon in the spray-dried material. This strategy is relevant whenever the subsequent heat treatment can be carried out in non-oxidizing atmosphere. For example, soluble precursors of carbon are commonly added to suspensions for the preparation of LiFePO$_4$/C composites. In a work by Liu et al. [283], LiFePO$_4$ with 2.5 wt % C was obtained by heat treatment in N$_2$ of a precursor prepared from an aqueous suspension of Li$_2$CO$_3$ and FePO$_4$ into which glucose had been dissolved. The authors compared spray-drying with microwave drying through testing of 14500-type cylindrical batteries with a graphite negative electrode and attributed the ~10% better performance of the spray-dried material to the higher compaction density of the electrode (2.55 g/cm^3) that could be reached thanks to the favorable microstructure.

In the previous example, the LiFePO$_4$ active material was formed during the heat treatment. In other cases, spray-drying is used to create a composite of carbon with an existing active material, such as silicon. As seen in Table 6, there is an impressive variety of carbon sources to choose from, but comparison is difficult because of the wide range of Si/C ratio in the final materials. In view of guiding the development of Si/C negative electrodes with high Si content, Ogata et al. [79] used two spray-dried Si/C composites (Si/flake graphite/CNT with 54 wt % Si and Si/flake graphite with 87 wt % Si—both are extensively characterized in the Methods section of ref. [79]) as the reference materials for a very detailed study of the phenomena governing coulombic efficiency. This was done by cycling the materials at different depth of discharge in order to probe the volume change of the amorphous phase and/or the amorphous-crystalline transformations. As shown in Figure 14 (reproduced from [79]), a broad range of techniques were used to characterize the (micro)structure and composition at different stages of individual cycles.

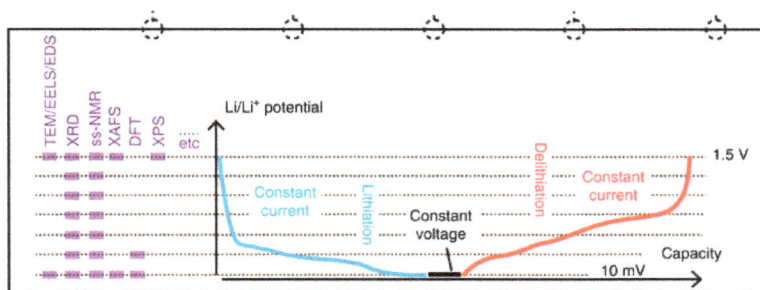

Figure 14. Overview of a structural characterization study conducted on spray-dried Si/C composites at different stages during individual cycles. The set of characterizations was repeated every 20 cycles. (Reproduced from reference [79] under CC BY 4.0—https://creativecommons.org/licenses/by/4.0/).

From a chemical point of view, the most complex case is probably when a solid form of carbon is dispersed in a solution of several inorganic salts. This is typically the case for the spray-drying synthesis of phosphates or fluorophosphates from solutions where carbon nanotubes or graphene oxide are added to provide electronic conductivity. In our work on Na$_3$V$_2$(PO$_4$)$_2$F$_3$/CNT [93], we found that an excess amount of NaF was necessary to prevent the formation of a small amount of fluorine-free Na$_3$V$_2$(PO$_4$)$_3$ secondary phase, suggesting that the addition of CNT to the solution interferes a little with the inorganic components. Conversely, the high concentration of several ions in the solution is supposed to affect the dispersion of carbons, although this effect has not yet been studied in such very complex situations. This might be one of the reasons why we observed an inhomogeneous distribution of carbon black (CB) in spray-dried granules of Na$_2$FePO$_4$F/CB [89], leading to a drop of 60% in discharge capacity compared to Na$_2$FePO$_4$F/CNT composites with similar carbon content [87,89]. A work by another group [342] on the fluorine-free alluaudite phosphate $Na_3V_2(PO_4)_3$ (with the drawback of a lower operating voltage) confirms that excellent rate capability is possible for a $Na_3V_2(PO_4)_3/CNT$ composite (Figure 15, reproduced from [342]). Along the same lines, Table 5 shows that graphene oxide (reduced during post-treatment) is becoming a popular choice for many phosphates, as exemplified by the results for $NaTi_2(PO_4)_3/RGO$ [340], where the discharge capacity decreases by less than 10% when going from 0.1 C to 30 C rate (130 mAh/g at 0.1 C, 118 mAh/g at 30 C).

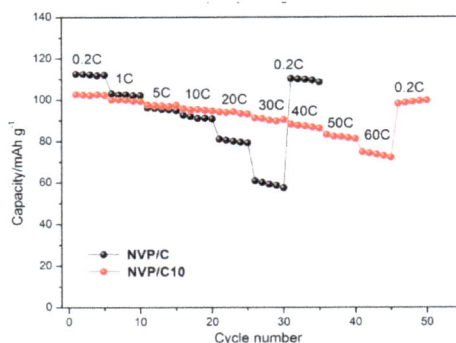

Figure 15. Rate capability of $Na_3V_2(PO_4)_3$ with 10 wt % CNT (NVP/C10) and without CNT (NVP/C). The electrodes were cycled vs. Na in the 2.0–3.8 V voltage range. Both samples were obtained by heat treatment of a spray-dried precursor prepared from a citric acid solution of NaHCO_3, NH_4VO_3 and NH_4H_2PO_4 into which CNT were dispersed in the case of the NVP/C10 sample. (Reproduced with permission from [342]. Copyright (2018) American Chemical Society).

8. Concluding Remarks

It should be clear from the preceding sections that the term "spray-drying" covers many different realities. Reasons for using spray-drying are varied since it can be used as a tool for mixing, shaping, or synthesizing (or combining several of these objectives simultaneously).

In many cases, spray-drying is not really a rival to other routes but rather a way to bring a laboratory-scale procedure to the next level in terms of production quantities, reproducibility, and control of agglomeration. This is true, for example, for many solid state reaction syntheses on the condition that the starting materials are not soluble in the liquid medium of the suspension. This can also be the case for sol(ution)-gel routes, taking into account that the increase in drying speed might modify some characteristics by comparison with a conventionally-dried gel. More generally, spray-drying can be considered in all cases where no problem comes from the fact that, except for volatiles, everything that is injected in the spray-dryer turns up in the as-spray-dried powder.

In other cases, spray-drying offers new opportunities, such as the dispersion of carbon in active material or the possibility offered by the droplet scale to use a simple solvent evaporation route (which, in other conditions, would result in unacceptably large composition inhomogeneities).

Spray-drying is commonly used in industry in many fields of applications. The 300+ publications referenced in this review demonstrate that the potential of spray-drying is increasingly recognized in the academic community for the synthesis of electrode materials from lab- to pilot-scale quantities.

However, the apparent simplicity of the spray-drying concept should not mask the fact that choices regarding the formulation of solutions/suspensions and the selection of experimental spray-drying parameters decisively affect the characteristics of the final material. Optimization of the parameters of the subsequent heat treatment is also very important but cannot alter drastically the microstructural properties. It is the hope of the authors that this review can contribute to a realization that making the

Materials **2018**, *11*, 1076

most of spray-drying requires a considered choice amongst possible strategies and careful consideration of the solution/suspension formulation.

Author Contributions: N.E., C.P., J.B. and A.M. performed the bibliographic search and prepared the tables and figures. N.E. performed the original experiments corresponding to the results in Figure 2. F.B. and B.V. conceived and wrote the review. All authors read the draft, provided corrections and approved the final version.

Acknowledgments: The authors are grateful to University of Liege and FRS-FNRS for equipment grants. Part of this work was supported by the Walloon Region under the "PE PlanMarshall2.vert" program (BATWAL –1318146). NE thanks FRIA, Belgium for a Ph.D. fellowship [Grant 1.E118.16]. A. Mahmoud is grateful to the Walloon region (Belgium) for a Beware Fellowship Academia [Grant 2015-1, RESIBAT 1510399].

Conflicts of Interest: The authors declare no conflict of interest.

Appendix A

Table A1. Spray-drying parameters for layered oxides AMO_2 ($A = Li^+$, Na^+; $M = Li$, Ni, Mn, Co, Al, ...) Sections in the table are the same as in Table 3 (see main text) where compound stoichiometries and solution/suspension compositions can be found. Information about the spray-drying instruments is given as provided in the referenced papers. - = not available.

	T_{inlet} (°C)	T_{outlet} (°C)	Other Parameters	Spray-Drying Instrument
SPRAY-DRYING OF SOLUTIONS				
A. Spray-drying of aqueous solution of nitrates and/or acetates				
Duvigneaud et al. [145]	190	150	-	Buchi mini spray-dryer 190
He et al. [146] and He et al. [148]	200	-	400 mL/h	SD-2500 (Shanghai Triowin Lab Technology Company)
Kim et al. [152]	-	-	Bifluid nozzle 0.2 MPa	-
Kim et al. [187]	-	-	-	SD-1000, Tokyo Rikakikai Co. Ltd, Tokyo, Japan
Konstantinov et al. [153]	190–200	90–100	-	Yamato GA32
Li et al. [154]	-	-	-	Yamato GB32 pulvis mini-spray
Li et al. [156] and Li et al. [157]	-	-	-	Buchi mini spray-dryer B-290
Li et al. [160]	300	100	Bifluid nozzle 0.4 MPa 10 L/h	-
Liu et al. [166]	350	150	Bifluid nozzle 0.4 MPa	-
Wang et al. [263]	-	-	-	-
Wang et al. [172]	200	-	2.5 mol/L total cation concentration	-
Wang et al. [173]	210	110	-	-
Wu et al. [175]	220	110	Air pressure 0.2 MPa	-
Yue et al. [179,180]	220	110	-	-
Zhang et al. [183] and Zhang et al. [186] and Zhao et al. [264]	-	-	-	Pulvis mini-spray GB22, Yamato, Japan
B. Spray-drying of aqueous solution of salts dissolved in aqueous citric acid				
Li et al. [158]	180	65–70	-	-
Sun et al. [171]	-	-	2 mol/L concentration	Pulvis mini-spray GB22, Yamato, Japan
Watanabe et al. [174]	-	-	2 mol/L concentration	Buchi B-290
Zhang et al. [184,185]	-	-	-	Pulvis mini-spray GB22, Yamato, Japan
C. Spray-drying of aqueous solution of citrates				
Li et al. [155]	-	-	-	Yamato GB32 pulvis mini-spray
Qiao et al. [169]	-	-	-	L217, Lai Heng
Yuan et al. [178]	-	-	-	L217, Lai Heng

Table A1. *Cont.*

	T_{inlet} (°C)	T_{outlet} (°C)	Other Parameters	Spray-Drying Instrument
D. Spray-drying of aqueous solution (others)				
Li et al. [159]	200	-	Pumping 1.2 g/s Jet-air speed 6 kg/h 4 wt % solution	Spray-dryer Minor Niro A/S, Söborg, Denmark
Oh et al. [167]	-	-	-	-
SPRAY-DRYING OF SUSPENSIONS				
E. Spray-drying of an aqueous suspension to mix reactants				
Hou et al. [149]	-	-	-	-
Lin et al. [164]	200	-	-	-
Liu et al. [165]	-	-	-	-
Wang et al. [189]	-	-	-	-
Yue et al. [181]	-	-	-	-
F. Spray-drying of an ethanol suspension to mix reactants				
Hu et al. [150]	-	-	-	-
Lin et al. [161]	-	-	-	Niro 2108, Copenhagen
Lin et al. [162]	150	-	-	Niro 2108, Copenhagen
G. Mixing of AMO$_2$ active material with conductive carbon or conductive carbon precursor				
Cheng et al. [144]	200	-	Bifluid nozzle 0.2 MPa	SD-2500
Xia et al. [176]	-	-	-	SD-1500 laboratory scale spray-dryer (Tridwin Tech. Co. Shanghai, China)
Yang et al. [177]	220	-	Atomization pressure 0.5 MPa	-
Yue et al. [182]	-	-	-	-
H. Shaping of AMO$_2$ as spheres				
Chen et al. [143]	220	90	Compressed air pressure 0.2 MPa	-

Table A2. Inventory of the starting materials used in the publications referenced in this review.

Element	Precursor
Al	Nitrate [145,250,337]
B	H$_3$BO$_3$ [20], LiBO$_2$.8H$_2$O [21,22]
Ce	Nitrate [98]
Co	Acetate [146,148,151–157,160,171–175,177–180,186,187,200,201,215], nitrate [92,99–102,138,139,145,166,213,355], Co$_3$O$_4$ [7,161–163,189], Co(OH)$_2$ [139,167], (Co,Ni,Mn)OH$_x$ [149,165,188], (Co,Ni,Mn)O$_x$ [150]
Cr	Acetate [146,203], chloride [103], sulfate [203], Cr$_2$O$_3$ [202,203]
Cu	Acetate [140,346], nitrate [104–106,366]
F	NaF [87–89,91–94], HF [84], NH$_4$F [72], trifluoroacetic acid CF$_3$COOH [85,96]
Ge	GeO$_2$ dissolved in ammonia solution [111], GeO$_2$ from hydrolysis of GeCl$_4$ [112]
Fe	Fe [87–89]
Fe^{2+}	Oxalate [27,273–275,279,280,317,323,337,349–352,359], sulfate [135,281,296,304,310], acetate [86,307], chloride [310], (Fe,Mn)$_3$(PO$_4$)$_2$.xH$_2$O [312–314,319]
Fe^{3+}	Nitrate [21,22,110,136,158,212,268–270,285,286,300–303,310,347,354,361,363,366], phosphate [272,277,278,283,284,287,288,290–294,299,306–308,311], citrate [107,108,295,360,362], Fe$_2$O$_3$ [109,190,276],
La	Nitrate [113,357]
Li	Carbonate [7,20,84,149], [35],[35,161–164,167,188–190,202,203,216–218,220,221,227,228,230–232,237,243,247,248,251,256–261,271,273,277,283,286,294,295,296–303,308,317,323,325,329,333,334,362], hydroxide [150,159,165,168,169,172,177,178,184,185,192,193,199,210,222–226,229,233,235,239–241,249,253–257,262,272,278,279,281,284,285,287,288,290–293,296,304,306,311–314,319,327, 331,332,335–337,349,351,358–360,363], acetate [72,85,86,146,148,160,171,173–175,179,180,183,191,194–198,200,201,204,205,208,209,211,212,215,250,274,275,305,367,357], nitrate [145,151,152,154,156–158,166,186,187,244,324,364], oxalate [350,352], LiBO$_2$.8H$_2$O [21,22], LiH$_2$PO$_4$ [276,280,310,328,365]
Mg	Acetate [308]
Mn	Acetate [40,91,114,140,146,151,152,154–157,168,169,171–174,177–180,183,186,187,191–198,200,201,204–209,211–215,262–264,357,365], nitrate [158,166,199,310,311,364], carbonate [20,161–165,189,192], chloride [310], sulfate [310], MnC$_2$O$_4$.2H$_2$O [317,323], MnO$_2$ [7,191,202,203], Mn$_3$O$_4$ [164], (Co,Ni,Mn)OH$_x$ [149,165,188], (Ni,Mn) oxalate [211], (Fe,Mn)$_3$(PO$_4$)$_2$.xH$_2$O [312–314,319]
Mo	(NH$_4$)$_6$Mo$_7$O$_{24}$.4H$_2$O [115,141,367], MoS$_2$ [353]
Na	NaOH [87–89,97], acetate [263,265,343], Na$_2$CO$_3$ [339,340,344,346,351], NaHCO$_3$ [342], NaNO$_3$ [264,347], NaF [87–94], NaH$_2$PO$_4$ [91,345], sodium carboxymethylcellulose [333]

Table A2. *Cont.*

Element	Precursor
Ni	Acetate [22,85,117,146,148,151,152,154–157,169,171,172,174,175,177–180,183–187,204,205,207–209,211,212,215,263,264,357,358,375], nitrate [138,139,141,145,158,166,206,213,214,367], carbonate [164], Ni(OH)$_2$ [167], NiO [7,161–163,189], (Co,Ni,Mn)OH$_x$ [149,165,188], (Co,Ni,Mn)O$_x$ [150], (Ni,Mn) oxalate [210]
Nb	Nb$_2$O$_5$ [142,190], (NH$_4$)NbO(C$_2$O$_4$)$_2$·H$_2$O [131], ethoxide [251]
P	NH$_4$H$_2$PO$_4$ [87–90,93,94,268–271,273–275,279,285,286,295,301,302,305,317,323–325,327,329,332–335,337,339,340,342–344,346,359,360,362], NaH$_2$PO$_4$ [91,345], LiH$_2$PO$_4$ [276,280,310,328,365], H$_3$PO$_4$ [92,93,281,296,300,303,304,312–314,319,331,336,363,364], FePO$_4$(.xH$_2$O) [272,278,283,284,287,288,290–293,294,306–308,351], 1-hydroxyethane 1,1-diphosphonic acid HEDP (CH$_3$C(OH)(H$_2$PO$_3$)$_2$) [347,348], P [36], (Fe,Mn)$_3$(PO$_4$)$_2$.xH$_2$O [312–314,319]
Ru	Acetate [212]
S	Thiourea [114], sulfur [37–39], MoS$_2$ [353]
Sb	SbCl$_3$ [40]
Se	Se [83], H$_2$SeO$_3$ by dissolving SeO$_2$ in water [354], H$_2$Se gas for post-treatment of spray-dried precursor [367]
Si	Si [42–47,50,51,53–60,62–67,69,71,73–81,366], SiO$_2$ [48,49,120,121,349], SiO [52,118,119], tetraethyl ethoxysilane TEOS [350,352], Si/poly(acrylonitrile-divinylbenzene) hybrid microspheres [68], Si/poly(styrene-acrylonitrile) hybrid microspheres [70]
Sn^{2+}	Oxalate [122,124,355], chloride [169]
Sn^{4+}	Chloride [82,123,348]
Ti	TiO$_2$ [84,133,142,216–218,221,223–226,228,230–233,237,244,247,308,323,339,340], TiO$_2$ from basic hydrolysis of TiOSO$_4$·H$_2$SO$_4$·8H$_2$O [126], TiOSO$_4$·H$_2$SO$_4$·H$_2$O [131], Ti peroxo-carbonate solution [127], acidic solution of [NH$_4$]$_2$[Ti(C$_2$O$_4$)$_3$] [184,185], titania nanosheets [129,130], TiO(OH)$_2$(·xH$_2$O) [220,356], Ti tetraisopropoxide (C$_3$H$_7$O)$_4$Ti [128,132,212,227,235,243,248,251], Ti tetrabutoxide (C$_4$H$_9$O)$_4$Ti [211,222,229,238–241,249,250,265]
V	NH$_4$VO$_3$ [94,254,257,324,327,329,331–333,335–337,342,343,345,346,352,359,361,363,365], V$_2$O$_5$ [93,253,255,256,258–261,325,328,334,344,360,362,364]
Zn	Sulfate [135], nitrate [136]
Zr	ZrO$_2$ [161], Zr(NO$_3$)$_4$·5H$_2$O [357]

Table 3. Discharge specific capacity (in mAh/g) after 50 cycles at the indicated current density (in A/g or as a C-rate). For counter electrode, see first column.

Compound Type, Formulas and References	Discharge Capacity after 50 Cycles	
Borates		
LiFeBO$_3$ vs. Li [21]	127 mAh/g	after 30 cycles at 10 mA/g + 20 cycles at 20 mA/g
LiFe$_{0.94}$Ni$_{0.06}$BO$_3$ vs. Li [22]	132 mAh/g	after 35 cycles at 10 mA/g + 15 cycles at 20 mA/g
Elements		
C vs. Li [25]	355 mAh/g	after 50 cycles at 0.1 A/g
C vs. Li [27]	460 mAh/g	after 50 cycles at 0.37 A/g (1 C)
C vs. Li [31]	245 mAh/g	after 50 cycles at 0.1 A/g
C vs. Li [33]	460 mAh/g	after 50 cycles at 0.05 A/g
C (with 4 wt % Ni) vs. Li [35]	640 mAh/g	after 50 cycles at 0.5 A/g
P/C vs. Na [36]	2200 mAh/g	after 50 cycles at 0.1 A/g
S/C vs. Li [37]	980 mAh/g	after 50 cycles at 0.2 C
C/S vs. Li [38]	980 mAh/g	after 50 cycles at 0.1 C
S/C vs. Li [39]	840 mAh/g	after 50 cycles at 0.1 C
Sb/C vs. Na [40]	630 mAh/g	after 50 cycles at 0.2 A/g (0.33 C)
Si/C vs. Li [41]	1150 mAh/g	after 50 cycles at 0.45 A/g
Si/C vs. Li [42]	2200 mAh/g	after 50 cycles at 0.3 A/g
Si/C vs. Li [43]	1150 mAh/g	after 50 cycles at 0.1 A/g
Si/C vs. Li [44]	500 mAh/g	after 50 cycles at 0.1 A/g
Si/C vs. Li [46]	900 mAh/g	after 50 cycles at 0.2 A/g
Si/C vs. Li [47]	2450 mAh/g	after 50 cycles at 0.3 A/g
Si/C vs. Li [48]	1100 mAh/g	after 50 cycles at 0.3 A/g
Si/C vs. Li [49]	2200 mAh/g	after 50 cycles at 1 A/g
Si/C vs. Li [50]	420 mAh/g	after 50 cycles at 0.05 A/g
Si/C vs. Li [52]	600 mAh/g	after 50 cycles at 0.1 A/g
Si/C vs. Li [54]	1250 mAh/g	after 50 cycles at 1 A/g
Si/C vs. Li [55]	2100 mAh/g	after 50 cycles at 0.5 C
Si/C vs. Li [56]	570 mAh/g	after 50 cycles at 0.1 C
Si/C vs. Li [58]	650 mAh/g	after 50 cycles at 0.1 A/g
Si/C vs. Li [60]	1160 mAh/g	after 50 cycles at 0.1 A/g
Si/C vs. Li [61]	580 mAh/g	after 50 cycles at 0.1 A/g
Si/C vs. Li [63]	1800 mAh/g	after 50 cycles at 0.2 A/g
Si/C vs. Li [64]	560 mAh/g	after 50 cycles at 0.05 A/g
Si/C vs. Li [65]	500 mAh/g	after 50 cycles at 0.1 A/g
Si/C vs. Li [66]	500 mAh/g	after 50 cycles at 0.1 A/g
Si/C vs. Li [67]	950 mAh/g	after 50 cycles at 0.1 A/g
Si/C vs. Li [68]	500 mAh/g	after 50 cycles at 0.1 A/g
Si/C vs. Li [69]	2100 mAh/g	after 50 cycles at 0.5 A/g
Si/C vs. Li [70]	450 mAh/g	after 50 cycles at 0.1 A/g
Si/C vs. Li [71]	500 mAh/g	after 50 cycles at 5 C
Si/C vs. Li [73]	820 mAh/g	after 50 cycles at 0.1 A/g
Si/C vs. Li [74]	1400 mAh/g	after 50 cycles at 0.05 C
Si/C vs. Li [75]	500 mAh/g	after 50 cycles at 0.05 A/g
Si/C vs. Li [76]	1200 mAh/g	after 50 cycles at 0.3 A/g
Si/C vs. Li [77]	1100 mAh/g	after 50 cycles at 0.2 A/g
Si/C vs. Li [78]	780 mAh/g	after 50 cycles at 0.2 A/g
Si/C vs. Li [79]	1700 mAh/g	after 50 cycles at 1 C
Si/C vs. Li [80]	1550 mAh/g	after 50 cycles at 0.05 A/g
Si/C vs. Li [81]	1860 mAh/g	after 50 cycles at 0.1 A/g
Sn/C vs. Li [82]	670 mAh/g	after 50 cycles at 0.2 A/g
Sn/C vs. Na [82]	400 mAh/g	after 50 cycles at 0.05 A/g
Se/C vs. Li [83]	590 mAh/g	after 50 cycles at 0.1 C
Fluorides		
Li$_3$FeF$_6$ vs. Li [86]	85 mAh/g	after 50 cycles at 0.05 C
Fluorophosphates		
Na$_2$MnPO$_4$F/C vs. Na [90]	77 mAh/g	after 50 cycles at 6.2 mA/g
Na$_3$V$_2$(PO$_4$)$_2$F$_3$/C vs. Li [93]	100 mAh/g	after 50 cycles at 1 C
Na$_3$V$_2$O$_2$(PO$_4$)$_2$F/C vs. Na [94]	117 mAh/g	after 50 cycles at 0.5 C

Table 3. *Cont.*

Compound Type, Formulas and References	Discharge Capacity after 50 Cycles	
Organic salts		
$Li_2C_8H_4O_4$ vs. Li [95]	150 mAh/g	after 50 cycles at 0.05 C
$Na_2C_8H_4O_4$/C vs. Li [96]	210 mAh/g	after 50 cycles at 0.1 C
Oxides M_xO_y		
CoO/C vs. Li [100]	900 mAh/g	after 50 cycles at 1.4 A/g
Co_3O_4 vs. Li [100]	830 mAh/g	after 50 cycles at 1.4 A/g
Co_3O_4 vs. Li [101]	1020 mAh/g	after 50 cycles at 0.5 A/g
Co_3O_4 vs. Li [102]	1050 mAh/g	after 50 cycles at 1.4 A/g
Cr_2O_3/C vs. Li [103]	630 mAh/g	after 50 cycles at 0.1 A/g
CuO vs. Li [104]	690 mAh/g	after 50 cycles at 1 A/g
CuO/C vs. Li [105]	700 mAh/g	after 50 cycles at 2 A/g
CuO vs. Li [106]	760 mAh/g	after 50 cycles at 1 A/g
Fe_2O_3 vs. Li [107]	870 mAh/g	after 50 cycles at 0.4 A/g
Fe_2O_3/C vs. Li [108]	880 mAh/g	after 50 cycles at 0.4 A/g
Fe_2O_3/C vs. Li [109]	710 mAh/g	after 50 cycles at 0.8 A/g
Fe_2O_3 vs. Li [110]	1020 mAh/g	after 50 cycles at 0.4 A/g
GeO_x/C vs. Li [111]	975 mAh/g	after 50 cycles at 0.5 A/g
GeO_2/C vs. Li [112]	1060 mAh/g	after 50 cycles at 0.2 C
MnO/C vs. Li [114]	300 mAh/g	after 50 cycles at 0.5 A/g
MoO_3/C vs. Li [115]	1120 mAh/g	after 50 cycles at 0.5 A/g
NiO vs. Li [117]	590 mAh/g	after 50 cycles at 0.1 C
SnO_2/C vs. Li [122]	600 mAh/g	after 50 cycles at 2 A/g
SnO_2/C vs. Li [123]	1200 mAh/g	after 50 cycles at 0.1 A/g
SnO_2 vs. Li [124]	715 mAh/g	after 50 cycles at 2 A/g
SnO_2 vs. $LiMn_2O_4$ [124]	365 mAh/g	after 50 cycles at 1 A/g
TiO_2 vs. Li [126]	75 mAh/g	after 50 cycles from 0.1 C to 10 C
TiO_2/C vs. Li [127]	150 mAh/g	after 50 cycles at 0.94 A/g
TiO_2 vs. Li [130]	80 mAh/g	after 50 cycles at 0.02A/g
TiO_2 vs. Li [131]	190 mAh/g	after 50 cycles at 0.5 C
TiO_2/C vs. Na [133]	140 mAh/g	after 50 cycles at 0.2 C
V_2O_5/C vs. Li [134]	240 mAh/g	after 50 cycles at 0.2 C
Oxides $M_xM'_yO_z$		
$ZnFe_2O_4$ vs. Li [135]	1250 mAh/g	after 50 cycles at 0.1 A/g
$ZnFe_2O_4$ vs. Li [136]	750 mAh/g	after 50 cycles at 0.5 A/g
$Mn_{0.5}Co_{0.5}Fe_2O_4$/C vs. Li [137]	610 mAh/g	after 50 cycles at 0.1 A/g
$(Ni,Co)O_x$ vs. Li [139]	850 mAh/g	after 50 cycles at 1 A/g
$Cu_{1.5}Mn_{1.5}O_4$ vs. Li [140]	460 mAh/g	after 50 cycles at 0.1 A/g
$NiMoO_4$ vs. Li [141]	1000 mAh/g	after 50 cycles at 1 A/g
$TiNb_2O_7$/C vs. Li [142]	300 mAh/g	after 50 cycles at 0.25 C
Oxides $Li_xM_yO_z$ (layered)		
$LiCoO_2$ vs. graphite [153]	132 mAh/g	after 50 cycles at 0.3 mA/g
$LiNi_{0.8}Co_{0.2}O_2$ vs. Li [167]	160 mAh/g	after 50 cycles at 0.5 C
$LiNi_{0.8}Co_{0.15}Al_{0.05}O_2$ vs. Li [143]	151 mAh/g	after 50 cycles at 2 C
$LiNi_{0.6}Co_{0.2}Mn_{0.2}O_2$ vs. Li [179]	132 mAh/g at 50 °C	after 50 cycles at 0.16 A/g
$LiNi_{0.6}Co_{0.2}Mn_{0.2}O_2$ vs. Li [180]	135 mAh/g	after 50 cycles at 0.08 A/g
$LiNi_{0.6}Co_{0.2}Mn_{0.2}O_2$/C vs. Li [182]	154 mAh/g	after 50 cycles at 0.5 C
$LiNi_{0.425}Mn_{0.425}Co_{0.15}O_2$ vs. Li [155]	110 mAh/g	after 50 cycles at 1 C
$LiMn_{1/3}Ni_{1/3}Co_{1/3}O_2$ (ZrO_2-coated) vs. Li [156]	140 mAh/g	after 50 cycles at 0.5 C
$LiMn_{1/3}Ni_{1/3}Co_{1/3}O_2$-0.1 LiF vs. Li [157]	133 mAh/g	after 50 cycles at 0.32 A/g
$LiMn_{1/3}Ni_{1/3}Co_{1/3}O_2$ vs. Li [163]	180 mAh/g	after 50 cycles at 0.2 C
$LiMn_{1/3}Ni_{1/3}Co_{1/3}O_2$ vs. Li [165]	160 mAh/g	after 50 cycles at 1 C
0.98 $LiCoO_2$-0.02 Li_2MnO_3 vs. Li [173]	140 mAh/g	after 50 cycles at 1 C
$Li_{1.06}Ni_{0.3}Co_{0.4}Mn_{0.3}O_{2-d}$ vs. Li [187]	180 mAh/g	after 50 cycles at 0.03 A/g
$Li_{1.11}(Ni_{0.4}Co_{0.2}Mn_{0.4})_{0.89}O_2$ vs. Li [152]	187 mAh/g at 50 °C	after 50 cycles at 0.1 A/g
0.7 $LiMn_{0.337}Ni_{0.487}Co_{0.137}Cr_{0.04}O_2$ -0.3 Li_2MnO_3 vs. Li [146]	158 mAh/g	after 20 cycles at 0.05 A/g + 30 cycles at 0.25 A/g

Table 3. *Cont.*

Compound Type, Formulas and References	Discharge Capacity after 50 Cycles	
$0.7\,LiMn_{0.5}Ni_{0.4}Co_{0.1}O_2$ -$0.3\,Li_2MnO_3$ vs. Li [148]	200 mAh/g	after 50 cycles at 0.05 A/g (0.2 C)
$Li_{1.17}(Mn_{1/3}Ni_{1/3}Co_{1/3})_{0.83}O_2$ vs. Li [151]	177 mAh/g	after 50 cycles at 0.03 A/g
$Li_{1.17}Ni_{0.2}Co_{0.05}Mn_{0.58}O_2$ (CeO_2-coated) vs. Li [178]	212 mAh/g	after 50 cycles at 0.3 A/g
$Li_{1.17}Ni_{0.25}Mn_{0.58}O_2$ (Li-Mn-PO_4-coated) vs. Li [168]	265 mAh/g	after 50 cycles at 0.03 A/g
$Li_{1.17}Ni_{0.25}Mn_{0.55}Sn_{0.03}O_2$ vs. Li [169]	170 mAh/g	after 50 cycles at 0.3 A/g
$Li_{1.2}Mn_{0.54}Co_{0.13}Ni_{0.13}O_2$/C vs. Li [144]	160 mAh/g	after 20 cycles at 0.2 C + 30 cycles at 1 C
$Li_{1.2}Mn_{0.54}Ni_{0.13}Co_{0.13}O_2$/C vs. Li [147]	177 mAh/g	after 20 cycles at 0.05 A/g + 30 cycles at 0.125 A/g
$Li_{1.2}Ni_{0.13}Co_{0.13}Mn_{0.54}O_2$ vs. Li [188]	160 mAh/g	after 50 cycles from 0.1 C to 0.5 C
$Li_{1.2}Mn_{0.54}Ni_{0.13}Co_{0.13}O_2$ vs. Li [189]	200 mAh/g	after 50 cycles at 1 C
$Li_{1.2}Ni_{0.13}Co_{0.13}Mn_{0.54}O_2$/C vs. Li [177]	175 mAh/g	after 50 cycles from 0.2 C to 5 C
$Li_{1.2}Ni_{0.2}Mn_{0.6}O_2$ vs. Li [164]	150 mAh/g	after 50 cycles at 0.5 C
$0.5\,LiMn_{1/3}Ni_{1/3}Co_{1/3}O_2$ -$0.5\,Li_2MnO_3$ vs. Li [149]	189 mAh/g	after 50 cycles at 1 C
$0.5\,LiMn_{1/3}Ni_{1/3}Co_{1/3}O_2$ -$0.5\,Li_2MnO_3$ vs. soft C [172]	190 mAh/g	after 50 cycles at 1 C
$0.95\,LiNiO_2$-$0.05\,Li_2TiO_3$ vs. Li [184]	175 mAh/g	after 50 cycles at 0.02 A/g
Oxides $Li_xM_yO_z$ (others)		
$LiMn_2O_4$ vs. Li [191]	113 mAh/g	after 50 cycles at 1 C
$LiMn_2O_4$ vs. Li [192]	117 mAh/g	after 50 cycles at 0.2 C
$LiMn_2O_4$ vs. Li [193]	110 mAh/g	after 50 cycles at 0.2 C
$LiMn_2O_4$ vs. Li [194]	113 mAh/g	after 50 cycles at 1 C
$LiMn_2O_4$ vs. Li [198]	113 mAh/g	after 50 cycles at 1 C
$LiMn_2O_4$ vs. Li [199]	106 mAh/g	after 50 cycles at 0.5 C
$LiMn_{11/6}Co_{1/6}O_4$ vs. Li [201]	112 mAh/g	after 50 cycles at 0.2 C
$LiNi_{0.5}Mn_{1.5}O_4$ vs. Li [206]	135 mAh/g	after 50 cycles at 0.15 C
$LiNi_{0.5}Mn_{1.5}O_4$ vs. Li [207]	132 mAh/g	after 50 cycles at 0.1 C
$LiNi_{0.5}Mn_{1.5}O_4$ vs. Li [208]	118 mAh/g	after 50 cycles at 2 C
$LiNi_{0.5}Mn_{1.5}O_4$/C vs. Li [210]	130 mAh/g	after 50 cycles at 0.5 C
$LiNi_{0.5}Mn_{1.47}Ti_{0.03}O_4$ vs. Li [211]	125 mAh/g	after 50 cycles from 0.05 C to 5 C
$LiNi_{0.5}Mn_{1.4}Fe_{0.1}Ti_{0.03}O_4$ vs. Li [212]	170 mAh/g	after 50 cycles at 0.5 C
$LiNi_{0.5}Mn_{1.4}Ru_{0.1}Ti_{0.03}O_4$ vs. Li [212]	180 mAh/g	after 50 cycles at 0.5 C
$LiNi_{0.3}Mn_{1.5}Co_{0.2}O_4$ vs. Li [213]	115 mAh/g at 60 °C	after 50 cycles at 3.5 C
$LiNi_{0.45}Mn_{1.5}Co_{0.05}O_4$ vs. Li [214]	126 mAh/g	after 50 cycles at 0.15 C
$Li_4Ti_5O_{12}$ vs. Li [216]	147 mAh/g at 50 °C	after 50 cycles at 1 C
$Li_4Ti_5O_{12}$ vs. Li [217]	150 mAh/g	after 50 cycles at 1 C
$Li_4Ti_5O_{12}$/C vs. Li [219]	150 mAh/g	after 50 cycles at 2 C
$Li_4Ti_5O_{12}$ vs. Li [220]	150 mAh/g	after 50 cycles at 1 C
$Li_4Ti_5O_{12}$ vs. Li [222]	160 mAh/g	after 50 cycles at 1 C
$Li_4Ti_5O_{12}$ vs. Li [223]	175 mAh/g	after 50 cycles at 0.2 C
$Li_4Ti_5O_{12}$/C vs. Li [226]	165 mAh/g	after 50 cycles at 1 C
$Li_4Ti_5O_{12}$ vs. Li [229]	211 mAh/g	after 50 cycles at 2 C
$Li_4Ti_5O_{12}$/C vs. Li [230]	155 mAh/g	after 50 cycles at 1 C
$Li_4Ti_5O_{12}$ vs. Li [233]	162 mAh/g	after 50 cycles at 1 C
$Li_4Ti_5O_{12}$ vs. Li [234]	170 mAh/g	after 50 cycles at 1 C
$Li_4Ti_5O_{12}$/C vs. Li [235]	164 mAh/g	after 50 cycles at 1 C
$Li_4Ti_5O_{12}$/TiO_2 vs. Li [236]	168 mAh/g	after 50 cycles at 1 C
$Li_4Ti_5O_{12}$ vs. Li [239]	168 mAh/g	after 50 cycles at 1 C
$Li_4Ti_5O_{12}$ vs. Li [240]	172 mAh/g	after 50 cycles at 1 C
$Li_4Ti_5O_{12}$/C vs. Li [241]	142 mAh/g	after 50 cycles at 10 C
$Li_{4.3}Ti_5O_{12}$/C vs. Li [242]	132 mAh/g	after 50 cycles at 3 C
$Li_{4.3}Ti_5O_{12}$ vs. Li [243]	140 mAh/g	after 50 cycles at 1 C
$Li_4Ti_5O_{12}$/C vs. Li [245]	158 mAh/g	after 50 cycles at 5 C

Table 3. *Cont.*

Compound Type, Formulas and References	Discharge Capacity after 50 Cycles	
$Li_4Ti_5O_{12}/C$ vs. Li [246]	167 mAh/g	after 50 cycles at 0.1 C
$Li_4Ti_5O_{12}/C$ vs. Li [247]	143 mAh/g	after 50 cycles at 1 C
$Li_4Ti_5O_{12}/C$ vs. Li [248]	146 mAh/g	after 50 cycles at 2 C
$Li_4Ti_5O_{12}$ vs. Li [249]	168 mAh/g	after 50 cycles at 1 C
$Li_{3.98}Al_{0.06}Ti_{4.96}O_{12}/C$ vs. Li [250]	160 mAh/g	after 50 cycles at 1 C
$Li_{1.1}V_3O_8/C$ vs. Li [254]	225 mAh/g	after 50 cycles at 0.33 C
LiV_3O_8 vs. Li [255]	260 mAh/g	after 50 cycles at 0.125 A/g
Li_3VO_4/C vs. Li [258]	315 mAh/g	after 50 cycles at 10 C
Li_3VO_4/C vs. Li [259]	400 mAh/g	after 50 cycles at 0.2 C
Li_3VO_4/C vs. Li [260]	395 mAh/g	after 50 cycles at 0.5 C
$Li_4Mn_5O_{12}$ vs. Li [262]	128 mAh/g	after 50 cycles at 0.5 C
Oxides $Na_xM_yO_z$		
$Na_{2/3}Ni_{1/3}Mn_{2/3}O_2$ vs. Na [263]	102 mAh/g	after 50 cycles at 0.1 C
$Na_2Ti_3O_7$ vs. Na [265]	95 mAh/g	after 50 cycles from 0.1 C to 5 C
$Na_4Mn_9O_{18}/C$ in aqueous Na-ion battery [266]	85 mAh/g	after 50 cycles at 4 C
$Na_4Mn_9O_{18}/C$ in aqueous Na-ion battery [267]	50 mAh/g	after 50 cycles at 4 C
Phosphates		
$LiFePO_4/C$ vs. Li [271]	159 mAh/g	after 50 cycles at 1 C
$LiFePO_4/C$ vs. Li [273]	156 mAh/g	after 50 cycles at 1 C
$LiFePO_4/C$ vs. Li [275]	137 mAh/g	after 50 cycles at 1 C
$LiFePO_4/C$ vs. Li [276]	110 mAh/g	after 50 cycles at 1 C
$LiFePO_4/C$ vs. Li [278]	154 mAh/g	after 50 cycles at 1 C
$LiFePO_4/C$ vs. Li [281]	160 mAh/g	after 50 cycles at 0.1 C
$LiFePO_4/C$ vs. Li [282]	150 mAh/g	after 50 cycles at 1 C
$LiFePO_4/C$ vs. Li [283]	160 mAh/g	after 50 cycles at 1 C
$LiFePO_4/C$ vs. Li [284]	159 mAh/g	after 50 cycles at 0.1 C
$LiFePO_4/C$ vs. Li [285]	130 mAh/g	after 50 cycles at 5 C
$LiFePO_4/C$ vs. Li [286]	110 mAh/g	after 50 cycles from 0.1 C to 2 C
$LiFePO_4/C$ vs. Li [289]	110 mAh/g	after 50 cycles at 10 C
$LiFePO_4/C$ vs. Li [290]	123 mAh/g	after 50 cycles at 10 C
$LiFePO_4/C$ vs. Li [291]	162 mAh/g	after 50 cycles at 0.5 C
$LiFePO_4/C$ vs. Li [292]	156 mAh/g	after 50 cycles at 1 C
$LiFePO_4/C$ vs. Li [293]	120 mAh/g	after 50 cycles at 10 C
$LiFePO_4/C$ vs. Li [294]	140 mAh/g	after 50 cycles at 2 C
$LiFePO_4/C$ vs. Li [295]	137 mAh/g	after 50 cycles from 0.1 C to 4 C
$LiFePO_4/C$ vs. Li [296]	149 mAh/g	after 50 cycles at 1 C
$LiFePO_4/C$ vs. Li [298]	100 mAh/g	after 50 cycles at 3 C
$LiFePO_4/C$ vs. Li [299]	147 mAh/g	after 50 cycles at 3 C
$LiFePO_4/C$ vs. Li [300]	142 mAh/g	after 50 cycles at 0.1 C
$LiFePO_4/C$ vs. Li [304]	110 mAh/g	after 50 cycles at 10 C
$LiFePO_4/C$ vs. Li [305]	110 mAh/g	after 50 cycles at 10 C
$LiFePO_4/C$ vs. Li [306]	120 mAh/g	after 50 cycles at 10 C
$LiFePO_4/C$ vs. Li [307]	137 mAh/g	after 50 cycles at 1 C
$LiFePO_4/C$ vs. Li [308]	152 mAh/g	after 50 cycles at 1 C
$LiFePO_4/C$ vs. Li [309]	105 mAh/g	after 50 cycles at 1 C
$LiFe_{0.6}Mn_{0.4}PO_4/C$ vs. Li [315]	137 mAh/g	after 50 cycles at 2 C
$LiFe_{0.6}Mn_{0.4}PO_4/C$ vs. Li [316]	150 mAh/g	after 50 cycles at 0.5 C
$LiMn_{0.5}Fe_{0.5}PO_4/C$ vs. Li [318]	150 mAh/g at 55 °C	after 50 cycles at 1 C
$LiMn_{0.6}Fe_{0.4}PO_4/C$ vs. Li [312]	425 Wh/kg	after 50 cycles at 10 C
$LiMn_{0.7}Fe_{0.3}PO_4/C$ vs. Li [319]	145 mAh/g	after 50 cycles at 5 C
$LiMn_{0.75}Fe_{0.25}PO_4/C$ vs. Li [310]	120 mAh/g	after 50 cycles at 10 C
$LiMn_{0.8}Fe_{0.2}PO_4/C$ vs. Li [313]	138 mAh/g	after 50 cycles at 5 C
$LiMn_{0.8}Fe_{0.2}PO_4/C$ vs. $Li_4Ti_5O_{12}$ [313]	122 mAh/g	after 50 cycles at 1 C
$LiMn_{0.85}Fe_{0.2}PO_4/C$ vs. Li [314]	132 mAh/g	after 50 cycles at 5 C
$LiMn_{0.85}Fe_{0.15}PO_4/C$ vs. Li [317]	136 mAh/g	after 50 cycles at 1 C
$LiMn_{0.85}Fe_{0.15}PO_4/C$ vs. Li [320]	136 mAh/g	after 50 cycles at 1 C
$Li(Mn_{0.85}Fe_{0.15})_{0.92}Ti_{0.08}PO_4/C$ vs. Li [323]	144 mAh/g	after 50 cycles at 1 C

Table 3. *Cont.*

Compound Type, Formulas and References	Discharge Capacity after 50 Cycles	
$LiMn_{0.97}Fe_{0.03}PO_4$/C vs. Li [311]	158 mAh/g	after 50 cycles at 0.5 C
$LiMnPO_4$/C vs. Li [321]	96 mAh/g	after 50 cycles at 0.05 C
$LiVOPO_4$ vs. Li [324]	50 mAh/g	after 50 cycles at 0.2 C
$Li_3V_2(PO_4)_3$/C vs. Li [325]	143 mAh/g	after 50 cycles at 20 C
$Li_3V_2(PO_4)_3$/C vs. Li [326]	100 mAh/g	after 50 cycles from 0.2 C to 20 C
$Li_3V_2(PO_4)_3$/C vs. Li [327]	127 mAh/g	after 50 cycles at 0.1 C
$Li_3V_2(PO_4)_3$/C vs. Li [328]	131 mAh/g	after 50 cycles at 0.02 A/g
$Li_3V_2(PO_4)_3$/C vs. Li [329]	149 mAh/g	after 50 cycles at 10 C
$Li_3V_2(PO_4)_3$/C vs. Li [330]	118 mAh/g	after 50 cycles from 0.1 C to 5 C
$Li_3V_2(PO_4)_3$/C vs. Li [332]	123 mAh/g	after 50 cycles at 2 C
$Li_3V_2(PO_4)_3$/C vs. Li [333]	131 mAh/g	after 50 cycles at 0.1 C
$Li_3V_2(PO_4)_3$/C vs. Li [334]	138 mAh/g	after 50 cycles at 1 C
$Li_3V_2(PO_4)_3$/C vs. Li [335]	94 mAh/g	after 50 cycles at 1 C
$NaTi_2(PO_4)_3$/C vs. Na [339]	110 mAh/g	after 50 cycles from 0.2 C to 4 C
$NaTi_2(PO_4)_3$/C vs. Na [340]	128 mAh/g	after 50 cycles from 0.1 C to 5 C
$NaTi_2(PO_4)_3$/C vs. $Na_3V_2(PO_4)_3$/C [340]	98 mAh/g	after 50 cycles at 10 C
$Na_3V_2(PO_4)_3$/C vs. Na [342]	92 mAh/g	after 50 cycles at 10 C
$Na_3V_2(PO_4)_3$/C vs. Na [344]	103 mAh/g	after 50 cycles at 5 C
$Na_3V_2(PO_4)_3$/C vs. Na [345]	93 mAh/g	after 50 cycles at 5 C
$Na_3V_{1.95}Cu_{0.05}(PO_4)_3$/C vs. Na [346]	103 mAh/g	after 50 cycles at 20 C
Pyrophosphates		
$Na_2FeP_2O_7$/C vs. Na [347]	87 mAh/g	after 50 cycles at 0.1 C
$Na_2FeP_2O_7$/C vs. hard carbon [347]	62 mAh/g	after 50 cycles at 1 C
SnP_2O_7/C vs. Li [348]	645 mAh/g	after 50 cycles at 0.1 C
Silicates		
Li_2FeSiO_4/C vs. Li [349]	137 mAh/g	after 50 cycles at 1 C
Li_2FeSiO_4/C vs. Li [350]	140 mAh/g	after 50 cycles at 0.1 C
$Li_{1.95}Na_{0.05}FeSiO_4$/C vs. Li [351]	138 mAh/g	after 50 cycles at 2 C
$Li_2Fe_{0.5}V_{0.5}SiO_4$/C vs. Li [352]	157 mAh/g	after 50 cycles at 0.5 C
Sulfides and selenides		
MoS_2/C vs. Li [353]	800 mAh/g	after 50 cycles at 0.1 A/g
MoS_2/C vs. Na [353]	350 mAh/g	after 50 cycles at 0.1 A/g
$FeSe_2$/C vs. Na [354]	510 mAh/g	after 50 cycles at 0.5 A/g
MnS/C vs. Li [114]	700 mAh/g	after 50 cycles at 0.5 A/g
NiS/C vs. Na [375]	490 mAh/g	after 50 cycles at 0.3 A/g
Composites (not with carbon)		
$Sn–Sn_2Co_3@CoSnO_3–Co_3O_4$ vs. Li [355]	1050 mAh/g	after 50 cycles at 1 A/g
$0.5\ LiNi_{0.5}Mn_{1.5}O_4$-$0.5\ Li_7La_3Zr_2O_{12}$ vs. Li [357]	116 mAh/g	after 50 cycles at 1 C
$3Li_4Ti_5O_{12}.NiO$ [358]	240 mAh/g	after 50 cycles at 1 C
$9\ LiFePO_4$-$1\ Li_3V_2(PO_4)_3$/C vs. Li [362]	154 mAh/g	after 50 cycles at 1 C
$3\ LiFePO_4$-$1\ Li_3V_2(PO_4)_3$/C vs. Li [360]	152 mAh/g	after 50 cycles at 1 C
$0.7\ LiFePO_4$ -$0.3\ Li_3V_2(PO_4)_3$/C vs. Li [359]	120 mAh/g	after 50 cycles from 0.03 A/g to 1.5 A/g
$2\ LiFePO_4$-$1\ Li_3V_2(PO_4)_3$/C vs. Li [361]	143 mAh/g	after 50 cycles at 0.1 C
$1\ LiMnPO_4$-$1\ Li_3V_2(PO_4)_3$/C vs. Li [364]	123 mAh/g	after 50 cycles at 0.1 C
$1\ LiMnPO_4$-$2\ Li_3V_2(PO_4)_3$/C vs. Li [365]	130 mAh/g	after 50 cycles at 0.1 C
Si-$FeSi_2$-$Cu_{3.17}Si$ vs. Li [366]	410 mAh/g	after 50 cycles at 0.5 C
MoS_2-Ni_9S_8 vs. Na [367]	500 mAh/g	after 50 cycles at 0.5 A/g
$MoSe_2$-NiSe-C vs. Na [367]	390 mAh/g	after 50 cycles at 0.5 A/g

References

1. Tarascon, J.-M. Key challenges in future Li-battery research. *Philos. Trans. R. Soc. Math. Phys. Eng. Sci.* **2010**, *368*, 3227–3241. [CrossRef] [PubMed]

2. Croguennec, L.; Palacin, M.R. Recent Achievements on Inorganic Electrode Materials for Lithium-Ion Batteries. *J. Am. Chem. Soc.* **2015**, *137*, 3140–3156. [CrossRef] [PubMed]

3. Stunda-Zujeva, A.; Irbe, Z.; Berzina-Cimdina, L. Controlling the morphology of ceramic and composite powders obtained via spray-drying—A review. *Ceram. Int.* **2017**, *43*, 11543–11551. [CrossRef]

4. Jia, X.; Kan, Y.; Zhu, X.; Ning, G.; Lu, Y.; Wei, F. Building flexible $Li_4Ti_5O_{12}$/CNT lithium-ion battery anodes with superior rate performance and ultralong cycling stability. *Nano Energy* **2014**, *10*, 344–352. [CrossRef]

5. Ju, S.H.; Jang, H.C.; Kang, Y.C. Al-doped Ni-rich cathode powders prepared from the precursor powders with fine size and spherical shape. *Electrochim. Acta* **2007**, *52*, 7286–7292. [CrossRef]

6. Jung, D.S.; Hwang, T.H.; Park, S.B.; Choi, J.W. Spray-drying Method for Large-Scale and High-Performance Silicon Negative Electrodes in Li-Ion Batteries. *Nano Lett.* **2013**, *13*, 2092–2097. [CrossRef] [PubMed]

7. Chang, H.-Y.; Sheu, C.-I.; Cheng, S.-Y.; Wu, H.-C.; Guo, Z.-Z. Synthesis of $Li_{1.1}Ni_{1/3}Co_{1/3}Mn_{1/3}O_2$ cathode material using spray-microwave method. *J. Power Sources* **2007**, *174*, 985–989. [CrossRef]

8. Kim, J.H.; Kang, Y.C.; Choi, Y.J.; Kim, Y.S.; Lee, J.-H. Electrochemical properties of yolk-shell structured layered-layered composite cathode powders prepared by spray pyrolysis. *Electrochim. Acta* **2014**, *144*, 288–294. [CrossRef]

9. Nandiyanto, A.B.D.; Okuyama, K. Progress in developing spray-drying methods for the production of controlled morphology particles: From the nanometer to submicrometer size ranges. *Adv. Powder Technol.* **2011**, *22*, 1–19. [CrossRef]

10. Mezhericher, M.; Levy, A.; Borde, I. Theoretical Models of Single Droplet Drying Kinetics: A Review. *Dry. Technol.* **2010**, *28*, 278–293. [CrossRef]

11. Zbicinski, I. Modeling and Scaling Up of Industrial Spray-dryers: A Review. *J. Chem. Eng. Jpn.* **2017**, *50*, 757–767. [CrossRef]

12. Deshmukh, R.; Wagh, P.; Naik, J. Solvent evaporation and spray-drying technique for micro- and nanospheres/particles preparation: A review. *Dry. Technol.* **2016**, *34*, 1758–1772. [CrossRef]

13. Singh, A.; van den Mooter, G. Spray-drying formulation of amorphous solid dispersions. *Adv. Drug Deliv. Rev.* **2016**, *100*, 27–50. [CrossRef] [PubMed]

14. Gharsallaoui, A.; Roudaut, G.; Chambin, O.; Voilley, A.; Saurel, R. Applications of spray-drying in microencapsulation of food ingredients: An overview. *Food Res. Int.* **2007**, *40*, 1107–1121. [CrossRef]

15. Schuck, P.; Jeantet, R.; Bhandari, B.; Chen, X.D.; Perrone, Í.T.; de Carvalho, A.F.; Fenelon, M.; Kelly, P. Recent advances in spray-drying relevant to the dairy industry: A comprehensive critical review. *Dry. Technol.* **2016**, *34*, 1773–1790. [CrossRef]

16. Masquelier, C.; Croguennec, L. Polyanionic (Phosphates, Silicates, Sulfates) Frameworks as Electrode Materials for Rechargeable Li (or Na) Batteries. *Chem. Rev.* **2013**, *113*, 6552–6591. [CrossRef] [PubMed]

17. Kundu, D.; Talaie, E.; Duffort, V.; Nazar, L.F. The Emerging Chemistry of Sodium Ion Batteries for Electrochemical Energy Storage. *Angew. Chem. Int. Ed.* **2015**, *54*, 3431–3448. [CrossRef] [PubMed]

18. Nayak, P.K.; Yang, L.; Brehm, W.; Adelhelm, P. From Lithium-Ion to Sodium-Ion Batteries: Advantages, Challenges, and Surprises. *Angew. Chem. Int. Ed.* **2018**, *57*, 102–120. [CrossRef] [PubMed]

19. Toprakci, O.; Toprakci, H.A.K.; Ji, L.; Zhang, X. Fabrication and Electrochemical Characteristics of $LiFePO_4$ Powders for Lithium-Ion Batteries. *KONA Powder Part. J.* **2010**, *28*, 50–73. [CrossRef]

20. Lee, K.-J.; Kang, L.-S.; Uhm, S.; Yoon, J.S.; Kim, D.-W.; Hong, H.S. Synthesis and characterization of $LiMnBO_3$ cathode material for lithium ion batteries. *Curr. Appl. Phys.* **2013**, *13*, 1440–1443. [CrossRef]

21. Zhang, B.; Ming, L.; Zheng, J.; Zhang, J.; Shen, C.; Han, Y.; Wang, J.; Qin, S. Synthesis and characterization of multi-layer core-shell structural $LiFeBO_3$/C as a novel Li-battery cathode material. *J. Power Sources* **2014**, *261*, 249–254. [CrossRef]

22. Zhang, B.; Ming, L.; Tong, H.; Zhang, J.; Zheng, J.; Wang, X.; Li, H.; Cheng, L. Ni-doping to improve the performance of $LiFeBO_3$/C cathode material for lithium-ion batteries. *J. Alloys Compd.* **2018**, *740*, 382–388. [CrossRef]

23. Zhou, H.; Wang, D.; Fu, A.; Liu, X.; Wang, Y.; Li, Y.; Guo, P.; Li, H.; Zhao, X.S. Mesoporous carbon spheres with tunable porosity prepared by a template-free method for advanced lithium–sulfur batteries. *Mater. Sci. Eng. B* **2018**, *227*, 9–15. [CrossRef]

24. Ye, X.; Ma, J.; Hu, Y.-S.; Wei, H.; Ye, F. MWCNT porous microspheres with an efficient 3D conductive network for high performance lithium–sulfur batteries. *J. Mater. Chem. A* **2016**, *4*, 775–780. [CrossRef]

25. Chen, M.; Wang, Z.; Wang, A.; Li, W.; Liu, X.; Fu, L.; Huang, W. Novel self-assembled natural graphite based composite anodes with improved kinetic properties in lithium-ion batteries. *J. Mater. Chem. A* **2016**, *4*, 9865–9872. [CrossRef]

26. Deng, T.; Zhou, X. The preparation of porous graphite and its application in lithium ion batteries as anode material. *J. Solid State Electrochem.* **2016**, *20*, 2613–2618. [CrossRef]

27. Ma, Z.; Cui, Y.; Xiao, X.; Deng, Y.; Song, X.; Zuo, X.; Nan, J. A reconstructed graphite-like carbon micro/nano-structure with higher capacity and comparative voltage plateau of graphite. *J. Mater. Chem. A* **2016**, *4*, 11462–11471. [CrossRef]

28. Ma, Z.; Cui, Y.; Zuo, X.; Sun, Y.; Xiao, X.; Nan, J. Self-assembly flower-like porous carbon nanosheet powders for higher lithium-ion storage capacity. *Electrochim. Acta* **2015**, *184*, 308–315. [CrossRef]

29. Ma, Z.; Zhuang, Y.; Deng, Y.; Song, X.; Zuo, X.; Xiao, X.; Nan, J. From spent graphite to amorphous sp^2+sp^3 carbon-coated sp^2 graphite for high-performance lithium ion batteries. *J. Power Sources* **2018**, *376*, 91–99. [CrossRef]

30. Mei, R.; Song, X.; Hu, Y.; Yang, Y.; Zhang, J. Hollow reduced graphene oxide microspheres as a high-performance anode material for Li-ion batteries. *Electrochim. Acta* **2015**, *153*, 540–545. [CrossRef]

31. Wang, L.; Liu, Y.; Chong, C.; Wang, J.; Shi, Z.; Pan, J. Phenolic formaldehyde resin/graphene composites as lithium-ion batteries anode. *Mater. Lett.* **2016**, *170*, 217–220. [CrossRef]

32. Yuan, T.; Zhang, W.; Li, W.-T.; Song, C.; He, Y.-S.; Razal, J.M.; Ma, Z.-F.; Chen, J. N-doped pierced graphene microparticles as a highly active electrocatalyst for Li-air batteries. *2D Mater.* **2015**, *2*, 024002. [CrossRef]

33. Zhang, L.; Zhang, M.; Wang, Y.; Zhang, Z.; Kan, G.; Wang, C.; Zhong, Z.; Su, F. Graphitized porous carbon microspheres assembled with carbon black nanoparticles as improved anode materials in Li-ion batteries. *J. Mater. Chem. A* **2014**, *2*, 10161. [CrossRef]

34. Zhuang, H.; Deng, W.; Wang, W.; Liu, Z. Facile fabrication of nanoporous graphene powder for high-rate lithium–sulfur batteries. *RSC Adv.* **2017**, *7*, 5177–5182. [CrossRef]

35. Zhou, G.; Wang, D.-W.; Shan, X.; Li, N.; Li, F.; Cheng, H.-M. Hollow carbon cage with nanocapsules of graphitic shell/nickel core as an anode material for high rate lithium ion batteries. *J. Mater. Chem.* **2012**, *22*, 11252. [CrossRef]

36. Lee, G.-H.; Jo, M.R.; Zhang, K.; Kang, Y.-M. A reduced graphene oxide-encapsulated phosphorus/carbon composite as a promising anode material for high-performance sodium-ion batteries. *J. Mater. Chem. A* **2017**, *5*, 3683–3690. [CrossRef]

37. He, J.; Zhou, K.; Chen, Y.; Xu, C.; Lin, J.; Zhang, W. Wrinkled sulfur@graphene microspheres with high sulfur loading as superior-capacity cathode for LiS batteries. *Mater. Today Energy* **2016**, *1*, 11–16. [CrossRef]

38. Ma, J.; Fang, Z.; Yan, Y.; Yang, Z.; Gu, L.; Hu, Y.-S.; Li, H.; Wang, Z.; Huang, X. Novel Large-Scale Synthesis of a C/S Nanocomposite with Mixed Conducting Networks through a Spray-drying Approach for Li-S Batteries. *Adv. Energy Mater.* **2015**, *5*, 1500046. [CrossRef]

39. Tian, Y.; Sun, Z.; Zhang, Y.; Wang, X.; Bakenov, Z.; Yin, F. Micro-Spherical Sulfur/Graphene Oxide Composite via Spray-drying for High Performance Lithium Sulfur Batteries. *Nanomaterials* **2018**, *8*, 50. [CrossRef] [PubMed]

40. Wu, L.; Lu, H.; Xiao, L.; Ai, X.; Yang, H.; Cao, Y. Electrochemical properties and morphological evolution of pitaya-like Sb@C microspheres as high-performance anode for sodium ion batteries. *J. Mater. Chem. A* **2015**, *3*, 5708–5713. [CrossRef]

41. Bao, Q.; Huang, Y.-H.; Lan, C.-K.; Chen, B.-H.; Duh, J.-G. Scalable Upcycling Silicon from Waste Slicing Sludge for High-performance Lithium-ion Battery Anodes. *Electrochim. Acta* **2015**, *173*, 82–90. [CrossRef]

42. Bie, Y.; Yu, J.; Yang, J.; Lu, W.; Nuli, Y.; Wang, J. Porous microspherical silicon composite anode material for lithium ion battery. *Electrochim. Acta* **2015**, *178*, 65–73. [CrossRef]

43. Chen, H.; Hou, X.; Qu, L.; Qin, H.; Ru, Q.; Huang, Y.; Hu, S.; Lam, K. Electrochemical properties of core–shell nano-Si@carbon composites as superior anode materials for high-performance Li-ion batteries. *J. Mater. Sci. Mater. Electron.* **2017**, *28*, 250–258. [CrossRef]

44. Chen, H.; Wang, Z.; Hou, X.; Fu, L.; Wang, S.; Hu, X.; Qin, H.; Wu, Y.; Ru, Q.; Liu, X. Mass-producible method for preparation of a carbon-coated graphite@plasma nano-silicon@carbon composite with enhanced performance as lithium ion battery anode. *Electrochim. Acta* **2017**, *249*, 113–121. [CrossRef]

45. Chen, L.; Xie, X.; Wang, B.; Wang, K.; Xie, J. Spherical nanostructured Si/C composite prepared by spray-drying technique for lithium ion batteries anode. *Mater. Sci. Eng. B* **2006**, *131*, 186–190. [CrossRef]

46. Fan, X.; Jiang, X.; Wang, W.; Liu, Z. Green synthesis of nanoporous Si/C anode using NaCl template with improved cycle life. *Mater. Lett.* **2016**, *180*, 109–113. [CrossRef]

47. Feng, X.; Cui, H.; Miao, R.; Yan, N.; Ding, T.; Xiao, Z. Nano/micro-structured silicon@carbon composite with buffer void as anode material for lithium ion battery. *Ceram. Int.* **2016**, *42*, 589–597. [CrossRef]

48. Feng, X.; Ding, T.; Cui, H.; Yan, N.; Wang, F. A Low-Cost Nano/Micro Structured-Silicon-MWCNTs from Nano-Silica for Lithium Storage. *Nano* **2016**, *11*, 1650031. [CrossRef]

49. Feng, X.; Yang, J.; Bie, Y.; Wang, J.; Nuli, Y.; Lu, W. Nano/micro-structured Si/CNT/C composite from nano-SiO$_2$ for high power lithium ion batteries. *Nanoscale* **2014**, *6*, 12532–12539. [CrossRef] [PubMed]

50. Gan, L.; Guo, H.; Wang, Z.; Li, X.; Peng, W.; Wang, J.; Huang, S.; Su, M. A facile synthesis of graphite/silicon/graphene spherical composite anode for lithium-ion batteries. *Electrochim. Acta* **2013**, *104*, 117–123. [CrossRef]

51. He, Y.-S.; Gao, P.; Chen, J.; Yang, X.; Liao, X.-Z.; Yang, J.; Ma, Z.-F. A novel bath lily-like graphene sheet-wrapped nano-Si composite as a high performance anode material for Li-ion batteries. *RSC Adv.* **2011**, *1*, 958. [CrossRef]

52. Hou, X.; Wang, J.; Zhang, M.; Liu, X.; Shao, Z.; Li, W.; Hu, S. Facile spray-drying/pyrolysis synthesis of intertwined SiO@CNFs&G composites as superior anode materials for Li-ion batteries. *RSC Adv.* **2014**, *4*, 34615–34622.

53. Lai, J.; Guo, H.; Wang, Z.; Li, X.; Zhang, X.; Wu, F.; Yue, P. Preparation and characterization of flake graphite/silicon/carbon spherical composite as anode materials for lithium-ion batteries. *J. Alloys Compd.* **2012**, *530*, 30–35. [CrossRef]

54. Lee, J.; Moon, J.H. Spherical graphene and Si nanoparticle composite particles for high-performance lithium batteries. *Korean J. Chem. Eng.* **2017**, *34*, 3195–3199. [CrossRef]

55. Li, C.; Ju, Y.; Qi, L.; Yoshitake, H.; Wang, H. A micro-sized Si-CNT anode for practical application via a one-step, low-cost and green method. *RSC Adv.* **2017**, *7*, 54844–54851. [CrossRef]

56. Li, J.; Wang, J.; Yang, J.; Ma, X.; Lu, S. Scalable synthesis of a novel structured graphite/silicon/pyrolyzed-carbon composite as anode material for high-performance lithium-ion batteries. *J. Alloys Compd.* **2016**, *688*, 1072–1079. [CrossRef]

57. Li, J.; Yang, J.-Y.; Wang, J.-T.; Lu, S.-G. A scalable synthesis of silicon nanoparticles as high-performance anode material for lithium-ion batteries. *Rare Met.* **2017**. [CrossRef]

58. Li, M.; Hou, X.; Sha, Y.; Wang, J.; Hu, S.; Liu, X.; Shao, Z. Facile spray-drying/pyrolysis synthesis of core-shell structure graphite/silicon-porous carbon composite as a superior anode for Li-ion batteries. *J. Power Sources* **2014**, *248*, 721–728. [CrossRef]

59. Li, S.; Qin, X.; Zhang, H.; Wu, J.; He, Y.-B.; Li, B.; Kang, F. Silicon/carbon composite microspheres with hierarchical core-shell structure as anode for lithium ion batteries. *Electrochem. Commun.* **2014**, *49*, 98–102. [CrossRef]

60. Lin, J.; He, J.; Chen, Y.; Li, Q.; Yu, B.; Xu, C.; Zhang, W. Pomegranate-Like Silicon/Nitrogen-doped Graphene Microspheres as Superior-Capacity Anode for Lithium-Ion Batteries. *Electrochim. Acta* **2016**, *215*, 667–673. [CrossRef]

61. Liu, X.; Wang, Z.; Guo, H.; Li, X.; Zhou, R.; Zhou, Y. Chitosan: A N-doped carbon source of silicon-based anode material for lithium ion batteries. *Ionics* **2017**, *23*, 2311–2318. [CrossRef]

62. Paireau, C.; Jouanneau, S.; Ammar, M.-R.; Simon, P.; Béguin, F.; Raymundo-Piñero, E. Si/C composites prepared by spray-drying from cross-linked polyvinyl alcohol as Li-ion batteries anodes. *Electrochim. Acta* **2015**, *174*, 361–368. [CrossRef]

63. Pan, Q.; Zuo, P.; Lou, S.; Mu, T.; Du, C.; Cheng, X.; Ma, Y.; Gao, Y.; Yin, G. Micro-sized spherical silicon@carbon@graphene prepared by spray-drying as anode material for lithium-ion batteries. *J. Alloys Compd.* **2017**, *723*, 434–440. [CrossRef]

64. Ren, W.; Zhang, Z.; Wang, Y.; Tan, Q.; Zhong, Z.; Su, F. Preparation of porous silicon/carbon microspheres as high performance anode materials for lithium ion batteries. *J. Mater. Chem. A* **2015**, *3*, 5859–5865. [CrossRef]

65. Su, M.; Wang, Z.; Guo, H.; Li, X.; Huang, S.; Gan, L.; Xiao, W. Enhanced cycling performance of Si/C composite prepared by spray-drying as anode for Li-ion batteries. *Powder Technol.* **2013**, *249*, 105–109. [CrossRef]

66. Su, M.; Wang, Z.; Guo, H.; Li, X.; Huang, S.; Xiao, W.; Gan, L. Enhancement of the Cyclability of a Si/Graphite@Graphene composite as anode for Lithium-ion batteries. *Electrochim. Acta* **2014**, *116*, 230–236. [CrossRef]

67. Tao, H.; Xiong, L.; Zhu, S.; Zhang, L.; Yang, X. Porous Si/C/reduced graphene oxide microspheres by spray-drying as anode for Li-ion batteries. *J. Electroanal. Chem.* **2017**, *797*, 16–22. [CrossRef]

68. Wang, A.; Liu, F.; Wang, Z.; Liu, X. Self-assembly of silicon/carbon hybrids and natural graphite as anode materials for lithium-ion batteries. *RSC Adv.* **2016**, *6*, 104995–105002. [CrossRef]

69. Wang, J.; Hou, X.; Zhang, M.; Li, Y.; Wu, Y.; Liu, X.; Hu, S. 3-Aminopropyltriethoxysilane-Assisted Si@SiO$_2$/CNTs Hybrid Microspheres as Superior Anode Materials for Li-ion Batteries. *Silicon* **2017**, *9*, 97–104. [CrossRef]

70. Wang, Z.; Mao, Z.; Lai, L.; Okubo, M.; Song, Y.; Zhou, Y.; Liu, X.; Huang, W. Sub-micron silicon/pyrolyzed carbon@natural graphite self-assembly composite anode material for lithium-ion batteries. *Chem. Eng. J.* **2017**, *313*, 187–196. [CrossRef]

71. Xu, Q.; Li, J.-Y.; Sun, J.-K.; Yin, Y.-X.; Wan, L.-J.; Guo, Y.-G. Watermelon-Inspired Si/C Microspheres with Hierarchical Buffer Structures for Densely Compacted Lithium-Ion Battery Anodes. *Adv. Energy Mater.* **2017**, *7*, 1601481. [CrossRef]

72. Yang, Y.; Wang, Z.; Zhou, R.; Guo, H.; Li, X. Effects of lithium fluoride coating on the performance of nano-silicon as anode material for lithium-ion batteries. *Mater. Lett.* **2016**, *184*, 65–68. [CrossRef]

73. Yang, Y.; Wang, Z.; Zhou, Y.; Guo, H.; Li, X. Synthesis of porous Si/graphite/carbon nanotubes@C composites as a practical high-capacity anode for lithium-ion batteries. *Mater. Lett.* **2017**, *199*, 84–87. [CrossRef]

74. Zhang, H.; Xu, H.; Jin, H.; Li, C.; Bai, Y.; Lian, K. Flower-like carbon with embedded silicon nano particles as an anode material for Li-ion batteries. *RSC Adv.* **2017**, *7*, 30032–30037. [CrossRef]

75. Zhang, L.; Wang, Y.; Kan, G.; Zhang, Z.; Wang, C.; Zhong, Z.; Su, F. Scalable synthesis of porous silicon/carbon microspheres as improved anode materials for Li-ion batteries. *RSC Adv.* **2014**, *4*, 43114–43120. [CrossRef]

76. Zhang, M.; Hou, X.; Wang, J.; Li, M.; Hu, S.; Shao, Z.; Liu, X. Interweaved Si@C/CNTs&CNFs composites as anode materials for Li-ion batteries. *J. Alloys Compd.* **2014**, *588*, 206–211.

77. Zhang, Y.; Li, K.; Ji, P.; Chen, D.; Zeng, J.; Sun, Y.; Zhang, P.; Zhao, J. Silicon-multi-walled carbon nanotubes-carbon microspherical composite as high-performance anode for lithium-ion batteries. *J. Mater. Sci.* **2017**, *52*, 3630–3641. [CrossRef]

78. Zhou, Y.; Guo, H.; Wang, Z.; Li, X.; Zhou, R.; Peng, W. Improved electrochemical performance of Si/C material based on the interface stability. *J. Alloys Compd.* **2017**, *725*, 1304–1312. [CrossRef]

79. Ogata, K.; Jeon, S.; Ko, D.-S.; Jung, I.S.; Kim, J.H.; Ito, K.; Kubo, Y.; Takei, K.; Saito, S.; Cho, Y.-H. Evolving affinity between Coulombic reversibility and hysteretic phase transformations in nano-structured silicon-based lithium-ion batteries. *Nat. Commun.* **2018**, *9*, 479. [CrossRef] [PubMed]

80. Sun, Z.; Wang, X.; Ying, H.; Wang, G.; Han, W.-Q. Facial Synthesis of Three-Dimensional Cross-Linked Cage for High-Performance Lithium Storage. *ACS Appl. Mater. Interfaces* **2016**, *8*, 15279–15287. [CrossRef] [PubMed]

81. Wang, D.; Gao, M.; Pan, H.; Liu, Y.; Wang, J.; Li, S.; Ge, H. Enhanced cycle stability of micro-sized Si/C anode material with low carbon content fabricated via spray-drying and in situ carbonization. *J. Alloys Compd.* **2014**, *604*, 130–136. [CrossRef]

82. Ying, H.; Zhang, S.; Meng, Z.; Sun, Z.; Han, W.-Q. Ultrasmall Sn nanodots embedded inside N-doped carbon microcages as high-performance lithium and sodium ion battery anodes. *J. Mater. Chem. A* **2017**, *5*, 8334–8342. [CrossRef]

83. Youn, H.-C.; Jeong, J.H.; Roh, K.C.; Kim, K.-B. Graphene–Selenium Hybrid Microballs as Cathode Materials for High-performance Lithium–Selenium Secondary Battery Applications. *Sci. Rep.* **2016**, *6*, 30865. [CrossRef] [PubMed]

84. Gocheva, I.D.; Okada, S.; Yamaki, J. Electrochemical Properties of Trirutile-type Li$_2$TiF$_6$ as Cathode Active Material in Li-ion Batteries. *Electrochemistry* **2010**, *78*, 471–474. [CrossRef]

85. Lieser, G.; de Biasi, L.; Scheuermann, M.; Winkler, V.; Eisenhardt, S.; Glatthaar, S.; Indris, S.; Gesswein, H.; Hoffmann, M.J.; Ehrenberg, H. Sol-Gel Processing and Electrochemical Conversion of Inverse Spinel-Type Li$_2$NiF$_4$. *J. Electrochem. Soc.* **2015**, *162*, A679–A686. [CrossRef]

86. Lieser, G.; Schroeder, M.; Geßwein, H.; Winkler, V.; Glatthaar, S.; Yavuz, M.; Binder, J.R. Sol-gel processing and electrochemical characterization of monoclinic Li$_3$FeF$_6$. *J. Sol-Gel Sci. Technol.* **2014**, *71*, 50–59. [CrossRef]

87. Brisbois, M.; Caes, S.; Sougrati, M.T.; Vertruyen, B.; Schrijnemakers, A.; Cloots, R.; Eshraghi, N.; Hermann, R.P.; Mahmoud, A.; Boschini, F. Na$_2$FePO$_4$F/multi-walled carbon nanotubes for lithium-ion batteries: Operando Mössbauer study of spray-dried composites. *Sol. Energy Mater. Sol. Cells* **2016**, *148*, 67–72. [CrossRef]

88. Brisbois, M.; Krins, N.; Hermann, R.P.; Schrijnemakers, A.; Cloots, R.; Vertruyen, B.; Boschini, F. Spray-drying synthesis of Na$_2$FePO$_4$F/carbon powders for lithium-ion batteries. *Mater. Lett.* **2014**, *130*, 263–266. [CrossRef]

89. Mahmoud, A.; Caes, S.; Brisbois, M.; Hermann, R.P.; Berardo, L.; Schrijnemakers, A.; Malherbe, C.; Eppe, G.; Cloots, R.; Vertruyen, B. Spray-drying as a tool to disperse conductive carbon inside Na$_2$FePO$_4$F particles by addition of carbon black or carbon nanotubes to the precursor solution. *J. Solid State Electrochem.* **2018**, *22*, 103–112. [CrossRef]

90. Lin, X.; Hou, X.; Wu, X.; Wang, S.; Gao, M.; Yang, Y. Exploiting Na$_2$MnPO$_4$F as a high-capacity and well-reversible cathode material for Na-ion batteries. *RSC Adv.* **2014**, *4*, 40985–40993. [CrossRef]

91. Wu, L.; Hu, Y.; Zhang, X.; Liu, J.; Zhu, X.; Zhong, S. Synthesis of carbon-coated Na$_2$MnPO$_4$F hollow spheres as a potential cathode material for Na-ion batteries. *J. Power Sources* **2018**, *374*, 40–47. [CrossRef]

92. Zou, H.; Li, S.; Wu, X.; McDonald, M.J.; Yang, Y. Spray-Drying Synthesis of Pure Na$_2$CoPO$_4$F as Cathode Material for Sodium Ion Batteries. *ECS Electrochem. Lett.* **2015**, *4*, A53–A55. [CrossRef]

93. Eshraghi, N.; Caes, S.; Mahmoud, A.; Cloots, R.; Vertruyen, B.; Boschini, F. Sodium vanadium (III) fluorophosphate/carbon nanotubes composite (NVPF/CNT) prepared by spray-drying: Good electrochemical performance thanks to well-dispersed CNT network within NVPF particles. *Electrochim. Acta* **2017**, *228*, 319–324. [CrossRef]

94. Yin, Y.; Xiong, F.; Pei, C.; Xu, Y.; An, Q.; Tan, S.; Zhuang, Z.; Sheng, J.; Li, Q.; Mai, L. Robust three-dimensional graphene skeleton encapsulated Na$_3$V$_2$O$_2$(PO$_4$)$_2$F nanoparticles as a high-rate and long-life cathode of sodium-ion batteries. *Nano Energy* **2017**, *41*, 452–459. [CrossRef]

95. Zhang, H.; Deng, Q.; Zhou, A.; Liu, X.; Li, J. Porous Li$_2$C$_8$H$_4$O$_4$ coated with N-doped carbon by using CVD as an anode material for Li-ion batteries. *J. Mater. Chem. A* **2014**, *2*, 5696–5702. [CrossRef]

96. Deng, Q.; Wang, Y.; Zhao, Y.; Li, J. Disodium terephthalate/multiwall-carbon nanotube nanocomposite as advanced anode material for Li-ion batteries. *Ionics* **2017**, *23*, 2613–2619. [CrossRef]

97. Wu, X.; Ma, J.; Ma, Q.; Xu, S.; Hu, Y.-S.; Sun, Y.; Li, H.; Chen, L.; Huang, X. A spray-drying approach for the synthesis of a Na$_2$C$_6$H$_2$O$_4$/CNT nanocomposite anode for sodium-ion batteries. *J. Mater. Chem. A* **2015**, *3*, 13193–13197. [CrossRef]

98. Qian, X.; Zhao, D.; Jin, L.; Yao, S.; Rao, D.; Shen, X.; Zhou, Y.; Xi, X. A separator modified by spray-dried hollow spherical cerium oxide and its application in lithium sulfur batteries. *RSC Adv.* **2016**, *6*, 114989–114996. [CrossRef]

99. Hong, S.-H.; Song, M.Y. Syntheses of nano-sized Co-based powders by carbothermal reduction for anode materials of lithium ion batteries. *Ceram. Int.* **2018**, *44*, 4225–4229. [CrossRef]

100. Kim, J.H.; Kang, Y.C. Electrochemical properties of micron-sized, spherical, meso- and macro-porous Co$_3$O$_4$ and CoO–carbon composite powders prepared by a two-step spray-drying process. *Nanoscale* **2014**, *6*, 4789. [CrossRef] [PubMed]

101. Park, G.D.; Lee, J.-H.; Lee, J.-K.; Kang, Y.C. Effect of esterification reaction of citric acid and ethylene glycol on the formation of multi-shelled cobalt oxide powders with superior electrochemical properties. *Nano Res.* **2014**, *7*, 1738–1748. [CrossRef]

102. Son, M.Y.; Kim, J.H.; Kang, Y.C. Study of Co$_3$O$_4$ mesoporous nanosheets prepared by a simple spray-drying process and their electrochemical properties as anode material for lithium secondary batteries. *Electrochim. Acta* **2014**, *116*, 44–50. [CrossRef]

103. Xiang, Y.; Chen, Z.; Chen, C.; Wang, T.; Zhang, M. Design and synthesis of Cr$_2$O$_3$@C@G composites with yolk-shell structure for Li + storage. *J. Alloys Compd.* **2017**, *724*, 406–412. [CrossRef]

104. Jeon, K.M.; Kim, J.H.; Choi, Y.J.; Kang, Y.C. Electrochemical properties of hollow copper (II) oxide nanopowders prepared by salt-assisted spray-drying process applying nanoscale Kirkendall diffusion. *J. Appl. Electrochem.* **2016**, *46*, 469–477. [CrossRef]

105. Park, G.D.; Kang, Y.C. Superior Lithium-Ion Storage Properties of Mesoporous CuO-Reduced Graphene Oxide Composite Powder Prepared by a Two-Step Spray-Drying Process. *Chem. Eur. J.* **2015**, *21*, 9179–9184. [CrossRef] [PubMed]

106. Won, J.M.; Kim, J.H.; Choi, Y.J.; Cho, J.S.; Kang, Y.C. Electrochemical properties of CuO hollow nanopowders prepared from formless Cu-C composite via nanoscale Kirkendall diffusion process. *J. Alloys Compd.* **2016**, *671*, 74–83. [CrossRef]

107. Padashbarmchi, Z.; Hamidian, A.H.; Zhang, H.; Zhou, L.; Khorasani, N.; Kazemzad, M.; Yu, C. A systematic study on the synthesis of α-Fe$_2$O$_3$ multi-shelled hollow spheres. *RSC Adv.* **2015**, *5*, 10304–10309. [CrossRef]

108. Zhang, H.; Sun, X.; Huang, X.; Zhou, L. Encapsulation of α-Fe$_2$O$_3$ nanoparticles in graphitic carbon microspheres as high-performance anode materials for lithium-ion batteries. *Nanoscale* **2015**, *7*, 3270–3275. [CrossRef] [PubMed]

109. Zhou, G.-W.; Wang, J.; Gao, P.; Yang, X.; He, Y.-S.; Liao, X.-Z.; Yang, J.; Ma, Z.-F. Facile Spray-drying Route for the Three-Dimensional Graphene-Encapsulated Fe$_2$O$_3$ Nanoparticles for Lithium Ion Battery Anodes. *Ind. Eng. Chem. Res.* **2013**, *52*, 1197–1204. [CrossRef]

110. Zhou, L.; Xu, H.; Zhang, H.; Yang, J.; Hartono, S.B.; Qian, K.; Zou, J.; Yu, C. Cheap and scalable synthesis of α-Fe$_2$O$_3$ multi-shelled hollow spheres as high-performance anode materials for lithium ion batteries. *Chem. Commun.* **2013**, *49*, 8695. [CrossRef] [PubMed]

111. He, W.; Tian, H.; Wang, X.; Xin, F.; Han, W. Three-dimensional interconnected network GeO$_x$/multi-walled CNT composite spheres as high-performance anodes for lithium ion batteries. *J. Mater. Chem. A* **2015**, *3*, 19393–19401. [CrossRef]

112. Jia, H.; Kloepsch, R.; He, X.; Badillo, J.P.; Winter, M.; Placke, T. One-step synthesis of novel mesoporous three-dimensional GeO$_2$ and its lithium storage properties. *J Mater Chem A* **2014**, *2*, 17545–17550. [CrossRef]

113. Qian, X.; Zhao, D.; Jin, L.; Shen, X.; Yao, S.; Rao, D.; Zhou, Y.; Xi, X. ming Hollow spherical Lanthanum oxide coated separator for high electrochemical performance lithium-sulfur batteries. *Mater. Res. Bull.* **2017**, *94*, 104–112. [CrossRef]

114. Jeon, K.M.; Cho, J.S.; Kang, Y.C. Electrochemical properties of MnS-C and MnO-C composite powders prepared via spray-drying process. *J. Power Sources* **2015**, *295*, 9–15. [CrossRef]

115. Park, G.D.; Kim, J.H.; Choi, Y.J.; Kang, Y.C. Large-Scale Production of MoO$_3$-Reduced Graphene Oxide Powders with Superior Lithium Storage Properties by Spray-Drying Process. *Electrochim. Acta* **2015**, *173*, 581–587. [CrossRef]

116. Tao, Y.; Wei, Y.; Liu, Y.; Wang, J.; Qiao, W.; Ling, L.; Long, D. Kinetically-enhanced polysulfide redox reactions by Nb$_2$O$_5$ nanocrystals for high-rate lithium–sulfur battery. *Energy Environ. Sci.* **2016**, *9*, 3230–3239. [CrossRef]

117. Xiao, A.; Zhou, S.; Zuo, C.; Zhuan, Y.; Ding, X. Synthesis of nickel oxide nanospheres by a facile spray-drying method and their application as anode materials for lithium ion batteries. *Mater. Res. Bull.* **2015**, *70*, 200–203. [CrossRef]

118. Li, Y.; Hou, X.; Wang, J.; Mao, J.; Gao, Y.; Hu, S. Catalyst Ni-assisted synthesis of interweaved SiO/G/CNTs&CNFs composite as anode material for lithium-ion batteries. *J. Mater. Sci. Mater. Electron.* **2015**, *26*, 7507–7514.

119. Yang, X.; Zhang, P.; Shi, C.; Wen, Z. Porous Graphite/Silicon Micro-Sphere Prepared by In-Situ Carbothermal Reduction and Spray-drying for Lithium Ion Batteries. *ECS Solid State Lett.* **2012**, *1*, M5–M7. [CrossRef]

120. Wu, H.; Tang, Q.; Fan, H.; Liu, Z.; Hu, A.; Zhang, S.; Deng, W.; Chen, X. Dual-Confined and Hierarchical-Porous Graphene/C/SiO$_2$ Hollow Microspheres through Spray-drying Approach for Lithium-Sulfur Batteries. *Electrochim. Acta* **2017**, *255*, 179–186. [CrossRef]

121. Jiao, M.; Liu, K.; Shi, Z.; Wang, C. SiO$_2$/Carbon Composite Microspheres with Hollow Core-Shell Structure as a High-Stability Electrode for Lithium-Ion Batteries. *ChemElectroChem* **2017**, *4*, 542–549. [CrossRef]

122. Choi, S.H.; Kang, Y.C. Kilogram-Scale Production of SnO$_2$ Yolk-Shell Powders by a Spray-Drying Process Using Dextrin as Carbon Source and Drying Additive. *Chem. Eur. J.* **2014**, *20*, 5835–5839. [CrossRef] [PubMed]

123. Liu, D.; Kong, Z.; Liu, X.; Fu, A.; Wang, Y.; Guo, Y.-G.; Guo, P.; Li, H.; Zhao, X.S. Spray-Drying-Induced Assembly of Skeleton-Structured SnO₂/Graphene Composite Spheres as Superior Anode Materials for High-Performance Lithium-Ion Batteries. *ACS Appl. Mater. Interfaces* **2018**, *10*, 2515–2525. [CrossRef] [PubMed]

124. Cho, J.S.; Ju, H.S.; Kang, Y.C. Applying Nanoscale Kirkendall Diffusion for Template-Free, Kilogram-Scale Production of SnO₂ Hollow Nanospheres via Spray-drying System. *Sci. Rep.* **2016**, *6*, 23915. [CrossRef] [PubMed]

125. Chunju, L.; Hu, T.; Shu, K.; Chen, D.; Tian, G. Porous TiO₂ nanowire microsphere constructed by spray-drying and its electrochemical lithium storage properties. *Microsc. Res. Tech.* **2014**, *77*, 170–175. [CrossRef] [PubMed]

126. He, Y.-B.; Liu, M.; Xu, Z.-L.; Zhang, B.; Li, B.; Kang, F.; Kim, J.-K. Li-ion Reaction to Improve the Rate Performance of Nanoporous Anatase TiO₂ Anodes. *Energy Technol.* **2013**, *1*, 668–674. [CrossRef]

127. Mondal, A.; Maiti, S.; Singha, K.; Mahanty, S.; Panda, A.B. TiO₂-rGO nanocomposite hollow spheres: Large scale synthesis and application as an efficient anode material for lithium-ion batteries. *J. Mater. Chem. A* **2017**, *5*, 23853–23862. [CrossRef]

128. Park, G.D.; Lee, J.; Piao, Y.; Kang, Y.C. Mesoporous graphitic carbon-TiO₂ composite microspheres produced by a pilot-scale spray-drying process as an efficient sulfur host material for Li-S batteries. *Chem. Eng. J.* **2018**, *335*, 600–611. [CrossRef]

129. Sakao, M.; Kijima, N.; Akimoto, J.; Okutani, T. Synthesis and Electrochemical Properties of Porous Titania Prepared by Spray-drying of Titania Nanosheets. *Chem. Lett.* **2012**, *41*, 1515–1517. [CrossRef]

130. Sakao, M.; Kijima, N.; Yoshinaga, M.; Akimoto, J.; Okutani, T. Synthesis and Electrochemical Properties of Porous Titania Fabricated from Nanosheets. *Key Eng. Mater.* **2013**, *566*, 111–114. [CrossRef]

131. Ventosa, E.; Mei, B.; Xia, W.; Muhler, M.; Schuhmann, W. TiO₂(B)/Anatase Composites Synthesized by Spray-drying as High Performance Negative Electrode Material in Li-Ion Batteries. *ChemSusChem* **2013**, *6*, 1312–1315. [CrossRef] [PubMed]

132. Wilhelm, O.; Pratsinis, S.; de Chambrier, E.; Crouzet, M.; Exnar, I. Electrochemical performance of granulated titania nanoparticles. *J. Power Sources* **2004**, *134*, 197–201. [CrossRef]

133. Zhu, X.; Li, Q.; Fang, Y.; Liu, X.; Xiao, L.; Ai, X.; Yang, H.; Cao, Y. Graphene-Modified TiO₂ Microspheres Synthesized by a Facile Spray-Drying Route for Enhanced Sodium-Ion Storage. *Part. Part. Syst. Charact.* **2016**, *33*, 545–552. [CrossRef]

134. Li, Q.; Chen, Y.; He, J.; Fu, F.; Qi, F.; Lin, J.; Zhang, W. Carbon Nanotube Modified V₂O₅ Porous Microspheres as Cathodes for High-Performance Lithium-Ion Batteries. *Energy Technol.* **2017**, *5*, 665–669. [CrossRef]

135. Mao, J.; Hou, X.; Chen, H.; Ru, Q.; Hu, S.; Lam, K. Facile spray-drying synthesis of porous structured ZnFe₂O₄ as high-performance anode material for lithium-ion batteries. *J. Mater. Sci. Mater. Electron.* **2017**, *28*, 3709–3715. [CrossRef]

136. Won, J.M.; Choi, S.H.; Hong, Y.J.; Ko, Y.N.; Kang, Y.C. Electrochemical properties of yolk-shell structured ZnFe₂O₄ powders prepared by a simple spray-drying process as anode material for lithium-ion battery. *Sci. Rep.* **2014**, *4*, 5857. [CrossRef] [PubMed]

137. Zhang, Z.; Ren, W.; Wang, Y.; Yang, J.; Tan, Q.; Zhong, Z.; Su, F. Mn₀.₅Co₀.₅Fe₂O₄ nanoparticles highly dispersed in porous carbon microspheres as high performance anode materials in Li-ion batteries. *Nanoscale* **2014**, *6*, 6805. [CrossRef] [PubMed]

138. Mondal, A.; Maiti, S.; Mahanty, S.; Baran Panda, A. Large-scale synthesis of porous NiCo₂O₄ and rGO-NiCo₂O₄ hollow-spheres with superior electrochemical performance as a faradaic electrode. *J. Mater. Chem. A* **2017**, *5*, 16854–16864. [CrossRef]

139. Choi, S.H.; Park, S.K.; Lee, J.-K.; Kang, Y.C. Facile synthesis of multi-shell structured binary metal oxide powders with a Ni/Co mole ratio of 1:2 for Li-Ion batteries. *J. Power Sources* **2015**, *284*, 481–488. [CrossRef]

140. Quan, J.; Mei, L.; Ma, Z.; Huang, J.; Li, D. Cu₁.₅Mn₁.₅O₄ spinel: A novel anode material for lithium-ion batteries. *RSC Adv.* **2016**, *6*, 55786–55791. [CrossRef]

141. Park, J.-S.; Cho, J.S.; Kang, Y.C. Scalable synthesis of NiMoO₄ microspheres with numerous empty nanovoids as an advanced anode material for Li-ion batteries. *J. Power Sources* **2018**, *379*, 278–287. [CrossRef]

142. Zhu, G.; Li, Q.; Zhao, Y.; Che, R. Nanoporous TiNb₂O₇/C Composite Microspheres with Three-Dimensional Conductive Network for Long-Cycle-Life and High-Rate-Capability Anode Materials for Lithium-Ion Batteries. *ACS Appl. Mater. Interfaces* **2017**, *9*, 41258–41264. [CrossRef] [PubMed]

143. Chen, Y.; Li, P.; Zhao, S.; Zhuang, Y.; Zhao, S.; Zhou, Q.; Zheng, J. Influence of integrated microstructure on the performance of $LiNi_{0.8}Co_{0.15}Al_{0.05}O_2$ as a cathodic material for lithium ion batteries. *RSC Adv.* **2017**, *7*, 29233–29239. [CrossRef]

144. Cheng, J.; Li, X.; He, Z.; Wang, Z.; Guo, H.; Peng, W. Significant improved electrochemical performance of layered $Li_{1.2}Mn_{0.54}Co_{0.13}Ni_{0.13}O_2$ via graphene surface modification. *Mater. Technol.* **2016**, *31*, 658–665. [CrossRef]

145. Duvigneaud, P.H.; Segato, T. Synthesis and characterisation of $LiNi_{1-x-y}Co_xAl_yO_2$ cathodes for lithium-ion batteries by the PVA precursor method. *J. Eur. Ceram. Soc.* **2004**, *24*, 1375–1380. [CrossRef]

146. He, Z.; Wang, Z.; Cheng, L.; Zhu, Z.; Li, T.; Li, X.; Guo, H. Structural and electrochemical characterization of layered $0.3Li_2MnO_3 \cdot 0.7LiMn_{0.35-x/3}Ni_{0.5-x/3}Co_{0.15-x/3}Cr_xO_2$ cathode synthesized by spray-drying. *Adv. Powder Technol.* **2014**, *25*, 647–653. [CrossRef]

147. He, Z.; Wang, Z.; Guo, H.; Li, X.; Xianwen, W.; Yue, P.; Wang, J. A simple method of preparing graphene-coated $Li[Li_{0.2}Mn_{0.54}Ni_{0.13}Co_{0.13}]O_2$ for lithium-ion batteries. *Mater. Lett.* **2013**, *91*, 261–264. [CrossRef]

148. He, Z.; Wang, Z.; Guo, H.; Li, X.; Yue, P.; Wang, J.; Xiong, X. Synthesis and electrochemical performance of $xLi_2MnO_3 \cdot (1-x)LiMn_{0.5}Ni_{0.4}Co_{0.1}O_2$ for lithium ion battery. *Powder Technol.* **2013**, *235*, 158–162. [CrossRef]

149. Hou, M.; Guo, S.; Liu, J.; Yang, J.; Wang, Y.; Wang, C.; Xia, Y. Preparation of lithium-rich layered oxide micro-spheres using a slurry spray-drying process. *J. Power Sources* **2015**, *287*, 370–376. [CrossRef]

150. Hu, S.-K.; Cheng, G.-H.; Cheng, M.-Y.; Hwang, B.-J.; Santhanam, R. Cycle life improvement of ZrO_2-coated spherical $LiNi_{1/3}Co_{1/3}Mn_{1/3}O_2$ cathode material for lithium ion batteries. *J. Power Sources* **2009**, *188*, 564–569. [CrossRef]

151. Kim, J.-M.; Kumagai, N.; Kadoma, Y.; Yashiro, H. Synthesis and electrochemical properties of lithium non-stoichiometric $Li_{1+x}(Ni_{1/3}Co_{1/3}Mn_{1/3})O_{2+\delta}$ prepared by a spray-drying method. *J. Power Sources* **2007**, *174*, 473–479. [CrossRef]

152. Kim, J.-M.; Kumagai, N.; Cho, T.-H. Synthesis, Structure, and Electrochemical Characteristics of Overlithiated $Li_{[1+x]}(Ni_{[z]}Co_{[1-2z]}Mn_{[z]})_{[1-x]}O_2$ ($z = 0.1 - 0.4$ and $x = 0.0 - 0.1$) Positive Electrodes Prepared by Spray-Drying Method. *J. Electrochem. Soc.* **2008**, *155*, A82. [CrossRef]

153. Konstantinov, K.; Wang, G.X.; Yao, J.; Liu, H.K.; Dou, S.X. Stoichiometry-controlled high-performance $LiCoO_2$ electrode materials prepared by a spray solution technique. *J. Power Sources* **2003**, *119*, 195–200. [CrossRef]

154. Li, D.-C.; Muta, T.; Zhang, L.-Q.; Yoshio, M.; Noguchi, H. Effect of synthesis method on the electrochemical performance of $LiNi_{1/3}Mn_{1/3}Co_{1/3}O_2$. *J. Power Sources* **2004**, *132*, 150–155. [CrossRef]

155. Li, D.-C.; Noguchi, H.; Yoshio, M. Electrochemical characteristics of $LiNi_{0.5-x}Mn_{0.5-x}Co_{2x}O_2$ ($0 < x \le 0.1$) prepared by spray-dry method. *Electrochim. Acta* **2004**, *50*, 427–430.

156. Li, D.; Kato, Y.; Kobayakawa, K.; Noguchi, H.; Sato, Y. Preparation and electrochemical characteristics of $LiNi_{1/3}Mn_{1/3}Co_{1/3}O_2$ coated with metal oxides coating. *J. Power Sources* **2006**, *160*, 1342–1348. [CrossRef]

157. Li, D.; Sasaki, Y.; Kobayakawa, K.; Noguchi, H.; Sato, Y. Preparation, morphology and electrochemical characteristics of $LiNi_{1/3}Mn_{1/3}Co_{1/3}O_2$ with LiF addition. *Electrochim. Acta* **2006**, *52*, 643–648. [CrossRef]

158. Li, J.; Wang, L.; Chen, J.; He, X. Li Storage Properties of $(1-x-y)Li[Li_{1/3}Mn_{2/3}]O_2-xLiFeO_2-yLiNiO_2$ Solid Solution Cathode Materials. *ECS Trans.* **2014**, *62*, 79–87. [CrossRef]

159. Li, L.; Meyer, W.H.; Wegner, G.; Wohlfahrt-Mehrens, M. Synthesis of Submicrometer-Sized Electrochemically Active Lithium Cobalt Oxide via a Polymer Precursor. *Adv. Mater.* **2005**, *17*, 984–988. [CrossRef]

160. Li, Y.; Wan, C.; Wu, Y.; Jiang, C.; Zhu, Y. Synthesis and characterization of ultrafine $LiCoO_2$ powders by a spray-drying method. *J. Power Sources* **2000**, *85*, 294–298. [CrossRef]

161. Lin, B.; Wen, Z.; Gu, Z.; Xu, X. Preparation and electrochemical properties of $Li[Ni_{1/3}Co_{1/3}Mn_{1-x/3}Zr_{x/3}]O_2$ cathode materials for Li-ion batteries. *J. Power Sources* **2007**, *174*, 544–547. [CrossRef]

162. Lin, B.; Wen, Z.; Gu, Z.; Huang, S. Morphology and electrochemical performance of $Li[Ni_{1/3}Co_{1/3}Mn_{1/3}]O_2$ cathode material by a slurry spray-drying method. *J. Power Sources* **2008**, *175*, 564–569. [CrossRef]

163. Lin, B.; Wen, Z.; Wang, X.; Liu, Y. Preparation and characterization of carbon-coated $Li[Ni_{1/3}Co_{1/3}Mn_{1/3}]O_2$ cathode material for lithium-ion batteries. *J. Solid State Electrochem.* **2010**, *14*, 1807–1811. [CrossRef]

164. Lin, M.-H.; Cheng, J.-H.; Huang, H.-F.; Chen, U.-F.; Huang, C.-M.; Hsieh, H.-W.; Lee, J.-M.; Chen, J.-M.; Su, W.-N.; Hwang, B.-J. Revealing the mitigation of intrinsic structure transformation and oxygen evolution in a layered $Li_{1.2}Ni_{0.2}Mn_{0.6}O_2$ cathode using restricted charging protocols. *J. Power Sources* **2017**, *359*, 539–548. [CrossRef]

165. Liu, Y.; Qian, K.; He, J.; Chu, X.; He, Y.-B.; Wu, M.; Li, B.; Kang, F. In-situ polymerized lithium polyacrylate (PAALi) as dual-functional lithium source for high-performance layered oxide cathodes. *Electrochim. Acta* **2017**, *249*, 43–51. [CrossRef]

166. Liu, Z.; Hu, G.; Peng, Z.; Deng, X.; Liu, Y. Synthesis and characterization of layered $Li(Ni_{1/3}Mn_{1/3}Co_{1/3})O_2$ cathode materials by spray-drying method. *Trans. Nonferrous Met. Soc. China* **2007**, *17*, 291–295. [CrossRef]

167. Oh, S.H.; Jeong, W.T.; Cho, W.I.; Cho, B.W.; Woo, K. Electrochemical characterization of high-performance $LiNi_{0.8}Co_{0.2}O_2$ cathode materials for rechargeable lithium batteries. *J. Power Sources* **2005**, *140*, 145–150. [CrossRef]

168. Qiao, Q.Q.; Zhang, H.Z.; Li, G.R.; Ye, S.H.; Wang, C.W.; Gao, X.P. Surface modification of Li-rich layered $Li(Li_{0.17}Ni_{0.25}Mn_{0.58})O_2$ oxide with Li–Mn–PO$_4$ as the cathode for lithium-ion batteries. *J. Mater. Chem. A* **2013**, *1*, 5262. [CrossRef]

169. Qiao, Q.-Q.; Qin, L.; Li, G.-R.; Wang, Y.-L.; Gao, X.-P. Sn-stabilized Li-rich layered $Li(Li_{0.17}Ni_{0.25}Mn_{0.58})O_2$ oxide as a cathode for advanced lithium-ion batteries. *J. Mater. Chem. A* **2015**, *3*, 17627–17634. [CrossRef]

170. Qin, L.; Wen, Y.; Xue-Ping, G. Surface Modification of Li-rich Layered $Li(Li_{0.17}Ni_{0.2}Mn_{0.58}Co_{0.05})O_2$ Oxide with TiO$_2$(B) as the Cathode for Lithium-ion Batteries. *J. Inorg. Mater.* **2014**, *29*, 1257. [CrossRef]

171. Sun, Y.; Xia, Y.; Shiosaki, Y.; Noguchi, H. Preparation and electrochemical properties of $LiCoO_2$-$LiNi_{0.5}Mn_{0.5}O_2$-Li_2MnO_3 solid solutions with high Mn contents. *Electrochim. Acta* **2006**, *51*, 5581–5586. [CrossRef]

172. Wang, T.; Chen, Z.; Zhao, R.; Li, A.; Chen, H. A New High Energy Lithium ion Batteries Consisting of $0.5Li_2MnO_3 \cdot 0.5LiMn_{0.33}Ni_{0.33}Co_{0.33}O_2$ and Soft Carbon Components. *Electrochim. Acta* **2016**, *194*, 1–9. [CrossRef]

173. Wang, Z.; Wang, Z.; Guo, H.; Peng, W.; Li, X. Synthesis of Li_2MnO_3-stabilized $LiCoO_2$ cathode material by spray-drying method and its high-voltage performance. *J. Alloys Compd.* **2015**, *626*, 228–233. [CrossRef]

174. Watanabe, A.; Matsumoto, F.; Fukunishi, M.; Kobayashi, G.; Ito, A.; Hatano, M.; Ohsawa, Y.; Sato, Y. Relationship between Electrochemical Pre-Treatment and Cycle Performance of a Li-Rich Solid-Solution Layered $Li_{1-alpha}[Ni_{0.18}Li_{0.20+d}]_{alpha}Co_{0.03}Mn_{0.58}]O_2$ Cathode for Li-Ion Secondary Batteries. *Electrochemistry* **2012**, *80*, 561–565. [CrossRef]

175. Wu, H.M.; Tu, J.P.; Chen, X.T.; Yuan, Y.F.; Li, Y.; Zhao, X.B.; Cao, G.S. Synthesis and characterization of $LiNi_{0.8}Co_{0.2}O_2$ as cathode material for lithium-ion batteries by a spray-drying method. *J. Power Sources* **2006**, *159*, 291–294. [CrossRef]

176. Xia, L.; Li, S.-L.; Ai, X.-P.; Yang, H.-X.; Cao, Y.-L. Temperature-sensitive cathode materials for safer lithium-ion batteries. *Energy Environ. Sci.* **2011**, *4*, 2845. [CrossRef]

177. Yang, S.; Huang, G.; Hu, S.; Hou, X.; Huang, Y.; Yue, M.; Lei, G. Improved electrochemical performance of the $Li_{1.2}Ni_{0.13}Co_{0.13}Mn_{0.54}O_2$ wired by CNT networks for lithium-ion batteries. *Mater. Lett.* **2014**, *118*, 8–11. [CrossRef]

178. Yuan, W.; Zhang, H.Z.; Liu, Q.; Li, G.R.; Gao, X.P. Surface modification of $Li(Li_{0.17}Ni_{0.2}Co_{0.05}Mn_{0.58})O_2$ with CeO$_2$ as cathode material for Li-ion batteries. *Electrochim. Acta* **2014**, *135*, 199–207. [CrossRef]

179. Yue, P.; Wang, Z.; Guo, H.; Wu, F.; He, Z.; Li, X. Effect of synthesis routes on the electrochemical performance of $Li[Ni_{0.6}Co_{0.2}Mn_{0.2}]O_2$ for lithium ion batteries. *J. Solid State Electrochem.* **2012**, *16*, 3849–3854. [CrossRef]

180. Yue, P.; Wang, Z.; Peng, W.; Li, L.; Chen, W.; Guo, H.; Li, X. Spray-drying synthesized $LiNi_{0.6}Co_{0.2}Mn_{0.2}O_2$ and its electrochemical performance as cathode materials for lithium ion batteries. *Powder Technol.* **2011**, *214*, 279–282. [CrossRef]

181. Yue, P.; Wang, Z.; Peng, W.; Li, L.; Guo, H.; Li, X.; Hu, Q.; Zhang, Y. Preparation and electrochemical properties of submicron $LiNi_{0.6}Co_{0.2}Mn_{0.2}O_2$ as cathode material for lithium ion batteries. *Scr. Mater.* **2011**, *65*, 1077–1080. [CrossRef]

182. Yue, P.; Wang, Z.; Zhang, Q.; Yan, G.; Guo, H.; Li, X. Synthesis and electrochemical performance of $LiNi_{0.6}Co_{0.2}Mn_{0.2}O_2$/reduced graphene oxide cathode materials for lithium-ion batteries. *Ionics* **2013**, *19*, 1329–1334. [CrossRef]

183. Zhang, L.; Li, D.; Wang, X.; Noguchi, H.; Yoshio, M. Properties of Li-Ni-Mn-O electrode materials prepared from solution spray synthesized powders. *Mater. Lett.* **2005**, *59*, 2693–2697. [CrossRef]

184. Zhang, L.; Muta, T.; Noguchi, H.; Wang, X.; Zhou, M.; Yoshio, M. Peculiar electrochemical behaviors of $(1-x)LiNiO_2 \cdot xLi_2TiO_3$ cathode materials prepared by spray-drying. *J. Power Sources* **2003**, *117*, 137–142. [CrossRef]

185. Zhang, L.; Noguchi, H.; Li, D.; Muta, T.; Wang, X.; Yoshio, M.; Taniguchi, I. Synthesis and electrochemistry of cubic rocksalt Li–Ni–Ti–O compounds in the phase diagram of $LiNiO_2$–$LiTiO_2$–$Li[Li_{1/3}Ti_{2/3}]O_2$. *J. Power Sources* **2008**, *185*, 534–541. [CrossRef]

186. Zhang, L.; Wang, X.; Muta, T.; Li, D.; Noguchi, H.; Yoshio, M.; Ma, R.; Takada, K.; Sasaki, T. The effects of extra Li content, synthesis method, sintering temperature on synthesis and electrochemistry of layered $LiNi_{1/3}Mn_{1/3}Co_{1/3}O_2$. *J. Power Sources* **2006**, *162*, 629–635. [CrossRef]

187. Kim, J.-M.; Kumagai, N.; Komaba, S. Improved electrochemical properties of $Li_{1+x}(Ni_{0.3}Co_{0.4}Mn_{0.3})O_{2-\delta}$ ($x = 0, 0.03$ and 0.06) with lithium excess composition prepared by a spray-drying method. *Electrochim. Acta* **2006**, *52*, 1483–1490. [CrossRef]

188. Gao, J.; Huang, Z.; Li, J.; He, X.; Jiang, C. Preparation and characterization of $Li_{1.2}Ni_{0.13}Co_{0.13}Mn_{0.54}O_2$ cathode materials for lithium-ion battery. *Ionics* **2014**, *20*, 301–307. [CrossRef]

189. Wang, Z.; Yin, Y.; Ren, Y.; Wang, Z.; Gao, M.; Ma, T.; Zhuang, W.; Lu, S.; Fan, A.; Amine, K. High performance lithium-manganese-rich cathode material with reduced impurities. *Nano Energy* **2017**, *31*, 247–257. [CrossRef]

190. Ji, M.-J.; Kim, E.-K.; Ahn, Y.-T.; Choi, B.-H. Crystallinity and Battery Properties of Lithium Manganese Oxide Spinel with Lithium Titanium Oxide Spinel Coating Layer on Its Surface. *J. Korean Ceram. Soc.* **2010**, *47*, 633–637. [CrossRef]

191. Tu, J.P.; Wu, H.M.; Yang, Y.Z.; Zhang, W.K. Spray-drying technology for the synthesis of nanosized $LiMn_2O_4$ cathode material. *Mater. Lett.* **2007**, *61*, 864–867. [CrossRef]

192. Wan, C.; Cheng, M.; Wu, D. Synthesis of spherical spinel $LiMn_2O_4$ with commercial manganese carbonate. *Powder Technol.* **2011**, *210*, 47–51. [CrossRef]

193. Wan, C.; Wu, M.; Wu, D. Synthesis of spherical $LiMn_2O_4$ cathode material by dynamic sintering of spray-dried precursors. *Powder Technol.* **2010**, *199*, 154–158. [CrossRef]

194. Wu, H.M.; Tu, J.P.; Yang, Y.Z.; Shi, D.Q. Spray-drying process for synthesis of nanosized $LiMn_2O_4$ cathode. *J. Mater. Sci.* **2006**, *41*, 4247–4250. [CrossRef]

195. Wu, H.M.; Tu, J.P.; Chen, X.T.; Li, Y.; Zhao, X.B.; Cao, G.S. Electrochemical study on $LiMn_2O_4$ as cathode material for lithium ion batteries. *J. Electroanal. Chem.* **2006**, *586*, 180–183. [CrossRef]

196. Wu, H.M.; Tu, J.P.; Yuan, Y.F.; Li, Y.; Zhao, X.B.; Cao, G.S. Structural, morphological and electrochemical characteristics of spinel $LiMn_2O_4$ prepared by spray-drying method. *Scr. Mater.* **2005**, *52*, 513–517. [CrossRef]

197. Wu, H.M.; Tu, J.P.; Yuan, Y.F.; Li, Y.; Zhao, X.B.; Cao, G.S. Preparation of $LiMn_2O_4$ by two methods for lithium ion batteries. *Mater. Chem. Phys.* **2005**, *93*, 461–465. [CrossRef]

198. Wu, H.M.; Tu, J.P.; Yuan, Y.F.; Li, Y.; Zhang, W.K.; Huang, H. Electrochemical performance of nanosized $LiMn_2O_4$ for lithium-ion batteries. *Phys. B Condens. Matter* **2005**, *369*, 221–226. [CrossRef]

199. Wu, H.M.; Tu, J.P.; Yuan, Y.F.; Li, Y.; Zhao, X.B.; Cao, G.S. Synthesis and electrochemical characteristics of spinel $LiMn_2O_4$ via a precipitation spray-drying process. *Mater. Sci. Eng. B* **2005**, *119*, 75–79. [CrossRef]

200. Huang, H.; Wang, C.; Zhang, W.K.; Gan, Y.P.; Kang, L. Electrochemical study on $LiCo_{1/6}Mn_{11/6}O_4$ as cathode material for lithium ion batteries at elevated temperature. *J. Power Sources* **2008**, *184*, 583–588. [CrossRef]

201. Zhang, W.K.; Wang, C.; Huang, H.; Gan, Y.P.; Wu, H.M.; Tu, J.P. Synthesis and electrochemical properties of spinel $LiCo_{1/6}Mn_{11/6}O_4$ powders by a spray-drying method. *J. Alloys Compd.* **2008**, *465*, 250–254. [CrossRef]

202. Jiang, Q.; Hu, G.; Peng, Z.; Du, K.; Cao, Y.; Tang, D. Preparation of spherical spinel $LiCr_{0.04}Mn_{1.96}O_4$ cathode materials based on the slurry spray-drying method. *Rare Met.* **2009**, *28*, 618–623. [CrossRef]

203. Peng, Z.D.; Jiang, Q.L.; Du, K.; Wang, W.G.; Hu, G.R.; Liu, Y.X. Effect of Cr-sources on performance of $Li_{1.05}Cr_{0.04}Mn_{1.96}O_4$ cathode materials prepared by slurry spray-drying method. *J. Alloys Compd.* **2010**, *493*, 640–644. [CrossRef]

204. Wu, H.M.; Tu, J.P.; Chen, X.T.; Li, Y.; Zhao, X.B.; Cao, G.S. Effects of Ni-ion doping on electrochemical characteristics of spinel $LiMn_2O_4$ powders prepared by a spray-drying method. *J. Solid State Electrochem.* **2006**, *11*, 173–176. [CrossRef]

205. Wu, H.M.; Tu, J.P.; Chen, X.T.; Shi, D.Q.; Zhao, X.B.; Cao, G.S. Synthesis and characterization of abundant Ni-doped LiNi$_x$Mn$_{2-x}$O$_4$ (x = 0.1–0.5) powders by spray-drying method. *Electrochim. Acta* **2006**, *51*, 4148–4152. [CrossRef]

206. Li, D.; Ito, A.; Kobayakawa, K.; Noguchi, H.; Sato, Y. Electrochemical characteristics of LiNi$_{0.5}$Mn$_{1.5}$O$_4$ prepared by spray-drying and post-annealing. *Electrochim. Acta* **2007**, *52*, 1919–1924. [CrossRef]

207. He, S.; Zhang, Q.; Liu, W.; Fang, G.; Sato, Y.; Zheng, J.; Li, D. Influence of post-annealing in N$_2$ on structure and electrochemical characteristics of LiNi$_{0.5}$Mn$_{1.5}$O$_4$. *Chem. Res. Chin. Univ.* **2013**, *29*, 329–332. [CrossRef]

208. Risthaus, T.; Wang, J.; Friesen, A.; Wilken, A.; Berghus, D.; Winter, M.; Li, J. Synthesis of spinel LiNi$_{0.5}$Mn$_{1.5}$O$_4$ with secondary plate morphology as cathode material for lithium ion batteries. *J. Power Sources* **2015**, *293*, 137–142. [CrossRef]

209. Wu, H.M.; Tu, J.P.; Yuan, Y.F.; Li, Y.; Zhao, X.B.; Cao, G.S. Electrochemical and ex situ XRD studies of a LiMn$_{1.5}$Ni$_{0.5}$O$_4$ high-voltage cathode material. *Electrochim. Acta* **2005**, *50*, 4104–4108. [CrossRef]

210. Yang, W.; Dang, H.; Chen, S.; Zou, H.; Liu, Z.; Lin, J.; Lin, W. In Situ Carbon Coated LiNi$_{0.5}$Mn$_{1.5}$O$_4$ Cathode Material Prepared by Prepolymer of Melamine Formaldehyde Resin Assisted Method. *Int. J. Polym. Sci.* **2016**, *2016*, 1–5. [CrossRef]

211. Schroeder, M.; Glatthaar, S.; Geßwein, H.; Winkler, V.; Bruns, M.; Scherer, T.; Chakravadhanula, V.S.K.; Binder, J.R. Post-doping via spray-drying: A novel sol–gel process for the batch synthesis of doped LiNi$_{0.5}$Mn$_{1.5}$O$_4$ spinel material. *J. Mater. Sci.* **2013**, *48*, 3404–3414. [CrossRef]

212. Höweling, A.; Stoll, A.; Schmidt, D.O.; Geßwein, H.; Simon, U.; Binder, J.R. Influence of Synthesis, Dopants and Cycling Conditions on the Cycling Stability of Doped LiNi$_{0.5}$Mn$_{1.5}$O$_4$ Spinels. *J. Electrochem. Soc.* **2017**, *164*, A6349–A6358. [CrossRef]

213. Ito, A.; Li, D.; Lee, Y.; Kobayakawa, K.; Sato, Y. Influence of Co substitution for Ni and Mn on the structural and electrochemical characteristics of LiNi$_{0.5}$Mn$_{1.5}$O$_4$. *J. Power Sources* **2008**, *185*, 1429–1433. [CrossRef]

214. Li, D.; Ito, A.; Kobayakawa, K.; Noguchi, H.; Sato, Y. Structural and electrochemical characteristics of LiNi$_{0.5-x}$Co$_{2x}$Mn$_{1.5-x}$O$_4$ prepared by spray-drying process and post-annealing in O$_2$. *J. Power Sources* **2006**, *161*, 1241–1246. [CrossRef]

215. Wu, H.M.; Tu, J.P.; Yuan, Y.F.; Xiang, J.Y.; Chen, X.T.; Zhao, X.B.; Cao, G.S. Effects of abundant Co doping on the structure and electrochemical characteristics of LiMn$_{1.5}$Ni$_{0.5-x}$Co$_x$O$_4$. *J. Electroanal. Chem.* **2007**, *608*, 8–14. [CrossRef]

216. Alaboina, P.K.; Ge, Y.; Uddin, M.-J.; Liu, Y.; Lee, D.; Park, S.; Zhang, X.; Cho, S.-J. Nanoscale Porous Lithium Titanate Anode for Superior High Temperature Performance. *ACS Appl. Mater. Interfaces* **2016**, *8*, 12127–12133. [CrossRef] [PubMed]

217. Dai, C.; Ye, J.; Zhao, S.; He, P.; Zhou, H. Fabrication of High-Energy Li-Ion Cells with Li$_4$Ti$_5$O$_{12}$ Microspheres as Anode and 0.5 Li$_2$MnO$_3$ 0.5 LiNi$_{0.4}$Co$_{0.2}$Mn$_{0.4}$O$_2$ Microspheres as Cathode. *Chem. Asian J.* **2016**, *11*, 1273–1280. [CrossRef] [PubMed]

218. Deng, L.; Yang, W.-H.; Zhou, S.-X.; Chen, J.-T. Effect of carbon nanotubes addition on electrochemical performance and thermal stability of Li$_4$Ti$_5$O$_{12}$ anode in commercial LiMn$_2$O$_4$/Li$_4$Ti$_5$O$_{12}$ full-cell. *Chin. Chem. Lett.* **2015**, *26*, 1529–1534. [CrossRef]

219. Fleutot, B.; Davoisne, C.; Gachot, G.; Cavalaglio, S.; Grugeon, S.; Viallet, V. New chemical approach to obtain dense layer phosphate-based ionic conductor coating on negative electrode material surface: Synthesis way, outgassing and improvement of C-rate capability. *Appl. Surf. Sci.* **2017**, *400*, 139–147. [CrossRef]

220. Gao, J.; Jiang, C.; Wan, C. Influence of carbon additive on the properties of spherical Li$_4$Ti$_5$O$_{12}$ and LiFePO$_4$ materials for lithium-ion batteries. *Ionics* **2010**, *16*, 417–424. [CrossRef]

221. Han, S.-W.; Ryu, J.H.; Jeong, J.; Yoon, D.-H. Solid state synthesis of Li$_4$Ti$_5$O$_{12}$ for high power lithium ion battery applications. *J. Alloys Compd.* **2013**, *570*, 144–149. [CrossRef]

222. He, Z.; Wang, Z.; Wu, F.; Guo, H.; Li, X.; Xiong, X. Spherical Li$_4$Ti$_5$O$_{12}$ synthesized by spray-drying from a different kind of solution. *J. Alloys Compd.* **2012**, *540*, 39–45. [CrossRef]

223. Hsiao, K.-C.; Liao, S.-C.; Chen, J.-M. Microstructure effect on the electrochemical property of Li$_4$Ti$_5$O$_{12}$ as an anode material for lithium-ion batteries. *Electrochim. Acta* **2008**, *53*, 7242–7247. [CrossRef]

224. Hsieh, C.-T.; Chen, I.-L.; Jiang, Y.-R.; Lin, J.-Y. Synthesis of spinel lithium titanate anodes incorporated with rutile titania nanocrystallites by spray-drying followed by calcination. *Solid State Ion.* **2011**, *201*, 60–67. [CrossRef]

225. Hsieh, C.-T.; Lin, J.-Y. Influence of Li addition on charge/discharge behavior of spinel lithium titanate. *J. Alloys Compd.* **2010**, *506*, 231–236. [CrossRef]

226. Jung, H.-G.; Kim, J.; Scrosati, B.; Sun, Y.-K. Micron-sized, carbon-coated $Li_4Ti_5O_{12}$ as high power anode material for advanced lithium batteries. *J. Power Sources* **2011**, *196*, 7763–7766. [CrossRef]

227. Kadoma, Y.; Chiba, Y.; Yoshikawa, D.; Mitobe, Y.; Kumagai, N.; Ui, K. Influence of the Carbon Source on the Surface and Electrochemical Characteristics of Lithium Excess $Li_{4.3}Ti_5O_{12}$ Carbon Composite. *Electrochemistry* **2012**, *80*, 759–761. [CrossRef]

228. Lee, B.; Yoon, J.R. Synthesis of high-performance $Li_4Ti_5O_{12}$ and its application to the asymmetric hybrid capacitor. *Electron. Mater. Lett.* **2013**, *9*, 871–873. [CrossRef]

229. Li, C.; Li, G.; Wen, S.; Ren, R. Spray-drying synthesis and characterization of $Li_4Ti_5O_{12}$ anode material for lithium ion batteries. *J. Adv. Oxid. Technol.* **2017**, *20*. [CrossRef]

230. Liu, W.; Wang, Q.; Cao, C.; Han, X.; Zhang, J.; Xie, X.; Xia, B. Spray-drying of spherical $Li_4Ti_5O_{12}$/C powders using polyvinyl pyrrolidone as binder and carbon source. *J. Alloys Compd.* **2015**, *621*, 162–169. [CrossRef]

231. Wen, Z.; Gu, Z.; Huang, S.; Yang, J.; Lin, Z.; Yamamoto, O. Research on spray-dried lithium titanate as electrode materials for lithium ion batteries. *J. Power Sources* **2005**, *146*, 670–673. [CrossRef]

232. Lu, X.; Gu, L.; Hu, Y.-S.; Chiu, H.-C.; Li, H.; Demopoulos, G.P.; Chen, L. New Insight into the Atomic-Scale Bulk and Surface Structure Evolution of $Li_4Ti_5O_{12}$ Anode. *J. Am. Chem. Soc.* **2015**, *137*, 1581–1586. [CrossRef] [PubMed]

233. Nakahara, K.; Nakajima, R.; Matsushima, T.; Majima, H. Preparation of particulate $Li_4Ti_5O_{12}$ having excellent characteristics as an electrode active material for power storage cells. *J. Power Sources* **2003**, *117*, 131–136. [CrossRef]

234. Nowack, L.V.; Bunjaku, T.; Wegner, K.; Pratsinis, S.E.; Luisier, M.; Wood, V. Design and Fabrication of Microspheres with Hierarchical Internal Structure for Tuning Battery Performance. *Adv. Sci.* **2015**, *2*, 1500078. [CrossRef] [PubMed]

235. Ogihara, T.; Yamada, M.; Fujita, A.; Akao, S.; Myoujin, K. Effect of organic acid on the electrochemical properties of $Li_4Ti_5O_{12}$/C composite powders synthesized by spray pyrolysis. *Mater. Res. Bull.* **2011**, *46*, 796–800. [CrossRef]

236. Ren, J.; Ming, H.; Jia, Z.; Zhang, Y.; Ming, J.; Zhou, Q.; Zheng, J. High Tap Density $Li_4Ti_5O_{12}$ Microspheres: Synthetic Conditions and Advanced Electrochemical Performance. *Energy Technol.* **2017**, *5*, 1680–1686. [CrossRef]

237. Ruan, D.; Kim, M.-S.; Yang, B.; Qin, J.; Kim, K.-B.; Lee, S.-H.; Liu, Q.; Tan, L.; Qiao, Z. 700 F hybrid capacitors cells composed of activated carbon and $Li_4Ti_5O_{12}$ microspheres with ultra-long cycle life. *J. Power Sources* **2017**, *366*, 200–206. [CrossRef]

238. Wen, S.; Li, G.; Ren, R.; Li, C. Preparation of spherical $Li_4Ti_5O_{12}$ anode materials by spray-drying. *Mater. Lett.* **2015**, *148*, 130–133. [CrossRef]

239. Wu, F.; Li, X.; Wang, Z.; Guo, H.; He, Z.; Zhang, Q.; Xiong, X.; Yue, P. Low-temperature synthesis of nano-micron $Li_4Ti_5O_{12}$ by an aqueous mixing technique and its excellent electrochemical performance. *J. Power Sources* **2012**, *202*, 374–379. [CrossRef]

240. Wu, F.; Li, X.; Wang, Z.; Guo, H. Synthesis of chromium-doped lithium titanate microspheres as high-performance anode material for lithium ion batteries. *Ceram. Int.* **2014**, *40*, 13195–13204. [CrossRef]

241. Xu, G.; Quan, X.; Gao, H.; Li, J.; Cai, Y.; Cheng, X.; Guo, L. Facile spray-drying route for large scale nitrogen-doped carbon-coated $Li_4Ti_5O_{12}$ anode material in lithium-ion batteries. *Solid State Ion.* **2017**, *304*, 40–45. [CrossRef]

242. Yoshikawa, D.; Suzuki, N.; Kadoma, Y.; Ui, K.; Kumagai, N. Li excess $Li_{4+x}Ti_{5-x}O_{12-\delta}$/C composite using spray-drying method and its electrode properties. *Funct. Mater. Lett.* **2012**, *5*, 1250001. [CrossRef]

243. Yoshikawa, D.; Kadoma, Y.; Kim, J.-M.; Ui, K.; Kumagai, N.; Kitamura, N.; Idemoto, Y. Spray-drying synthesized lithium-excess $Li_{4+x}Ti_{5-x}O_{12-\delta}$ and its electrochemical property as negative electrode material for Li-ion batteries. *Electrochim. Acta* **2010**, *55*, 1872–1879. [CrossRef]

244. Yuan, T.; Li, W.-T.; Zhang, W.; He, Y.-S.; Zhang, C.; Liao, X.-Z.; Ma, Z.-F. One-Pot Spray-Dried Graphene Sheets-Encapsulated Nano-$Li_4Ti_5O_{12}$ Microspheres for a Hybrid BatCap System. *Ind. Eng. Chem. Res.* **2014**, *53*, 10849–10857. [CrossRef]

245. Zhang, Q.; Peng, W.; Wang, Z.; Li, X.; Xiong, X.; Guo, H.; Wang, Z.; Wu, F. $Li_4Ti_5O_{12}$/Reduced Graphene Oxide composite as a high rate capability material for lithium ion batteries. *Solid State Ion.* **2013**, *236*, 30–36. [CrossRef]

246. Zheng, X.; Dong, L.; Dong, C. Easy synthesis of $Li_4Ti_5O_{12}$/C microspheres containing nanoparticles as anode material for high-rate lithium batteries. *Surf. Rev. Lett.* **2014**, *21*, 1450023. [CrossRef]

247. Zhu, G.-N.; Liu, H.-J.; Zhuang, J.-H.; Wang, C.-X.; Wang, Y.-G.; Xia, Y.-Y. Carbon-coated nano-sized $Li_4Ti_5O_{12}$ nanoporous micro-sphere as anode material for high-rate lithium-ion batteries. *Energy Environ. Sci.* **2011**, *4*, 4016. [CrossRef]

248. Zhu, W.; Zhuang, Z.; Yang, Y.; Zhang, R.; Lin, Z.; Lin, Y.; Huang, Z. Synthesis and electrochemical performance of hole-rich $Li_4Ti_5O_{12}$ anode material for lithium-ion secondary batteries. *J. Phys. Chem. Solids* **2016**, *93*, 52–58. [CrossRef]

249. Wu, F.; Wang, Z.; Li, X.; Guo, H.; Yue, P.; Xiong, X.; He, Z.; Zhang, Q. Characterization of spherical-shaped $Li_4Ti_5O_{12}$ prepared by spray-drying. *Electrochim. Acta* **2012**, *78*, 331–339. [CrossRef]

250. Dong, G.-H.; Liu, H.-J.; Zhou, L.; Chong, L.; Yang, J.; Qiao, Y.-M.; Zhang, D.-H. Investigation of various synthetic conditions for large-scale synthesis and electrochemical properties of $Li_{3.98}Al_{0.06}Ti_{4.96}O_{12}$/C as anode material. *J. Alloys Compd.* **2014**, *615*, 817–824. [CrossRef]

251. Kumagai, N.; Yoshikawa, D.; Kadoma, Y.; Ui, K. Spray-Drying Synthesized Lithium-excess $Li_{4+x}Ti_{4.95-x}Nb_{0.05}O_{12-d}$ and its Electrochemical Property as Negative Electrode Material for Li-ion Batteries. *Electrochemistry* **2010**, *78*, 754–756. [CrossRef]

252. Ng, S.-H.; Tran, N.; Bramnik, K.G.; Hibst, H.; Novák, P. A Feasibility Study on the Use of $Li_4V_3O_8$ as a High Capacity Cathode Material for Lithium-Ion Batteries. *Chem. Eur. J.* **2008**, *14*, 11141–11148. [CrossRef] [PubMed]

253. West, K. Comparison of LiV_3O_8 Cathode Materials Prepared by Different Methods. *J. Electrochem. Soc.* **1996**, *143*, 820. [CrossRef]

254. Tran, N.; Bramnik, K.G.; Hibst, H.; Prölß, J.; Mronga, N.; Holzapfel, M.; Scheifele, W.; Novák, P. Spray-Drying Synthesis and Electrochemical Performance of Lithium Vanadates as Positive Electrode Materials for Lithium Batteries. *J. Electrochem. Soc.* **2008**, *155*, A384. [CrossRef]

255. Xiong, X.; Wang, Z.; Guo, H.; Li, X.; Wu, F.; Yue, P. High performance LiV_3O_8 cathode materials prepared by spray-drying method. *Electrochim. Acta* **2012**, *71*, 206–212. [CrossRef]

256. Xiong, X.; Wang, Z.; Li, X.; Guo, H. Study on ultrafast synthesis of LiV_3O_8 cathode material for lithium-ion batteries. *Mater. Lett.* **2012**, *76*, 8–10. [CrossRef]

257. Gao, J.; Jiang, C.; Wan, C. Preparation and characterization of spherical $Li_{1+x}V_3O_8$ cathode material for lithium secondary batteries. *J. Power Sources* **2004**, *125*, 90–94. [CrossRef]

258. Yang, Y.; Li, J.; Chen, D.; Zhao, J. Spray-drying-Assisted Synthesis of Li_3VO_4/C/CNTs Composites for High-Performance Lithium Ion Battery Anodes. *J. Electrochem. Soc.* **2017**, *164*, A6001–A6006. [CrossRef]

259. Yang, Y.; Li, J.; He, X.; Wang, J.; Sun, D.; Zhao, J. A facile spray-drying route for mesoporous Li_3VO_4/C hollow spheres as an anode for long life lithium ion batteries. *J. Mater. Chem. A* **2016**, *4*, 7165–7168. [CrossRef]

260. Yang, Y.; Li, J.; Huang, J.; Huang, J.; Zeng, J.; Zhao, J. Polystyrene-template-assisted synthesis of Li_3VO_4/C/rGO ternary composite with honeycomb-like structure for durable high-rate lithium ion battery anode materials. *Electrochim. Acta* **2017**, *247*, 771–778. [CrossRef]

261. Zeng, J.; Yang, Y.; Li, C.; Li, J.; Huang, J.; Wang, J.; Zhao, J. Li_3VO_4: An insertion anode material for magnesium ion batteries with high specific capacity. *Electrochim. Acta* **2017**, *247*, 265–270. [CrossRef]

262. Jiang, Y.P.; Xie, J.; Cao, G.S.; Zhao, X.B. Electrochemical performance of $Li_4Mn_5O_{12}$ nano-crystallites prepared by spray-drying-assisted solid state reactions. *Electrochim. Acta* **2010**, *56*, 412–417. [CrossRef]

263. Wang, H.; Yang, B.; Liao, X.-Z.; Xu, J.; Yang, D.; He, Y.-S.; Ma, Z.-F. Electrochemical properties of P2-$Na_{2/3}[Ni_{1/3}Mn_{2/3}]O_2$ cathode material for sodium ion batteries when cycled in different voltage ranges. *Electrochim. Acta* **2013**, *113*, 200–204. [CrossRef]

264. Zhao, W.; Yamamoto, S.; Tanaka, A.; Noguchi, H. Synthesis of Li-excess layered cathode material with enhanced reversible capacity for Lithium ion batteries through the optimization of precursor synthesis method. *Electrochim. Acta* **2014**, *143*, 347–356. [CrossRef]

265. Zou, W.; Li, J.; Deng, Q.; Xue, J.; Dai, X.; Zhou, A.; Li, J. Microspherical $Na_2Ti_3O_7$ prepared by spray-drying method as anode material for sodium-ion battery. *Solid State Ion.* **2014**, *262*, 192–196. [CrossRef]

266. Yin, F.; Liu, Z.; Yang, S.; Shan, Z.; Zhao, Y.; Feng, Y.; Zhang, C.; Bakenov, Z. Na$_4$Mn$_9$O$_{18}$/Carbon Nanotube Composite as a High Electrochemical Performance Material for Aqueous Sodium-Ion Batteries. *Nanoscale Res. Lett.* **2017**, *12*, 569. [CrossRef] [PubMed]

267. Yin, F.; Liu, Z.; Zhao, Y.; Feng, Y.; Zhang, Y. Electrochemical Properties of an Na$_4$Mn$_9$O$_{18}$-Reduced Graphene Oxide Composite Synthesized via Spray-drying for an Aqueous Sodium-Ion Battery. *Nanomaterials* **2017**, *7*, 253. [CrossRef] [PubMed]

268. Yang, F.; Zhang, H.; Shao, Y.; Song, H.; Liao, S.; Ren, J. Formic acid as additive for the preparation of high-performance FePO$_4$ materials by spray-drying method. *Ceram. Int.* **2017**, *43*, 16652–16658. [CrossRef]

269. Yang, X.; Zhang, S.M.; Zhang, J.X. Synthesis and Modification of Iron-based Cathode Materials: Iron Phosphate for Lithium Secondary Batteries. *Arab. J. Sci. Eng.* **2014**, *39*, 6687–6691. [CrossRef]

270. Yang, X.; Zhang, J.X.; Zhang, S.M.; Yan, L.C.; Mei, Y.; Geng, G. Preparation of Spherical FePO$_4$ Cathode Material for Lithium Ion Batteries. *Adv. Mater. Res.* **2012**, *347*, 576–581.

271. Cao, F.; Pan, G.X.; Zhang, Y.J. Construction of ultrathin N-doped carbon shell on LiFePO$_4$ spheres as enhanced cathode for lithium ion batteries. *Mater. Res. Bull.* **2017**, *96*, 325–329. [CrossRef]

272. Chen, L.; Lu, C.; Chen, Q.A.; Gu, Y.J.; Wang, M.; Chen, Y.B. Preparation and Characterization of Nano-LiFePO$_4$/C Using Two-Fluid Spray-dryer. *Appl. Mech. Mater.* **2014**, *563*, 62–65. [CrossRef]

273. Chen, Z.; Zhao, Q.; Xu, M.; Li, L.; Duan, J.; Zhu, H. Electrochemical properties of self-assembled porous micro-spherical LiFePO$_4$/PAS composite prepared by spray-drying method. *Electrochim. Acta* **2015**, *186*, 117–124. [CrossRef]

274. Gao, F.; Tang, Z. Kinetic behavior of LiFePO$_4$/C cathode material for lithium-ion batteries. *Electrochim. Acta* **2008**, *53*, 5071–5075. [CrossRef]

275. Gao, F.; Tang, Z.; Xue, J. Preparation and characterization of nano-particle LiFePO$_4$ and LiFePO$_4$/C by spray-drying and post-annealing method. *Electrochim. Acta* **2007**, *53*, 1939–1944. [CrossRef]

276. Gu, Y.J.; Hao, F.X.; Chen, Y.B.; Liu, H.Q.; Wang, Y.M.; Liu, P.; Zhang, Q.G.; Li, S.Q. Electrochemical Properties of LiFePO$_4$/C Composite by Spray-Drying Method. *Adv. Mater. Res.* **2013**, *643*, 96–99. [CrossRef]

277. Gu, Y.; Zhang, X.; Lu, S.; Jiang, D.; Wu, A. High rate performance of LiF modified LiFePO$_4$/C cathode material. *Solid State Ion.* **2015**, *269*, 30–36. [CrossRef]

278. Guan, X.; Li, G.; Li, C.; Ren, R. Synthesis of porous nano/micro structured LiFePO$_4$/C cathode materials for lithium-ion batteries by spray-drying method. *Trans. Nonferrous Met. Soc. China* **2017**, *27*, 141–147. [CrossRef]

279. Huang, B.; Zheng, X.; Jia, D.; Lu, M. Design and synthesis of high-rate micron-sized, spherical LiFePO$_4$/C composites containing clusters of nano/microspheres. *Electrochim. Acta* **2010**, *55*, 1227–1231. [CrossRef]

280. Huang, B.; Zheng, X.; Fan, X.; Song, G.; Lu, M. Enhanced rate performance of nano–micro structured LiFePO$_4$/C by improved process for high-power Li-ion batteries. *Electrochim. Acta* **2011**, *56*, 4865–4868. [CrossRef]

281. Kim, J.-K. Supercritical synthesis in combination with a spray process for 3D porous microsphere lithium iron phosphate. *CrystEngComm* **2014**, *16*, 2818–2822. [CrossRef]

282. Kim, M.-S.; Lee, G.-W.; Lee, S.-W.; Jeong, J.H.; Mhamane, D.; Roh, K.C.; Kim, K.-B. Synthesis of LiFePO$_4$/graphene microspheres while avoiding restacking of graphene sheet's for high-rate lithium-ion batteries. *J. Ind. Eng. Chem.* **2017**, *52*, 251–259. [CrossRef]

283. Liu, H.; Liu, Y.; An, L.; Zhao, X.; Wang, L.; Liang, G. High Energy Density LiFePO$_4$/C Cathode Material Synthesized by Wet Ball Milling Combined with Spray-drying Method. *J. Electrochem. Soc.* **2017**, *164*, A3666–A3672. [CrossRef]

284. Liu, J.; Wang, J.; Yan, X.; Zhang, X.; Yang, G.; Jalbout, A.F.; Wang, R. Long-term cyclability of LiFePO$_4$/carbon composite cathode material for lithium-ion battery applications. *Electrochim. Acta* **2009**, *54*, 5656–5659. [CrossRef]

285. Liu, Q.-B.; Liao, S.-J.; Song, H.-Y.; Liang, Z.-X. High-performance LiFePO$_4$/C materials: Effect of carbon source on microstructure and performance. *J. Power Sources* **2012**, *211*, 52–58. [CrossRef]

286. Liu, Q.; Liao, S.; Song, H.; Zeng, J. LiFePO$_4$/C Microspheres with Nano-micro Structure, Prepared by Spray-drying Method Assisted with PVA as Template. *Curr. Nanosci.* **2012**, *8*, 208–214. [CrossRef]

287. Lu, C.; Chen, L.; Chen, Y.B.; Gu, Y.J.; Wang, M.; Zuo, L.L.; Liu, H.Q.; Wang, Y.M.; Sun, X.F. Effects of Different Granularity Control Methods on Morphology, Structure and Electrochemical Performance of LiFePO$_4$/C. *Adv. Mater. Res.* **2014**, *893*, 830–833. [CrossRef]

288. Lu, C.; Chen, L.; Chen, Y.B.; Gu, Y.J.; Wang, M.; Zuo, L.L.; Zhang, Z.; Chen, Q.A.; Liu, H.Q.; Wang, Y.M. Effects of Different Organic Carbon Sources on Properties of LiFePO$_4$/C Synthesized by Spray-Drying. *Appl. Mech. Mater.* **2014**, *535*, 725–728. [CrossRef]

289. Luo, W.; Wen, L.; Luo, H.; Song, R.; Zhai, Y.; Liu, C.; Li, F. Carbon nanotube-modified LiFePO$_4$ for high rate lithium ion batteries. *New Carbon Mater.* **2014**, *29*, 287–294. [CrossRef]

290. Lv, Y.-J.; Su, J.; Long, Y.-F.; Lv, X.-Y.; Wen, Y.-X. Effect of milling time on the performance of bowl-like LiFePO$_4$/C prepared by wet milling-assisted spray-drying. *Ionics* **2014**, *20*, 471–478. [CrossRef]

291. Lv, Y.-J.; Long, Y.-F.; Su, J.; Lv, X.-Y.; Wen, Y.-X. Synthesis of bowl-like mesoporous LiFePO$_4$/C composites as cathode materials for lithium ion batteries. *Electrochim. Acta* **2014**, *119*, 155–163. [CrossRef]

292. Mei, R.; Yang, Y.; Song, X.; An, Z.; Zhang, J. Triple carbon coated LiFePO$_4$ composite with hierarchical conductive architecture as high-performance cathode for Li-ion batteries. *Electrochim. Acta* **2015**, *153*, 523–530. [CrossRef]

293. Ni, L.; Zheng, J.; Qin, C.; Lu, Y.; Liu, P.; Wu, T.; Tang, Y.; Chen, Y. Fabrication and characteristics of spherical hierarchical LiFePO$_4$/C cathode material by a facile method. *Electrochim. Acta* **2014**, *147*, 330–336. [CrossRef]

294. Ren, J.; Pu, W.; He, X.; Jiang, C.; Wan, C. A carbon-LiFePO$_4$ nanocomposite as high-performance cathode material for lithium-ion batteries. *Ionics* **2011**, *17*, 581–586. [CrossRef]

295. Wu, L.; Zhong, S.-K.; Liu, J.-Q.; Lv, F.; Wan, K. High tap-density and high performance LiFePO$_4$/C cathode material synthesized by the combined sol spray-drying and liquid nitrogen quenching method. *Mater. Lett.* **2012**, *89*, 32–35. [CrossRef]

296. Yang, C.-C.; Hsu, Y.-H.; Shih, J.-Y.; Wu, Y.-S.; Karuppiah, C.; Liou, T.-H.; Lue, S.J. Preparation of 3D micro/mesoporous LiFePO$_4$ composite wrapping with porous graphene oxide for high-power lithium ion battery. *Electrochim. Acta* **2017**, *258*, 773–785. [CrossRef]

297. Yang, C.-C.; Jang, J.-H.; Jiang, J.-R. Comparison Electrochemical Performances of Spherical LiFePO$_4$/C Cathode Materials at Low and High Temperatures. *Energy Procedia* **2014**, *61*, 1402–1409. [CrossRef]

298. Yang, C.-C.; Jang, J.-H.; Jiang, J.-R. Preparation of carbon and oxide co-modified LiFePO$_4$ cathode material for high performance lithium-ion battery. *Mater. Chem. Phys.* **2015**, *165*, 196–206. [CrossRef]

299. Yang, X.; Tu, J.; Lei, M.; Zuo, Z.; Wu, B.; Zhou, H. Selection of Carbon Sources for Enhancing 3D Conductivity in the Secondary Structure of LiFePO$_4$/C Cathode. *Electrochim. Acta* **2016**, *193*, 206–215. [CrossRef]

300. Yu, F.; Zhang, J.-J.; Yang, Y.-F.; Song, G.-Z. Up-scalable synthesis, structure and charge storage properties of porous microspheres of LiFePO$_4$@C nanocomposites. *J. Mater. Chem.* **2009**, *19*, 9121. [CrossRef]

301. Yu, F.; Zhang, J.; Yang, Y.; Song, G. Preparation and characterization of mesoporous LiFePO$_4$/C microsphere by spray-drying assisted template method. *J. Power Sources* **2009**, *189*, 794–797. [CrossRef]

302. Yu, F.; Zhang, J.; Yang, Y.; Song, G. Reaction mechanism and electrochemical performance of LiFePO$_4$/C cathode materials synthesized by carbothermal method. *Electrochim. Acta* **2009**, *54*, 7389–7395. [CrossRef]

303. Yu, F.; Zhang, J.; Yang, Y.; Song, G. Porous micro-spherical aggregates of LiFePO$_4$/C nanocomposites: A novel and simple template-free concept and synthesis via sol-gel-spray-drying method. *J. Power Sources* **2010**, *195*, 6873–6878. [CrossRef]

304. Zhou, X.; Wang, F.; Zhu, Y.; Liu, Z. Graphene modified LiFePO$_4$ cathode materials for high power lithium ion batteries. *J. Mater. Chem.* **2011**, *21*, 3353. [CrossRef]

305. Sun, X.; Zhang, L. Outstanding Li-storage performance of LiFePO$_4$@MWCNTs cathode material with 3D network structure for lithium-ion batteries. *J. Phys. Chem. Solids* **2018**, *116*, 216–221. [CrossRef]

306. Wang, B.; Wang, Y.; Wu, H.; Yao, L.; Yang, L.; Li, J.; Xiang, M.; Zhang, Y.; Liu, H. Ultrafast and Durable Lithium Storage Enabled by Porous Bowl-Like LiFePO$_4$/C Composite with Na + Doping. *ChemElectroChem* **2017**, *4*, 1141–1147. [CrossRef]

307. Zou, B.; Wang, Y.; Zhou, S. Spray-drying-assisted synthesis of LiFePO$_4$/C composite microspheres with high performance for lithium-ion batteries. *Mater. Lett.* **2013**, *92*, 300–303. [CrossRef]

308. Tu, J.; Wu, K.; Tang, H.; Zhou, H.; Jiao, S. Mg–Ti co-doping behavior of porous LiFePO$_4$ microspheres for high-rate lithium-ion batteries. *J. Mater. Chem. A* **2017**, *5*, 17021–17028. [CrossRef]

309. Yang, C.-C.; Jang, J.-H.; Jiang, J.-R. Study of electrochemical performances of lithium titanium oxide-coated LiFePO$_4$/C cathode composite at low and high temperatures. *Appl. Energy* **2016**, *162*, 1419–1427. [CrossRef]

310. Kim, M.-S.; Kim, H.-K.; Lee, S.-W.; Kim, D.-H.; Ruan, D.; Chung, K.Y.; Lee, S.H.; Roh, K.C.; Kim, K.-B. Synthesis of Reduced Graphene Oxide-Modified LiMn$_{0.75}$Fe$_{0.25}$PO$_4$ Microspheres by Salt-Assisted Spray-drying for High-Performance Lithium-Ion Batteries. *Sci. Rep.* **2016**, *6*. [CrossRef] [PubMed]

311. Li, C.; Li, G.; Guan, X. Synthesis and electrochemical performance of micro-nano structured LiFe$_{1-x}$Mn$_x$PO$_4$/C (0 ≤ x ≤ 0.05) cathode for lithium-ion batteries. *J. Energy Chem.* **2017**. [CrossRef]

312. Li, J.; Wang, Y.; Wu, J.; Zhao, H.; Wu, H.; Zhang, Y.; Liu, H. Preparation of Enhanced-Performance LiMn$_{0.6}$Fe$_{0.4}$PO$_4$/C Cathode Material for Lithium-Ion Batteries by using a Divalent Transition-Metal Phosphate as an Intermediate. *ChemElectroChem* **2017**, *4*, 175–182. [CrossRef]

313. Li, J.; Wang, Y.; Wu, J.; Zhao, H.; Liu, H. CNT-embedded LiMn$_{0.8}$Fe$_{0.2}$PO$_4$/C microsphere cathode with high rate capability and cycling stability for lithium ion batteries. *J. Alloys Compd.* **2018**, *731*, 864–872. [CrossRef]

314. Li, J.; Xiang, M.; Wang, Y.; Wu, J.; Zhao, H.; Liu, H. Effects of adhesives on the electrochemical performance of monodisperse LiMn$_{0.8}$Fe$_{0.2}$PO$_4$/C microspheres as cathode materials for high power lithium-ion batteries. *J. Mater. Chem. A* **2017**, *5*, 7952–7960. [CrossRef]

315. Liu, W.; Gao, P.; Mi, Y.; Chen, J.; Zhou, H.; Zhang, X. Fabrication of high tap density LiFe$_{0.6}$Mn$_{0.4}$PO$_4$/C microspheres by a double carbon coating–spray-drying method for high rate lithium ion batteries. *J Mater Chem A* **2013**, *1*, 2411–2417. [CrossRef]

316. Mi, Y.; Gao, P.; Liu, W.; Zhang, W.; Zhou, H. Carbon nanotube-loaded mesoporous LiFe$_{0.6}$Mn$_{0.4}$PO$_4$/C microspheres as high performance cathodes for lithium-ion batteries. *J. Power Sources* **2014**, *267*, 459–468. [CrossRef]

317. Xu, S.; Lv, X.-Y.; Wu, Z.; Long, Y.-F.; Su, J.; Wen, Y.-X. Synthesis of porous-hollow LiMn$_{0.85}$Fe$_{0.15}$PO$_4$/C microspheres as a cathode material for lithium-ion batteries. *Powder Technol.* **2017**, *308*, 94–100. [CrossRef]

318. Yang, C.-C.; Chen, W.-H. Microsphere LiFe$_{0.5}$Mn$_{0.5}$PO$_4$/C composite as high rate and long-life cathode material for lithium-ion battery. *Mater. Chem. Phys.* **2016**, *173*, 482–490. [CrossRef]

319. Yang, L.; Wang, Y.; Wu, J.; Xiang, M.; Li, J.; Wang, B.; Zhang, Y.; Wu, H.; Liu, H. Facile synthesis of micro-spherical LiMn$_{0.7}$Fe$_{0.3}$PO$_4$/C cathodes with advanced cycle life and rate performance for lithium-ion battery. *Ceram. Int.* **2017**, *43*, 4821–4830. [CrossRef]

320. Lei, Z.; Wang, J.; Yang, J.; Nuli, Y.; Ma, Z. Nano/micro-hierarchical-structured LiMn$_{0.85}$Fe$_{0.15}$PO$_4$ cathode material for advanced lithium ion battery. *ACS Appl. Mater. Interfaces* **2017**. [CrossRef] [PubMed]

321. Jiang, Y.; Liu, R.; Xu, W.; Jiao, Z.; Wu, M.; Chu, Y.; Su, L.; Cao, H.; Hou, M.; Zhao, B. A novel graphene modified LiMnPO$_4$ as a performance-improved cathode material for lithium-ion batteries. *J. Mater. Res.* **2013**, *28*, 2584–2589. [CrossRef]

322. Zhang, Y.J.; Wang, X.Y.; Gao, Y. The Synthesis and SEM Characterization of Spherical LiMnPO$_4$/C Composite Prepared by Spray-drying. *Adv. Mater. Res.* **2013**, *631*, 472–475.

323. Huang, Q.-Y.; Wu, Z.; Su, J.; Long, Y.-F.; Lv, X.-Y.; Wen, Y.-X. Synthesis and electrochemical performance of Ti-Fe co-doped LiMnPO$_4$/C as cathode material for lithium-ion batteries. *Ceram. Int.* **2016**, *42*, 11348–11354. [CrossRef]

324. Zheng, J.; Han, Y.; Zhang, B.; Shen, C.; Ming, L.; Zhang, J. Comparative investigation of microporous and nanosheet LiVOPO$_4$ as cathode materials for lithium-ion batteries. *RSC Adv.* **2014**, *4*, 41076–41080. [CrossRef]

325. Hu, Y.; Ma, X.; Guo, P.; Jaeger, F.; Wang, Z. 3D graphene-encapsulated Li$_3$V$_2$(PO$_4$)$_3$ microspheres as a high-performance cathode material for energy storage. *J. Alloys Compd.* **2017**, *723*, 873–879. [CrossRef]

326. Huang, B.; Fan, X.; Zheng, X.; Lu, M. Synthesis and rate performance of lithium vanadium phosphate as cathode material for Li-ion batteries. *J. Alloys Compd.* **2011**, *509*, 4765–4768. [CrossRef]

327. Jiang, Y.; Xu, W.; Chen, D.; Jiao, Z.; Zhang, H.; Ma, Q.; Cai, X.; Zhao, B.; Chu, Y. Graphene modified Li$_3$V$_2$(PO$_4$)$_3$ as a high-performance cathode material for lithium ion batteries. *Electrochim. Acta* **2012**, *85*, 377–383. [CrossRef]

328. Liu, Q.; Ren, L.; Cong, C.; Ding, F.; Guo, F.; Song, D.; Guo, J.; Shi, X.; Zhang, L. Study on Li$_3$V$_2$(PO$_4$)$_3$/C cathode materials prepared using pitch as a new carbon source by different approaches. *Electrochim. Acta* **2016**, *187*, 264–276. [CrossRef]

329. Wang, X.; Dong, S.; Wang, H. Three-dimensional CNTs wrapped Li$_3$V$_2$(PO$_4$)$_3$ microspheres cathode with high-rate capability and cycling stability for Li-ion batteries. *Solid State Ion.* **2017**, *309*, 146–151. [CrossRef]

330. Wu, L.; Zhong, S.; Lu, J.; Lv, F.; Liu, J. Li$_3$V$_2$(PO$_4$)$_3$/C microspheres with high tap density and high performance synthesized by a two-step ball milling combined with the spray-drying method. *Mater. Lett.* **2014**, *115*, 60–63. [CrossRef]

331. Yu, F.; Zhang, J.; Yang, Y.; Song, G. Preparation and electrochemical performance of Li$_3$V$_2$(PO$_4$)$_3$/C cathode material by spray-drying and carbothermal method. *J. Solid State Electrochem.* **2010**, *14*, 883–888. [CrossRef]

332. Zhang, B.; Zheng, J. Synthesis of $Li_3V_2(PO_4)_3$/C with high tap-density and high-rate performance by spray-drying and liquid nitrogen quenching method. *Electrochim. Acta* **2012**, *67*, 55–61. [CrossRef]

333. Zhang, L.-L.; Peng, G.; Liang, G.; Zhang, P.-C.; Wang, Z.-H.; Jiang, Y.; Huang, Y.-H.; Lin, H. Controllable synthesis of spherical $Li_3V_2(PO_4)_3$/C cathode material and its electrochemical performance. *Electrochim. Acta* **2013**, *90*, 433–439. [CrossRef]

334. Zhang, X.; Guo, H.; Li, X.; Wang, Z.; Wu, L. High tap-density $Li_3V_2(PO_4)_3$/C composite material synthesized by sol spray-drying and post-calcining method. *Electrochim. Acta* **2012**, *64*, 65–70. [CrossRef]

335. Zuo, Z.L.; Wang, J.; Deng, J.Q.; Yao, Q.R.; Wang, Z.M.; Zhou, H.Y. Electrochemical Performance of Spherical $Li_3V_2(PO_4)_3$/C Synthesized by Spray-drying Method. *Key Eng. Mater.* **2017**, *727*, 738–743. [CrossRef]

336. Yang, G.; Jiang, C.Y.; He, X.M.; Ying, J.R.; Gao, J. Preparation of $Li_3V_2(PO_4)_3$/$LiFePO_4$ composite cathode material for lithium ion batteries. *Ionics* **2013**, *19*, 1247–1253. [CrossRef]

337. Kee, Y.; Dimov, N.; Kobayashi, E.; Kitajou, A.; Okada, S. Structural and electrochemical properties of Fe- and Al-doped $Li_3V_2(PO_4)_3$ for all-solid state symmetric lithium ion batteries prepared by spray-drying-assisted carbothermal method. *Solid State Ion.* **2015**, *272*, 138–143. [CrossRef]

338. Yang, B.; Li, X.; Guo, H.; Wang, Z.; Xiao, W. Preparation and properties of $Li_{1.3}Al_{0.3}Ti_{1.7}(PO_4)_3$ by spray-drying and post-calcining method. *J. Alloys Compd.* **2015**, *643*, 181–185. [CrossRef]

339. Bian, M.; Tian, L. Design and synthesis of three-dimensional $NaTi_2(PO_4)_3$@CNT microspheres as advanced anode materials for rechargeable sodium-ion batteries. *Ceram. Int.* **2017**, *43*, 9543–9546. [CrossRef]

340. Fang, Y.; Xiao, L.; Qian, J.; Cao, Y.; Ai, X.; Huang, Y.; Yang, H. 3D Graphene Decorated $NaTi_2(PO_4)_3$ Microspheres as a Superior High-Rate and Ultracycle-Stable Anode Material for Sodium Ion Batteries. *Adv. Energy Mater.* **2016**, *6*, 1502197. [CrossRef]

341. Huang, C.; Zuo, Z.; Deng, J.; Yao, Q.; Wang, Z.; Zhou, H. Electrochemical Properties of Hollow Spherical $Na_3V_2(PO_4)_3$/C Cathode Materials for Sodium-ion Batteries. *Int. J. Electrochem. Sci.* **2017**, *12*, 9456–9464. [CrossRef]

342. Chen, H.; Zhang, B.; Wang, X.; Dong, P.; Tong, H.; Zheng, J.; Yu, W.; Zhang, J. CNT-Decorated $Na_3V_2(PO_4)_3$ Microspheres as a High-Rate and Cycle-Stable Cathode Material for Sodium Ion Batteries. *ACS Appl. Mater. Interfaces* **2018**, *10*, 3590–3595. [CrossRef] [PubMed]

343. Zeng, J.; Yang, Y.; Lai, S.; Huang, J.; Zhang, Y.; Wang, J.; Zhao, J. A Promising High-Voltage Cathode Material Based on Mesoporous $Na_3V_2(PO_4)_3$/C for Rechargeable Magnesium Batteries. *Chem. Eur. J.* **2017**, *23*, 16898–16905. [CrossRef] [PubMed]

344. Zhang, J.; Fang, Y.; Xiao, L.; Qian, J.; Cao, Y.; Ai, X.; Yang, H. Graphene-Scaffolded $Na_3V_2(PO_4)_3$ Microsphere Cathode with High Rate Capability and Cycling Stability for Sodium Ion Batteries. *ACS Appl. Mater. Interfaces* **2017**, *9*, 7177–7184. [CrossRef] [PubMed]

345. Zheng, W.; Huang, X.; Ren, Y.; Wang, H.; Zhou, S.; Chen, Y.; Ding, X.; Zhou, T. Porous spherical $Na_3V_2(PO_4)_3$/C composites synthesized via a spray-drying -assisted process with high-rate performance as cathode materials for sodium-ion batteries. *Solid State Ion.* **2017**, *308*, 161–166. [CrossRef]

346. Zhang, D.; Feng, P.; Xu, B.; Li, Z.; Qiao, J.; Zhou, J.; Chang, C. High Rate Performance of $Na_3V_{2-x}Cu_x(PO_4)_3$/C Cathodes for Sodium Ion Batteries. *J. Electrochem. Soc.* **2017**, *164*, A3563–A3569. [CrossRef]

347. Cao, J.; Ni, L.; Qin, C.; Tang, Y.; Chen, Y. Synthesis of hierarchical $Na_2FeP_2O_7$ spheres with high electrochemical performance via spray-drying. *Ionics* **2017**, *23*, 1783–1791. [CrossRef]

348. Wu, T.; Dai, G.; Qin, C.; Cao, J.; Tang, Y.; Chen, Y. A novel method to synthesize SnP_2O_7 spherical particles for lithium-ion battery anode. *Ionics* **2016**, *22*, 2315–2319. [CrossRef]

349. Huang, X.; You, Y.; Ren, Y.; Wang, H.; Chen, Y.; Ding, X.; Liu, B.; Zhou, S.; Chu, F. Spray-drying-assisted synthesis of hollow spherical Li_2FeSiO_4/C particles with high performance for Li-ion batteries. *Solid State Ion.* **2015**, *278*, 203–208. [CrossRef]

350. Zhang, Z.; Liu, X.; Wang, L.; Wu, X.; Zhao, H.; Chen, B.; Xiong, W. Fabrication and characterization of carbon-coated Li_2FeSiO_4 nanoparticles reinforced by carbon nanotubes as high performance cathode materials for lithium-ion batteries. *Electrochim. Acta* **2015**, *168*, 8–15. [CrossRef]

351. Ren, Y.; Lu, P.; Huang, X.; Ding, J.; Wang, H.; Zhou, S.; Chen, Y.; Ding, X. Spherical $Li_{1.95}Na_{0.05}FeSiO_4$/C composite as nanoporous cathode material exhibiting high rate capability. *Mater. Lett.* **2016**, *173*, 207–210. [CrossRef]

352. Zhang, Z.; Liu, X.; Wu, Y.; Zhao, H.; Chen, B.; Xiong, W. Synthesis and Characterization of Spherical $Li_2Fe_{0.5}V_{0.5}SiO_4$/C Composite for High-Performance Cathode Material of Lithium-Ion Secondary Batteries. *J. Electrochem. Soc.* **2015**, *162*, A737–A742. [CrossRef]

353. Kalluri, S.; Seng, K.H.; Guo, Z.; Du, A.; Konstantinov, K.; Liu, H.K.; Dou, S.X. Sodium and Lithium Storage Properties of Spray-Dried Molybdenum Disulfide-Graphene Hierarchical Microspheres. *Sci. Rep.* **2015**, *5*. [CrossRef] [PubMed]

354. Park, G.D.; Kim, J.H.; Kang, Y.C. Large-scale production of spherical $FeSe_2$ -amorphous carbon composite powders as anode materials for sodium-ion batteries. *Mater. Charact.* **2016**, *120*, 349–356. [CrossRef]

355. Park, G.D.; Kang, Y.C. Design and Synthesis of Spherical Multicomponent Aggregates Composed of Core-Shell, Yolk-Shell, and Hollow Nanospheres and Their Lithium-Ion Storage Performances. *Small* **2018**, *14*, 1703957. [CrossRef] [PubMed]

356. Kijima, N.; Yomono, H.; Manabe, T.; Akimoto, J.; Igarashi, K. Microwave Synthesis of Fe_2O_3/SnO_2 Nanocomposites and Its Lithium Storage Performance. *Chem. Lett.* **2017**, *46*, 886–888. [CrossRef]

357. Liu, W.; Shi, Q.; Qu, Q.; Gao, T.; Zhu, G.; Shao, J.; Zheng, H. Improved Li-ion diffusion and stability of a $LiNi_{0.5}Mn_{1.5}O_4$ cathode through in situ co-doping with dual-metal cations and incorporation of a superionic conductor. *J. Mater. Chem. A* **2017**, *5*, 145–154. [CrossRef]

358. Kong, X.W.; Zhang, R.L.; Zhong, S.K.; Wu, L. Synthesis and characterisation of high-performance $3Li_4Ti_5O_{12}$·NiO composite anode material for lithium-ion batteries. *Mater. Res. Innov.* **2015**, *19*, 418–422. [CrossRef]

359. Ma, P.; Hu, P.; Liu, Z.; Xia, J.; Xia, D.; Chen, Y.; Liu, Z.; Lu, Z. Structural and electrochemical characterization of $0.7LiFePO_4$·$0.3Li_3V_2(PO_4)_3$/C cathode materials using PEG and glucose as carbon sources. *Electrochim. Acta* **2013**, *106*, 187–194. [CrossRef]

360. Wu, L.; Lu, J.; Zhong, S. Studies of $xLiFePO_4$·$yLi_3V_2(PO_4)_3$/C composite cathode materials with high tap density and high performance prepared by sol spray-drying method. *J. Solid State Electrochem.* **2013**, *17*, 2235–2241. [CrossRef]

361. Zhang, J.; Shen, C.; Zhang, B.; Zheng, J.; Peng, C.; Wang, X.; Yuan, X.; Li, H.; Chen, G. Synthesis and performances of $2LiFePO_4$·$Li_3V_2(PO_4)_3$/C cathode materials via spray-drying method with double carbon sources. *J. Power Sources* **2014**, *267*, 227–234. [CrossRef]

362. Zhong, S.; Wu, L.; Zheng, J.; Liu, J. Preparation of high tap-density $9LiFePO_4$·$Li_3V_2(PO_4)_3$/C composite cathode material by spray-drying and post-calcining method. *Powder Technol.* **2012**, *219*, 45–48. [CrossRef]

363. Yu, F.; Qi, P.; An, Y.; Wang, G.; Xia, L.; Zhu, M.; Dai, B. Up-Scaled Microspherical Aggregates of $LiFe_{0.4}V_{0.4}PO_4$/C Nanocomposites as Cathode Materials for High-Rate Li-Ion Batteries. *Energy Technol.* **2015**, *3*, 496–502. [CrossRef]

364. Wang, F.; Yang, J.; NuLi, Y.; Wang, J. Composites of $LiMnPO_4$ with $Li_3V_2(PO_4)_3$ for cathode in lithium-ion battery. *Electrochim. Acta* **2013**, *103*, 96–102. [CrossRef]

365. Zhang, J.; Wang, X.; Zhang, B.; Tong, H. Porous spherical $LiMnPO_4$·$2Li_3V_2(PO_4)_3$/C cathode material synthesized via spray-drying route using oxalate complex for lithium-ion batteries. *Electrochim. Acta* **2015**, *180*, 507–513. [CrossRef]

366. Chae, S.; Ko, M.; Park, S.; Kim, N.; Ma, J.; Cho, J. Micron-sized Fe–Cu–Si ternary composite anodes for high energy Li-ion batteries. *Energy Environ. Sci.* **2016**, *9*, 1251–1257. [CrossRef]

367. Park, J.-S.; Chan Kang, Y. Multicomponent (Mo, Ni) metal sulfide and selenide microspheres with empty nanovoids as anode materials for Na-ion batteries. *J. Mater. Chem. A* **2017**, *5*, 8616–8623. [CrossRef]

368. Arpagaus, C.; Collenberg, A.; Rütti, D. Laboratory spray-drying of materials for batteries, lasers, and bioceramics. *Dry. Technol.* **2018**, *30*, 1–9. [CrossRef]

369. Arpagaus, C. A Novel Laboratory-Scale Spray-dryer to Produce Nanoparticles. *Dry. Technol.* **2012**, *30*, 1113–1121. [CrossRef]

370. Anandharamakrishnan, C.; Ishwarya, S.P. Introduction to spray-drying. In *Spray-Drying Techniques for Food Ingredient Encapsulation*; The IFT Press Series; John Wiley & Sons, Ltd.: Chichester, UK; Hoboken, NJ, USA, 2015; pp. 14–15, ISBN 978-1-118-86419-7.

371. Feng, X.; Cui, H.; Li, Z.; Miao, R.; Yan, N. Scalable Synthesis of Dual-Carbon Enhanced Silicon-Suboxide/Silicon Composite as Anode for Lithium Ion Batteries. *Nano* **2017**, *12*, 1750084. [CrossRef]

372. Das, A.; Sen, D.; Mazumder, S.; Ghosh, A.K.; Basak, C.B.; Dasgupta, K. Formation of nano-structured core-shell micro-granules by evaporation induced assembly. *RSC Adv.* **2015**, *5*, 85052–85060. [CrossRef]

373. Fu, N.; Wu, W.D.; Wu, Z.; Moo, F.T.; Woo, M.W.; Selomulya, C.; Chen, X.D. Formation process of core-shell microparticles by solute migration during drying of homogenous composite droplets. *AIChE J.* **2017**, *63*, 3297–3310. [CrossRef]

374. Zellmer, S.; Garnweitner, G.; Breinlinger, T.; Kraft, T.; Schilde, C. Hierarchical Structure Formation of Nanoparticulate Spray-Dried Composite Aggregates. *ACS Nano* **2015**, *9*, 10749–10757. [CrossRef] [PubMed]

375. Park, G.D.; Cho, J.S.; Kang, Y.C. Sodium-ion storage properties of nickel sulfide hollow nanospheres/reduced graphene oxide composite powders prepared by a spray-drying process and the nanoscale Kirkendall effect. *Nanoscale* **2015**, *7*, 16781–16788. [CrossRef] [PubMed]

376. Du, K.; Xie, H.; Hu, G.; Peng, Z.; Cao, Y.; Yu, F. Enhancing the Thermal and Upper Voltage Performance of Ni-Rich Cathode Material by a Homogeneous and Facile Coating Method: Spray-Drying Coating with Nano-Al_2O_3. *ACS Appl. Mater. Interfaces* **2016**, *8*, 17713–17720. [CrossRef] [PubMed]

377. Wang, J.; Yin, L.; Jia, H.; Yu, H.; He, Y.; Yang, J.; Monroe, C.W. Hierarchical Sulfur-Based Cathode Materials with Long Cycle Life for Rechargeable Lithium Batteries. *ChemSusChem* **2014**, *7*, 563–569. [CrossRef] [PubMed]

378. Liu, L.; Wei, Y.; Zhang, C.; Zhang, C.; Li, X.; Wang, J.; Ling, L.; Qiao, W.; Long, D. Enhanced electrochemical performances of mesoporous carbon microsphere/selenium composites by controlling the pore structure and nitrogen doping. *Electrochim. Acta* **2015**, *153*, 140–148. [CrossRef]

379. Kim, J.H.; Lee, J.-H.; Kang, Y.C. Electrochemical properties of cobalt sulfide-carbon composite powders prepared by simple sulfidation process of spray-dried precursor powders. *Electrochim. Acta* **2014**, *137*, 336–343. [CrossRef]

380. Wang, Y.; Shen, Y.; Du, Z.; Zhang, X.; Wang, K.; Zhang, H.; Kang, T.; Guo, F.; Liu, C.; Wu, X. A lithium-carbon nanotube composite for stable lithium anodes. *J. Mater. Chem. A* **2017**, *5*, 23434–23439. [CrossRef]

381. Shui, J.L.; Lin, B.; Liu, W.L.; Yang, P.H.; Jiang, G.S.; Chen, C.H. Li-Mn-Co-O shelled $LiMn_2O_4$ spinel powder as a positive electrode material for lithium secondary batteries. *Mater. Sci. Eng. B* **2004**, *113*, 236–241. [CrossRef]

382. Shi, J.-L.; Peng, H.-J.; Zhu, L.; Zhu, W.; Zhang, Q. Template growth of porous graphene microspheres on layered double oxide catalysts and their applications in lithium–sulfur batteries. *Carbon* **2015**, *92*, 96–105. [CrossRef]

383. Zhang, Z.; Wang, Y.; Ren, W.; Zhong, Z.; Su, F. Synthesis of porous microspheres composed of graphitized carbon@amorphous silicon/carbon layers as high performance anode materials for Li-ion batteries. *RSC Adv.* **2014**, *4*, 55010–55015. [CrossRef]

384. Bahadur, J.; Sen, D.; Mazumder, S.; Bhattacharya, S.; Frielinghaus, H.; Goerigk, G. Origin of Buckling Phenomenon during Drying of Micrometer-Sized Colloidal Droplets. *Langmuir* **2011**, *27*, 8404–8414. [CrossRef] [PubMed]

385. Wang, D.; Fu, A.; Li, H.; Wang, Y.; Guo, P.; Liu, J.; Zhao, X.S. Mesoporous carbon spheres with controlled porosity for high-performance lithium–sulfur batteries. *J. Power Sources* **2015**, *285*, 469–477. [CrossRef]

Review

Fe-Based Nano-Materials in Catalysis

Stavros Alexandros Theofanidis [1], Vladimir V. Galvita [1,*], Christos Konstantopoulos [2],
Hilde Poelman [1] and Guy B. Marin [1]

[1] Laboratory for Chemical Technology, Ghent University, Technologiepark 914, B-9052 Ghent, Belgium;
StavrosAlexandros.Theofanidis@Ugent.be (S.A.T.); Hilde.Poelman@UGent.be (H.P.);
Guy.Marin@UGent.be (G.B.M.)

[2] Department of Engineering, University of Campania "Luigi Vanvitelli", Via Roma 29, 81031 Aversa (CE), Italy;
konstantopoulos.christos@gmail.com

* Correspondence: Vladimir.Galvita@UGent.be

Received: 5 April 2018; Accepted: 10 May 2018; Published: 17 May 2018

Abstract: The role of iron in view of its further utilization in chemical processes is presented, based on current knowledge of its properties. The addition of iron to a catalyst provides redox functionality, enhancing its resistance to carbon deposition. FeO_x species can be formed in the presence of an oxidizing agent, such as CO_2, H_2O or O_2, during reaction, which can further react via a redox mechanism with the carbon deposits. This can be exploited in the synthesis of active and stable catalysts for several processes, such as syngas and chemicals production, catalytic oxidation in exhaust converters, etc. Iron is considered an important promoter or co-catalyst, due to its high availability and low toxicity that can enhance the overall catalytic performance. However, its operation is more subtle and diverse than first sight reveals. Hence, iron and its oxides start to become a hot topic for more scientists and their findings are most promising. The scope of this article is to provide a review on iron/iron-oxide containing catalytic systems, including experimental and theoretical evidence, highlighting their properties mainly in view of syngas production, chemical looping, methane decomposition for carbon nanotubes production and propane dehydrogenation, over the last decade. The main focus goes to Fe-containing nano-alloys and specifically to the Fe–Ni nano-alloy, which is a very versatile material.

Keywords: role of iron; CO_2 utilization; chemical looping; nano-alloys; carbon; hydrocarbon conversion; dehydrogenation

1. Introduction and Motivation

Iron is one of the most abundant elements in the earth's crust composing 5% of it, and iron oxides have proven to be valuable materials to mankind over the years, starting from the pre-historic age where iron oxide containing ochre pigments were used to decorate cave walls (Figure 1). Fe_3O_4 containing rocks were man's first experience with magnetism, while compass-like instruments based on Fe_3O_4 were already exploited for religious purposes in China around 200 BC [1]. The development of Fe_3O_4-based compasses for navigation occurred in Europe approximately around 850 AD. Throughout the 20th century, iron oxides were at the forefront of discovery in science. For example, Fe_3O_4 as $Fe^{2+}Fe^{3+}_2O^{2-}_4$ was one of the first spinel structures solved by Bragg in 1915 [2] and Verwey discovered one of the first metal–insulator transitions in Fe_3O_4 in 1939.

Iron is involved in several biological processes. Proteins containing iron can be found in all living organisms [3,4]. In humans, an iron–protein, hemoglobin, is responsible for oxygen transport from the lungs to the rest of the body and for the blood color (Figure 1). Iron oxides, like Fe_3O_4, aid the navigation of magnetotactic bacteria [5], and it is thought that they play a similar role in the beaks of homing pigeons, while they have also been discovered in the human brain and other body tissues in unknown amounts.

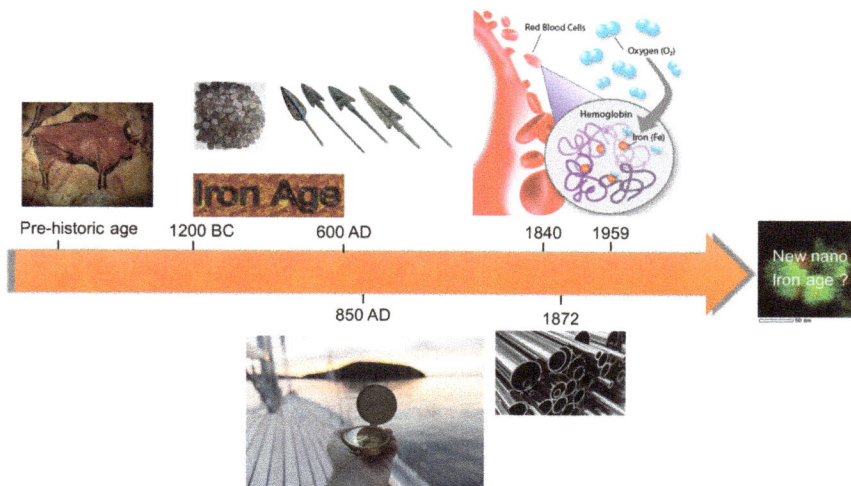

Figure 1. Use of iron/iron oxides throughout mankind [6–8].

Recently, there has been a resurgence of research into iron oxide materials for chemical/catalytic application [9–15]. Tartaj and co-workers [16] describe in their article entitled "The Iron Oxides Strike Back: . . . " how the exciting properties of iron oxides, coupled to their low toxicity, stability and economic viability, make them ideal for applications in a broad range of emerging fields. As one of the most significant earth oxides, iron oxide can be employed in the development of active and stable catalytic materials for reforming reactions to produce syngas [17–21], for production of chemicals [22–24], such as allyl alcohol [25,26], as an active component for catalytic oxidation in exhaust converters [27,28], for hydrodeoxygenation [29,30] and hydrogenation [31,32] reactions, for hydrogen sulfide removal from sewage [33], for electrochemical reduction of CO_2 [34], in batteries [35], in chemical looping processes [36–40], in water gas shift reaction [41–44], etc. The use of iron in the proton exchange membrane (PEM) fuel cells [43,45] has also attracted special interest. Sebastian and co-workers [46] utilized Fe–N–C based catalyst as cathode in a direct methanol fuel cell (DMFC) in order to efficiently produce power. They reported an outstanding performance even at high methanol concentration, while at high temperature the catalyst displayed a similar current–time behavior to a membrane–electrode assembly based on a Pt cathode. Galvita and co-workers [47] suggested the use of iron-based materials for energy storage. Their concept includes a reactor configuration consisting of two chambers, both utilizing iron-based materials. Initially, the materials in the two chambers are reduced to metallic form, thus "charging" the reactor. In the second "discharging" step, steam is fed to the inner chamber, while air is sent to the outer. Hydrogen is produced by the inner chamber, whereas the external chamber is used for heat generation. Apart from iron, the external chamber contains a Ni-based layer, which is pyrophoric, in order to enable the startup of heat generation at room temperature under air flow.

In many of the aforementioned applications, the interest in iron is associated with the unique ability of the oxides to be reduced and then re-oxidized by H_2O/CO_2 [48]. Based on these iron oxide redox properties, a new reforming process has been developed by Buelens and co-workers [49], termed as "super-dry reforming". The authors efficiently transformed CO_2 from waste product to CO. They used Fe_2O_3 supported on $MgAl_2O_4$ as a solid oxygen carrier material (OCM), where three molecules of CO_2 are consumed per one CH_4, resulting in an enhanced CO production.

All of the above highlight the importance of iron/iron oxide systems, especially in the field of catalysis. Scientists consider iron-based materials as promising candidates to be employed in various chemical applications, like syngas production, chemical looping, methane decomposition for carbon nanotubes production and propane dehydrogenation. Therefore, this work focuses on reviewing the progress that has been made in the past few years, trying to unravel the role of iron in Fe-containing materials for sustainable application in chemical processes.

2. Fe in CeO_2 for Chemical Looping

Chemical looping is a cyclic process where in the first half cycle, the materials undergo reduction through release of lattice oxygen producing e.g., CO, CO_2 or H_2O. In the second half cycle the oxygen vacancy in the lattice is refilled due to the reaction with an oxidizing gas, such as O_2, CO_2 or H_2O, resulting in the production of CO or H_2 [50]. Key properties for chemical looping are the reducibility of the carrier, its cost, toxicity, thermal stability and attrition resistance. Oxides of Ni, Cu, Mo, and Fe, are typically used as oxygen carriers [51,52]. Among these, iron oxides stand out because of their natural abundance and high reoxidation capacity with CO_2 or H_2O over a wide range of operating conditions (700–1000 °C). However, pure iron oxides tend to deactivate rapidly [53,54]. The major factor for deactivation in pure iron oxide materials is sintering. To overcome this challenge, iron oxides are often modified with other oxide materials, e.g., MgO, TiO_2 [55], Al_2O_3 [56], CeO_2 [57], ZrO_2 [58,59], $CeZrO_2$ [60], SiO_2 [61], and $MgAl_2O_4$ [57,62–64]. Certain promoters contribute towards the redox reaction, alongwith iron oxide. These are therefore termed chemically active promoters, e.g., CeO_2, $CeZrO_2$ among the latter, CeO_2 stands out as it has high activity toward methane oxidation by lattice oxygen, as well as reasonable H_2O or CO_2 reoxidation capacity [65].

The interaction between Ce and Fe was found to induce structural modification and stabilization of iron oxides, making it an ideal candidate for promoting iron oxide in a chemical looping process. CeO_2 improves the activity of Fe_2O_3 toward selective CH_4 oxidation by lattice oxygen, as well as the re-oxidation capacity by H_2O or CO_2 [50,57,66]. The interaction can be established through the formation of a solid solution where Fe^{3+} cations dissolve in the ceria structure. The evolution of the Fe_2O_3–CeO_2 structure as a function of composition is shown in Figure 2. In general, the formation of a solid solution between CeO_2 and MeO_x (Me = Mn, Fe, or Cu) is responsible for enhancing the CeO_2 reducibility compared with pure CeO_2 [48,65,67]. CeO_2 has a fluorite structure, with each Ce^{4+} cation surrounded by eight equivalent nearest O^{2-}, that form the corners of a cube. When Ce^{4+} ions are replaced by lower valence cations, an oxygen vacancy or lattice defect can be created, which is considered to be the most reactive site. Both surface and bulk oxygen vacancies tend to form within CeO_2, the former being suitable for adsorption purposes.

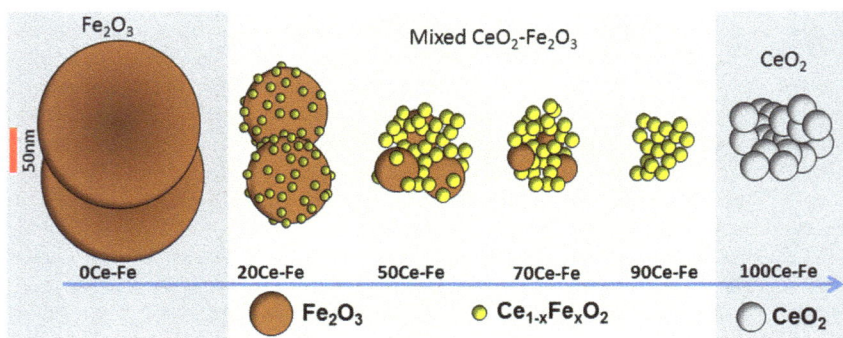

Figure 2. Schematic illustration of mixed CeO_2–Fe_2O_3 samples, based upon ICP composition, XRD patterns, STEM, EDX, and EELS. Obtained from [65].

Although the reduction at the surface of Fe_2O_3–CeO_2 is independent of whether CeO_2 is present or not, after consuming the available surface oxygen for CH_4 oxidation, oxygen can be transferred from bulk to surface more rapidly in Fe_2O_3–CeO_2 than in Fe_2O_3. This was ascribed to the CeO_2 additive, creating oxygen vacancies in the solid solution. These vacancies are able to quickly transfer oxygen from the bulk to the surface of the oxygen carrier material through vacancy diffusion or even oxygen tunnels formed by vacancies. According to reported CH_4-TPR profiles [50] for both CeO_2 and Fe_2O_3–CeO_2 samples, the removal of the most reactive oxygen mainly occurs at lower temperatures (773–823 K), giving rise to deep oxidation of CH_4 to CO_2, while CO is the main product at higher temperature (>873 K). Pure CeO_2 shows a high CH_4 oxidation activity at a temperature around 923 K due to the high consumption rate of surface lattice oxygen. In comparison to the profile of CeO_2, $Ce_{0.9}Fe_{0.1}O_{2-\delta}$ shows a dramatic decrease of CO_2 production with an increase in production of partial oxidation reaction products. The Fe_2O_3–CeO_2 mixed oxides with Fe content above 0.5 fail to increase the conversion of CH_4 and show a decline in CO selectivity, which is due to the increasing amount of pure Fe yielding deep oxidation products. Therefore, an equal weight loading of Fe and Ce can maximally promote the reactivity for redox reactions of the material [48,50].

Overall, three types of deactivation were identified for the Fe_2O_3–CeO_2 materials: (1) Fe extraction from the solid solution $Ce_{1-x}Fe_xO_2$, (2) perovskite formation ($CeFeO_3$) and (3) sintering. The extraction of Fe from the $Ce_{1-x}Fe_xO_{2-x}$ occurs very fast. It leads to lower reducibility of CeO_2, but at the same time provides more iron oxide storage capacity by setting free extra Fe. $CeFeO_3$ perovskite formation leads to loss of oxygen storage capacity as it is non-reducible at temperatures lower than 1073 K. Finally, sintering is a slow process which continues throughout cyclic operation. It causes crystallites to grow in size, thereby increasing the diffusion time of bulk oxygen to the surface. Hence, a lower degree of reduction is reached in a given reduction time and upon re-oxidation with CO_2, a lower CO yield is obtained. The relative importance of these deactivation types depends on the composition of the oxygen storage materials. In iron rich samples deactivation is predominantly caused by sintering of iron oxides. Fe extraction is of minor importance given the composition of this material. Similarly, perovskite formation may occur, but will hardly affect the cycling productivity. In ceria rich samples, all three types of deactivation occur. Compared to pure Fe_2O_3, sintering as the main deactivation type is tempered by the strategy of decorating Fe_2O_3 with CeO_2 nanoparticles.

3. Fe in Spinels for Chemical Looping

One of the most common iron containing chemical compounds with spinel structure is Fe_3O_4. It naturally occurs as the mineral magnetite, containing Fe^{2+} and Fe^{3+} ions. Nano-Fe_3O_4 has recently gained attention as heterogeneous catalyst due to its environmental compatibility, simple handling and ease of recovery using an external magnetic field [68–70]. There are many reports in literature using Fe_3O_4-based materials in environmental applications [69,71], in Fenton-like processes [68,72] and in wastewater treatment [9,73,74].

Iron can also form spinel phases with aluminum and magnesium, depending on the applied conditions during the catalyst synthesis, e.g., calcination temperature, resulting in $FeAl_2O_4$ and $MgFe_2O_4$ structures, respectively [75,76]. These materials have been used as oxygen storage during chemical looping processes, preventing the sintering of Fe particles and thus increasing the process stability. Ferrites have also been utilized for oxidation of alcohols to the corresponding ketones or aldehydes [77,78]. However, the aforementioned iron spinel structures require higher reduction/oxidation temperature, resulting in more severe operating conditions [79,80]. On the other hand, Dharanipragada and co-workers [81] synthesized a novel material, combining Al^{3+}, Fe^{3+} and Mg^{2+} in one spinel structure, forming a $MgFe_xAl_{2-x}O_4$ material that was used for oxygen storage during chemical looping for CO_2 to CO conversion. They concluded that at low Fe loading (<30 wt %), most of the iron is in a spinel structure with magnesium aluminate. Even though Fe incorporated inside the spinel has lower oxygen storage capacity compared to Fe_2O_3 supported on the $MgFe_xAl_{2-x}O_4$ material (Figure 3), the stabilization of Fe in the spinel structure results in an improved performance.

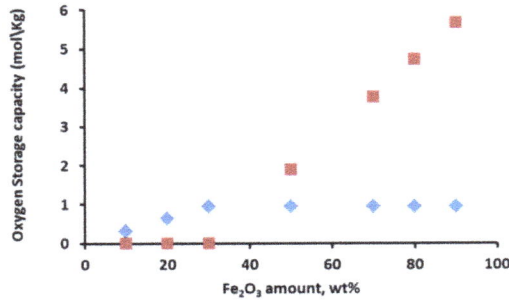

Figure 3. Oxygen storage capacity of $MgFe_xAl_{2-x}O_4$ materials as a function of the Fe_2O_3 content. Note that when Fe_2O_3 loading is less than 30 wt %, it is completely incorporated into the spinel structure without separate Fe_2O_3 phases. ♦: iron incorporated in spinel structure; ■: separate Fe_2O_3 phase. Obtained from [81].

The occurrence of a separate Fe_2O_3 phase will drastically increase the oxygen storage capacity of the material. When Fe is fully incorporated into the spinel, redox cycling proceeds between Fe^{3+} and Fe^{2+}, based on Mossbauer spectra [81]. On the other hand, for the materials with higher Fe_2O_3 loadings, the cycling of $MgFe_xAl_{2-x}O_4 + Fe_2O_3$ will change the Fe oxidation state between Fe^{3+} and Fe^{2+} in the spinel and between Fe^{3+} and Fe^0 in the separate iron oxides. However, the latter materials do suffer from severe sintering. Figure 4 shows that already after five isothermal cycles under H_2/CO_2 at 1023 K, the crystallite size for Fe_2O_3 in $MgFe_xAl_{2-x}O_4$ with 50 wt % Fe_2O_3 increased from 60 to 80 nm, while the size of the $MgFe_xAl_{2-x}O_4$ remained stable at 10–22 nm [81]. This implies that the incorporation of Fe inside the lattice of the magnesium aluminate spinel structure can greatly improve the stability of the material during chemical looping, alternating between reducing and oxidizing environment. And this stability of the material determines the economics of the process [48,82–84].

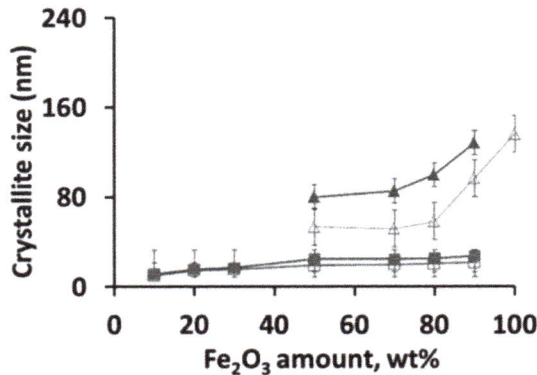

Figure 4. Crystallite size of Fe_2O_3 and $MgFe_xAl_{2-x}O_4$ phases in the samples, as calculated based on XRD using the Scherrer equation. As-prepared: (□) $MgFe_xAl_{2-x}O_4$ and (Δ) Fe_2O_3; (■) $MgFe_xAl_{2-x}O_4$ and (▲) Fe_3O_4 after 5 isothermal redox cycles of H_2/CO_2 at 1023 K. Obtained from [81].

The $MgFe_{0.14}Al_{1.86}O_4$ spinel structure with 10 wt % Fe_2O_3 (x = 0.14) shows the highest stability during isothermal H_2/CO_2 cycles without any Fe_2O_3 phase segregation (Figure 5). Dharanipragada and co-workers [85] further examined the reduction kinetics of this $MgFe_{0.14}Al_{1.86}O_4$ using XRD and in-situ

QXANES at the Fe–K edge. They found that Fe is incorporated in the octahedral sites of the spinel, replacing Al in the lattice.

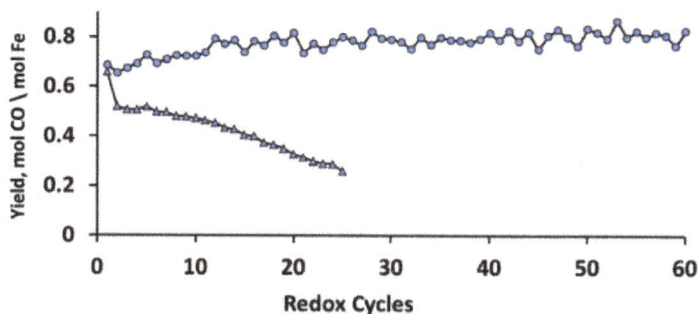

Figure 5. CO yield in CO_2 to CO conversion as a function of isothermal H_2-CO_2 redox cycles for $MgFe_xAl_{2-x}O_4$ and $Fe_2O_3/MgFe_xAl_{2-x}O_4$ with (●) 10 wt % Fe_2O_3 ($MgFe_{0.14}Al_{1.86}O_4$) and (▲) 90 wt % Fe_2O_3. Each cycle (16 min) is composed of 4 min H_2 (5% in Ar), 4 min He, 4 min CO_2 (100%) and 4 min He at 1123 K. All the gas flows were 1.1 NmL/s. Obtained from [81].

During reduction, 55% of Fe could be reduced from 3+ to 2+, with the rest remaining identical to the "as-prepared" state. A shrinking core model was proposed [85], where initially the external surface of the solid is involved in the reaction (reduction). The reduced layer then thickens, depending on the exposure time under reducing environment, enclosing a shrinking core of unreacted solid (Figure 6). This shrinking core model provided an adequate description for the transition from Fe^{3+} to Fe^{2+} in the $MgFe_{0.14}Al_{1.86}O_4$ material (top right inset of Figure 6).

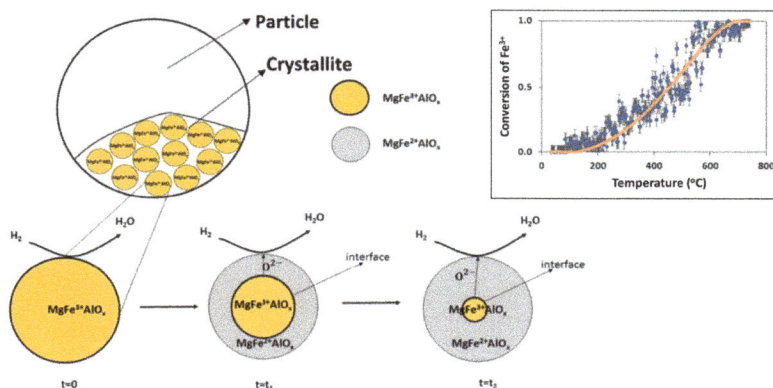

Figure 6. Schematic representation of the shrinking core model in a $MgFe_xAl_{2-x}O_4$ crystallite. Top right inset: Observed and calculated conversion profile of Fe^{3+} based on pre-edge fitting of QXANES spectra from $MgFe_{0.14}Al_{1.86}O_4$. Obtained from [85].

On the crystallite scale, the solid-solid transformations are governed by three phenomena according to Dharanipragada and co-workers [85]: (1) Reaction of surface oxygen with H_2, forming H_2O, (2) reduction of $MgFe_x^{3+}Al_{2-x}O_4$ to $MgFe_x^{2+}Al_{2-x}O_4$ at the interface between unreacted core of the crystallite and reduced material and (3) oxygen diffusion from the core through the reduced layer to the surface, where the reaction takes place.

Fe-based spinel materials and more specifically the $MgFe_xAl_{2-x}O_4$, are recently receiving more attention as they combine good redox properties and thermal stability. They can be applied to many processes varying from pollutants removal, e.g., SO_2 [86] and chemical looping [81] to syngas production via catalytic steam and/or dry reforming.

4. Fe in Nano-Alloys for Catalysis

Many applications of iron use it in alloyed form, steel being the most famous Fe-containing alloy. In catalysis, Fe-containing nano-alloys are often used, e.g., in bimetallic nano-alloys combined with a noble metal or non-noble element. Yfanti and co-workers [70] used Fe–Pt catalysts for hydrodeoxygenation of glycerol. They reported an electronic interaction between Fe and Pt, which increased the glycerol conversion, compared to the monometallic Pt. By increasing the Fe content, the catalyst surface structure was changed, as the iron oxide clusters started to cover the Pt particles. This resulted in a slight decrease in the main product selectivity, 1,2-propanediol, but at the same time, the stability of the catalyst was increased. The improved catalytic performance was attributed to the Fe addition as it enhanced the carbon-resistance of the catalyst and prevented the sintering of Pt particles. Saravanan and co-workers [87] used Fe–Pt catalysts for the oxidation of indoor pollutants, such as CO and benzene, in a temperature range of 298–473 K, demonstrating that they can be a possible alternative for the existing monometallic Pt catalysts. The authors concluded that the intermetallic phase $PtFe_3$ is more active than the Pt_3Fe. On the other hand, Jiang and co-workers [88] used Fe–Pd bimetallic catalysts with a core-shell structure for the oxygen reduction reaction (ORR) as an alternative to Pt-based catalysts. They demonstrated that Fe–Pd had a robust catalytic activity and durability in ORR.

There is a number of studies in literature indicating the promoting effect of Fe to Rh-based catalysts for syngas conversion to C_{2+} oxygenates, such as ethanol [23,24,89–92]. An alloy based on Fe and Rh has been reported by Palomino and co-workers [91], who investigated the effect of alloying on syngas conversion. They found that the addition of Fe increased the selectivity towards ethanol, but partially suppressed the catalytic activity due to blocking or modifying of Rh active sites depending on the Fe content. Similarly, Liu and co-workers [93] used Rh supported on SiO_2 catalysts promoted with Mn and Fe for CO hydrogenation towards light hydrocarbons and oxygenates. A trimetallic Rh-Fe-Mn alloy was formed, with molar ratio of 1:0.15:0.10, that resulted in higher selectivities than the bimetallic counterparts.

A synergetic effect of Fe and Ru supported on TiO_2 was reported by Phan and co-workers [94] during anisole hydrodeoxygenation reaction (HDO). The addition of Fe to the Ru/TiO_2 catalyst altered the surface properties, changing the reaction pathway. More specifically, the anisole conversion and product distribution were affected by the Fe loading (Figure 7). The combination of Ru and Fe lead to a higher selectivity of benzene and a lower selectivity of methoxycyclohexane, indicating that direct deoxygenation (DDO) is the main reaction pathway. The enhanced performance with Fe was attributed to the increased number of oxygen vacancies on the surface of the TiO_2 support.

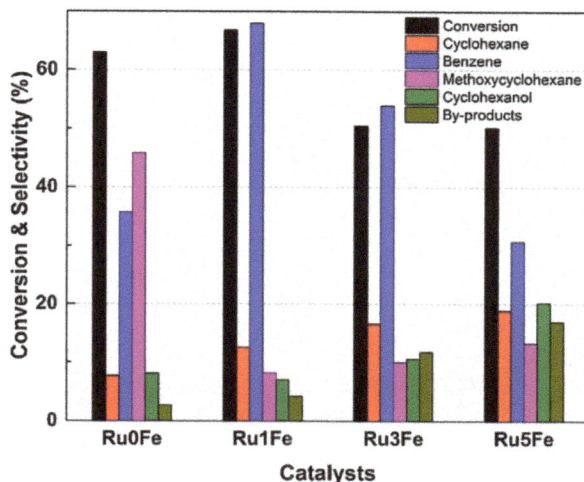

Figure 7. Conversion and product distribution of the Ru_xFe/TiO_2 catalysts, during anisole hydrodeoxygenation (HDO), in a stainless-steel batch reactor. Reaction conditions: 10 wt % anisole (40 mL), catalyst (1.0 g), 200 °C and 10 bar H_2 for 3 h. Obtained from [94].

Tungsten forms alloy with Fe and retains functional properties (mechanical, magnetic, etc.) even at elevated temperature, having a variety of applications in the industrial sector [95]. Tharamani and co-workers used the Fe–W alloy as an anode in a methanol oxidative fuel cell with a H_2SO_4 medium [95]. Shi and co-workers [96] used Cu–Fe bimetallic catalysts supported on carbon nanotubes for the synthesis of higher alcohols from syngas. They found that the selectivity toward methanol decreased, and the formation of C_{2+}–OH alcohols increased, reaching a selectivity of 68.8% for the best candidate with a Fe:Cu atomic ratio of 1. A Fe–Co alloy phase was reported to form after reduction in hydrogen by Koike and co-workers [97]. This Fe–Co catalyst was active for toluene steam reforming, but deactivated due to oxidation of the alloy phase. The addition of hydrogen in the feed stream resulted in higher activity.

In what follows, the bimetallic Ni-containing Fe nano-alloys will be discussed in detail as they have an outstanding ability to limit surface carbon accumulation.

4.1. Fe–Ni Nano-Alloy

The preparation of a Fe–Ni alloy generally involves impregnation of their precursors on a support material, calcination under air and reduction [20,98]. However, this might result in large and non-uniform Fe–Ni particles [20]. According to the Fe–Ni phase diagram (Figure 8) [99], at least one regular Ni-rich alloy with $FeNi_3$ composition is known. Other Fe–Ni alloy structures with composition NiFe, Ni_3Fe_2 and Ni_2Fe have also been reported [100]. However, a bimetallic Fe–Ni system will most likely contain a wide range of different structures of the nano-alloy, depending on the Fe/Ni ratio and the applied temperature. Figure 8 shows that Ni and Fe, as well as their alloys, have similar melting points. This implies that the surface migration and aggregation phenomena, which are correlated with the Tammann temperature (=0.52·melting point), will be within the same temperature range.

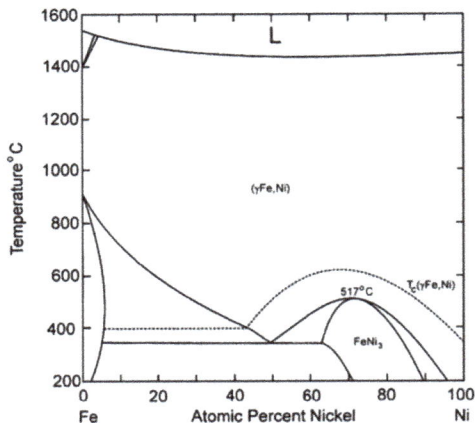

Figure 8. Phase diagram of the bimetallic Fe–Ni system. Obtained from [99].

Co-impregnation was used by Theofanidis and co-workers [17] to prepare Fe–Ni catalysts supported on MgAl$_2$O$_4$. A surface area of 84.7 $\pm$ 5.8 and 47.6 $\pm$ 11.4 m$^2 \cdot$g^{-1} was measured for 8 wt %Ni-5 wt %Fe and 8 wt %Ni-8 wt %Fe (Ni/(Ni + Fe) ratios of 0.6 and 0.5), respectively, after the calcination step under air flow (named as "as-prepared"). Similar values, in the range of 53–71 m$^2 \cdot$g^{-1}, were obtained by Kustov and co-workers for Fe–Ni catalysts with different total metal loading and a Ni/(Ni + Fe) ratio varying from 0 to 0.8, supported on MgAl$_2$O$_4$ [100]. On the other hand, a Fe–Ni catalyst supported on Mg$_x$Al$_y$O$_z$ hydrotalcite has been reported to have higher surface area, in the range of 172–175 m$^2 \cdot$g^{-1} [101]. Li and co-workers [19] also prepared Fe–Ni, as a steam reforming catalyst, using a hydrotalcite type of precursor. They obtained uniform Fe–Ni nanoparticles, with particle size varying from 8.1 to 10.2 nm depending on the Ni/(Ni + Fe) ratio (from 0.4 to 0.9).

The crystalline phases of the Fe-Ni/MgAl$_2$O$_4$ samples were determined by X-ray diffraction (XRD). In the "as-prepared" state, NiO, NiAl$_2$O$_4$, NiFe$_2$O$_4$ and Fe oxides were detected, depending on the used support material [17,18,101]. Upon reduction, a bimetallic Fe–Ni nano-alloy with a crystallite size of approximately 5–20 nm is formed (Figure 9), depending on the metal (Ni and Fe) loading, shifting the main 2θ angle position to lower values than for metallic Ni [18,21,101]. The XRD pattern after oxidation by CO$_2$ (Figure 9) shows that the Fe–Ni alloy was decomposed to Ni and Fe$_3$O$_4$, while the NiAl$_2$O$_4$ and MgAl$_2$O$_4$ support diffractions remained stable.

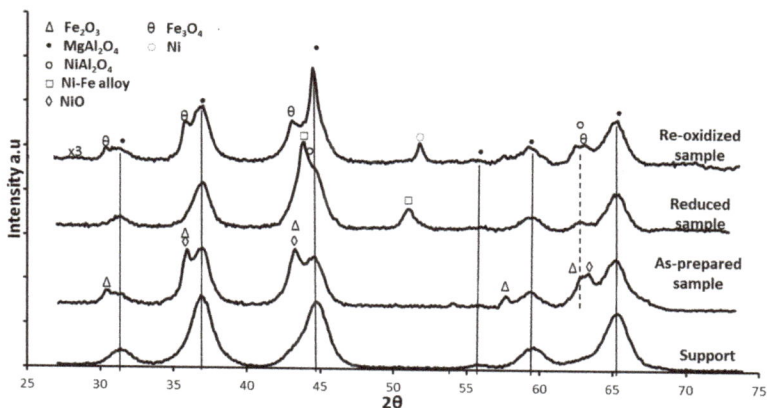

Figure 9. Full XRD scans of $MgAl_2O_4$, as-prepared, reduced and re-oxidized 8 wt %Ni-5 wt %Fe/$MgAl_2O_4$ (1 mL/s of 10%H_2/He mixture or CO_2 at a total pressure of 101.3 kPa and 1123 K). The $NiFe_2O_4$ phase cannot be distinguished due to overlapping with Fe_2O_3. Reproduced from [17].

The Ni and Fe elements are uniformly distributed in the nano-alloy (Figure 10A) after reduction. In contrast, after CO_2 oxidation Ni and Fe particles are segregated (Figure 10B) and Fe is oxidized to Fe_3O_4 [17,22,102].

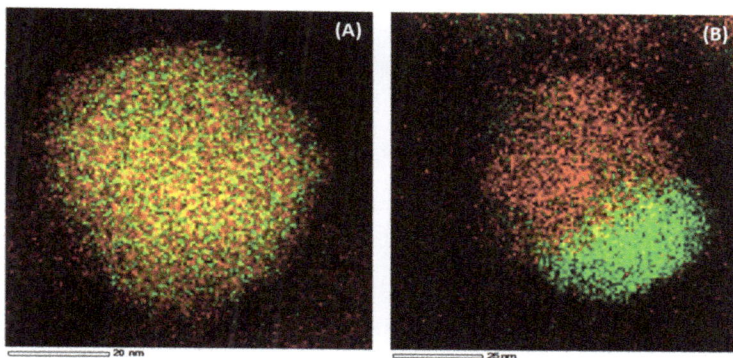

Figure 10. EDX element mapping of 8 wt %Ni-5 wt %Fe/$MgAl_2O_4$. (**A**) After H_2-reduction (1 mL/s of 5%H_2/Ar mixture at a total pressure of 101.3 kPa and 1123 K). (**B**) After CO_2 oxidation (1 mL/s of CO_2 at a total pressure of 101.3 kPa and 1123 K). Red and green colors correspond to Fe and Ni elements respectively. Obtained from [17].

A schematic illustration of the Fe–Ni nano-alloy formation and decomposition is presented in Figure 11. The alloy is decomposed during CO_2 oxidation between 850 K and 1123 K yielding two separate phases of Ni and Fe_3O_4 (see the EDX elemental mapping image Figure 10B). Metallic Ni in the bulk cannot be oxidized to NiO under CO_2 flow up to 1123 K. A subsequent H_2 reduction step leads again to the formation of a Fe–Ni nano-alloy [17].

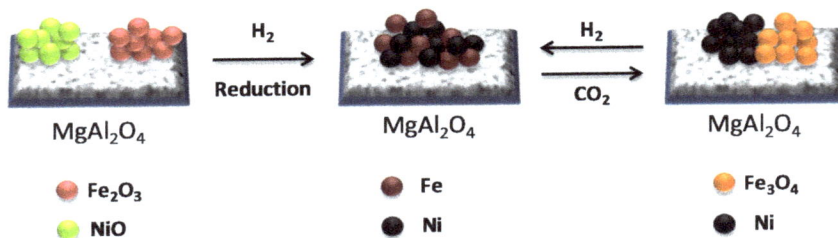

Figure 11. Schematic diagram of Fe–Ni nano-alloy formation and decomposition, depending on the applied environment. Obtained from [17].

4.1.1. Activity during Methane Decomposition

Monometallic [103,104] and bimetallic Fe-based catalysts were extensively used for carbon formation [105–107]. Even if the carbon formation and growth on catalysts is an undesired phenomenon in reforming reactions, the synthesis of carbon nanotubes (CNT), a type of carbon material with graphite layers and tubular structure, plays a very important role in the field of nanotechnology [105–107]. Carbon nanotubes were first identified by Lijma [108]. They require a source of elemental carbon, such as methane, and energy in order to be formed. The CNTs have numerous properties like high surface area, electronic and thermal conductivity, tensile strength, resistance to acidic/basic chemicals, making them ideal to be used in a variety of applications such as catalyst supports, air and water filtration, conductive adhesive, fibers and fabrics, etc. [109].

Methane is often used as a carbon source and the understanding of its activation step, which typically occurs over metals, is essential. The activation of CH_4 only, without co-feed of other reagents, under methane decomposition (MD) reaction conditions, at 1023 K and 1 bar under the flow of 1 mL/s 50%CH_4-50%Ar, over monometallic Ni, Fe and bimetallic Fe–Ni, was investigated by Theofanidis and co-workers [17]. Carbon accumulated according to the methane decomposition reaction ($CH_4 \rightarrow C + 2H_2$) [102]. After oxidation by CO_2, it was found that more carbon was deposited on the bimetallic catalyst than on the monometallic ones, implying that the Fe–Ni alloy does not suppress carbon formation. Wang and co-workers used Fe–Ni catalysts with different Ni/(Ni + Fe) ratios for methane decomposition (Figure 12) in order to produce hydrogen and carbon nanotubes (Figure 13) [110]. They also found that the Fe–Ni alloy is active for methane decomposition. Figure 12A shows the methane conversion as a function of time-on-stream (TOS) for three catalysts with Ni/(Ni + Fe) ratio of 1.0, 0.7 and 0.3 respectively. The monometallic Ni (Ni/(Ni + Fe) of 1.0) deactivated after 16 h TOS, while the Fe-rich sample (Ni/(Ni + Fe) of 0.3) displayed almost no activity, as it was completely deactivated after less than 2 h TOS. On the other hand, the bimetallic Fe–Ni catalyst with a Ni/(Ni + Fe) ratio of 0.7 had a stable performance throughout 20 h TOS. They further examined the best candidate for the same reaction for longer TOS (Figure 12B). The conversion dropped from 72% to 40% in the first 50 h, while hereafter the catalyst remained stable, even up to 210 h TOS. 56.2 g of carbon were produced, Figure 12B, which equals 562 g of C/g of catalyst during the 210 h.

According to many researchers, the carbon accumulation follows the deposition-diffusion-precipitation mechanism (or bulk diffusion mechanism) [110–113], where the properties of the metal play a crucial role. The modification of the Ni catalyst with Fe may increase the carbon diffusion rate, thereby decreasing the surface carbon accumulation. Indeed, the diffusion of carbon atoms in Fe is 3 orders of magnitude faster than in Ni [114]. The fast removal of carbon atoms from the surface can suppress the reverse reaction of methane formation ($C + 2H_2 \rightarrow CH_4$), thus compensating for the lower methane decomposition rate of bimetallic Fe–Ni catalysts compared to monometallic Ni. Indeed, Ni is more active than Fe for methane decomposition and hence the addition of Fe is likely to reduce the carbon formation rate. As a result, the balance among carbon formation, diffusion and precipitation as carbon nanotube is maintained in Fe–Ni catalysts leading to improved catalytic performance [110].

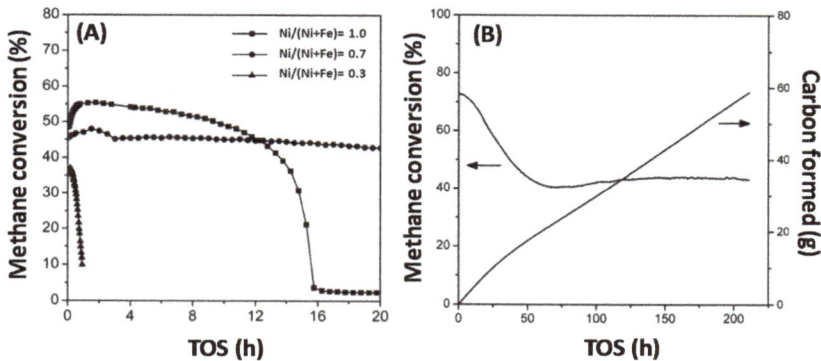

Figure 12. (**A**) Methane conversion over Fe–Ni catalysts with Ni/(Ni + Fe) ratio of 1.0, 0.7 and 0.3 as a function of TOS at 873 K. (**B**) Long term test of a Fe–Ni catalyst with Ni/(Ni + Fe) ratio of 0.7 during methane decomposition at 923 K. Reproduced from [110].

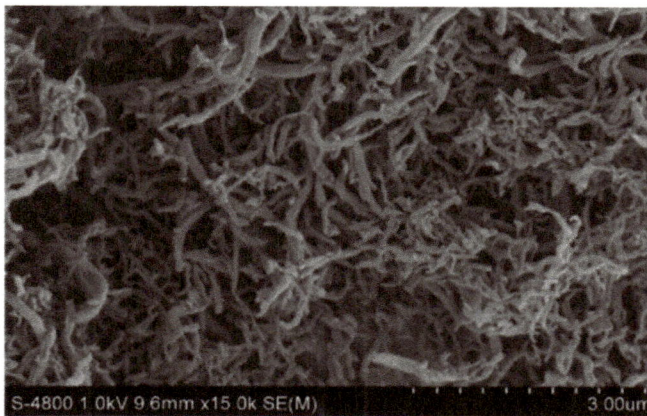

Figure 13. SEM image of carbon nanotubes produced over a Fe–Ni catalyst with Ni/(Ni + Fe) of 0.7 after 210 h TOS under methane decomposition at 923 K. Obtained from [110].

4.1.2. Activity during Syngas Production

Syngas production over Fe–Ni catalysts strongly depends on the composition of the nano-alloy that is formed after the reduction process [17,21,101]. More specifically, the Fe–Ni catalysts are sensitive to the Fe content and their activity is related to the employed Ni/Fe [17,21] or Ni/(Ni + Fe) ratio (Figure 14A) [101].Wang and co-workers [21] found that the addition of Fe promoted the steam reforming reaction in the range of Ni/Fe $\geq$ 2. On the other hand, Theofanidis and co-workers [17] found a slight improvement in the activity of Ni-Fe catalysts in the same range of Ni/Fe ratio, while the carbon deposition was suppressed remarkably. Pure Fe is twenty times less active than a pure Ni catalyst for methane dry reforming (DRM) at 923 K, Figure 14B, with a CH_4 consumption rate of 0.022 mol·s^{-1}·kg^{-1}cat and 0.34 mol·s^{-1}·kg^{-1}cat, respectively. However, pure Ni loses 30% of its activity after only 10 h TOS. On the other hand, the bimetallic Ni-rich Fe catalysts, with Ni/(Ni + Fe) ratios of

0.8 and 0.75 show an activity similar to pure Ni at 923 K, 0.32 and 0.25 mol·s^{-1}·kg$^{-1}_{cat}$, respectively. Their stable performance is emphasized by their modest activity loss during 10 h TOS, by only 6.4% and 4.0%, respectively [101].

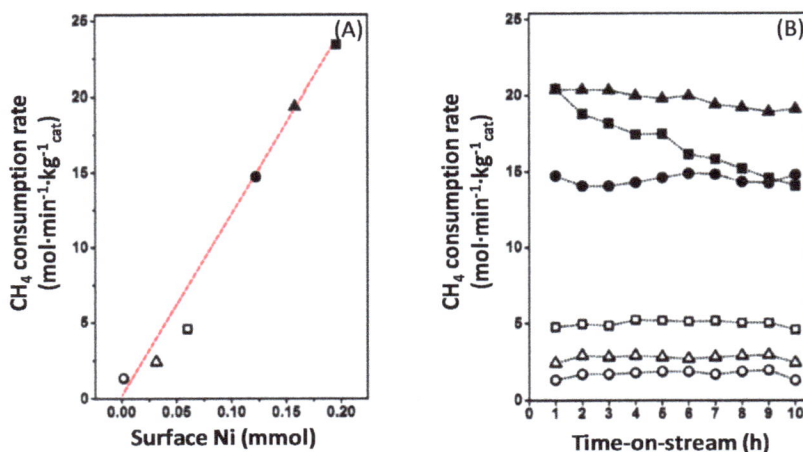

Figure 14. (**A**) Rate of methane consumption (mol·min^{-1}·kg$^{-1}_{cat}$) as a function of the amount of surface Ni (mmol), during methane dry reforming. ■: Pure Ni; ▲: Ni/(Ni + Fe) = 0.80; ●: Ni/(Ni + Fe) = 0.75; □: Ni/(Ni + Fe) = 0.5; △: Ni/(Ni + Fe) = 0.25 and ○: Pure Fe, all supported on Mg$_x$Al$_y$O$_z$ and (**B**) rate of methane consumption (mol·min^{-1}·kg$^{-1}_{cat}$) as a function of time-on-stream (TOS) during DRM at 923 K. Reproduced from [101].

The deposited carbon as a function of Ni/(Ni + Fe) ratio can be seen in Figure 15. Carbon filaments start to grow as the Ni/(Ni + Fe) ratio approaches 1 (pure Ni) [115] after 4 h TOS. On the other hand, a negligible amount of carbon was accumulated on bimetallic Fe–Ni with Ni/(Ni + Fe) ratio ≤ 0.6 (Figure 15).

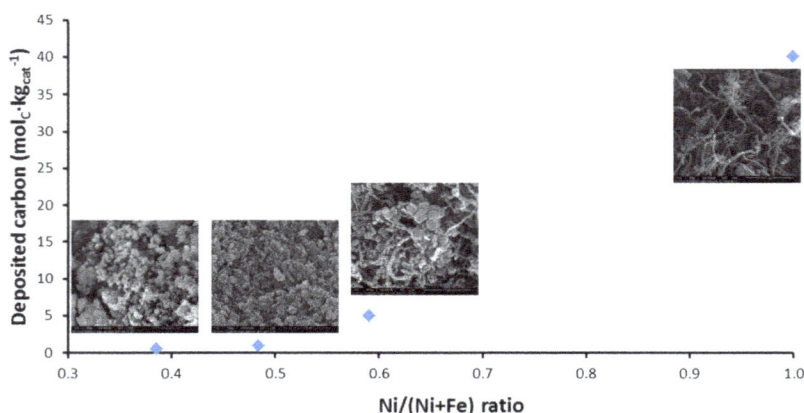

Figure 15. Deposited carbon as a function of Ni/(Ni + Fe) ratio along with the SEM micrographs of "spent" catalysts supported on MgAl$_2$O$_4$. Temperature 1023 K, CH$_4$:CO$_2$ = 1:1, reaction time 4 h. Reproduced from [17].

During a stability test over longer time-on-stream for DRM at 1023 K (Figure 16), Theofanidis and co-workers [116] observed a loss of 62% in the CH_4 consumption rate of a bimetallic Fe–Ni catalyst supported on $MgAl_2O_4$ with Ni/(Ni + Fe) of 0.65. They examined carbon formation as a possible reason for the deactivation. However, the deposited carbon was below detection limits after 24 h TOS, implying that the addition of Fe increased the carbon-resistance of the catalyst during reforming reactions. They also evaluated the reversibility of the observed deactivation. As much as 76% of the catalyst initial activity could be restored [116]. Since no carbon was deposited, it was concluded that sintering was at the origin of the irreversible deactivation that accounted for the persisting 24% of activity loss. The reversible deactivation was attributed to Fe segregation from the Fe–Ni nano-alloy structure. Indeed, an increase in CO/H_2 ratio from 1.3 after 1 h TOS to 2.5 after 24 h TOS (Figure 16) was observed, indicating a modification in the nature of active sites during the reaction. As Fe is more active for the reverse water-gas-shift reaction (RWGS: $CO_2 + H_2$ $H_2O + CO$) than Ni, its segregation from the alloy leads to consumption of H_2 and hence an increase in CO/H_2 ratio. The Fe–Ni nano-alloy can however be reconstructed upon regeneration and reduction steps.

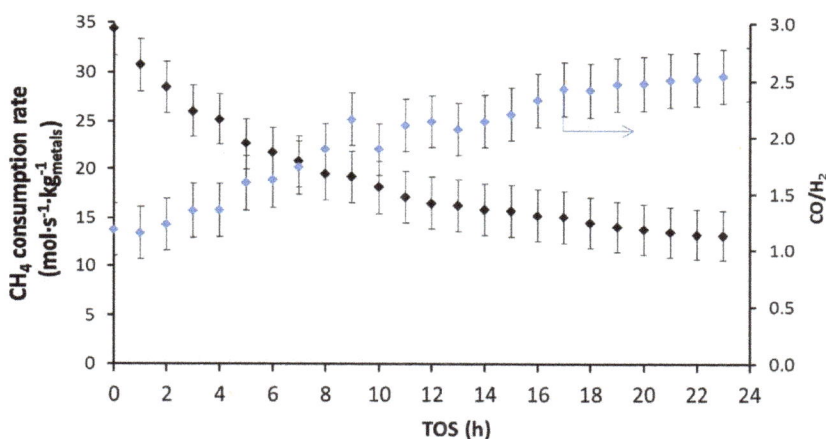

Figure 16. CH_4 consumption rate ($mol_{CH4} \cdot s^{-1} \cdot kg^{-1}_{metals}$) and the produced CO/H_2 ratio over a bimetallic Fe–Ni catalyst with Ni/(Ni + Fe) = 0.65 during DRM at 1023 K (total pressure of 101.3 kPa and CH_4:CO_2 = 1:1). W_{metals}/F^0_{CH4} = 0.025 $mol_{CH4} \cdot s^{-1} \cdot kg^{-1}_{metals}$, X_{CH4}: from 62% to 24%. Reproduced from [116].

The ratio between reducing and oxidizing gases determines the material's position in the iron/iron oxides system and is as such very important for the stability of Fe containing alloys (Figure 17) [22,49,117]. The outlet gas of a reforming reaction contains syngas, a mixture of CO and H_2, both reducing gases, as well as unreacted CO_2 and H_2O, from the reverse water-gas-shift reaction, as oxidizing gases. The reduction potential of this gas mixture strongly depends on the ratio between reducing and oxidizing gases. Indeed, the presence of CO_2 or H_2O in the reaction mixture significantly decreases the achieved reduction degree of iron oxide because they both act as oxidizing agents. The ratio R_c, or reduction capacity, which indicates the reducing strength of the gas composition, can be expressed as follows:

$$R_c = (CO + H_2)/(CO_2 + H_2O) \tag{1}$$

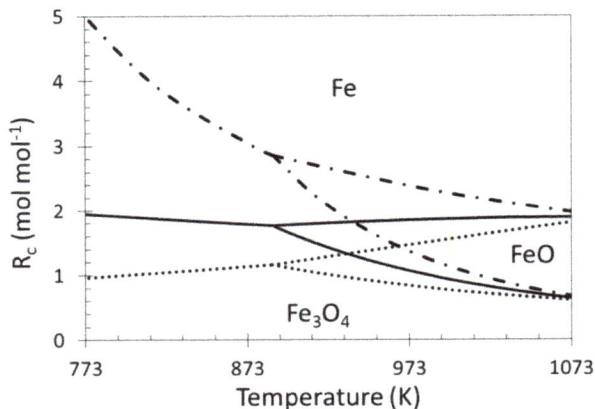

Figure 17. Phase diagram showing the equilibrium lines between Fe_3O_4, FeO and Fe as a function of temperature and reduction capacity in presence of: $-\cdot-$ H_2/H_2O; —— $H_2/CO/H_2O/CO_2$ (equimolar amount of C and H_2 corresponding with a feed of $CH_4 + CO_2$); ········ CO/CO_2. Obtained from [49].

However, during methane reforming, iron involved in the CO_2 or H_2O activation will be segregated from the Fe–Ni alloy [118], even under an overall reducing environment ($R_c > 1$). This redistribution of elements could eventually result in Fe species located on top of alloy particles [101,116]. Wang and co-workers examined Fe–Ni catalysts supported on Al_2O_3 for steam reforming of tars and used Extended X-ray absorption fine structure (EXAFS) spectroscopy to analyze the local structure of the Fe–Ni nano-alloys [21]. They found a lower coordination number for Fe than for Ni, suggesting that Fe/Fe oxide species are enriched in the outer layers of the alloy particles. These iron species can further interact with the C, CH_x and H species at the surface. A similar mechanism of deactivation can be invoked for any high concentration Fe containing alloy: It can decompose at high temperature under H_2O/CO_2 [17,22,102], resulting in segregation of Fe from the alloy (Figure 18). The deactivation can then be attributed to the lowered surface Ni/Fe ratio, since Fe is less active in reforming than Ni [17,101]. All of the above implies that even if R_c can determine the oxidation state of Fe under reaction conditions, the local interaction of Fe with oxidizing gases will lead to iron segregation, independent from the reduction capacity R_c.

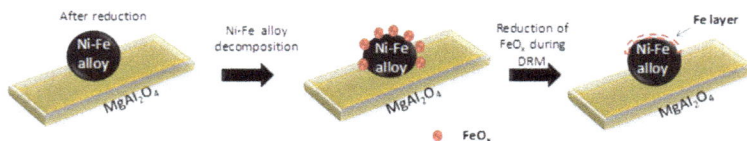

Figure 18. Deactivation due to Fe segregation from the Fe–Ni surface alloy during DRM at high temperature (1023 K). Obtained from [116].

Theofanidis and co-workers [116] evaluated the thermodynamic tendency of Fe to move towards the alloy surface using Density Functional Theory. They compared this tendency of Fe in a bimetallic Fe–Ni and a trimetallic catalyst, containing a noble metal, Pd, Fe–Ni–Pd (Table 1). The DFT calculations reveal that (i) the segregation behavior of Fe is a very strong function of the adsorbate layer present, and (ii) the presence of Pd in a Fe–Ni alloy will reduce the tendency of Fe to segregate to the surface for coverages that are close to what can be expected during DRM conditions.

Table 1. Segregation energies without (ΔE_{seg}, kJ/mol) and with adsorbates (ΔE_{seg}^{ads}) for the exchange of Fe in the subsurface layer of a (111) surface of Ni$_3$Fe or Ni$_2$FePd, with Ni or Pd from the surface layer, for various coverages, representative for methane dry reforming (DRM). All coverages refer to the species adsorbed on the fcc sites of a periodically repeated unit cell with 4 surface atoms. Obtained from [116].

ΔE_{seg} (kJ/mol)	Ni$_3$Fe	Ni$_2$PdFe	
Adsorbate Overlayer	**Fe ↔ Ni**	**Fe ↔ Ni**	**Fe ↔ Pd**
0% (vacuum)	+55	+53	+104
100% H	+49	+52	+43
100% CO	+29	+42	+3
100% O	−94	−76	−218
50% CO, 50% O	−25	−8	−92
50% CO, 25% O, 25% H	+1	+11	+38
25% of CH, CO, O, H	+2	+9	+143

4.1.3. Catalyst Regeneration: Carbon Removal by CO$_2$

Despite the different ways to control catalyst deactivation due to carbon deposition, carbon accumulation will eventually occur during reforming reactions and thus regeneration will be required in order to remove all carbon species [119,120]. Therefore, it is important to understand the catalyst regeneration mechanisms. The rate of carbon removal depends on its structure [121], location [122] and on the nature of the catalyst [123–125].

The existence of two different carbon species structures, graphitic and amorphous, was observed by Guo and co-workers [126], who performed Raman spectroscopy over Ni/MgAl$_2$O$_4$ after coking via CH$_4$ temperature programmed decomposition. Raman spectroscopy is widely used in order to investigate the structure and crystallite size of carbon species [127]. It provides information about the electronic properties and can detect the presence of ordered carbon species [126]. The Raman spectrum of a single crystal graphene sample only shows the G band at approximately 1581 cm^{-1} Raman shift. However, in case of imperfect, polycrystalline graphite and other carbonaceous materials [128], additional bands are detected at 1355 cm^{-1} (D band) and 1620 cm^{-1} (D' band). The ratio of areas I_D/I_G has been correlated to the inverse crystallite size of graphite [129].

In alignment with Guo, Theofanidis and co-workers found the presence of amorphous and graphitic-like carbon using Raman (Figure 19) and TEM (Figure 20). Figure 19 shows the Raman spectra for graphite, a spent Fe–Ni catalyst (with Ni/(Ni + Fe) ratio of 0.6) after 1 h TOS during DRM at 1023 K, the same catalyst after CO$_2$-TPO to 950 K and after CO$_2$-TPO to 1123 K. The analysis for the spent Ni–Fe catalyst (black line in Figure 18) confirmed the existence of two types of carbon species structures. The G band of single crystal graphene, shifted from 1581 cm^{-1} to 1584 cm^{-1}, implies the presence of graphitic-like carbon species on the catalyst (more graphene layers). According to literature, the G Raman peak changes in position, shape and intensity as a function of the number of graphene layers [130]. The D and D' bands at 1350 and 1619 cm^{-1} were also observed and attributed to a defective and disordered structure [128,130]. This disordered carbon species structure, following from the D band, can be amorphous. The Raman spectrum of the Ni–Fe catalyst after CO$_2$-TPO at 950 K (grey line in Figure 19) showed the same peaks as the spent Ni–Fe catalyst, implying the existence of the same types of carbon. Finally, the same types of carbon were observed on the Ni-Fe catalyst after CO$_2$ treatment at 1123 K [118].

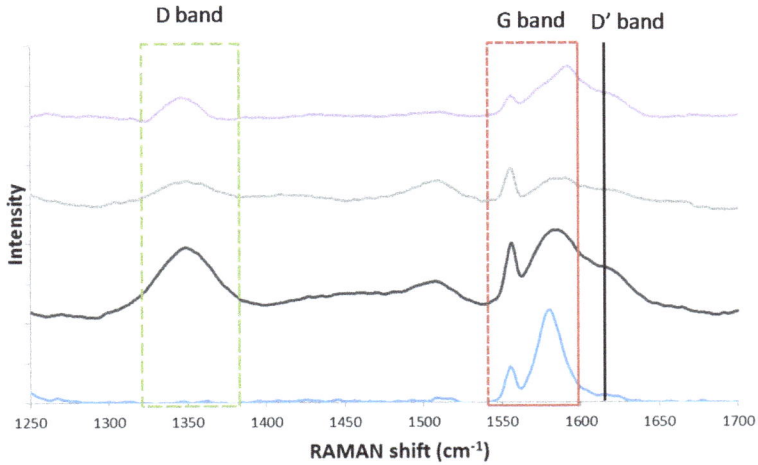

Figure 19. Raman spectrum of the spent Fe–Ni catalyst, with Ni/(Ni + Fe) ratio of 0.6 (DRM for 1 h, 1023 K, $CH_4:CO_2$ = 1.1, total pressure of 101.3 kPa). Blue line: pure graphite as a reference, black line: spent Fe–Ni catalyst, grey line: spent Fe–Ni catalyst after CO_2-TPO up to 950 K, purple line: Spent Fe–Ni catalyst after CO_2-TPO up to 1123 K. Obtained from [118].

Figure 20A shows a TEM image of a spent Fe–Ni catalyst with Ni/(Ni + Fe) ratio of 0.6. The presence of filamentous carbon with Fe–Ni nano-alloy particles on top is observed, which can be verified by the EDX mapping (Figure 20B–D).

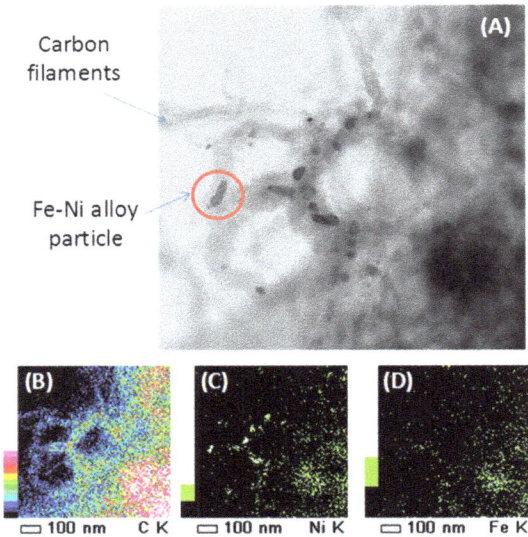

Figure 20. (**A**) HRTEM image of a spent Fe–Ni catalyst with Ni/(Ni + Fe) ratio of 0.6 (after DRM at 1023 K, $CH_4:CO_2:He$ = 1.1:1:1, total pressure of 101.3 kPa, reaction time 1 h). EDX element mapping of (**B**) carbon, (**C**) Ni and (**D**) Fe. Obtained from [118].

CO$_2$-regeneration resulted in the removal of carbon on the active metals of the catalysts [118]. However, EDX-STEM (Energy-dispersive X-ray spectroscopy Scanning Transmission Electron Microscope) mapping (Figure 21) showed the persistence of carbon species located far from the catalyst active metals, implying the absence of direct interaction between carbon species and CO$_2$ from the gas phase.

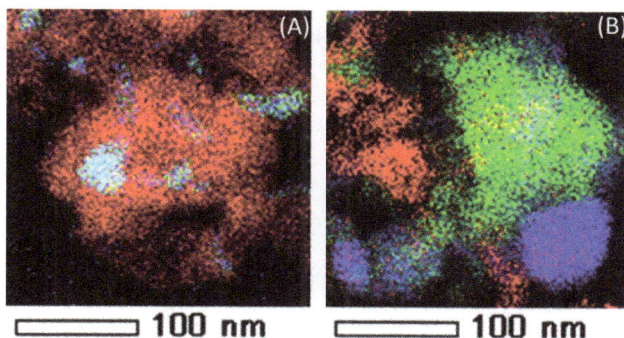

Figure 21. EDX element mapping of Fe–Ni. (**A**) After DRM (1023 K, CH$_4$/CO$_2$/He = 1.1/1/1, total pressure of 101.3 kPa, reaction time 1 h). (**B**) After CO$_2$ oxidation (1 mL/s of CO$_2$ at a total pressure of 101.3 kPa and 1123 K). Red, green and blue colors correspond to carbon, Fe and Ni elements respectively. Obtained from [118].

Theofanidis and co-workers [118] used operando XRD and isothermal experiments in a Temporal Analysis of Products (TAP) reactor, in order to unravel the major mechanistic aspects of carbon species removal by CO$_2$ over a spent Fe–Ni catalyst. They reported that the process could be described by two parallel contributions (Figure 22): (1) Dissociation of CO$_2$ over Ni followed by the oxidation of carbon species by surface oxygen; (2) Fe oxidation by CO$_2$ and subsequent carbon species oxidation by Fe oxide lattice oxygen (Fe oxide reduction step).

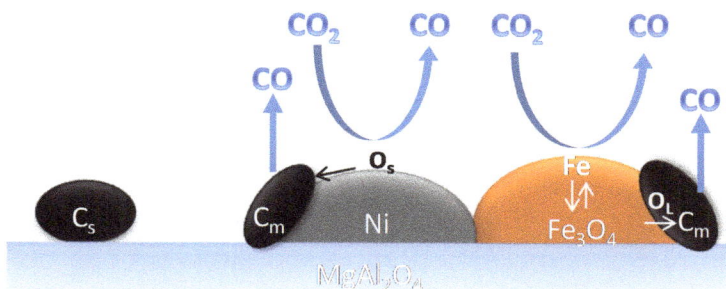

Figure 22. Schematic representation of carbon species removal by CO$_2$ over Fe–Ni catalyst. C$_s$: deposited carbon. O$_s$: surface oxygen, O$_L$: lattice oxygen. C$_m$: carbon deposited on metals, C$_s$: Carbon deposited far from metals, O$_s$: surface oxygen, O$_L$: lattice oxygen. The carbon illustration is not corresponding to the real carbon structure. Obtained from [118].

4.2. Trimetallic Fe-Containing Alloys for Hydrocarbon Conversion

A trimetallic Fe-containing alloy, along with Ni and Pd supported on $MgAl_2O_4$, forming upon H_2-Temperature programmed reduction (TPR), was also reported by Theofanidis and co-workers [116]. Time-resolved in situ XRD (Figure 23) was used to follow up on the phases. The diffraction peaks associated to Fe_2O_3 were not detected due to the low concentration and their overlapping with $MgAl_2O_4$ peaks. However, during reduction, PdO peaks disappeared at 400 K and NiO peaks above 800 K. The metallic Pd related diffraction shifted from 40.1° to an angle of 42.4°, above 820 K, higher than that for Ni–Pd alloy (41.9°), which was hence attributed to a trimetallic Fe–Ni–Pd alloy diffraction peak [116].

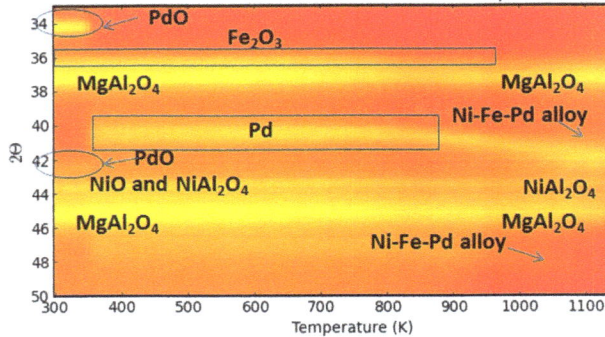

Figure 23. 2D in situ XRD pattern during H_2-TPR for Fe–Ni–Pd. Heating rate: 30 K/min, maximum temperature 1123 K, flow rate: 1 NmL/s, 10%H_2/He. Obtained from [116].

The elemental distribution of "as-prepared" and reduced Fe–Ni–Pd catalyst is indicated in Figure 24, using energy-dispersive X-ray spectroscopy (EDX)-STEM mapping. Oxide clusters are detected in the as-prepared sample, Ni (green), Fe (red) and Pd (blue), while upon reduction the elements get redistributed, resulting in the formation of a trimetallic alloy in the outer shell. Based upon element loadings, the core of the alloy will be close to bimetallic Fe–Ni, while the surface contains truly trimetallic Fe–Ni–Pd [116]. The trimetallic Fe–Ni–Pd alloy with low Pd concentration, less than 0.5 wt %, has been utilized for syngas production [116], displaying promising results in terms of suppressing carbon formation due to Fe presence. The stability of Fe–Ni catalyst increases due to Pd addition by means of a thin Fe–Ni–Pd shell surface layer in the alloy. The latter acts as a barrier for Fe segregation from the core during syngas production [116].

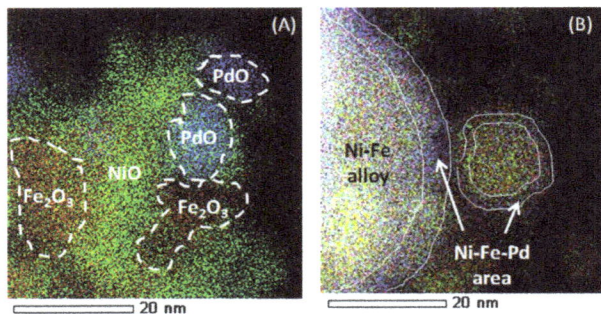

Figure 24. EDX element mapping of a Fe–Ni–Pd catalyst supported on $MgAl_2O_4$. (**A**) as-prepared (**B**) reduced (1 NmL/s of 5%H_2/He mixture at a total pressure of 101.3 kPa and 1123 K). Red, green and blue colors correspond to Fe, Ni and Pd elements, respectively. Obtained from [116].

Noble metals like Pt and Pd are good dehydrogenation catalysts that have been widely used [131–136]. The property of the aforementioned Fe–Ni–Pd catalyst to form a core-shell alloy structure after reduction, where small concentrations of Pd are mainly located in the shell, in combination with the carbon-resistance of the catalyst due to the presence of Fe, can be exploited during propane dehydrogenation (PDH) and oxidative propane dehydrogenation (OPDH). The dehydrogenation of light alkanes (ethane, propane, butane) obtained from natural gas sources is considered an important route for the selective production of high-purity alkenes, which are basic chemicals for the industry. An important industrial propylene production is based on selective, non-oxidative propane dehydrogenation resulting in catalyst deactivation, low conversion. Oxidative dehydrogenation (ODH) provides a promising alternative route based on elimination of thermodynamic limitations and avoiding of catalyst regeneration. Indeed, co-feeding an oxidant such as CO_2 can offer a myriad of opportunities, especially for catalysts containing Fe, which has proven to suppress carbon deposition. Furthermore, the oxidant CO_2 will react with product H_2, thereby shifting the equilibrium and enhancing the catalyst selectivity. The by-products of the CO_2-ODH reaction are CO and H_2O, via the reverse water gas shift reaction. Catalysts with redox properties, such as Fe-based catalysts, could possess high catalytic activity for the various ODH reactions of hydrocarbons.

Our preliminary results show that the addition of Pd to Fe–Ni slightly increase the selectivity of the catalyst towards the main product of C_3H_6, while the C_3H_8 conversion during propane dehydrogenation at 873 K was slightly higher compared to bimetallic Fe–Ni. On the other hand, during oxidative propane dehydrogenation, the trimetallic Fe–Ni–Pd showed slightly higher C_3H_8 conversion, but lower selectivity compared to Fe–Ni.

Further optimization of the catalysts is needed in order to fine-tune the catalytic properties through alloying. Nano-alloys synthesized by mixing elements, can produce intermetallic compounds with significantly modified properties compared to the monometallic counterparts, due to "synergistic effects". Their chemical reactivity can be changed by modifying the composition and atomic ordering, as well as the size of the clusters. This ability to modify and fine-tune properties through alloying is the reason why the field of nano-alloys in catalysis is increasingly attracting scientific attention.

5. Summary and Outlook: The Role of Fe

Significant progress has been achieved in the past few years on understanding the role of Fe in nano-materials, in view of further utilization in chemical processes as a promoter or catalyst. In this review, the role of Fe, the current challenges and the future opportunities of using Fe in catalytic systems have been presented and discussed.

(1) The addition of Fe, either in bimetallic catalysts or incorporated into the support lattice, can provide redox functionality to the catalyst, which helps to suppress carbon formation.

The bimetallic Fe–Ni catalyst showed higher activity and stability compared to the monometallic samples, as the FeO_x species which form under reaction conditions in the presence of an oxidizing agent (CO_2, H_2O or O_2), react via a redox mechanism with the carbon deposits. On the other hand, the Fe concentration is a crucial parameter for the catalytic stability, because of Fe segregation from the Fe–Ni alloy under reaction conditions. Therefore, Ni-rich catalysts with Ni/(Ni + Fe) ratio equal to or higher than 0.8 are preferred. The dosed amount of Fe can still increase the carbon-resistance of the catalyst, while, at the same time avoiding deactivation due to blocking of Ni sites.

(2) The mechanism of carbon species removal by CO_2 over bimetallic Fe–Ni is different from that over a monometallic Ni catalyst.

Carbon deposits close to active metals can be removed by CO_2, a process that can be described by two parallel contributions. One contains the dissociation of CO_2 over Ni and subsequent oxidation of carbon species by the surface oxygen. The second consists of the Fe oxidation by CO_2 followed by carbon species oxidation by Fe oxide lattice oxygen, i.e., Fe oxide reduction.

(3) The redox properties of Fe can be exploited in different processes.

The use of Fe is not limited to the processes described in this review. The super-dry reforming process was developed based on Fe redox properties. Fe_2O_3 supported on $MgAl_2O_4$ was used as a solid oxygen carrier material and three molecules of CO_2 were consumed per one CH_4, resulting in an enhanced CO production. Because of the multiple oxidation states of Fe, Fe–Ni alloys were also exploited as oxygen carriers during chemical looping dry reforming, tuning the product selectivities when CH_4 is used as a fuel.

The novel $MgFe_xAl_{2-x}O_4$ support, where Fe is incorporated in the octahedral sites of the magnesium aluminate spinel structure can be further optimized and exploited as a new, low cost support material for different processes. The redox functionality acquired by the Fe addition to magnesium aluminate combined with enhanced thermal stability are required properties that a support material should offer. Further insight in catalyst optimization, in terms of activity and stability, can be obtained by investigating the oxygen mobility of this material when a metal, such as Ni, is deposited on top of the $MgFe_xAl_{2-x}O_4$ support.

Author Contributions: S.A.T. made the literature analysis and wrote the manuscript. C.K. made literature analysis, performed the oxidative dehydrogenation experiments and wrote 10% of the manuscript. V.V.G. participated in discussions related to the content of the manuscript. H.P. participated in discussions related to the content of the manuscript and wrote 10%. G.B.M. participated in discussions related to the content of the manuscript.

Acknowledgments: This work was supported by the "Long Term Structural Methusalem Funding by the Flemish Government" and the Fund for Scientific Research Flanders (FWO-Vlaanderen) in supplying financing of travel costs and beam time at the DUBBLE beam line of the ESRF. The authors acknowledge the assistance from Alessandro Longo (DUBBLE beamline, ESRF), support from C. Detavernier with the in situ XRD equipment (Department of Solid State Sciences, Ghent University) and from Vitaliy Bliznuk (Department of Materials Science and Engineering, Ghent University) and Lukas Buelens (Laboratory for Chemical Technology, Ghent University) for the HRTEM measurements.

Conflicts of Interest: The authors declare no conflicts of interest.

References

1. Parkinson, G.S. Iron oxide surfaces. *Surf. Sci. Rep.* **2016**, *71*, 272–365. [CrossRef]
2. Bragg, W.H. The structure of magnetite and the spinels. *Nature* **1915**, *95*, 561. [CrossRef]
3. Coltrain, B.K.; Herron, N.; Busch, D.H. Oxygen Activation by Transition Metal Complexes of Macrobicyclic Cyclidene Ligands. In *The Activation of Dioxygen and Homogeneous Catalytic Oxidation*; Barton, D.H.R., Martell, A.E., Sawyer, D.T., Eds.; Springer: Boston, MA, USA, 1993; pp. 359–380.
4. Dlouhy, A.C.; Outten, C.E. The Iron Metallome in Eukaryotic Organisms. *Met. Ions Life Sci.* **2013**, *12*, 241–278. [PubMed]
5. Yan, L.; Zhang, S.; Chen, P.; Liu, H.; Yin, H.; Li, H. Magnetotactic bacteria, magnetosomes and their application. *Microbiol. Res.* **2012**, *167*, 507–519. [CrossRef] [PubMed]
6. Encyclopedia, A. Prehistoric Colour Palette. Available online: http://www.visual-arts-cork.com/index.htm (accessed on 14 May 2018).
7. Bellis, M. The Compass and Other Magnetic Innovations. Available online: https://www.thoughtco.com/ (accessed on 14 May 2018).
8. Davis, C.P. Hemoglobin (Low and High Range Causes). Available online: https://www.medicinenet.com/script/main/hp.asp (accessed on 14 May 2018).
9. Xu, P.; Zeng, G.M.; Huang, D.L.; Feng, C.L.; Hu, S.; Zhao, M.H.; Lai, C.; Wei, Z.; Huang, C.; Xie, G.X.; et al. Use of iron oxide nanomaterials in wastewater treatment: A review. *Sci. Total Environ.* **2012**, *424*, 1–10. [CrossRef] [PubMed]
10. Jagadeesh, R.V.; Surkus, A.-E.; Junge, H.; Pohl, M.-M.; Radnik, J.; Rabeah, J.; Huan, H.; Schünemann, V.; Brückner, A.; Beller, M. Nanoscale Fe_2O_3-Based Catalysts for Selective Hydrogenation of Nitroarenes to Anilines. *Science* **2013**, *342*, 1073–1076. [CrossRef] [PubMed]
11. Bagheri, S.; Julkapli, N.M. Modified iron oxide nanomaterials: Functionalization and application. *J. Magn. Magn. Mater.* **2016**, *416*, 117–133. [CrossRef]

12. Bykova, E.; Dubrovinsky, L.; Dubrovinskaia, N.; Bykov, M.; McCammon, C.; Ovsyannikov, S.V.; Liermann, H.P.; Kupenko, I.; Chumakov, A.I.; Ruffer, R.; et al. Structural complexity of simple Fe$_2$O$_3$ at high pressures and temperatures. *Nat. Commun.* **2016**, *7*, 10661. [CrossRef] [PubMed]

13. Chen, G.; Zhao, Y.; Fu, G.; Duchesne, P.N.; Gu, L.; Zheng, Y.; Weng, X.; Chen, M.; Zhang, P.; Pao, C.-W.; et al. Interfacial Effects in Iron-Nickel Hydroxide–Platinum Nanoparticles Enhance Catalytic Oxidation. *Science* **2014**, *344*, 495–499. [CrossRef] [PubMed]

14. Handa, S.; Wang, Y.; Gallou, F.; Lipshutz, B.H. Sustainable Fe–ppm Pd nanoparticle catalysis of Suzuki-Miyaura cross-couplings in water. *Science* **2015**, *349*, 1087–1091. [CrossRef] [PubMed]

15. Zhao, M.; Yuan, K.; Wang, Y.; Li, G.; Guo, J.; Gu, L.; Hu, W.; Zhao, H.; Tang, Z. Metal–organic frameworks as selectivity regulators for hydrogenation reactions. *Nature* **2016**, *539*, 76–80. [CrossRef] [PubMed]

16. Tartaj, P.; Morales, M.P.; Gonzalez-Carreño, T.; Veintemillas-Verdaguer, S.; Serna, C.J. The Iron Oxides Strike Back: From Biomedical Applications to Energy Storage Devices and Photoelectrochemical Water Splitting. *Adv. Mater.* **2011**, *23*, 5243–5249. [CrossRef] [PubMed]

17. Theofanidis, S.A.; Galvita, V.V.; Poelman, H.; Marin, G.B. Enhanced Carbon-Resistant Dry Reforming Fe-Ni Catalyst: Role of Fe. *ACS Catal.* **2015**, *5*, 3028–3039. [CrossRef]

18. Ashok, J.; Kawi, S. Nickel–Iron Alloy Supported over Iron–Alumina Catalysts for Steam Reforming of Biomass Tar Model Compound. *ACS Catal.* **2014**, *4*, 289–301. [CrossRef]

19. Li, D.; Koike, M.; Wang, L.; Nakagawa, Y.; Xu, Y.; Tomishige, K. Regenerability of Hydrotalcite-Derived Nickel–Iron Alloy Nanoparticles for Syngas Production from Biomass Tar. *ChemSusChem* **2014**, *7*, 510–522. [CrossRef] [PubMed]

20. Koike, M.; Li, D.; Nakagawa, Y.; Tomishige, K. A Highly Active and Coke-Resistant Steam Reforming Catalyst Comprising Uniform Nickel–Iron Alloy Nanoparticles. *ChemSusChem* **2012**, *5*, 2312–2314. [CrossRef] [PubMed]

21. Wang, L.; Li, D.; Koike, M.; Koso, S.; Nakagawa, Y.; Xu, Y.; Tomishige, K. Catalytic performance and characterization of Ni-Fe catalysts for the steam reforming of tar from biomass pyrolysis to synthesis gas. *Appl. Catal. A* **2011**, *392*, 248–255. [CrossRef]

22. Hu, J.; Buelens, L.; Theofanidis, S.-A.; Galvita, V.V.; Poelman, H.; Marin, G.B. CO$_2$ conversion to CO by auto-thermal catalyst-assisted chemical looping. *J. CO2 Util.* **2016**, *16*, 8–16. [CrossRef]

23. Haider, M.A.; Gogate, M.R.; Davis, R.J. Fe-promotion of supported Rh catalysts for direct conversion of syngas to ethanol. *J. Catal.* **2009**, *261*, 9–16. [CrossRef]

24. Burch, R.; Petch, M.I. Investigation of the synthesis of oxygenates from carbon monoxide/hydrogen mixtures on supported rhodium catalysts. *Appl. Catal. A* **1992**, *88*, 39–60. [CrossRef]

25. Liu, Y.; Tuysuz, H.; Jia, C.-J.; Schwickardi, M.; Rinaldi, R.; Lu, A.-H.; Schmidt, W.; Schuth, F. From glycerol to allyl alcohol: Iron oxide catalyzed dehydration and consecutive hydrogen transfer. *Chem. Commun.* **2010**, *46*, 1238–1240. [CrossRef] [PubMed]

26. Konaka, A.; Tago, T.; Yoshikawa, T.; Nakamura, A.; Masuda, T. Conversion of glycerol into allyl alcohol over potassium-supported zirconia–iron oxide catalyst. *Appl. Catal. B* **2014**, *146*, 267–273. [CrossRef]

27. Grossale, A.; Nova, I.; Tronconi, E. Study of a Fe–zeolite-based system as NH$_3$-SCR catalyst for diesel exhaust aftertreatment. *Catal. Today* **2008**, *136*, 18–27. [CrossRef]

28. Colombo, M.; Nova, I.; Tronconi, E.; Schmeißer, V.; Bandl-Konrad, B.; Zimmermann, L. NO/NO$_2$/N$_2$O–NH$_3$ SCR reactions over a commercial Fe-zeolite catalyst for diesel exhaust aftertreatment: Intrinsic kinetics and monolith converter modelling. *Appl. Catal. B* **2012**, *111–112*, 106–118. [CrossRef]

29. Kandel, K.; Anderegg, J.W.; Nelson, N.C.; Chaudhary, U.; Slowing, I.I. Supported iron nanoparticles for the hydrodeoxygenation of microalgal oil to green diesel. *J. Catal.* **2014**, *314*, 142–148. [CrossRef]

30. Li, X.; Zhai, Z.; Tang, C.; Sun, L.; Zhang, Y.; Bai, W. Production of propionic acid via hydrodeoxygenation of lactic acid over Fe$_x$O$_y$ catalysts. *RSC Adv.* **2016**, *6*, 62252–62262. [CrossRef]

31. Luska, K.L.; Bordet, A.; Tricard, S.; Sinev, I.; Grünert, W.; Chaudret, B.; Leitner, W. Enhancing the Catalytic Properties of Ruthenium Nanoparticle-SILP Catalysts by Dilution with Iron. *ACS Catal.* **2016**, *6*, 3719–3726. [CrossRef]

32. Tamura, M.; Yonezawa, D.; Oshino, T.; Nakagawa, Y.; Tomishige, K. In Situ Formed Fe Cation Modified Ir/MgO Catalyst for Selective Hydrogenation of Unsaturated Carbonyl Compounds. *ACS Catal.* **2017**, *7*, 5103–5111. [CrossRef]

33. Zeng, D.; Liu, S.; Gong, W.; Wang, G.; Qiu, J.; Chen, H. Effect of Surface Properties of Iron Oxide Sorbents on Hydrogen Sulfide Removal from Odor. *CLEAN* **2015**, *43*, 975–979. [CrossRef]

34. Pérez-Rodríguez, S.; Pastor, E.; Lázaro, M.J. Noble metal-free catalysts supported on carbon for CO_2 electrochemical reduction. *J. CO2 Util.* **2017**, *18*, 41–52. [CrossRef]

35. Zhang, L.; Wu, H.B.; Lou, X.W. Iron-Oxide-Based Advanced Anode Materials for Lithium-Ion Batteries. *Adv. Energy Mater.* **2014**, *4*, 1300958. [CrossRef]

36. Galvita, V.V.; Filez, M.; Poelman, H.; Bliznuk, V.; Marin, G.B. The Role of Different Types of CuO in CuO-CeO_2/Al_2O_3 for Total Oxidation. *Catal. Lett.* **2014**, *144*, 32–43. [CrossRef]

37. Meledina, M.; Turner, S.; Galvita, V.V.; Poelman, H.; Marin, G.B.; Van Tendeloo, G. Local environment of Fe dopants in nanoscale Fe: CeO_{2-x} oxygen storage material. *Nanoscale* **2015**, *7*, 3196–3204. [CrossRef] [PubMed]

38. Najera, M.; Solunke, R.; Gardner, T.; Veser, G. Carbon capture and utilization via chemical looping dry reforming. *Chem. Eng. Res. Des.* **2011**, *89*, 1533–1543. [CrossRef]

39. Cho, W.C.; Kim, C.G.; Jeong, S.U.; Park, C.S.; Kang, K.S.; Lee, D.Y.; Kim, S.D. Activation and Reactivity of Iron Oxides as Oxygen Carriers for Hydrogen Production by Chemical Looping. *Ind. Eng. Chem. Res.* **2015**, *54*, 3091–3100. [CrossRef]

40. Abad, A.; Mattisson, T.; Lyngfelt, A.; Johansson, M. The use of iron oxide as oxygen carrier in a chemical-looping reactor. *Fuel* **2007**, *86*, 1021–1035. [CrossRef]

41. Zhu, M.; Wachs, I.E. Iron-Based Catalysts for the High-Temperature Water–Gas Shift (HT-WGS) Reaction: A Review. *ACS Catal.* **2016**, *6*, 722–732. [CrossRef]

42. Meshkani, F.; Rezaei, M. A highly active and stable chromium free iron based catalyst for H_2 purification in high temperature water gas shift reaction. *Int. J. Hydrog. Energy* **2014**, *39*, 18302–18311. [CrossRef]

43. Galvita, V.; Sundmacher, K. Cyclic water gas shift reactor (CWGS) for carbon monoxide removal from hydrogen feed gas for PEM fuel cells. *Chem. Eng. J.* **2007**, *134*, 168–174. [CrossRef]

44. Galvita, V.; Hempel, T.; Lorenz, H.; Rihko-Struckmann, L.K.; Sundmacher, K. Deactivation of Modified Iron Oxide Materials in the Cyclic Water Gas Shift Process for CO-Free Hydrogen Production. *Ind. Eng. Chem. Res.* **2008**, *47*, 303–310. [CrossRef]

45. Galvita, V.; Schröder, T.; Munder, B.; Sundmacher, K. Production of hydrogen with low CO_x-content for PEM fuel cells by cyclic water gas shift reactor. *Int. J. Hydrog. Energy* **2008**, *33*, 1354–1360. [CrossRef]

46. Sebastián, D.; Serov, A.; Artyushkova, K.; Gordon, J.; Atanassov, P.; Aricò, A.S.; Baglio, V. High Performance and Cost-Effective Direct Methanol Fuel Cells: Fe-N-C Methanol-Tolerant Oxygen Reduction Reaction Catalysts. *ChemSusChem* **2016**, *9*, 1986–1995. [CrossRef] [PubMed]

47. Galvita, V.V.; Poelman, H.; Marin, G.B. Combined Chemical Looping: New Possibilities for Energy Storage and Conversion. *Energy Fuels* **2017**, *31*, 11509–11514. [CrossRef]

48. Galvita, V.V.; Poelman, H.; Fornero, E.; Saeys, M.; Marin, G.B. *Development and Performance of Iron Based Oxygen Carriers for Chemical Looping*; Wiley-VCH Verlag GmbH & Co. KGaA: Weinheim, Germany, 2017.

49. Buelens, L.C.; Galvita, V.V.; Poelman, H.; Detavernier, C.; Marin, G.B. Super-dry reforming of methane intensifies CO_2 utilization via Le Chatelier's principle. *Science* **2016**, *354*, 449–452. [CrossRef] [PubMed]

50. Miller, D.D.; Siriwardane, R. Mechanism of Methane Chemical Looping Combustion with Hematite Promoted with CeO_2. *Energy Fuels* **2013**, *27*, 4087–4096. [CrossRef]

51. Luo, M.; Yi, Y.; Wang, S.; Wang, Z.; Du, M.; Pan, J.; Wang, Q. Review of hydrogen production using chemical-looping technology. *Renew. Sustain. Energy Rev.* **2018**, *81*, 3186–3214. [CrossRef]

52. Protasova, L.; Snijkers, F. Recent developments in oxygen carrier materials for hydrogen production via chemical looping processes. *Fuel* **2016**, *181*, 75–93. [CrossRef]

53. Urasaki, K.; Tanimoto, N.; Hayashi, T.; Sekine, Y.; Kikuchi, E.; Matsukata, M. Hydrogen production via steam–iron reaction using iron oxide modified with very small amounts of palladium and zirconia. *Appl. Catal. A Gen.* **2005**, *288*, 143–148. [CrossRef]

54. Otsuka, K.; Kaburagi, T.; Yamada, C.; Takenaka, S. Chemical storage of hydrogen by modified iron oxides. *J. Power Sources* **2003**, *122*, 111–121. [CrossRef]

55. Corbella, B.M.; Palacios, J.M. Titania-supported iron oxide as oxygen carrier for chemical-looping combustion of methane. *Fuel* **2007**, *86*, 113–122. [CrossRef]

56. Rihko-Struckmann, L.K.; Datta, P.; Wenzel, M.; Sundmacher, K.; Dharanipragada, N.V.R.A.; Poelman, H.; Galvita, V.V.; Marin, G.B. Hydrogen and Carbon Monoxide Production by Chemical Looping over Iron-Aluminium Oxides. *Energy Technol.* **2016**, *4*, 304–313. [CrossRef]

57. Galinsky, N.L.; Shafiefarhood, A.; Chen, Y.; Neal, L.; Li, F. Effect of support on redox stability of iron oxide for chemical looping conversion of methane. *Appl. Catal. B Environ.* **2015**, *164*, 371–379. [CrossRef]

58. Hu, J.; Galvita, V.V.; Poelman, H.; Detavernier, C.; Marin, G.B. A core-shell structured $Fe_2O_3/ZrO_2@ZrO_2$ nanomaterial with enhanced redox activity and stability for CO_2 conversion. *J. CO2 Util.* **2017**, *17*, 20–31. [CrossRef]

59. Dharanipragada, N.V.R.A.; Galvita, V.V.; Poelman, H.; Buelens, L.C.; Detavernier, C.; Marin, G.B. Bifunctional Co- and Ni- ferrites for catalyst-assisted chemical looping with alcohols. *Appl. Catal. B Environ.* **2018**, *222*, 59–72. [CrossRef]

60. Galvita, V.; Sundmacher, K. Redox behavior and reduction mechanism of Fe_2O_3–$CeZrO_2$ as oxygen storage material. *J. Mater. Sci.* **2007**, *42*, 9300–9307. [CrossRef]

61. Zafar, Q.; Mattisson, T.; Gevert, B. Integrated Hydrogen and Power Production with CO_2 Capture Using Chemical-Looping ReformingRedox Reactivity of Particles of CuO, Mn_2O_3, NiO, and Fe_2O_3 Using SiO_2 as a Support. *Ind. Eng. Chem. Res.* **2005**, *44*, 3485–3496. [CrossRef]

62. Leion, H.; Mattisson, T.; Lyngfelt, A. The use of petroleum coke as fuel in chemical-looping combustion. *Fuel* **2007**, *86*, 1947–1958. [CrossRef]

63. Shulman, A.; Linderholm, C.; Mattisson, T.; Lyngfelt, A. High Reactivity and Mechanical Durability of $NiO/NiAl_2O_4$ and $NiO/NiAl_2O_4/MgAl_2O_4$ Oxygen Carrier Particles Used for more than 1000 h in a 10 kW CLC Reactor. *Ind. Eng. Chem. Res.* **2009**, *48*, 7400–7405. [CrossRef]

64. Johansson, M.; Mattisson, T.; Lyngfelt, A. Investigation of Fe_2O_3 with $MgAl_2O_4$ for Chemical-Looping Combustion. *Ind. Eng. Chem. Res.* **2004**, *43*, 6978–6987. [CrossRef]

65. Galvita, V.V.; Poelman, H.; Bliznuk, V.; Detavernier, C.; Marin, G.B. CeO_2-Modified Fe_2O_3 for CO_2 Utilization via Chemical Looping. *Ind. Eng. Chem. Res.* **2013**, *52*, 8416–8426. [CrossRef]

66. Tang, M.; Xu, L.; Fan, M. Progress in oxygen carrier development of methane-based chemical-looping reforming: A review. *Appl. Energy* **2015**, *151*, 143–156. [CrossRef]

67. Zhu, X.; Li, K.; Wei, Y.; Wang, H.; Sun, L. Chemical-Looping Steam Methane Reforming over a CeO_2–Fe_2O_3 Oxygen Carrier: Evolution of Its Structure and Reducibility. *Energy Fuels* **2014**, *28*, 754–760. [CrossRef]

68. Zeng, T.; Chen, W.-W.; Cirtiu, C.M.; Moores, A.; Song, G.; Li, C.-J. Fe_3O_4 nanoparticles: A robust and magnetically recoverable catalyst for three-component coupling of aldehyde, alkyne and amine. *Green Chem.* **2010**, *12*, 570–573. [CrossRef]

69. Hudson, R.; Riviere, A.; Cirtiu, C.M.; Luska, K.L.; Moores, A. Iron-iron oxide core-shell nanoparticles are active and magnetically recyclable olefin and alkyne hydrogenation catalysts in protic and aqueous media. *Chem. Commun.* **2012**, *48*, 3360–3362. [CrossRef] [PubMed]

70. Yfanti, V.-L.; Vasiliadou, E.S.; Sklari, S.; Lemonidou, A.A. Hydrodeoxygenation of glycerol with in situ H_2 formation over Pt catalysts supported on Fe modified Al_2O_3: Effect of Fe loading. *J. Chem. Technol. Biotechnol.* **2017**, *92*, 2236–2245. [CrossRef]

71. De Vos, Y.; Jacobs, M.; Van Driessche, I.; Van Der Voort, P.; Snijkers, F.; Verberckmoes, A. Processing and characterization of Fe-based oxygen carriers for chemical looping for hydrogen production. *Int. J. Greenhouse Gas Control* **2018**, *70*, 12–21. [CrossRef]

72. Tu, Y.; Tian, S.; Kong, L.; Xiong, Y. Co-catalytic effect of sewage sludge-derived char as the support of Fenton-like catalyst. *Chem. Eng. J.* **2012**, *185–186*, 44–51. [CrossRef]

73. Luo, M.; Yuan, S.; Tong, M.; Liao, P.; Xie, W.; Xu, X. An integrated catalyst of Pd supported on magnetic Fe_3O_4 nanoparticles: Simultaneous production of H_2O_2 and Fe^{2+} for efficient electro-Fenton degradation of organic contaminants. *Water Res.* **2014**, *48*, 190–199. [CrossRef] [PubMed]

74. Amanatidou, E.; Samiotis, G.; Trikoilidou, E.; Tsikritzis, L. Particulate organics degradation and sludge minimization in aerobic, complete SRT bioreactors. *Water Res.* **2016**, *94*, 288–295. [CrossRef] [PubMed]

75. Bolt, P.H.; Habraken, F.H.P.M.; Geus, J.W. Formation of Nickel, Cobalt, Copper, and Iron Aluminates from α- and γ-Alumina-Supported Oxides: A Comparative Study. *J. Solid State Chem.* **1998**, *135*, 59–69. [CrossRef]

76. Reshetenko, T.V.; Avdeeva, L.B.; Khassin, A.A.; Kustova, G.N.; Ushakov, V.A.; Moroz, E.M.; Shmakov, A.N.; Kriventsov, V.V.; Kochubey, D.I.; Pavlyukhin, Y.T.; et al. Coprecipitated iron-containing catalysts (Fe-Al$_2$O$_3$, Fe-Co-Al$_2$O$_3$, Fe-Ni-Al$_2$O$_3$) for methane decomposition at moderate temperatures: I. Genesis of calcined and reduced catalysts. *Appl. Catal. A* **2004**, *268*, 127–138. [CrossRef]

77. Martins, N.; Martins, L.; Amorim, C.; Amaral, V.; Pombeiro, A. Solvent-Free Microwave-Induced Oxidation of Alcohols Catalyzed by Ferrite Magnetic Nanoparticles. *Catalysts* **2017**, *7*, 222. [CrossRef]

78. Martins, N.; Martins, L.; Amorim, C.; Amaral, V.; Pombeiro, A. First-Row-Transition Ion Metals(II)-EDTA Functionalized Magnetic Nanoparticles as Catalysts for Solvent-Free Microwave-Induced Oxidation of Alcohols. *Catalysts* **2017**, *7*, 335. [CrossRef]

79. Baldychev, I.; Vohs, J.M.; Gorte, R.J. The effect of thermodynamic properties of zirconia-supported Fe$_3$O$_4$ on water-gas shift activity. *Appl. Catal. A* **2009**, *356*, 225–230. [CrossRef]

80. Cabello, A.; Dueso, C.; García-Labiano, F.; Gayán, P.; Abad, A.; de Diego, L.F.; Adánez, J. Performance of a highly reactive impregnated Fe$_2$O$_3$/Al$_2$O$_3$ oxygen carrier with CH$_4$ and H$_2$S in a 500Wth CLC unit. *Fuel* **2014**, *121*, 117–125. [CrossRef]

81. Dharanipragada, N.V.R.A.; Buelens, L.C.; Poelman, H.; De Grave, E.; Galvita, V.V.; Marin, G.B. Mg-Fe-Al-O for advanced CO$_2$ to CO conversion: Carbon monoxide yield vs. oxygen storage capacity. *J. Mater. Chem. A* **2015**, *3*, 16251–16262. [CrossRef]

82. Fan, L.-S.; Zeng, L.; Luo, S. Chemical-looping technology platform. *AICHE J.* **2015**, *61*, 2–22. [CrossRef]

83. Galvita, V.V.; Poelman, H.; Marin, G.B. Combined chemical looping for energy storage and conversion. *J. Power Sources* **2015**, *286*, 362–370. [CrossRef]

84. Iliuta, I.; Tahoces, R.; Patience, G.S.; Rifflart, S.; Luck, F. Chemical-looping combustion process: Kinetics and mathematical modeling. *AICHE J.* **2010**, *56*, 1063–1079. [CrossRef]

85. Dharanipragada, N.V.R.A.; Galvita, V.V.; Poelman, H.; Buelens, L.C.; Marin, G.B.; Longo, A. Insight in kinetics from pre-edge features using time resolved in situ XAS. *AICHE J.* **2017**. [CrossRef]

86. Wang, J.; Li, C. A study of surface and inner layer compositions of Mg-Fe-Al-O mixed spinel sulfur-transfer catalyst using Auger electron spectroscopy. *Mater. Lett.* **1997**, *32*, 223–227. [CrossRef]

87. Saravanan, G.; Jayasree, K.P.; Divya, Y.; Pallavi, M.; Nitin, L. Ordered intermetallic Pt-Fe nano-catalysts for carbon monoxide and benzene oxidation. *Intermetallics* **2018**, *94*, 179–185. [CrossRef]

88. Jiang, G.; Li, X.; Lv, X.; Chen, L. Core/shell FePd/Pd catalyst with a superior activity to Pt in oxygen reduction reaction. *Sci. Bull.* **2016**, *61*, 1248–1254. [CrossRef]

89. Wang, J.; Zhang, Q.; Wang, Y. Rh-catalyzed syngas conversion to ethanol: Studies on the promoting effect of FeO$_x$. *Catal. Today* **2011**, *171*, 257–265. [CrossRef]

90. Dimitrakopoulou, M.; Huang, X.; Krohnert, J.; Teschner, D.; Praetz, S.; Schlesiger, C.; Malzer, W.; Janke, C.; Schwab, E.; Rosowski, F.; et al. Insights into structure and dynamics of (Mn,Fe)O$_x$-promoted Rh nanoparticles. *Faraday Discuss.* **2017**. [CrossRef]

91. Palomino, R.M.; Magee, J.W.; Llorca, J.; Senanayake, S.D.; White, M.G. The effect of Fe–Rh alloying on CO hydrogenation to C$_{2+}$ oxygenates. *J. Catal.* **2015**, *329*, 87–94. [CrossRef]

92. Yang, N.; Medford, A.J.; Liu, X.; Studt, F.; Bligaard, T.; Bent, S.F.; Nørskov, J.K. Intrinsic Selectivity and Structure Sensitivity of Rhodium Catalysts for C$_{2+}$ Oxygenate Production. *J. Am. Chem. Soc.* **2016**, *138*, 3705–3714. [CrossRef] [PubMed]

93. Liu, Y.; Göeltl, F.; Ro, I.; Ball, M.R.; Sener, C.; Aragão, I.B.; Zanchet, D.; Huber, G.W.; Mavrikakis, M.; Dumesic, J.A. Synthesis Gas Conversion over Rh-Based Catalysts Promoted by Fe and Mn. *ACS Catal.* **2017**, *7*, 4550–4563. [CrossRef]

94. Phan, T.N.; Ko, C.H. Synergistic effects of Ru and Fe on titania-supported catalyst for enhanced anisole hydrodeoxygenation selectivity. *Catal. Today* **2017**. [CrossRef]

95. Tharamani, C.N.; Beera, P.; Jayaram, V.; Begum, N.S.; Mayanna, S.M. Studies on electrodeposition of Fe–W alloys for fuel cell applications. *Appl. Surf. Sci.* **2006**, *253*, 2031–2037. [CrossRef]

96. Shi, X.; Yu, H.; Gao, S.; Li, X.; Fang, H.; Li, R.; Li, Y.; Zhang, L.; Liang, X.; Yuan, Y. Synergistic effect of nitrogen-doped carbon-nanotube-supported Cu–Fe catalyst for the synthesis of higher alcohols from syngas. *Fuel* **2017**, *210*, 241–248. [CrossRef]

97. Koike, M.; Hisada, Y.; Wang, L.; Li, D.; Watanabe, H.; Nakagawa, Y.; Tomishige, K. High catalytic activity of Co-Fe/α-Al$_2$O$_3$ in the steam reforming of toluene in the presence of hydrogen. *Appl. Catal. B* **2013**, *140–141*, 652–662. [CrossRef]

98. Unmuth, E.E.; Schwartz, L.H.; Butt, J.B. Iron alloy Fischer-Tropsch catalysts: I. Oxidation-reduction studies of the Fe-Ni system. *J. Catal.* **1980**, *61*, 242–255. [CrossRef]

99. Okamoto, H. *Handbook of Binary Alloy Phase Diagrams*; ASM International®: Materials Park, OH, USA, 1990.

100. Kustov, A.L.; Frey, A.M.; Larsen, K.E.; Johannessen, T.; Nørskov, J.K.; Christensen, C.H. CO methanation over supported bimetallic Ni–Fe catalysts: From computational studies towards catalyst optimization. *Appl. Catal. A* **2007**, *320*, 98–104. [CrossRef]

101. Kim, S.M.; Abdala, P.M.; Margossian, T.; Hosseini, D.; Foppa, L.; Armutlulu, A.; van Beek, W.; Comas-Vives, A.; Copéret, C.; Müller, C. Cooperativity and Dynamics Increase the Performance of NiFe Dry Reforming Catalysts. *J. Am. Chem. Soc.* **2017**, *139*, 1937–1949. [CrossRef] [PubMed]

102. Galvita, V.V.; Poelman, H.; Detavernier, C.; Marin, G.B. Catalyst-assisted chemical looping for CO_2 conversion to CO. *Appl. Catal. B* **2015**, *164*, 184–191. [CrossRef]

103. Zhou, L.; Enakonda, L.R.; Saih, Y.; Loptain, S.; Gary, D.; Del-Gallo, P.; Basset, J.M. Catalytic Methane Decomposition over $Fe-Al_2O_3$. *ChemSusChem* **2016**, *9*, 1243–1248. [CrossRef] [PubMed]

104. Reshetenko, T.V.; Avdeeva, L.B.; Ushakov, V.A.; Moroz, E.M.; Shmakov, A.N.; Kriventsov, V.V.; Kochubey, D.I.; Pavlyukhin, Y.T.; Chuvilin, A.L.; Ismagilov, Z.R. Coprecipitated iron-containing catalysts ($Fe-Al_2O_3$, $Fe-Co-Al_2O_3$, $Fe-Ni-Al_2O_3$) for methane decomposition at moderate temperatures: Part II. Evolution of the catalysts in reaction. *Appl. Catal. A* **2004**, *270*, 87–99. [CrossRef]

105. Kong, J.; Soh, H.T.; Cassell, A.M.; Quate, C.F.; Dai, H. Synthesis of individual single-walled carbon nanotubes on patterned silicon wafers. *Nature* **1998**, *395*, 878–881. [CrossRef]

106. Boskovic, B.O.; Stolojan, V.; Khan, R.U.A.; Haq, S.; Silva, S.R.P. Large-area synthesis of carbon nanofibres at room temperature. *Nat. Mater.* **2002**, *1*, 165–168. [CrossRef] [PubMed]

107. Wang, D.; Yang, G.; Ma, Q.; Wu, M.; Tan, Y.; Yoneyama, Y.; Tsubaki, N. Confinement Effect of Carbon Nanotubes: Copper Nanoparticles Filled Carbon Nanotubes for Hydrogenation of Methyl Acetate. *ACS Catal.* **2012**, *2*, 1958–1966. [CrossRef]

108. Iijima, S. Helical microtubules of graphitic carbon. *Nature* **1991**, *354*, 56–58. [CrossRef]

109. Pan, X.; Bao, X. Reactions over catalysts confined in carbon nanotubes. *Chem. Commun.* **2008**. [CrossRef] [PubMed]

110. Wang, G.; Jin, Y.; Liu, G.; Li, Y. Production of Hydrogen and Nanocarbon from Catalytic Decomposition of Methane over a $Ni-Fe/Al_2O_3$ Catalyst. *Energy Fuels* **2013**, *27*, 4448–4456. [CrossRef]

111. Lobo, L.S.; Figueiredo, J.L.; Bernardo, C.A. Carbon formation and gasification on metals. Bulk diffusion mechanism: A reassessment. *Catal. Today* **2011**, *178*, 110–116. [CrossRef]

112. Lobo, L.S. Carbon Formation from Hydrocarbons on Metals. Ph.D. Thesis, Imperial College of Science and Technology, London, UK, 1971.

113. Rostrup-Nielsen, J.; Trimm, D.L. Mechanisms of carbon formation on nickel-containing catalysts. *J. Catal.* **1977**, *48*, 155–165. [CrossRef]

114. Chesnokov, V.V.; Buyanov, R.A. The formation of carbon filaments upon decomposition of hydrocarbons catalysed by iron subgroup metals and their alloys. *Russ. Chem. Rev.* **2000**, *69*, 623–638. [CrossRef]

115. Shah, N.; Panjala, D.; Huffman, G.P. Hydrogen Production by Catalytic Decomposition of Methane. *Energy Fuels* **2001**, *15*, 1528–1534. [CrossRef]

116. Theofanidis, S.A.; Galvita, V.V.; Sabbe, M.; Poelman, H.; Detavernier, C.; Marin, G.B. Controlling the stability of a Fe–Ni reforming catalyst: Structural organization of the active components. *Appl. Catal. B* **2017**, *209*, 405–416. [CrossRef]

117. Heidebrecht, P.; Sundmacher, K. Thermodynamic analysis of a cyclic water gas-shift reactor (CWGSR) for hydrogen production. *Chem. Eng. Sci.* **2009**, *64*, 5057–5065. [CrossRef]

118. Theofanidis, S.A.; Batchu, R.; Galvita, V.V.; Poelman, H.; Marin, G.B. Carbon gasification from Fe–Ni catalysts after methane dry reforming. *Appl. Catal. B* **2016**, *185*, 42–55. [CrossRef]

119. Lobo, L.S. Intrinsic kinetics in carbon gasification: Understanding linearity, "nanoworms" and alloy catalysts. *Appl. Catal. B Environ.* **2014**, *148–149*, 136–143. [CrossRef]

120. Lobo, L.S. Catalytic Carbon Gasification: Review of Observed Kinetics and Proposed Mechanisms or Models—Highlighting Carbon Bulk Diffusion. *Catal. Rev.* **2013**, *55*, 210–254. [CrossRef]

121. Zong, N.; Liu, Y. Learning about the mechanism of carbon gasification by CO_2 from DSC and TG data. *Thermochim. Acta* **2012**, *527*, 22–26. [CrossRef]

122. Pakhare, D.; Spivey, J. A review of dry (CO_2) reforming of methane over noble metal catalysts. *Chem. Soc. Rev.* **2014**, *43*, 7813–7837. [CrossRef] [PubMed]

123. Gardner, T.H.; Spivey, J.J.; Kugler, E.L.; Pakhare, D. CH_4–CO_2 reforming over Ni-substituted barium hexaaluminate catalysts. *Appl. Catal. A* **2013**, *455*, 129–136. [CrossRef]

124. Gac, W.; Denis, A.; Borowiecki, T.; Kępiński, L. Methane decomposition over Ni–MgO–Al_2O_3 catalysts. *Appl. Catal. A* **2009**, *357*, 236–243. [CrossRef]

125. Al–Fatish, A.S.A.; Ibrahim, A.A.; Fakeeha, A.H.; Soliman, M.A.; Siddiqui, M.R.H.; Abasaeed, A.E. Coke formation during CO_2 reforming of CH_4 over alumina-supported nickel catalysts. *Appl. Catal. A* **2009**, *364*, 150–155. [CrossRef]

126. Guo, J.; Lou, H.; Zheng, X. The deposition of coke from methane on a Ni/$MgAl_2O_4$ catalyst. *Carbon* **2007**, *45*, 1314–1321. [CrossRef]

127. Espinat, D.; Dexpert, H.; Freund, E.; Martino, G.; Couzi, M.; Lespade, P.; Cruege, F. Characterization of the coke formed on reforming catalysts by laser raman spectroscopy. *Appl. Catal.* **1985**, *16*, 343–354. [CrossRef]

128. Darmstadt, H.; Sümmchen, L.; Ting, J.M.; Roland, U.; Kaliaguine, S.; Roy, C. Effects of surface treatment on the bulk chemistry and structure of vapor grown carbon fibers. *Carbon* **1997**, *35*, 1581–1585. [CrossRef]

129. Jawhari, T.; Roid, A.; Casado, J. Raman spectroscopic characterization of some commercially available carbon black materials. *Carbon* **1995**, *33*, 1561–1565. [CrossRef]

130. Ferrari, A.C. Raman spectroscopy of graphene and graphite: Disorder, electron–phonon coupling, doping and nonadiabatic effects. *Solid State Commun.* **2007**, *143*, 47–57. [CrossRef]

131. De Miguel, S.R.; Jablonski, E.L.; Castro, A.A.; Scelza, O.A. Highly selective and stable multimetallic catalysts for propane dehydrogenation. *J. Chem. Technol. Biotechnol.* **2000**, *75*, 596–600. [CrossRef]

132. Siddiqi, G.; Sun, P.; Galvita, V.; Bell, A.T. Catalyst performance of novel Pt/Mg(Ga)(Al)O catalysts for alkane dehydrogenation. *J. Catal.* **2010**, *274*, 200–206. [CrossRef]

133. Jablonski, E.L.; Castro, A.A.; Scelza, O.A.; de Miguel, S.R. Effect of Ga addition to Pt/Al_2O_3 on the activity, selectivity and deactivation in the propane dehydrogenation. *Appl. Catal. A* **1999**, *183*, 189–198. [CrossRef]

134. Nawaz, Z.; Wei, F. Hydrothermal study of Pt–Sn-based SAPO-34 supported novel catalyst used for selective propane dehydrogenation to propylene. *J. Ind. Eng. Chem.* **2010**, *16*, 774–784. [CrossRef]

135. Zhang, Y.; Zhou, Y.; Huang, L.; Zhou, S.; Sheng, X.; Wang, Q.; Zhang, C. Structure and catalytic properties of the Zn-modified ZSM-5 supported platinum catalyst for propane dehydrogenation. *Chem. Eng. J.* **2015**, *270*, 352–361. [CrossRef]

136. Coq, B.; Tijani, A.; Figuéras, F. Influence of alloying platinum for the hydrogenation of p-chloronitrobenzene over PtM/Al_2O_3 catalysts with M: Sn, Pb, Ge, Al, Zn. *J. Mol. Catal.* **1992**, *71*, 317–333. [CrossRef]

materials

MDPI

Review

Advanced Chemical Looping Materials for CO₂ Utilization: A Review

Jiawei Hu, Vladimir V. Galvita *, Hilde Poelman and Guy B. Marin

Laboratory for Chemical Technology, Ghent University, Technologiepark 914, B-9052 Ghent, Belgium; Jiawei.Hu@UGent.be (J.H.); Hilde.Poelman@UGent.be (H.P.); Guy.Marin@UGent.be (G.B.M.)
* Correspondence: Vladimir.Galvita@UGent.be; Tel.: +32-468-10-6004; Fax: +32-9331-1759

Received: 18 May 2018; Accepted: 6 July 2018; Published: 10 July 2018

Abstract: Combining chemical looping with a traditional fuel conversion process yields a promising technology for low-CO₂-emission energy production. Bridged by the cyclic transformation of a looping material (CO₂ carrier or oxygen carrier), a chemical looping process is divided into two spatially or temporally separated half-cycles. Firstly, the oxygen carrier material is reduced by fuel, producing power or chemicals. Then, the material is regenerated by an oxidizer. In chemical looping combustion, a separation-ready CO₂ stream is produced, which significantly improves the CO₂ capture efficiency. In chemical looping reforming, CO₂ can be used as an oxidizer, resulting in a novel approach for efficient CO₂ utilization through reduction to CO. Recently, the novel process of catalyst-assisted chemical looping was proposed, aiming at maximized CO₂ utilization via the achievement of deep reduction of the oxygen carrier in the first half-cycle. It makes use of a bifunctional looping material that combines both catalytic function for efficient fuel conversion and oxygen storage function for redox cycling. For all of these chemical looping technologies, the choice of looping materials is crucial for their industrial application. Therefore, current research is focused on the development of a suitable looping material, which is required to have high redox activity and stability, and good economic and environmental performance. In this review, a series of commonly used metal oxide-based materials are firstly compared as looping material from an industrial-application perspective. The recent advances in the enhancement of the activity and stability of looping materials are discussed. The focus then proceeds to new findings in the development of the bifunctional looping materials employed in the emerging catalyst-assisted chemical looping technology. Among these, the design of core-shell structured Ni-Fe bifunctional nanomaterials shows great potential for catalyst-assisted chemical looping.

Keywords: CH₄ reforming; catalyst-assisted chemical looping; oxygen carrier; metal oxides; structured nanomaterials; core-shell; bifunctional materials

---

1. Introduction

To bolster the rapid economic growth of modern society, the entire world is facing increasing demands on its energy sources. Although the consumption of non-fossil fuels (such as renewable energy and nuclear power) is projected to grow faster than fossil fuels (coal, petroleum and natural gas), the latter are expected to account for almost 80% of global energy use through 2040 [1,2]. Therefore, the continued dependence on fossil fuels for energy supply in the foreseeable future is inevitable. However, fossil fuel-based energy production is widely considered to be a leading cause for greenhouse gas emissions such as CO₂, which is the largest contributor towards the global warming effect [3–8]. With the adoption of the Paris Agreement at the 21st Conference of the Parties of the United Nations Framework Convention on Climate Change (UNFCCC) in December 2015, scientists from all over the world are now being challenged with a new goal: To limit the global temperature increase to 1.5 °C above pre-industrial levels [9].

Hence, there is an urgent need for strategies to reduce CO_2 emissions in various related fields, especially industrial production and electricity generation, which contribute almost half of total emissions [10]. Carbon Capture and Storage (CCS) is a central strategy, as it not only reduces the associated greenhouse gas emissions in conformity with global targets, but also provides the opportunity to allow for a more sustainable use of fossil fuel energy [11]. The applications of the CCS scheme are reflected in two groups of technologies: One is the capture of CO_2 from power plants, followed by compression, transportation and permanent storage [12]; another is the combination with strategies to improve energy efficiency for the separation of CO_2, or the conversion of CO_2 to value added chemicals [13–15]. The recent progress in CO_2 capture and separation technologies with respect to the feasible processes, as well as new materials have been reported in some reviews [3,7,16]. In the past few decades, the large-scale capture of CO_2 using amine absorbers became commercially available and was mainly applied in CO_2 separation from flue gases [17]. This conventional amine scrubbing for CO_2 capture however increases the energy requirements of a plant by 25–40% [18].

From an economic point of view, there is need to improve current technologies for lowering the overall cost of CO_2 capture and separation processes, which can be implemented from two aspects: Upgrading the efficiency of capture materials to decrease material expense, or designing novel process concepts to lower equipment cost. Figure 1 shows a series of important developments of CO_2 capture technologies [16]. Evidently, more time and funding investments are required to commercialize the innovative techniques that possess higher cost reduction benefits. Among the innovating technologies, chemical looping technology has the potential to achieve inherent separation of CO_2 from a fuel feed, thus eliminating energy intensive gas-gas separation costs, expectedly providing highest cost reduction benefits once it is fully commercialized. Furthermore, this technique can also provide an approach to convert CO_2 into value added chemicals and energy sources (syngas, CO) [19–23].

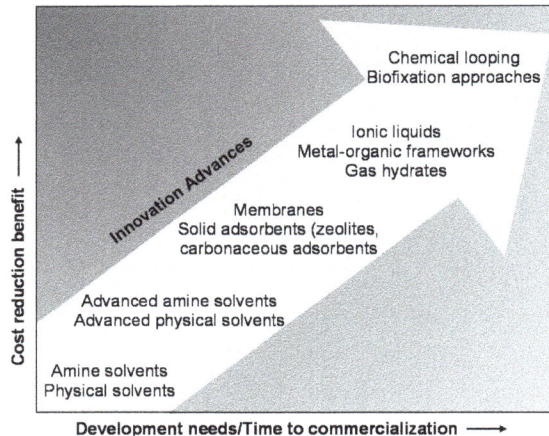

Figure 1. Innovation advances in CO_2 capture technologies: Their cost reduction benefits vs. the remaining development needs/time to commercialization. Reproduced with permission from D'Alessandro, D.M. et al. [16], Angewandte Chemie International Edition; published by John Wiley and Sons, 2010.

The concept chemical looping was first used by Ishida et al. [24] to describe a high-exergy efficiency combustion process that employs redox reactions of chemical intermediates. Later, this concept was extended to a broad range of applications, depending on the nature of a recyclable reaction scheme. Herein, a given reaction is decomposed into multiple sub-reactions, based on the reaction and regeneration of a solid mediator, called looping material (generally a metal oxide) performed in separate reactors or as subsequent steps in a single reactor [22]. Currently, the chemical looping processes can be categorized into two groups according to whether CO_2 generation or separation is involved or not, as displayed in Figure 2 [21].

Extensive studies on group A have been reported [25]. The type A-1 process, also known as chemical looping combustion (CLC) [26], can be divided into two separate half-cycles: In the reduction half-cycle, the metal oxide is reduced, whilst the carbonaceous fuel is fully oxidized to CO_2 and H_2O; the reduced metal oxide is regenerated in the re-oxidation half-cycle by different re-oxidizing agents, such as air, steam or CO_2, simultaneously producing power, H_2 and CO, respectively. In addition, the gaseous O_2 released from some metal oxides upon heating in the reduction half-cycle (i.e., oxygen uncoupling) can directly react with the fuel, which avoids the gas-solid reactions between the fuel and oxygen carrier. Based on this concept, an essential chemical looping process, i.e., chemical looping with oxygen uncoupling (CLOU), was established. Solid fuels, such as coal and biomass, can be used directly for combustion in CLOU process without gasification, showing advantage over a typical CLC process.

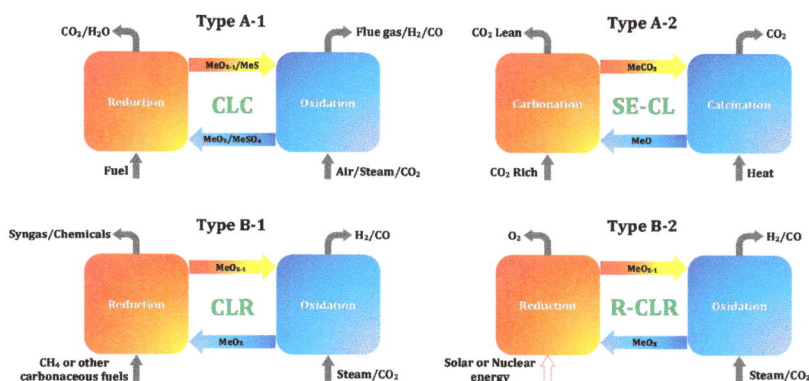

Figure 2. Chemical looping reactions involving CO_2 generation (**top group A**) and no CO_2 generation (**bottom group B**). Note: MeS: Metal sulfide; $MeSO_4$: Metal sulfate; $MeCO_3$: Metal carbonate; MeO: Metal oxide; MeO_X and MeO_{X-1}: Oxidized and reduced form of metal oxide, respectively [21].

The type A-2 process uses a regenerable solid sorbent to achieve the in-situ capture of CO_2 through a cyclic carbonation-calcination reaction before, during or after the carbonaceous fuel combustion or reforming [27]. It provides a viable CO_2 control technology, i.e., sorption-enhanced chemical looping (SE-CL), for process intensification [15] or high-purity chemicals (CO_2-lean) production [28]. The SE-CL process embodies two steps: CO_2 is in-situ absorbed through the carbonation of a metal oxide, forming metal carbonate in the carbonator, afterwards the latter is calcined at higher temperature to regenerate the metal oxide, thereby releasing CO_2.

In group B, a similar metal oxide regeneration process is involved as in group A, but the reactions in the former are not intended to produce CO_2. The type B-1, chemical looping reforming (CLR), aims to reform carbonaceous fuels to syngas (a mixture of H_2 and CO) or chemicals (e.g., ethylene) [29]. Utilization of CO_2 can be achieved if it is used as an oxidizing agent in the re-oxidation half-cycle, turning it into chemical looping dry reforming (CLDR). Compared to conventional dry reforming (DR), application of CLDR brings two main advantages: First, since the looping material is regenerated in each cycle, deposited carbon can be removed by reacting with CO_2. Second, as CO_2 is used as an oxidizing agent for material regeneration in the re-oxidation half-cycle, this process can utilize more CO_2 than conventional dry reforming; in case of CH_4 reforming, the CO_2 converted in CLDR is three times higher than in DR [23].

When using hydrocarbon fuels such as natural gas and biogas in CLDR, the kinetics of the reduction half-cycle is limited. The natural mixture of CH_4 with CO_2 (e.g., in biogas) significantly decreases the rate and degree of metal oxide reduction because CO_2 also acts as an oxidizing agent. Both CO_2 in the feed and as a product have this retarding effect on the metal oxide reduction, thus limiting the conversion of CO_2 to CO in the re-oxidation half-cycle. In order to overcome these drawbacks, an extra catalyst component can be introduced to convert CH_4 and CO_2 into CO and H_2, which then both reduce the metal oxide during the reduction half-cycle. This not only increases the CH_4 and CO_2 conversions, but also diminishes the oxidizing atmosphere because both CO and H_2 are reducing agents. Based on this concept, catalyst-assisted chemical looping dry reforming (CCDR) was recently proposed as an intensified process with the goal of maximum utilization of two main greenhouse gases, CH_4 and CO_2, in terms of CO yield. This process is generally implemented over a bifunctional reactor bed, composed of either a physical mixture of a reforming catalyst and a metal oxide looping material, or a bifunctional looping material [15,23,30]. The entire CCDR process yields the dry reforming of CH_4 into CO as an overall reaction, being a strongly endothermic process. Thermodynamic calculations indicate that an increase of temperature is necessary to obtain high CH_4 and CO_2 conversion, leading to larger energy requirements for heating the reactor, which highly increases the operating cost. Additionally, material deactivation due to active metal sintering will be aggravated under high-temperature conditions during dry reforming [31]. To counter these problems, CCDR was adjusted to yield a new process, called catalyst-assisted chemical looping auto-thermal dry reforming (CCAR) [20]. Here, a small amount of O_2 is added to CH_4 and CO_2 to partially convert CH_4 in an exothermic reaction (partial oxidation or combustion) [31–35], so that heat is generated in-situ, thereby compensating the strong endothermicity of the dry reforming reaction. Based on thermodynamic analysis, the whole CCAR process can run in thermoneutral conditions, i.e., $\Delta H = 0$ kJ·mol^{-1}, with molar ratios of CH_4:CO_2:O_2 = 1:2.4:0.3 at 600 °C or with molar ratios of CH_4:CO_2:O_2 = 1:2:0.5 at 750 °C [20]. Owing to the inherent composition of biogas, typically containing CH_4 (45~75%), CO_2 (25~55%) and O_2 (0.01~5%) [36], the CCAR can efficiently utilize biogas as fuel feed in the reduction half-cycle.

Compared to type B-1, the operation of type B-2 uses carbon neutral energy sources like solar and nuclear instead of carbonaceous fuels, providing a renewable chemical looping reforming (R-CLR) process. In R-CLR, considerable heat from solar or nuclear origin is imposed on the metal oxide to decompose it to metal and oxygen in the reducer. Subsequently, the metal is re-oxidized to metal oxide by steam or CO_2, forming pure H_2 or CO [37].

In light of the above overview of chemical looping processes, the key to success for chemical looping technology are the looping materials. Their adequacy in reactivity at temperatures below the melting point ensures the energy conversion efficiency during chemical looping. On the other hand, the contamination resistance and physical strength of a looping material at high redox reaction temperatures determine the durability of a chemical looping process. Further, the low price and environmental compatibility of the looping materials play a crucial role in the sustainability of the chemical looping technology. Therefore, major efforts have been made to develop the ideal chemical looping materials in the past decade. During the early stages of development, research focused on the metal oxide materials with a single function as CO_2 accepter or oxygen storage, which are mainly used for conventional chemical looping combustion and reforming processes. The advances were reported in a few review papers [1,2,14,21,26,38,39]. More recently, a novel process, so-called catalyst-assisted chemical looping, was proposed with the aim of maximal CO_2 utilization. Since then, the development of chemical looping materials has been directed towards the bifunctional material that integrates catalyst and oxygen storage material in one unit, achieving both catalytic fuel conversion and chemical looping CO_2 utilization. In this review article, an update of the previous overview of chemical looping materials is given, together with an extension focusing on new insights in the development of bifunctional looping materials based on the latest studies.

2. Materials for Chemical Looping

Depending on the chemical looping process performed, the looping materials are mainly classified in two types: CO_2 carrier and oxygen carrier, which are required to have high reactivity and sufficient stability, and to be cost-effective and environmentally friendly [40].

2.1. CO_2 Carrier

Among all the solid CO_2 sorbents, CaO is a proper material to act as CO_2 carrier because of numerous advantages [41,42]: (1) low cost since it can be derived from limestone that is readily available; (2) non-toxic to the environment; (3) the spent materials can be further used in the cement industry after decarburization, supporting sustainable development in economy; (4) CaO may also serve as an alkaline catalyst in biomass conversion processes; and (5) the heat from the exothermic carbonation of CaO can be retrieved to compensate the high heat requirement during chemical looping. However, it is well known that the reactivity decay over cycles of CaO-based CO_2 sorbents, particularly natural material-based sorbents, is inevitable. The main causes for the deactivation are particle sintering and attrition. Hence, many efforts have been made to improve their sintering resistance ability and mechanical strength, including doping of the material to suppress sintering, thermal or chemical pretreatment to achieve superior structural properties, development of new synthetic sorbents with good CO_2 capture capacity and stability. Additionally, the development of feasible technologies for the reactivation of spent sorbents is of equal importance. The recent progress in CaO-based looping materials has been summarized by numerous review articles [43–46].

2.2. Oxygen Carrier

The nature of the oxygen carrier is a crucial matter, because a chemical looping process requires it to possess a certain redox activity and be able to withstand multiple cycles without large loss in the integrity of physical and chemical properties. Ideally, the oxygen carrier should function with reproducible kinetics at high temperature, and the materials that only show reactivity near or even above their melting point should be eliminated [1]. Oxygen transport capacity is another important property of an oxygen carrier. It is defined as the usable oxygen in the oxygen carrier during one redox cycle, which determines the fuel conversion and the circulation rate [2,47]. If chemical looping process is performed in a fluidized bed, the oxygen carriers additionally need to have sufficient mechanical strength to avoid attrition, which is quite common during their lifetime. In this sense, the attrition behavior of materials should be evaluated according to standard crushing tests under chemical looping operation conditions [39]. When carbonaceous and sulfur-containing fuels (fossil fuels and biomass) are used as feedstocks in chemical looping, issues like coke deposition, sulfur poisoning and high operation temperature inevitably result in deactivation of the oxygen carrier. Hence, properties like coke resistance, sulfur tolerance and thermal stability are of great importance for developing a proper oxygen carrier. In addition, the effect of ash formation on the performance of an oxygen carrier cannot be ignored, particularly in the utilization of biomass. Ash with low melting point alkali metals such as Na or K will promote agglomeration of the oxygen carrier [48,49], while high-acid ash (e.g., SiO_2-rich ash) leads to the production of potassium silicates that facilitate particle sintering, suppressing the fuel conversion and CO_2 capture efficiency [50]. Therefore, the resistance of an oxygen carrier to agglomeration is also a critical factor for its practical application. Eventually, the reactivity of oxygen carrier includes not only the theoretical conversion of fuel, relying on the thermodynamic properties of the materials, but also transport and reaction kinetics, which depend on composition, structure as well as texture properties like particle size, porosity and specific surface area [1].

Among various transition metals, Ni, Fe, Cu, Mn and Co-based materials are commonly considered to serve as oxygen carriers, although they present some disadvantages: Low reactivity and low oxygen transport capacity of Fe and Mn, low melting point and high agglomeration tendency of

Cu, as well as high cost and negative health effects of Ni and Co. Figure 3 compares key information for these traditional oxygen carriers [39].

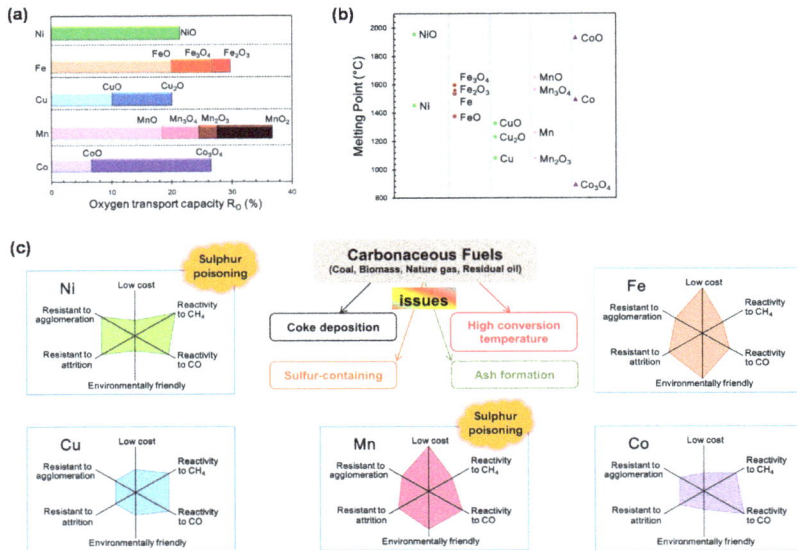

Figure 3. Comparison of Ni, Fe, Cu, Mn and Co-based oxygen carriers in term of oxygen transport capacity (**a**), melting point (**b**) as well as cost, reactivity, agglomeration or attrition resistance (**c**). The oxygen transport capacity R_o (%) is used to evaluate the maximum oxygen transport between the fully reduced (OC_r) and oxidized (OC_o) forms of the oxygen carrier (OC): $R_o = (OC_o - OC_r)/OC_o$. Adapted with permission from Zhao X. et al. [39], Energy & Environmental Science; published by The Royal Society of Chemistry, 2017.

Despite of the toxicity and high cost, Ni-based material still receives major attention as a reliable reforming catalyst due to its high catalytic activity with CH_4 and other light hydrocarbons [51]. Almost complete CH_4 conversion can be obtained over Ni-based oxygen carriers during a CLR process. However, particle sintering and the loss of metallic Ni, owing to partial re-oxidation into NiO, along with the repetition of cycles cause a significant loss of catalytic performance [52]. In addition, Ni-based materials are sensitive to deactivation by sulfur poisoning and coke deposition. Improved Ni-based materials are necessary to be designed to tackle these drawbacks for further application.

Fe-based materials are commonly known as a good option for oxygen carriers due to their low price, high mechanical strength and environmentally friendly nature [39]. They almost have no tendency for carbon or sulfur deposition [53], making them promising candidates in all chemical looping processes involving carbonaceous and sulfur-containing fuels. Further, iron oxide presents the highest oxygen storage capacity from CO_2 (0.7 mol CO_2/mol Fe) over a broad range of operating temperatures (600~1800 °C) [22]. Thus, chemical looping dry reforming over Fe-based oxygen carriers has been one of the target technologies for CO_2 utilization. However, within the multiple oxidized states of Fe system, only the transformation from Fe_2O_3 to Fe_3O_4 is suitable for CLC process due to thermodynamic limitations, leading to low oxygen transport capacity of Fe-based oxygen carriers. Other drawbacks also impede the application of Fe-based oxygen carriers, including relatively low reactivity towards gaseous fuels and the agglomeration related to magnetite formation [54]. The development of new-generation Fe-based oxygen carriers should focus on solving these issues.

The flexible redox behavior between the oxidized (CuO) and reduced (Cu or Cu_2O) forms of Cu ensures that Cu-based materials have both high reactivity and oxygen transfer capacity. The advantages of Cu also include relatively low cost and toxicity, so Cu-based oxygen carriers have received greater attention from both industry and academia. Moreover, the negative influence of sulfur impurities on the performance of Cu-based oxygen carriers during a chemical looping process is not significant when the oxygen carrier-to-fuel ratio lies above 1.5 [53]. There are two main disadvantages that limit the application of Cu-based materials as oxygen carrier: One is attrition owing to their weak mechanical strength, another is agglomeration as a result of the low melting point of Cu. The former can be relieved through stabilizing Cu by inert promoters, such as Al_2O_3, ZrO_2, TiO_2 and SiO_2 [55]. In order to solve the melting problem, keeping the operation temperature lower than 800 °C could be effective [39].

Similar to Fe, Mn has various oxidation states including MnO_2, Mn_2O_3, Mn_3O_4 and MnO, among which Mn_3O_4 is the only oxide phase of Mn that exists at a temperature above 800 °C. Hence, only the redox transformation between Mn_3O_4 and MnO is applicable for the CLC process, showing a relatively low oxygen transport capacity. Additionally, Mn oxides were found to show low reactivity towards CH_4. Similar to Ni-based materials, deactivation of Mn-based materials caused by sulfur contaminants cannot be ignored [56,57]. Nevertheless, Mn-based materials are still accepted to serve as oxygen carriers because of their low toxicity and cheap price. Some related research indicates that Mn promoted with typical materials like SiO_2, Al_2O_3, TiO_2, ZrO_2 and $MgAl_2O_4$, can interact to irreversibly form highly unreactive phases [58,59], making them unsuitable as oxygen carriers. However, Mn-based materials using dopant-stabilized ZrO_2 (applicable dopants such as MgO, CaO, Y_2O_3, La_2O_3 and CeO_2 [60]) as support proved to possess not only high reactivity but also high structural stability and agglomeration resistance during redox cycles, which is prospective as promising Mn-based oxygen carriers [61].

Owing to the high reactivity and oxygen transport capacity, Co-based materials have been increasingly considered as good oxygen carriers, even though they also suffer from cost and environmental concerns. As the highest oxidation state of Co, Co_3O_4 is thermodynamically unstable above 900 °C and easily transformed into CoO. In this respect, the redox transformation between Co and CoO is commonly adopted in chemical looping processes [2]. However, CoO shows strong interaction with common promoter materials such as Al_2O_3, MgO and TiO_2, forming unreactive phases like $CoAl_2O_4$, $Mg_{0.4}Co_{0.6}O$ and $CoTiO_3$, respectively. As a result, these oxygen carriers generally undergo fast deactivation during repeated cycles [62,63]. In contrast, the Co-based oxygen carrier with YSZ (yttria-stabilized zirconia) as support was found to exhibit good reactivity and high resistance to carbon formation in CLC process [62].

In summary, all of the above mentioned factors should be considered and balanced based on the involved feedstock and the desired chemical looping process when selecting an oxygen carrier. The total cost of the typical metal-based oxygen carriers follows the order: Co > Ni > Cu > Fe > Mn-based material (Figure 3), which includes the cost of active materials, promoter materials and the preparation section [2]. Overall, Fe- and Ni-based oxygen carriers are the most popular carriers for chemical looping, particularly for carbonaceous fuels-driven processes, because Fe oxides hardly suffer from carbon deposition and sulfur poisoning, while Ni shows excellent reactivity towards hydrocarbon fuels, comparable with the activity of noble metals (Pt, Pd and Rh).

3. Improvements to Oxygen Carriers

To meet the requirements of industrial chemical looping application, further improvements should be made to oxygen carriers to tackle their shortcomings. There are three main pathways: (1) the introduction of promoter and/or support; (2) the mixing of active metal oxides to integrate various advantages; and (3) the activation and stabilization of oxygen carriers by design of special structures. Here, we focus on reporting the advances related to improved Ni- and Fe-based oxygen carriers. The major improvement approaches, based on the promoters used, the material structure and the applied chemical looping processes, are summarized in Table 1.

Table 1. Summary of the improvement approaches for the different types of oxygen carriers.

Oxygen Carrier (Active Metal)	Promoters/Support	Preparation	Structure	Process	Reactant Gas Reducing	Reactant Gas Oxidizing	Ref.
Monometallic-Based Material							
	α-Al$_2$O$_3$	incipient wetness impregnation	supported	CLC	CH$_4$ or 30%CH$_4$/N$_2$	air or 5%O$_2$/N$_2$	[54,55]
	La-Co/γ-Al$_2$O$_3$	impregnation	supported	CLC	CH$_4$	air	[56]
	Al$_2$O$_3$, sepiolite, SiO$_2$, TiO$_2$, ZrO$_2$	mechanical mixing	mixed	CLC	70%CH$_4$/H$_2$O	air	[58]
	SiO$_2$, MgAl$_2$O$_4$	dry impregnation	supported	CLC and CLR	10%CH$_4$ 10%H$_2$O 5%CO$_2$ 75%N$_2$	5%O$_2$/N$_2$	[59]
	TiO$_2$, ZrO$_2$, TiO$_2$, SiO$_2$, Al$_2$O$_3$, NiAl$_2$O$_4$	incipient wetness impregnation	supported	CLC	20%CH$_4$/N$_2$	air	[67]
NiO		wet impregnation	supported	CLR	CH$_4$/H$_2$O (1/3)	air	[68]
	ZrO$_2$, γ-Al$_2$O$_3$	wet impregnation	supported	CLR	CH$_4$/H$_2$O (1/3)	air	[69]
	Mg-stabilized ZrO$_2$	freeze granulation	supported	CLC	CH$_4$/N$_2$ CH$_4$/H$_2$O	5%O$_2$/N$_2$	[70]
	Yttria-stabilized ZrO$_2$	sol-gel and dissolution	mixed	CLR	H$_2$	air	[71]
	NiAl$_2$O$_4$	sol-gel	spinel	CLC	17%CH$_4$/He	air	[72]
	NiAl$_2$O$_4$	spray-drying	mixed	CLC	50%CH$_4$/H$_2$O	5%O$_2$/N$_2$	[73]
	MgAl$_2$O$_4$	freeze granulation	supported	CLR	Nature gas	air	[74]
	α-Al$_2$O$_3$, MgAl$_2$O$_4$, CaAl$_2$O$_4$	dry impregnation	supported	CLC	10%CH$_4$ 20%H$_2$O 70%N$_2$	air	[75]
	CaAl$_2$O$_4$	impregnation	supported	CLC	15%CH$_4$/N$_2$	10%O$_2$/N$_2$	[76]
	CeO$_2$, La$_2$O$_3$	incipient wetness impregnation	supported	CLC	17%CH$_4$/Ar	air	[77]
	γ-Al$_2$O$_3$	impregnation	supported	CLC	CH$_4$	air	[78]
	α-Al$_2$O$_3$	sol-gel combustion	mixed	CLC	H$_2$	air	[79]
	Al$_2$O$_3$	sol-gel	mixed	CLC	lignite	air	[80]
	γ-Al$_2$O$_3$	impregnation	supported	CLC	ppm/CH$_4$ H$_2$S	air	[81]
	Al$_2$O$_3$	co-precipitation	mixed	CL ^{a}	H$_2$	CO$_2$	[82]
	MgO	theoretical study	supported	CLC	CO	O$_2$	[83]
	TiO$_2$	incipient wetness impregnation	supported	CLC	CH$_4$	O$_2$	[84]
	TiO$_2$	solid-state mixing	mixed	CLC	3%H$_2$/Ar	air	[85]
	TiO$_2$	mechanical mixing	mixed	CLR	syngas	H$_2$O	[86]
Fe$_2$O$_3$	SiO$_2$	dry impregnation	supported	CLC	50%CH$_4$/H$_2$O	5%O$_2$/N$_2$	[87]
	MgAl$_2$O$_4$	freeze granulation	supported	CLC	50%CH$_4$/H$_2$O	air	[88]
	MgFeAlO$_x$	co-precipitation	spinel	CL ^{a}	5%H$_2$/He	CO$_2$	[89]
	MgAl$_2$O$_4$	sequential wetness impregnation	supported	CLR	CO	H$_2$O	[90]
	MgAl$_2$O$_4$	spray-drying	doped	CLR	CH$_4$	H$_2$O	[91]
	CaO	wet impregnation	mixed	CLC	10%CO/N$_2$	20%CO$_2$/N$_2$	[92]
	CaO	mechanical mixing	mixed	CLC	10%CO/N$_2$	20%CO$_2$/N$_2$	[93]
	ZrO$_2$	co-precipitation	mixed	CLC	10%CO/N$_2$	20%CO$_2$/N$_2$	[94]
	ZrO$_2$	theoretical study	supported	CLC	CO	O$_2$	[95]
	ZrO$_2$	co-precipitation	mixed	CLC	CO	O$_2$ H$_2$O	[96]
	Ce, Ca, Mg-stabilized ZrO$_2$	freeze granulation	supported	CLC	CH$_4$/CO$_2$/CO	O$_2$	[5]

Table 1. *Cont.*

Oxygen Carrier (Active Metal)	Promotors/Support	Preparation	Structure	Process	Reactant Gas Reducing	Reactant Gas Oxidizing	Ref.
Monometallic-Based Material							
	CeO_2	co-precipitation	mixed	CL[a]	5%H_2/He	CO_2	[22]
	CeO_2	co-precipitation	mixed	CLR	CH_4	H_2O/N_2	[97]
	CeO_2	nanometric colloidal sol technique	Surfacegrafted	CL[a]	CO or H_2/N_2	O_2/N_2	[98]
	CeO_2	sol-gel	mixed	CLR	CO/H_2/N_2	H_2O/N_2	[99]
	CeO_2	co-precipitation	solid solution	CL[a]	H_2	CO_2	[100]
Mixed oxide-based material							
CoO-NiO	Yttria-stabilized ZrO_2	dissolution	supported	CLC	CH_4/H_2O (1/2)	air	[101]
	Al_2O_3	incipient wetness impregnation	supported	CLC	CH_4	air	[102]
Fe_2O_3-CuO	$MgAl_2O_4$	mechanical mixing	mixed	CLC	syngas or nature gas	air	[103]
$Fe_xMn_{(1-x)}O$	-	co-precipitation	mixed	CLC	5%CH_4/He	air	[104]
	Al_2O_3	spray-drying	supported	CLOU	CH_4 or syngas	5%O_2/N_2	[105]
$Ni_xMn_{(1-x)}O$	-	spray-drying	mixed	CLOU	CH_4 or syngas	5%O_2/N_2	[106]
	Al_2O_3	spray-drying	supported	CLOU	CH_4 or syngas	5%O_2/N_2	[107]
	CeO_2, Al_2O_3	sequential impregnation	supported	CLC	25%CH_4/N_2	10%O_2/N_2	[108]
Fe_2O_3-NiO	Al_2O_3	incipient wetness impregnation	supported	CLC	17%CH_4/Ar	20%O_2/He	[109]
	Al_2O_3	mechanical mixing and impregnation	supported	CLC[b]	biomass	air	[110]
	Al_2O_3	co-precipitation	mixed	CLC[b]	biomass	air	[111]
	$La_{0.8}Sn_{0.2}FeO_3$	sol-gel	supported	CLR	30%CH_4/N_2	30%CO_2/N_2	[112]
Perovskite-structured material							
$LaFeO_3$	-	Solution combustion	perovskite	CLR	CH_4/Ar	O_2/Ar	[113]
$La_{1-x}Sr_xFeO_3$	-	co-precipitation	perovskite	CLR	CH_4	H_2O	[114]
	-	solution combustion	perovskite	CLR	40%CH_4/N_2	air	[115]
$La_{1-x}Sr_xMn_yFe_{1-y}O_3$	M = Ni, Co, Cr, Cu	citrate method	perovskite	CLR	CH_4	O_2 or H_2O	[116]
Core-shell structured material							
NiO (core)	SiO_2 (shell)	SiO_2 coating	core-shell	DR	CO_2:CH_4:N_2 = 1:1:1		[117]
Fe_3O_4 (core)	Al_2O_3 (shell)	atomic layer deposition	core-shell	DR	CO_2:CH_4:He = 1:18		[118]
Fe_2O_3 (core)	SiO_2 (shell)	micremulsion-based synthesis	core-shell	CLC	50%CH_4/N_2	50%O_2/N_2	[119]
	$La_{1-x}Sr_xFeO_3$ (shell)	reverse microemulsion	core-shell	CL[a]	H_2	CO_2	[120]
MeO_x (core) (Me = Mn, Co, Fe)	$La_{1-x}Sr_xFeO_3$ (shell)	sequential sol-gel	core-shell	CLR	CH_4	O_2	[121]
$LaMn_{0.7}Fe_{0.3}O_{3.15}$ (core)	SiO_2 (shell)	modified Pechini method	core-shell	CLR	10%CH_4/He	10%O_2/He	[122]
CuO (shell)		surfactant-templating	core-shell	CLC	CH_4	air	[123]
	TiO_2-Al_2O_3 (core)	self-assembly template combustion	core-shell	CLOU	N_2	O_2	[124]
Fe_2O_3/ZrO_2 (core)	ZrO_2 (shell)	impregnation and nanocoating	core-shell	CL[a]	5%H_2/Ar	O_2 or H_2O	[40]
Bifunctional material							
Ni-Fe_2O_3	CeO_2	co-precipitation and impregnation	supported	CCDR	CO_2:CH_4 = 1:1	CO_2	[23]
$NiFe_2O_4$ or $CoFe_2O_4$	$CoZrO_2$	co-precipitation	spinel	CCDR	methanol or ethanol-H_2O	CO_2	[125]
Ni-Fe_2O_3	$MgAl_2O_4$	co-impregnation	supported	CCAR	CO_2:CH_4:O_2 = 1:1:0.5	CO_2	[20]
$Ni_{(shell)}$-$Fe_2O_{3(core)}$	ZrO_2	impregnation, nanocoating and SiO_2-template assisted	core-shell	CCAR	CO_2:CH_4:O_2 = 1:1:0.2	CO_2	[126]

[a] chemical looping; [b] chemical looping gasification.

3.1. Monometallic Materials

The promoter and support of monometallic materials can be classified as physically or chemically active based on their functions in oxygen carrier. The former usually serve as a physical barrier to improve thermal stability, while the latter contribute their redox properties to enhance reactivity [127].

Regarding to Ni-based oxygen carriers, series of heat-resistant oxides such as Al_2O_3 [64–66], SiO_2 [58,59], TiO_2 [67,68], ZrO_2 and stabilized ZrO_2 [69–71], as well as spinels like $NiAl_2O_4$ [72,73], $MgAl_2O_4$ and $CaAl_2O_4$ [74–76] are usually considered as adequate promoter materials. Among these, NiO/Al_2O_3 as an oxygen carrier demonstrates excellent reactivity, high thermal stability and strong resistance to carbon formation during chemical looping operation. However, the formation of an inactive $NiAl_2O_4$ spinel, owing to the strong interaction between NiO and Al_2O_3, leads to a partial loss of reactivity. As a result of $NiAl_2O_4$ formation, an excess of Ni should be added during material preparation to retain free NiO in the oxygen carrier. Accordingly, the $NiO/NiAl_2O_4$ oxygen carrier still shows high reactivity, although now a larger amount of Ni is required. Alternatively, the chemical passivation (i.e., the spinel formation behavior) can be used to enhance the inert function of Al_2O_3 promoter. Addition of MgO or CaO during the preparation of the NiO/Al_2O_3 oxygen carrier will form a more stable spinel like $MgAl_2O_4$ or $CaAl_2O_4$. The presence of such spinel in the oxygen carrier effectively suppresses the formation of $NiAl_2O_4$, maintaining the high reactivity of NiO during redox cycles and eliminating the need for more Ni. Generally, the use of other promoter materials shows problems with reactivity and mechanical strength: The use of TiO_2 for Ni-based oxygen carrier was found to show slow reaction rate or even absence of reaction, because of the formation of the stable phase $NiTiO_3$; the NiO/SiO_2 oxygen carrier gives much lower reactivity, poor mechanical strength and deactivation with progress of cycling; the NiO/ZrO_2 or $NiO/$stabilized ZrO_2 oxygen carriers present good reactivity but their mechanical strength is weak.

The use of a reducible oxide, particularly CeO_2, as support for a Ni-based oxygen carrier, presents an obvious potential for enhanced redox properties of the oxygen carrier through a synergistic effect between Ni and CeO_2 [77]. This synergistic effect may increase the utilization of Ni species due to the enhanced oxygen transfer from the CeO_2 support to the Ni particles via the Ni-CeO_2 interface, as shown in Figure 4: After the initial part of the reduction half-cycle, a Ni particle consists of two phases, i.e., metallic Ni (the outer layer of the particle) and NiO (the core of the particle). At the same time, CH_4 is cracked on the surface of metallic Ni to form Ni-bound carbonaceous species, which afterwards can be gasified by lattice oxygen derived from nearby CeO_2. The lost lattice oxygen in the support is refilled by oxygen from the core of the Ni particle (where the NiO phase is present), resulting in the continued reduction of the Ni particle.

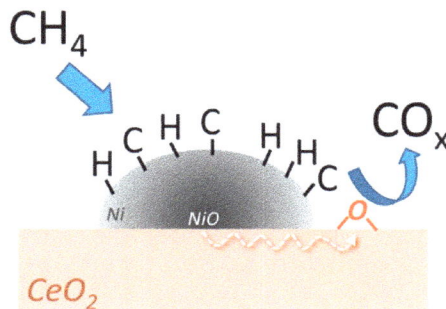

Figure 4. Schematic representation of a proposed mechanism for the synergistic effect between metal and reducible support in a NiO/CeO_2 oxygen carrier for enhanced metal utilization in a chemical looping process. Obtained with permission from Bhavsar S. et al. [77]. Copyright (2013) American Chemical Society.

Although Ni/CeO_2 oxygen carrier shows high reactivity towards CH_4, high carrier utilization and high resistance to carbon deposition in CLC process, a significant loss in surface area owing to sintering of CeO_2 over repeated cycles indicates that the CeO_2 support must be stabilized to make it applicable for industry. It has been confirmed that doping CeO_2 with metals of different valency (such as La and Zr) can not only increase the thermal stability of CeO_2 by maintaining the high surface area and preserving oxygen vacancies, but also further enhance the reducibility by increasing the oxygen mobility [128,129].

Regarding Fe-based oxygen carriers, promoter materials such as Al_2O_3 [78–82], MgO [83,130], TiO_2 [84–86], SiO_2 [87], $MgAl_2O_4$ [88–91,103], CaO [92,93], ZrO_2 [94–96] and CeO_2 [22,97–100] have received extensive scientific attention in the past decades. However, these promoters can easily lead to solid-solid transformation with iron oxides during a continuous cycling process [127]. Depending on the specific promoter, new phases like $AlFe_2O_4$, $MgFe_2O_4$, $FeTiO_3$, and Fe_2SiO_4, $Ca_2Fe_2O_5$ and $CeFeO_3$ have a strong tendency to develop during the redox reactions. The so formed metal-ferrites can continue to serve as a physical barrier to prevent sintering of the iron oxide, but the overall oxygen carrier becomes less reactive in the presence of these phases, since the regeneration of the active iron oxides from the aforementioned metal-ferrites requires much stronger reducing and oxidizing conditions. As a result, the oxygen storage capacity decreases significantly [94,120]. Especially metal-ferrites such as Fe_2SiO_4 and $Ca_2Fe_2O_5$ are unreactive and hard to decompose, and they usually cause a substantial decrease in the oxygen carrier reactivity and selectivity. Because the $MgAl_2O_4$ spinel forms more easily, the addition of an appropriate amount of Mg (the atomic ratio of Mg:Al = 1:2) to Fe_2O_3/Al_2O_3 could successfully alleviate the formation of $FeAl_2O_4$. Further, $MgAl_2O_4$ itself is also favored to act as an oxygen carrier support owing to its high thermal and chemical stability, as well as to its specific heat capacity. While using $Fe_2O_3/MgAl_2O_4$ as an oxygen carrier in a CLR process, an increase in activity and stability with the number of cycles was found. Marked by a strong inertness, the use of ZrO_2 as a promoter has received a great deal of attention. An extensive study on the Fe_2O_3/ZrO_2 oxygen carrier for chemical looping indicated that the physical structure of ZrO_2 was relatively stable during cycling and that no new phase formed [94]. Additionally, it was observed that the use of Ca-, Mg- and Ce-stabilized ZrO_2 as supports for Fe-based oxygen carriers yields a further enhancement of redox properties and thermal stability [54].

CeO_2 is an ideal candidate for promoting iron oxides in a chemical looping process, not only as a physical promoter, but also as a chemically active promoter due to its high activity towards methane or syngas oxidation by lattice oxygen, as well as its reasonable CO_2 and H_2O re-oxidation capacity [22,127]. The redox couple Ce^{4+} and Ce^{3+} facilitates oxygen storage and release from its bulk fluorite lattice. When the Ce^{4+} ions are replaced by lower valence cations (such as Fe^{3+}), oxygen vacancies are created, which are considered to be active sites. After consuming the available surface oxygen for oxidation, extra oxygen can be transferred from bulk to surface more rapidly in a CeO_2-Fe_2O_3 solid solution than in pure Fe_2O_3. The fast oxygen transfer from the bulk of Fe_2O_3 to its surface is enabled by the creation of oxygen vacancies owing to the CeO_2 additive [1]. Therefore, the addition of CeO_2 in Fe-based oxygen carriers enhances the reducibility of Fe_2O_3 by means of CeO_2-Fe_2O_3 interaction. Machida et al. [98] proposed the concept that CeO_2 and Fe_2O_3 can respectively serve as an oxygen gateway and an oxygen reservoir in a combined system, opening up a new type of Ce_2O_3-grafted Fe_2O_3 oxygen carrier that is different from the conventional Fe_2O_3/Ce_2O_3 solid solution materials. To validate this idea, they grafted CeO_2 nanoparticles onto the surface of Fe_2O_3 particles by a nanometric colloidal sol technique (Figure 5a,b). A possible reaction mechanism for the oxygen release and storage over the CeO_2-grafted Fe_2O_3 material (Figure 5c) was elaborated: Under reducing conditions, the surface oxygen of CeO_2 (O_S) is readily removed by a reducing agent (R), yielding an oxygen vacancy (V_O). The latter is instantaneously re-filled by lattice oxygen (O_O) supplied from the Fe_2O_3 bulk, actuating the reduction of Fe_2O_3. Subsequently, the reduced Fe species are quickly re-oxidized by dissociated O_2 transferred from the CeO_2 surface.

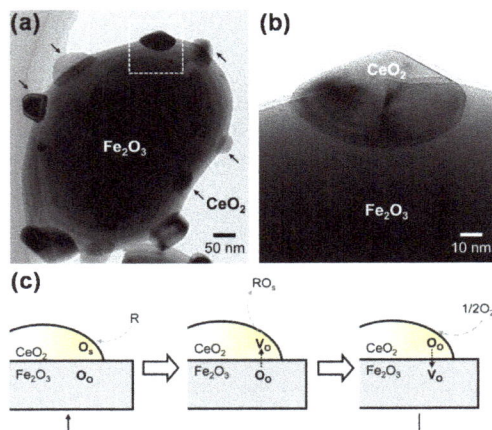

Figure 5. CeO_2-grafted Fe_2O_3 oxygen carrier: Morphological structure: TEM image (**a**) and an enlargement of the square region (**b**); Schematic representation of its oxygen release and storage mechanism (**c**). V_O: oxygen vacancy, O_O: lattice oxygen, O_s: surface oxygen. Adapted with permission from Machida M. et al. [98]. Copyright (2015) American Chemical Society.

3.2. Mixed Oxide Materials

It is well accepted that complex metal oxides may provide better properties than the individual metal oxides. Therefore, oxygen carriers based on mixed oxides have been investigated by combining different active metal oxides in a single particle or mixing different oxygen carriers, each composed of a single metal oxide. The main objective is to combine the complementary advantages of different oxygen carriers for the practical application in a chemical looping process.

The CoO-NiO supported YSZ oxygen carrier was found to have high reactivity towards CH_4, complete absence of coke formation and good regenerative ability in natural gas-fueled CLC, even though the formation of a $NiCoO_2$ solid solution decreased the redox reaction rate to some extent [101]. Hossain et al. [102] reported that the addition of Co into a Ni/Al_2O_3 material not only enhanced the reducibility of the oxygen carrier due to the suppression of $NiAl_2O_4$ formation, but also inhibited the agglomeration of particles by maintaining consistent Ni dispersion throughout the redox cycles.

The incorporation of Fe into Cu-based oxygen carriers can relieve their agglomeration problem, owing to the high melting point of Fe_2O_3 [131]. Wang et al. [103] evaluated the performance of series of Fe_2O_3-$CuO/MgAl_2O_4$ oxygen carriers for coke-oven-gas-based CLC in a laboratory pressurized circulating fluidized bed system. The best working oxygen carrier was found to be $Fe_2O_{3(45)}$-$CuO_{(15)}/MgAl_2O_{4(40)}$, which showed high reactivity and circulation stability. It was also found that the introduction of a NiO-based oxygen carrier into the Fe_2O_3/CuO-based oxygen carrier bed facilitated the reduction of the operating temperature in CLC process, due to the high reactivity of Ni towards CH_4. Further, a layer-by-layer mixing mode of a NiO-based oxygen carrier and a Fe_2O_3/CuO-based oxygen carrier in a fixed-bed reactor avoids the formation of $NiFe_2O_4$ spinel during the reduction half-cycle, ensuring a good redox rate, thereby providing a promising reactor configuration candidate for an enhanced-CH_4-reforming chemical looping process [132].

The bimetallic Fe-Mn-O system shows higher CH_4 combustion (close to 100%) in the absence of oxygen or air, higher CO_2 selectivity (almost no toxic CO byproduct) and more stable multicycle CLC performance, compared to the individual Fe or Mn oxides. Additionally, owing to the low cost of both Fe and Mn, the bimetallic Fe-Mn oxide can be considered as a very competitive oxygen carrier for the chemical looping process [104]. Moreover, both Fe-Mn-O and Ni-Mn-O systems have oxygen uncoupling properties, as a result, $Fe_xMn_{(1-x)}O$ and $NiMn_2O_4$ can be used as oxygen carriers for the CLOU process [105–107].

Continuing the comparison of Ni- and Fe-based oxygen carriers in Section 2.2: Ni shows best activity in CH_4 conversion, but low resistance to carbon formation, whilst Fe has almost no trend to carbon deposition, with high oxygen storage capacity from CO_2, but low reactivity towards CH_4. Due to its low price and non-toxicity, Fe is advantageous from an economic and environmental point of view compared to Ni. Aiming to combine the complementary advantages of these two constituent metals, many efforts have been made to develop bimetallic Fe-Ni oxygen carriers via two different main strategies, physical [108,133] and chemical [23,108–112] mixing. Johansson et al. [133] reported that the addition of only 3 wt % NiO in the bed of a Fe-based oxygen carrier was sufficient to reach a very high CH_4 conversion in CLC process. This is probably due to a synergy effect, i.e., Ni catalyzes the reforming of CH_4 to CO and H_2, which then both react with iron oxide at a considerably higher rate than CH_4. In the case of chemically mixed oxygen carriers, in which the $NiFe_2O_4$ spinel usually forms, the CH_4 conversion and reaction rate in CLC process follow the ranking: $NiO > NiFe_2O_4 > Fe_2O_3$ [109], further indicating that alloying a less reactive metal—Fe—with a highly reactive one—Ni—can significantly improve the activity of the resulting bimetallic oxygen carrier. Compared to pure NiO and Fe_2O_3, $NiFe_2O_4$ is more suitable to serve as oxygen carrier for the CLR process, because the limited oxygen transfer capacity of $NiFe_2O_4$ avoids the complete oxidation of CH_4, which results in an increased selectivity to syngas [112]. Alternatively, the addition of small amount of Fe into a Ni-based oxygen carrier or catalyst can obviously enhance the carbon resistance. Theofanidis et al. [51] prepared a series of bimetallic Fe-Ni/$MgAl_2O_4$ catalysts with different Fe/Ni molar ratios (from 0 to 1.5) to investigate the role of Fe on enhanced carbon resistance during CH_4 dry reforming. They found that the process of dry reforming over bimetallic Fe-Ni catalyst follows the Mars—van Krevelen mechanism, where CO_2 oxidizes Fe to iron oxide and CH_4 is decomposed on Ni sites to H_2 and surface carbon. The latter, simultaneously, can be oxidized by lattice oxygen from iron oxide, yielding CO (Figure 6a). During dry reforming, a Fe-Ni alloy is formed, which still constitutes an active phase and can be regenerated to metallic Ni and Fe_3O_4 by re-oxidizing by steam or CO_2. Compared to monometallic Ni/$MgAl_2O_4$, the carbon deposition on the bimetallic Fe-Ni/$MgAl_2O_4$ is significantly suppressed (Figure 6b).

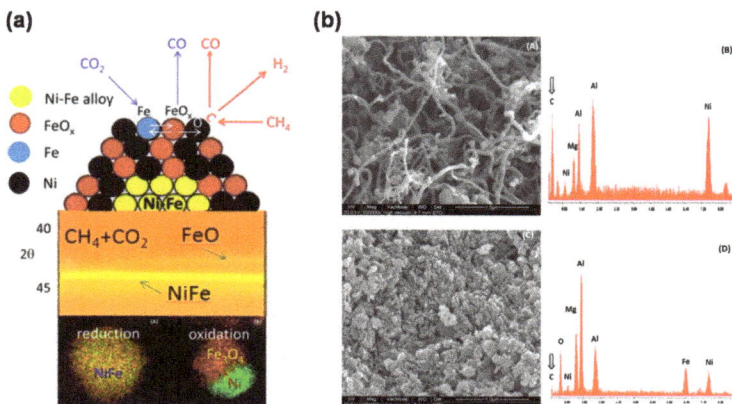

Figure 6. The bimetallic Fe-Ni/$MgAl_2O_4$ catalyst for enhanced carbon-resistant dry reforming: (**a**) The role of Fe for in-situ carbon removal from the catalyst surface during dry reforming; (**b**) Determination of the carbon resistance ability of the bimetallic catalyst by comparison of the spent samples of Ni/$MgAl_2O_4$ (A and B) and $Fe_{1.1}$-Ni/$MgAl_2O_4$ (C and D) through SEM micrographs (left) and EDX analysis (right). Reproduced with permission from Theofanidis S.A. et al. [51]. Copyright (2015) American Chemical Society.

3.3. Structured Materials

3.3.1. Perovskite Structure

Oxygen carriers with perovskite structure are receiving increasing attention owing to their enhanced redox properties, high oxygen mobility and thermal stability. Perovskite-structured oxide materials are typically formulated as ABO_3, where A is a large alkaline earth or rare earth element and B is a smaller transition metal cation [134]. Where the A-site cation is responsible for the thermal resistance, the B-site cation will be the contributor to the reactivity, e.g., the catalytic activity in chemical looping reforming of CH_4. As a result, the selection of B-site cations is crucial in the design of promising perovskite-structured oxygen carriers. In addition, the interaction between the A and B cations can affect the reactivity and selectivity during fuels conversion as well as the stability of the structure [1].

As well-established transition metal for oxygen carriers, Fe^{3+} has certainly been considered as a B-site cation. The performance of $AFeO_3$ (A = La, Nd and Eu) was first investigated for selective oxidation of CH_4 by Dai et al. [135]. It was found that, among the three $AFeO_3$ candidates, $LaFeO_3$ showed the best performance with respect to syngas production and stability during the redox cycling between CH_4 and air at 900 °C. Because of the kinetic limitations, the reduction of $LaFeO_3$ by CH_4 is basically performed from Fe^{3+} to Fe^{2+}, while deeper reduction is very difficult to achieve [113]. Further, partial substitution of the A-site cation by a low oxidation-state metal such as Sr leads to the formation of oxygen vacancies and a fraction of B-site cations in a higher valence state. A highly variable oxygen non-stoichiometry in the perovskite structure is hence introduced that can readily provide lattice oxygen, while subsequently the removed oxygen species can be replenished by gaseous oxygen. Substitution of La^{3+} by Sr^{2+} could cause electronic imbalance, but it can be compensated by the transformation of a fraction of Fe^{3+} to Fe^{4+}. Therefore, $La_{1-x}Sr_xFeO_3$ (x = 0 to ~1) perovskite oxides have been widely investigated as oxygen carriers in CLR processes [114,115,136]. Li et al. [136] proposed that there are two kinds of oxygen species on such perovskite oxides: One is the weakly-bound oxygen species which is active in complete oxidation of CH_4 to CO_2 and H_2O, another is strongly-bound oxygen which is responsible for partial oxidation of CH_4 to CO and H_2. The activity of such perovskite materials was found to show a decrease when x increased above 0.5, probably owing to the suppression of CH_4 decomposition by the Sr substitution [115]. The best performance was presented by $La_{0.7}Sr_{0.3}FeO_3$, and during the reduction half-cycle of chemical looping steam reforming a maximum H_2 yield was obtained over a simple mechanical mixture of $La_{0.7}Sr_{0.3}FeO_3$ with an additional 5 wt % NiO [114]. On the other hand, the partial substitution of active B-site cations by aliovalent transition metal cations can lead to the enhancement of catalytic activity due to synergistic valence changes and introduction of non-stoichiometry–related microstructural defects into the lattice. Lori et al. [116] modified the "base case" material, $La_{0.7}Sr_{0.3}FeO_3$, by transition metals such as Ni, Co, Cu and Cr. All the resulting perovskite-structured materials, with the general formula $La_{0.7}Sr_{0.3}M_yFe_{1-y}O_3$ (M is the additional transition metal), show the capability to act as suitable oxygen carriers in CLR process. The modification with partial substitution of Fe by Cr shows the best improvement compared to others. Additionally, $La_{0.7}Sr_{0.3}Cr_{0.1}Fe_{0.9}O_3$ was found to have the optimum composition, showing highest CH_4 conversion and syngas production, zero carbon deposition as well as good redox properties.

The application of CH_4-based CLR over the perovskite-structured oxygen carriers, that provide lattice oxygen for CH_4 conversion, represents an alternative in terms of high selectivity to syngas and elimination of the air separation process. However, the oxygen storage capacity of perovskite-structured metal oxides is relatively low compared to traditional metal oxides, and the oxygen transport capacity of the perovskites is less due to its slow reduction rate. Both disadvantages limit the application of the perovskite-structured oxide materials as industrial oxygen carriers.

3.3.2. Core-Shell Structure

Fast redox kinetics and hence high oxygen transport rates of oxygen carriers are considered to be obstructed by the solid-state diffusion limitation due to the formation of a dense oxide overlayer during oxidation [137]. The transition from a kinetically controlled regime to a diffusion limited regime occurs when the overlayer reaches a critical thickness. Hence, an oxygen carrier with smaller particle size (i.e., nano-sized oxygen carrier) is expected to suffer less from such diffusion limitations by shifting the transition point towards close to complete oxygen carrier conversion, because, from a geometrical point of view, the fraction of the total particle volume constituted by a layer with constant thickness strongly increases with decreasing particle diameter [14]. Moreover, owing to the relatively slower oxidation kinetics with CO_2 compared to air or steam, the use of nano-sized metal particles is expected to maximize the reactivity when CO_2 is used as oxidizing agent [120]. In addition, it is believed that carbon deposition is not energetically favored over small metal particles, especially for Ni. As a consequence, the use of nano-sized Ni as oxygen carrier is very beneficial to chemical looping dry reforming of CH_4 [138]. However, the fast deactivation of nanoparticles, due to severe sintering during high-temperature processes like chemical looping, poses a big issue for practical application. To counter this, oxygen carriers with a core-shell structure are designed, allowing stabilization of nano-size active metals by a thin and porous encapsulating shell (i.e., metal$_{core}$@oxide$_{shell}$ nanomaterial). This approach provides a promising strategy, not only to address the stability requirement, but also to improve the performance in terms of reactivity, redox properties and coke resistance, depending on the employed shell materials. Such core-shell structured nanomaterials have shown excellent performance in methane reforming [117,118,138–140] and chemical looping processes [14,40,119–124,141].

The first-generation metal$_{core}$@oxide$_{shell}$ nanomaterial of particular interest is a metal@SiO$_2$ core-shell material, because of its easy preparation and low cost, providing great potential to be developed as commercial material, as well as the easy adjustment of its structural properties, such as porosity, specific surface area and shell thickness through modification of the synthesis method of the SiO_2 shell [117,119]. Ni@SiO$_2$ core-shell materials have received broad interest for CH_4 reforming, as they turned out to be good redox catalysts or oxygen carriers, owing to their high activity and stability as well as their potential to resist coke deposition. Fe$_2$O$_3$@SiO$_2$ core-shell materials (Figure 7a) however, show poor utilization of the oxygen carrier due to the formation of Fe_2SiO_4 during the reduction half-cycle. The active Fe cannot easily be restored from this Fe_2SiO_4 phase in the CO_2 re-oxidation half-cycle, resulting in deactivation of material along with the proceeding cycles. To counter this deactivation, an additional air oxidation step is required to fully liberate active Fe from the Fe-silicate (Figure 7b), which leads to an increasing cost with process complexity [14,120].

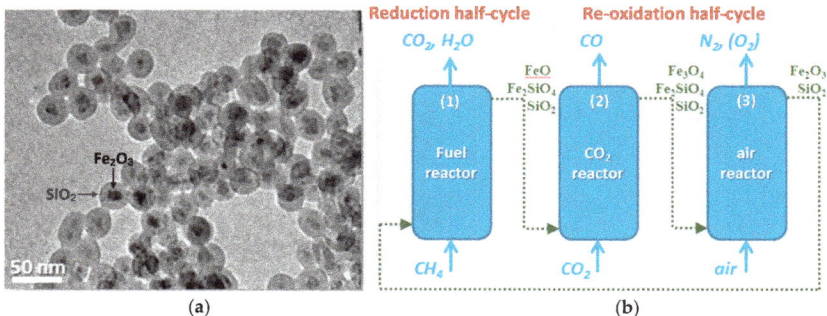

Figure 7. TEM image of a core-shell structured Fe$_2$O$_3$@SiO$_2$ oxygen carrier (**a**) and schematic representation of a three-stage chemical looping dry reforming process for CO_2 utilization (**b**). Reproduced with permission from Bhavsar S. et al. [14], Catalysis Today; published by Elsevier, 2014.

Although the above three-stage chemical looping dry reforming process allows sufficient utilization of the oxygen carrier via regaining Fe_2O_3 from the Fe_2SiO_4 phase in the air reactor, the low CO_2 conversion in the CO_2 reactor still limits the process efficiency in terms of CO_2 utilization. Therefore, there is need to further explore reproducible core-shell structures. Li's group designed a core–shell $Fe_2O_3@La_xSr_{1-x}FeO_{3-\delta}$ material ($Fe_2O_3@LSF$), composed of a nanoscale Fe_2O_3 core (~ 50 nm) and a perovskite-structured $La_xSr_{1-x}FeO_{3-\delta}$ shell (thickness of ~10 nm), which was successfully synthesized by using a sequential sol-gel method [121]. The performance of such material was investigated in CH_4 partial oxidation: In this well-defined structure (Figure 8), syngas production from CH_4 is enhanced by the highly selective LSF shell; the activity is improved with nano-size Fe_2O_3 which can provide more readily accessible lattice oxygen; the coating of Fe_2O_3 with a stable perovskite-structured shell does not hinder the recyclability of the material, but rather prevents Fe_2O_3 from sintering and carbon deposition, while improving the oxygen ion and electron fluxes. Thus, the proposed $Fe_2O_3@LSF$ core-shell material was suggested to be a promising oxygen carrier for CLR processes. Later on, this group investigated the effect of core and shell compositions of $MeO_x@LSF$ (Me = Mn, Co and Fe) materials on the performance for CLR [122]. They found that the presence of Sr in the shell is favorable to stabilize Fe in the core-shell structure (i.e., prevents irreversible Fe leaching to the surface) by forming strontium hexaferrite ($SrFe_{12}O_{19}$) or a Sr deficient perovskite. However, these phases are more difficult to reduce, leading to a decrease of the overall reduction rate of the $Fe_2O_3@LSF$ material. As a result, its redox activity is weakened.

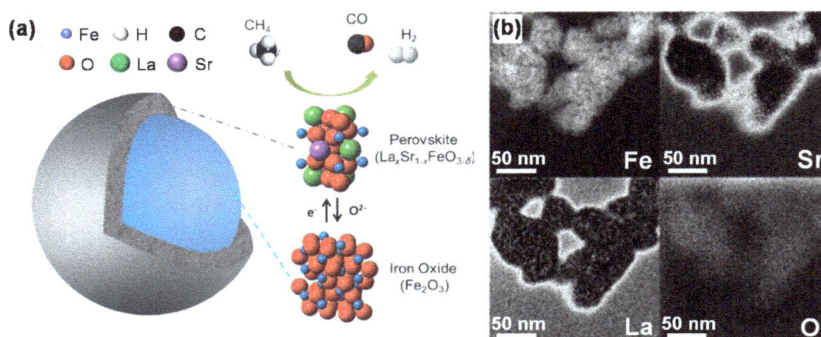

Figure 8. Schematics of the core-shell $Fe_2O_3@La_xSr_{1-x}FeO_{3-\delta}$ material (**a**) and the EFTEM mappings of individual Fe, Sr, La and O elements (**b**), the brighter areas represent regions enriched with the element of interest. La and Sr are shown concentrated on the particle surface, while Fe is homogeneously distributed throughout, suggesting the formation of a core-shell structure. Reproduced with permission from Shafiefarhood A. et al. [121], ChemCatChem; published by John Wiley and Sons, 2014.

More suitable shell materials with chemical stability should be considered in order to maintain the long-term activity of core-shell structured Fe-based oxygen carriers. ZrO_2 is of great interest as shell material due to its strong thermal and chemical stability. Li et al. [142] proposed a $Ni@ZrO_2$ nanocomposite, constructed from nano-Ni particles encapsulated in a nano-ZrO_2 framework, showing a strong ability to resistant sintering and coke deposition under reforming conditions. ZrO_2 exists in different phases, among which the t-ZrO_2 phase is dominant. It easily forms in nanoparticles and ultrathin films [143], where it contributes to a high specific surface area and oxygen conductivity. In a similar way, a core-shell material $Fe_2O_3@ZrO_2$, with a thin, porous ZrO_2 shell surrounding a nano-sized Fe_2O_3 core, can exhibit high reactivity and stability in chemical looping processes. The $Fe_2O_3@ZrO_2$ material can be synthesized via nanocoating of Fe_2O_3 with ZrO_2 through hydrolysis with a ZrO_2 precursor [144,145]. In this process, two challenges need to be addressed: (1) the high tendency of nanoparticles to aggregate during the coating process, resulting in multiple cores within one shell [144]; (2) generally, the as-prepared Fe_2O_3 particles are non-densified, providing an unstable support for the shell, which easily collapses upon high-temperature reaction. To counter these drawbacks, a novel core-shell structured nanomaterial was proposed, $Fe_2O_3/ZrO_2@ZrO_2$ [40]. The core consists of Fe_2O_3 nanoparticles decorating a stable ZrO_2 and it is coated afterwards with a porous thin ZrO_2 layer (Figure 9).

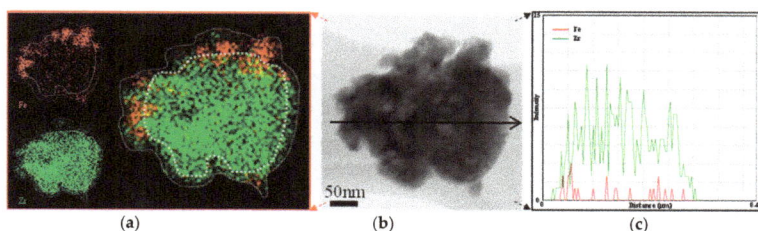

Figure 9. Morphology of $Fe_2O_3/ZrO_2@ZrO_2$ as core-shell nanomaterial: EDX element mappings of Zr and Fe (**a**), STEM image (**b**), and EDX line-scan (**c**). Contour lines of core and shell (dotted white line) in the mapping serve as guide to the eye. The black arrow in the STEM image represents the scanning route through the particle.

Applying Fe_2O_3 nanoparticles supported on stable crystalline ZrO_2 as core material is advantageous; the Fe_2O_3/ZrO_2 core avoids nanoparticle aggregation during the coating process, while at the same time providing a dense, stable support for the shell. The core material is covered with a ZrO_2 shell by means of a general nanocoating process [146,147], using P-123 (a nonionic amphiphilic surfactant) as an active template layer on the core surface, upon which the shell is formed. Eventually, the template layer is calcined to decompose, leaving a mesoporous ZrO_2 shell, permeable to reduction and oxidation gases to reach the iron oxide [148]. Compared to non-coated material, the core-shell $Fe_2O_3/ZrO_2@ZrO_2$ nanomaterial, designed to possess a stable Fe_2O_3/ZrO_2 core and a mesoporous t-ZrO_2 shell, exhibited excellent redox activity (higher CO_2 conversion to CO at 650 °C) and stability (stable CO production, thermal-stable structure and high capacity to resist particle sintering) during 100 cycles of chemical looping conversion of CO_2 (Figure 10).

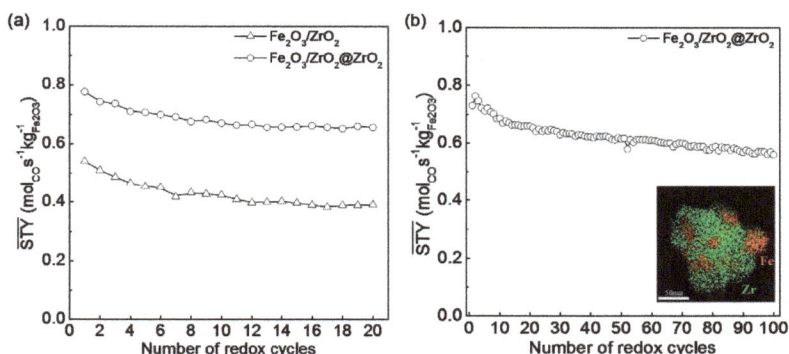

Figure 10. Time-averaged space-time yield ($\overline{STY}$) of CO vs. number of redox cycles: (**a**) Comparison of CO $\overline{STY}$ for non-coated Fe_2O_3/ZrO_2 material and core-shell structured $Fe_2O_3/ZrO_2@ZrO_2$ nanomaterial during 20 redox cycles at 650 °C; (**b**) Stability of CO $\overline{STY}$ over $Fe_2O_3/ZrO_2@ZrO_2$ during 100 prolonged redox cycles at 650 °C, the insert figure is the EDX element mapping of the spent core-shell material. Each cycle (10 min) is composed of 2 min H_2 (5% in Ar) reduction, 2 min CO_2 (pure) re-oxidation and 3 min intermediate He purging. Gas flow rates: 0.04 mmol·s^{-1}.

4. Bifunctional Materials for Catalyst-Assisted Chemical Looping

When using Fe-based oxygen carrier as looping material, the reduction rate of iron oxide by hydrocarbon fuel is very slow, especially in the presence of CO_2 (originating from feed and product). As a result, the conversion of fuel and CO_2 is limited. Through addition of a catalyst component (commonly Ni), a bifunctional material, possessing an extra catalytic function for decomposition of hydrocarbon fuel into syngas (the latter ensures the formation of metallic Fe for adequate CO_2 cyclic conversion), brings chemical looping towards a new stage—catalyst-assisted chemical looping dry (or auto-thermal dry) reforming (CCDR or CCAR), two processes which are designed for efficient conversion of hydrocarbon fuels accompanied by maximum CO_2 utilization.

Figure 11 sketches the CH_4-based catalyst-assisted chemical looping processes over a Ni-Fe bifunctional material. In the reduction half-cycle, the Ni catalyst converts CH_4 and CO_2 in syngas (a mixture of H_2 and CO), which reduces the iron oxide oxygen carrier. Reduction of iron oxide by H_2 and CO is significantly faster and deeper than by CH_4 [23,149], which emphasizes the importance of the Ni catalyzed CH_4 reforming process. Upon re-oxidation, the oxygen carrier is regenerated by feeding extra CO_2, thereby producing CO. In such processes, the Ni-Fe bifunctional material is required to provide not only an optimum activity towards conversion of CH_4 and CO_2 into CO and H_2, but also provide redox properties for interaction with both reducing gases (CO and H_2) and oxidizing gases (CO_2 and H_2O). Therefore, the material must show functions of both reforming catalyst and oxygen carrier, i.e., exhibit a bifunctional character.

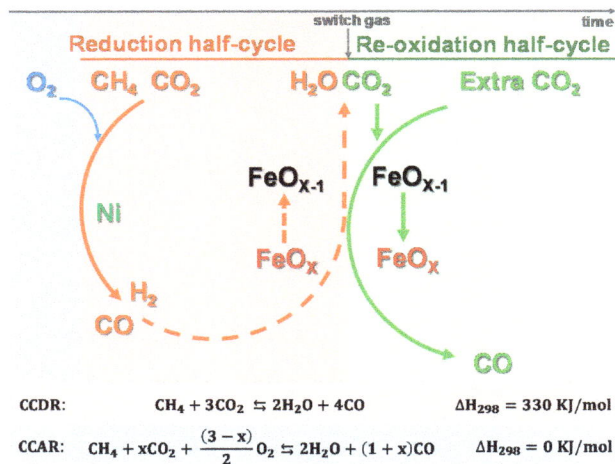

Figure 11. Schematic representations of catalyst-assisted chemical looping processes. Ni represents the Ni catalyst. FeO_{X-1} and FeO_X are the reduced and oxidized state of the iron oxide oxygen carrier, respectively. The equations in the bottom are the global reactions of the corresponding processes.

In previous work [23], a bifunctional 5 wt % Ni/CeO_2-Fe_2O_3 material was prepared by impregnation of the required amount of Ni on a 50 wt % CeO_2-Fe_2O_3 oxygen carrier. The working principle of the bifunctional material in CCDR is shown in Figure 12a. During the reduction half-cycle, the mixture feed of $CH_4 + CO_2$ is converted on Ni sites to CO and H_2, which in turn reduce the oxygen carrier component (from Fe_3O_4 to metallic Fe), yielding CO_2 and H_2O. Subsequently, the reduced oxygen carrier is re-oxidized (from Fe to Fe_3O_4) by extra CO_2, producing CO in the re-oxidation half-cycle. During H_2-TPR of the Ni/CeO_2-Fe_2O_3 bifunctional material, Fe_2O_3 is reduced to Fe_3O_4, and then FeO, finally transforming into metallic Fe. At around 350 °C metallic Ni appears due to the reduction of NiO. Interaction between Ni and Fe leads to Ni-Fe alloy formation from 580 °C onwards. This alloy remains stable up to 600 °C during CO_2-TPO but is completely decomposed into Ni and Fe_3O_4 at 700 °C (Figure 12b). The addition of Ni to CeO_2-Fe_2O_3 shows great promise towards generation of CO from CO_2 in CCDR. As seen in Figure 12c, the position of the CeO_2 (111) peak shows a downward shift indicating partial reduction. This shift is quite significant in Ni/CeO_2-Fe_2O_3, while it remains limited in CeO_2-Fe_2O_3 and doesn't occur at these temperatures in pure CeO_2, meaning the partial reduction is enhanced by the addition of Ni, leading to deeper reduction of the oxygen carrier. As a result, a higher CO yield can be obtained during the reduction half-cycle when feeding CO_2. The CO yield over the Ni/CeO_2-Fe_2O_3 bifunctional material is proven to be 10 times higher than over the CeO_2-Fe_2O_3 oxygen carrier (Figure 12d).

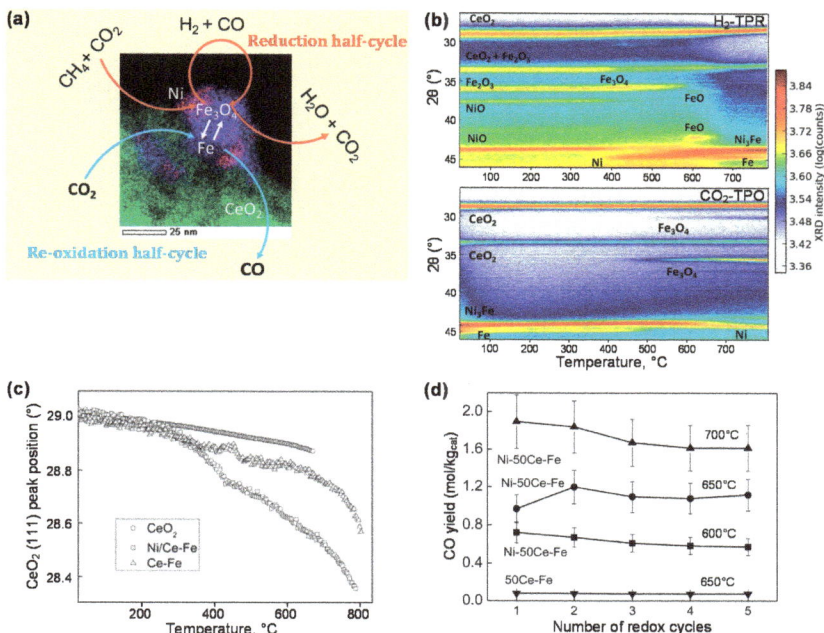

Figure 12. Catalyst-assisted chemical looping dry reforming over a 5 wt % Ni/CeO_2-Fe_2O_3 (1:1) bifunctional material: (**a**) Schematic representation of the reactions in the two separated half-cycles, the STEM-EDX element mappings of Ni (red), Fe (blue) and Ce (green) show the morphological structure of the bifunctional material; (**b**) The evolution of crystal structure of the bifunctional material during in-situ H_2-TPR and CO_2-TPO. Measurement conditions include: temperature ramp rate: 20 °C/min, reducing gas for TPR: 5% H_2/He, oxidizing gas for TPO: 100% CO_2; (**c**) Variation of peak position of CeO_2 (111) at 2θ = 29° during TPR for pure CeO_2 (○), CeO_2-Fe_2O_3 (△) and Ni/CeO_2-Fe_2O_3 (□). Same measurement conditions; (**d**) Comparison of CO yield during 5 redox cycles over CeO_2-Fe_2O_3 oxygen carrier and Ni/CeO_2-Fe_2O_3 bifunctional material at different temperatures. The error bar indicates twice the standard deviation. Adapted with permission from Galvita V.V. et al. [23], Applied Catalysis B: Environmental; published by Elsevier, 2015.

An alternative towards the bifunctional material for CCDR processes is provided by spinel ferrites (AFe_2O_4), possessing both a catalytic element (A) and an oxygen carrier element (Fe). This ferrite-based material is readily prepared via co-precipitation of all elements, unlike the additional modification of a Fe-based oxygen carrier by a catalyst component. Various Me-ferrites (Me = Ni, Co, Cu, Mn, Zn) have been investigated in chemical looping processes [112,150,151], among which the $NiFe_2O_4$ and $CoFe_2O_4$ show certain catalytic activity and redox stability during chemical looping dry reforming of CH_4. Given the increase of alcohol feedstocks from chemical conversion of biomass, these ferrite-based bifunctional materials were further tested in chemical looping reforming of alcohol fuels, such as methanol and ethanol, for the production of value added fuels such as syngas. Dharanipragada et al. [125] investigated the performance of $NiFe_2O_4$- or $CoFe_2O_4$-$CeZrO_2$ bifunctional materials in a CCDR process with two different alcohol fuels, methanol and an ethanol-water mixture (1:1 molar ratio). In this case (Figure 13), methanol or ethanol acts as feed in the reduction half-cycle, which is first converted into CO and H_2 (as well as some CH_4 when using the ethanol-water mixture), that simultaneously reduce the material, yielding CO_2 and H_2O. In the re-oxidation half-cycle, CO_2 is fed to regenerate the material and produce CO.

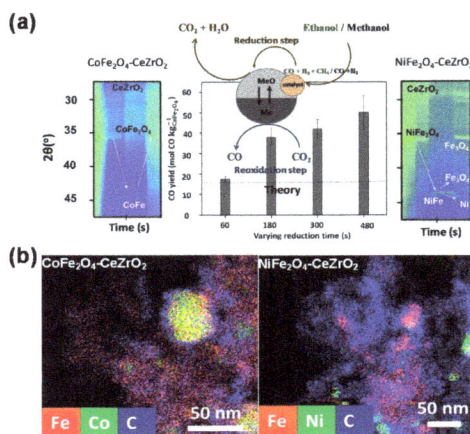

Figure 13. Catalyst-assisted chemical looping dry reforming with alcohol fuels as feed over Co- and Ni-ferrites modified with $CeZrO_2$ bifunctional materials: (**a**) Overview of the transformation behavior of gas and solid phases; (**b**) STEM-EDX mappings of Co, Ni, Fe and C in 20 wt % $CoFe_2O_4$- or $NiFe_2O_4$-$CeZrO_2$ after methanol redox cycles. Reproduced with permission from Dharanipragada N.V.R.A. et al. [125], Applied Catalysis B: Environmental; published by Elsevier, 2018.

During prolonged cycling, the material suffers from deactivation due to phase segregation and particle sintering. It was found that $CoFe_2O_4$-$CeZrO_2$ with low ferrite content (20 wt %) had good redox activity and stability, but the material with 80 wt % $CoFe_2O_4$ underwent deactivation owing to phase segregation into Co and Fe_3O_4. On the other hand, all $NiFe_2O_4$-$CeZrO_2$ materials suffered from phase segregation as well as sintering, and the separated Ni and Fe_3O_4 phases could not be restored into the original spinel phase ($NiFe_2O_4$). In the methanol-based CCDR process, carbon formation was observed on both $NiFe_2O_4$- and $CoFe_2O_4$-$CeZrO_2$, but significantly more intense on the former (Figure 13b), indicating a lower carbon-resistance ability of the Ni-ferrite. The deposited carbon in turn enhanced the CO yield during the re-oxidation half-cycle, because of the reaction between CO_2 and carbon. This phenomenon was verified experimentally, showing a more elevated CO yield than the theoretical value after material re-oxidation (as seen in Figure 13a). Furthermore, it was suggested that air or O_2 would be needed for complete gasification of the deposited carbon.

Of course, the deactivation of such bifunctional materials due to carbon deposition during CCDR cannot be ignored. In order to tackle this drawback, the upgraded technology of CCAR can provide a prospective solution. The feasibility of the CCAR process over a Ni-Fe/$MgAl_2O_4$ bifunctional material was studied through both thermodynamic analysis and prolonged cycling experiments [20]. In the reduction half-cycle, a mixture of CH_4, CO_2 and O_2 in a 1:1:0.5 molar ratio was first converted over the Ni catalyst into a mixture of CO and H_2, which both reduce iron oxide to metallic Fe (Figure 14a). The high endothermicity of dry reforming as well as the coke formation could be mitigated in this half-cycle under auto-thermal conditions. The steady-state reforming of the $CH_4:CO_2:O_2$ =1:1:0.5 mixture over Ni-Fe/MgAl$_2$O$_4$ at various temperatures showed the auto-thermal dry reforming reduction regime was established in the temperature range from 550 °C to 750 °C, where the O_2 is completely consumed and the conversion of CH_4 and CO_2 respectively reaches 90% and 80%, yielding syngas with a H_2:CO ratio close to unity (Figure 14b). During the re-oxidation half-cycle, CO is produced alongside the regeneration of iron oxide by CO_2 (Figure 14a). The overall CO yield depends on the degree of Fe_3O_4 reduction reached in the reduction half-cycle, which strongly depends on the ratio (R_c) between the present reducing (CO + H_2) and oxidizing (CO_2 + H_2O) gases. Based on thermodynamic analysis, high conversion of Fe_3O_4 to metallic Fe can be reached if $R_c > 2$ and T > 600 °C in the present study (Figure 14c).

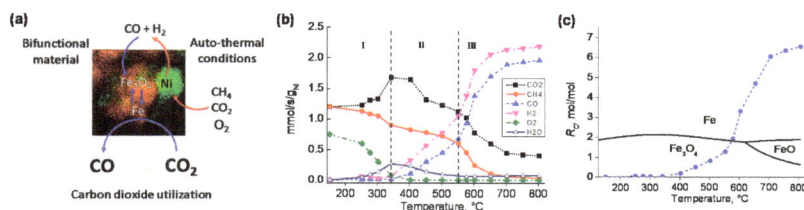

Figure 14. Catalyst-assisted chemical looping auto-thermal dry reforming over a 9 wt % NiO-16 wt % Fe$_2$O$_3$/MgAl$_2$O$_4$ bifunctional material: (**a**) Schematic representation of the reactions in the two half-cycles; (**b**) Space-time yield of reagents and products as a function of temperature during reforming of a mixture with molar ratio $CH_4:CO_2:O_2$ = 1:1:0.5 over the bifunctional material. Region I: total oxidation, region II: dry and steam reforming, region III: auto-thermal dry reforming; (**c**) The ratio R_c between reducing gases (CO + H_2) and oxidizing gases (CO_2 + H_2O) as a function of reaction temperature during catalyst-assisted chemical looping auto-thermal dry reforming process, based on the experimental data of (b). Solid circles with dashed line as guide to the eye. Solid lines: Equilibrium lines of the iron/iron oxide system for each R_c at different temperatures.

The stability of the Ni-Fe/MgAl$_2$O$_4$ bifunctional material was determined by examining the changes in morphological structure by SEM (Figure 15a), showing no obvious carbon deposition on the material surface, but an increase in the particle size and a collapse in the pore structure after 25 cycles of CCAR. Therefore, particle sintering results in a decrease of the CO time-averaged space-time yield, which reaches a stable level after the first 5 cycles (Figure 15b), suggesting that severe sintering of particles only occurs in the initial stage of redox cycling.

Obviously, deactivation of bifunctional materials through particle sintering during chemical looping remains a challenge. As mentioned in Section 3.3.2, a possible countering strategy could be the construction of material with a core-shell structure. Hence, a Ni-Fe bifunctional nanomaterial (Fe/Zr@Zr-Ni@Zr) was designed, specifically for the catalyst-assisted chemical looping process. It consists of a ZrO$_2$-coated Ni outer shell, encapsulating a Fe$_2$O$_3$/ZrO$_2$@ZrO$_2$ core (Figure 16a), showing multiple benefits: (1) the Ni-based shell and the Fe-based core serve as reforming catalyst and oxygen carrier, respectively, and achieve catalytic reforming and oxygen transfer in a single nanoscale unit; (2) the hollow sphere surrounding the core prevents the aggregation of core particles [152], while providing space for gas-solid contact; (3) at the same time, the sphere offers a large specific surface area, leading to fine dispersion of Ni particles and the adequate deposition of the outer ZrO$_2$ protective layer, both of which contribute to a high catalytic activity [153]. To design such a structure, an extra step is introduced in the synthesis pathway, where SiO$_2$ is typically used as a template [154]: after synthesis of the Fe$_2$O$_3$/ZrO$_2$@ZrO$_2$ core, the latter is coated with a dense SiO$_2$ layer using a modified Stöber process [155]. After loading the Ni@ZrO$_2$ shell, the SiO$_2$ template is selectively removed by dissolution with an adequate solvent to free the core, so that it can contact with reactant gases. The result is an eccentric core-shell structure as the core is now no longer supported within the hollow shell (Figure 16b).

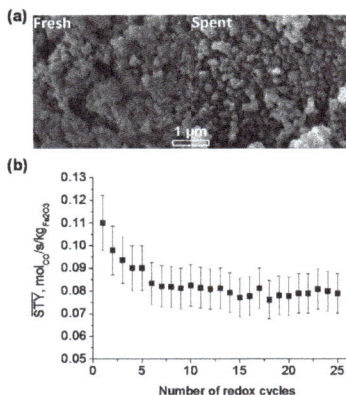

Figure 15. Stability of catalyst-assisted chemical looping auto-thermal dry reforming over the Ni-Fe/MgAl$_2$O$_4$ bifunctional material: (**a**) SEM micrographs of the fresh sample and the spend sample after 25 cycles; (**b**) Comparison of CO space-time yield ($\overline{\text{STY}}$) during 25 cycles at 750 °C.

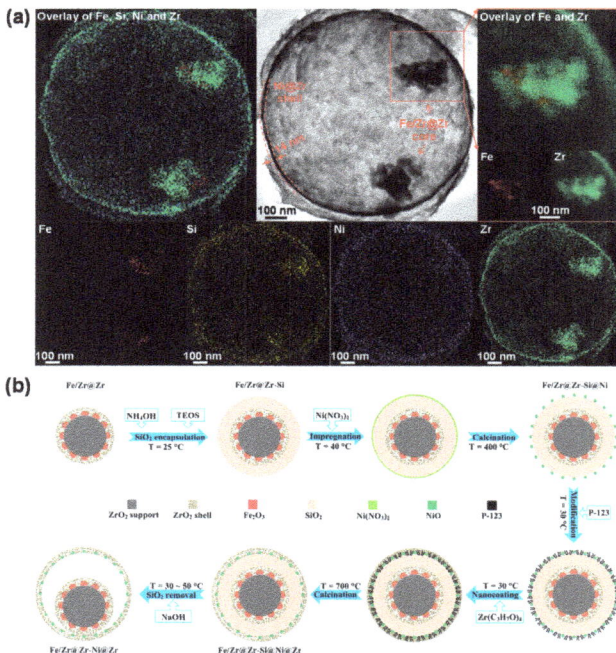

Figure 16. Core-shell bifunctional nanomaterial Fe/Zr@Zr-Ni@Zr: (**a**) STEM micrograph and EDX element mapping of Zr, Si, Ni and Fe, and the overlay mapping of these elements for the as-prepared material; (**b**) Schematics of the synthesis procedure, combining methods of incipient wetness impregnation, SiO$_2$ template-assistance and nanocoating.

The bifunctional material Fe/Zr@Zr-Ni@Zr, made according to this recipe, is active for reforming of a feed mixture of CH_4, CO_2 and O_2 (molar ratio of 1:1:0.2) as well as for conversion of CO_2 to CO in a CCAR process (Figure 17a). O_2 conversion during reduction is always 100% in these conditions. Methane dry reforming is the dominant reaction, yielding syngas with a H_2:CO ratio of unity, together with some methane combustion into CO_2 and H_2O. In the second half-cycle, the re-oxidation of Fe into Fe_3O_4 rapidly converts CO_2 to CO. Hence, the CO space-time yield shows a sharp peak in the first 20 s of the re-oxidation half-cycle and then steadily decreases towards zero, indicating the majority of Fe_3O_4 has been regenerated. Moreover, when averaged over 25 CCAR cycles, a stable CO time-averaged space-time yield is obtained over Fe/Zr@Zr-Ni@Zr (Figure 17b). Characterizations (XPS, STEM-EDX mapping and XRD) of the spent sample after 25 cycles indicate no obvious carbon deposits on the core-shell bifunctional material and no sintering of the Ni and Fe particles.

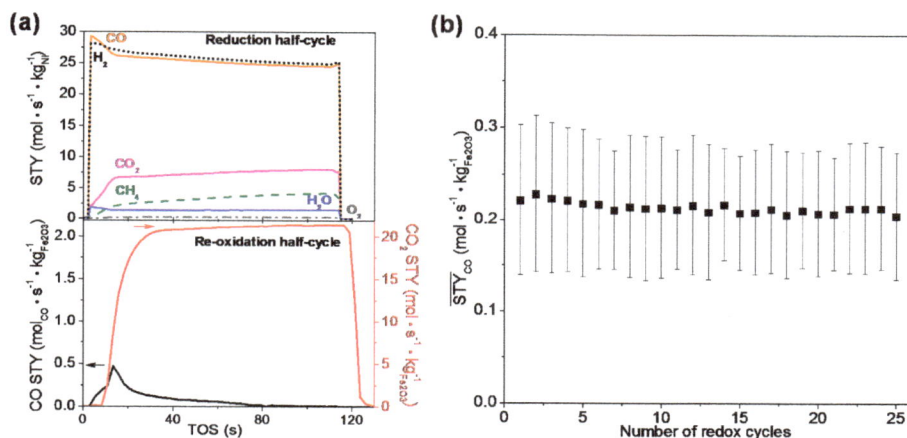

Figure 17. Reactivity and stability of the Fe/Zr@Zr-Ni@Zr bifunctional material during the catalyst-assisted chemical looping auto-thermal dry reforming process at 750 °C: (**a**) Space-time yield (STY) of products vs. time on stream (TOS) during 2 min reduction half-cycle (reactant: mixture of CH_4:CO_2:O_2 = 1:1:0.2, 150 mL·min^{-1}) and 2 min re-oxidation half-cycle (reactant: 90% CO_2/Ar, 150 mL·min^{-1}); (**b**) CO time-averaged space-time yield ($\overline{STY}$) of the re-oxidation half-cycle during 25 redox cycles.

The CO_2 utilization during this process can be further enhanced through the construction of a double-zone reactor bed configuration [126], composed of a core-shell structured Fe/Zr@Zr-Ni@Zr bifunctional material (the first zone) followed by a core-shell structured Fe_2O_3/ZrO_2@ZrO_2 oxygen carrier (the second zone). In this bed configuration, it is expected that much more CO_2 can be converted to CO due to the additional oxygen storage capacity provided by the second oxygen carrier zone.

5. Conclusions

Chemical looping technology has recently emerged as a promising "clean energy conversion technology" due to the advantages of manageable inherent CO_2 capture. In a chemical looping scheme, a single reaction is broken down into two, spatially or temporally separated half-cycle reactions, coupled through the cyclic transformation of a solid looping material. Depending on their functions, the looping materials can be divided into two main types, correspondingly defining two groups of chemical looping processes: As an oxygen carrier, to transport oxygen through cyclic reduction and re-oxidation during chemical looping combustion or reforming; as a CO_2 carrier, to remove CO_2 in-situ through cyclic carbonation and calcination during sorption-enhanced chemical looping.

The flexibility of chemical looping processes allows the utilization of various feedstocks (fossil fuels, CO_2 and renewables) and the production of various value-added chemicals (H_2, CO, syngas and liquid fuels) via tailoring of reactor feed and looping materials.

At the heart of chemical looping technology is the selection of looping materials. The employment of a suitable looping material determines the sustainability and economics of the technology. From a practical point of view, some commonly used looping materials were compared. CaO is an ideal material to serve as CO_2 carrier owing to its low cost, non-toxicity, acceptable CO_2 capacity, regenerability and heat conduction. Further efforts for a CaO-based carrier should focus on increasing its CO_2 capacity and adsorption rate, as well as improving its thermal stability and mechanical strength. Transition metals, such as Ni, Fe, Cu, Mn and Co-based materials, are commonly used as oxygen carriers, among which Fe- and Ni-based materials are most employed, especially for carbonaceous feedstock-driven chemical looping processes. Iron oxide is highly tolerant to carbon deposition, environmentally friendly, cheap, and has a high oxygen storage capacity from CO_2. A future task in the development of Fe-based oxygen carriers is the enhancement of its ability to resist sintering. Ni then again has excellent reactivity towards hydrocarbon fuels, making Ni-based material a promising redox catalyst for chemical looping reforming, but the drawbacks with regard to particle sintering, carbon deposition and sulfur poisoning of the material need to be further addressed.

Based on different purposes of improvement of oxygen carriers, approaches have been investigated, mainly focusing on compositional and structural optimization. The stabilization of active components with promoter materials is a widely used method. Some physical promoters (e.g., Al_2O_3) can react with the active component to form spinel phases (such as $NiAl_2O_4$ and $FeAl_2O_4$), which can continue to act as inert binders but resulting in the partial loss of reactivity. Adding an appropriate amount of MgO into NiO/Al_2O_3 or Fe_2O_3/Al_2O_3 systems can mitigate the formation of Ni- or Fe-Al_2O_4 spinel through the competition of $MgAl_2O_4$ formation. Owing to the thermal and chemical stability, both $MgAl_2O_4$ and ZrO_2 are considered as a good physical promoter materials for oxygen carriers. In addition, reducible CeO_2 can serve not only as a physical, but also a chemically active promoter. The CeO_2-modified oxygen carriers show high reactivity, good reducibility, high oxygen transport capacity as well as strong carbon resistance.

Another strategy is the use of an oxygen carrier with mixed active metals, combining the complementary advantages of individual metals. The mixing of Co into a Ni-based oxygen carrier can improve its reducibility and carbon resistance. The addition of Fe into a Cu-based oxygen carrier increases its melting point to resist agglomeration. Combining advantages of both Ni and iron oxide, the bimetallic Ni/Fe-based material—containing an optimized molar ratio of Ni/Fe and supported by a stable promoter material—gradually becomes a highly promising oxygen carrier or redox catalyst for chemical looping reforming, showing high reactivity towards hydrocarbon fuels, good thermal stability, high ability to resist carbon deposition and low cost.

Structural optimization is of equal importance for an ideal oxygen carrier. Perovskite-structured oxide (ABO_3-type) materials show remarkable results. Owing to the unique structure of perovskite, its lattice oxygen is readily utilized for syngas production with high selectivity, and phase transformation can be avoided due to its thermal stability. The partial substitution of A- and B-site metals by an aliovalent metal can improve the stability and reactivity, respectively. One optimum composition of perovskite-structured oxygen carriers is $La_{0.7}Sr_{0.3}Cr_{0.1}Fe_{0.9}O_3$. Future work must focus on improvement of the oxygen storage and transport capacities of perovskite-structured oxygen carriers. Furthermore, a core-shell structure, i.e., an active core encapsulated by a thin and porous shell, has become the most competitive structure, due to its strong adjustability: Control of the particle size of the active core enables to control the reaction kinetics; Tailoring of the composition, thickness and pore properties of the shell can change the stability of the material and its selectivity towards specific products. An ideal core-shell structured oxygen carrier applied in chemical looping should show high oxygen transport capacity and strong thermal stability under the harsh operation conditions. In this sense, a nano-size active metal core encapsulated by a chemically inert and porous shell is required to construct a core-shell structured oxygen carrier. A proposed $Fe_2O_3/ZrO_2@ZrO_2$ oxygen carrier, composed of a stable Fe_2O_3/ZrO_2 core and a thin and mesoporous ZrO_2 shell, proved to possess excellent redox activity and stability for CO_2 utilization during a long-term chemical looping process.

Efficient conversion of hydrocarbon fuels to syngas accompanied by a goal of maximum CO_2 utilization has been achieved through a novel process—catalyst-assisted chemical looping dry or auto-thermal dry reforming—implemented over a bifunctional material. The latter must show functions of both reforming catalyst and oxygen carrier. The performances of series of Ni-Fe bifunctional materials were investigated in terms of syngas production, CO_2 conversion and material stability. A core-shell structured Fe/Zr@Zr-Ni@Zr bifunctional material, providing catalyst and oxygen storage functions from a Ni-based shell and a Fe-based core, respectively, shows a stable CO yield during prolonged cycling without significant carbon deposition and particle sintering. On the reactor scale, careful distribution of the Ni-based catalyst and Fe-based oxygen carrier in the reactor bed could further enhance CO_2 utilization.

6. Outlook

Although chemical looping materials have now been developed as core-shell structured bifunctional nanomaterial, applicable in the new process of chemical looping technology—catalyst-assisted chemical looping, the research on the industrially-applicable looping materials has not yet reached its final destination. Further outlooks could be as follows:

(1) Considering the economic effectiveness, the establishment of low-cost and facile synthesis methods is crucial for the development of the industrial looping materials, especially for the core-shell structured nanomaterials.

(2) High-pressure operation of the chemical looping reactors is essential to maintain the circulation efficiency of an industrial platform. Hence, holistic evaluation of the effect of pressure on the behavior of the core-shell structured materials is necessary for further improvement of the materials.

(3) Research towards the enhancement of the long-term stability, for example maintaining reactivity and structural integrity for thousands of redox cycles, of the looping materials is crucial. The choice of the developed bifunctional looping materials should be screened by taking into account of their lifecycle and cost of production.

(4) Most of the fuel feedstocks contain sulfur-based contaminates, which easily bind to the active metal sites to form stable surface oxy-sulfides and/or bulk metal sulfides, resulting in the material deactivation. Therefore, study on the sulfur-poisoning mechanism of the materials by using feed compositions that are close to real feedstocks is indispensable.

Author Contributions: J.H. made the literature analysis and wrote the manuscript. V.V.G. participated in discussions related to the content of the manuscript. H.P. participated in discussions related to the content of the manuscript, G.B.M. participated in discussions related to the content of the manuscript.

Funding: Long Term Structural Methusalem Funding by the Flemish Government; CSC Funding by China Scholarship Council; Interuniversity Attraction Poles Programme (IAP7/5), Belgian State, Belgain Science Policy.

References

1. Tang, M.; Xu, L.; Fan, M. Progress in oxygen carrier development of methane-based chemical-looping reforming: A review. *Appl. Energy* **2015**, *151*, 143–156. [CrossRef]
2. Adanez, J.; Abad, A.; Garcia-Labiano, F.; Gayan, P.; De Diego, L.F. Progress in chemical-looping combustion and reforming technologies. *Prog. Energy Combust. Sci.* **2012**, *38*, 215–282. [CrossRef]
3. Li, B.; Duan, Y.; Luebke, D.; Morreale, B. Advances in CO_2 capture technology: A patent review. *Appl. Energy* **2013**, *102*, 1439–1447. [CrossRef]

4. Das, A.; D'Alessandro, D.M.; Peterson, V.K. Carbon Dioxide Separation, Capture, and Storage in Porous Materials. In *Neutron Applications in Materials for Energy. Neutron Scattering Applications and Techniques*; Kearley, G., Peterson, V., Eds.; Springer: Cham, Switzerland, 2015; pp. 33–60.

5. Guo, L.; Yang, J.; Hu, G.; Hu, X.; DaCosta, H.; Fan, M. CO_2 removal from flue gas with amine-impregnated titanate nanotubes. *Nano Energy* **2016**, *25*, 1–8. [CrossRef]

6. He, Y.; Zhang, L.; Teng, B.; Fan, M. New application of Z-scheme Ag_3PO_4/g-C_3N_4 composite in converting CO_2 to fuel. *Environ. Sci. Technol.* **2015**, *49*, 649–656. [CrossRef] [PubMed]

7. Mondal, M.K.; Balsora, H.K.; Varshney, P. Progress and trends in CO_2 capture/separation technologies: A review. *Energy* **2012**, *46*, 431–441. [CrossRef]

8. Kenarsari, S.D.; Yang, D.; Jiang, G.; Zhang, S.; Wang, J.; Russell, A.G.; Wei, Q.; Fan, M. Review of recent advances in carbon dioxide separation and capture. *RSC Adv.* **2013**, *3*, 22739–22773. [CrossRef]

9. Rogelj, J.; Knutti, R. Geosciences after Paris. *Nat. Geosci.* **2016**, *9*, 187–189. [CrossRef]

10. Environmental Protection Agency. Inventory of U.S. Greenhouse Gas Emissions and Sinks: 1990–2015. Greenhouse Gas Emissions. 2017. Available online: www.epa.gov/ghgemissions/inventory-us-greenhouse-gas-emissions-and-sinks-1990-2015 (accessed on 13 April 2017).

11. Jacobson, M.Z. Review of solutions to global warming, air pollution, and energy security. *Energy Environ. Sci.* **2009**, *2*, 148–173. [CrossRef]

12. Cebrucean, D.; Cebrucean, V.; Ionel, I. CO_2 capture and storage from fossil fuel power plants. *Energy Procedia* **2014**, *63*, 18–26. [CrossRef]

13. Roy, S.C.; Varghese, O.K.; Paulose, M.; Grimes, C.A. Toward solar fuels: Photocatalytic conversion of carbon dioxide to hydrocarbons. *ACS Nano* **2010**, *4*, 1259–1278. [CrossRef] [PubMed]

14. Bhavsar, S.; Najera, M.; Solunke, R.; Veser, G. Chemical looping: To combustion and beyond. *Catal. Today* **2014**, *228*, 96–105. [CrossRef]

15. Buelens, L.C.; Galvita, V.V.; Poelman, H.; Detavernier, C.; Marin, G.B. Super-dry reforming of methane intensifies CO_2 utilization via le chatelier's principle. *Science* **2016**, *354*, 449–452. [CrossRef] [PubMed]

16. D'Alessandro, D.M.; Smit, B.; Long, J.R. Carbon dioxide capture: Prospects for new materials. *Angew. Chem. Int. Ed. Engl.* **2010**, *49*, 6058–6082. [CrossRef] [PubMed]

17. Rochelle, G.T. Amine scrubbing for CO_2 capture. *Science* **2009**, *325*, 1652–1654. [CrossRef] [PubMed]

18. Haszeldine, R.S. Carbon capture and storage: How green can black be? *Science* **2009**, *325*, 1647–1652. [CrossRef] [PubMed]

19. De Diego, L.F.; Ortiz, M.; Adánez, J.; García-Labiano, F.; Abad, A.; Gayán, P. Synthesis gas generation by chemical-looping reforming in a batch fluidized bed reactor using Ni-based oxygen carriers. *Chem. Eng. J.* **2008**, *144*, 289–298. [CrossRef]

20. Hu, J.; Buelens, L.; Theofanidis, S.-A.; Galvita, V.V.; Poelman, H.; Marin, G.B. CO_2 conversion to CO by auto-thermal catalyst-assisted chemical looping. *J. CO2 Util.* **2016**, *16*, 8–16. [CrossRef]

21. Fan, L.-S.; Zeng, L.; Luo, S. Chemical-looping technology platform. *AIChE J.* **2015**, *61*, 2–22. [CrossRef]

22. Galvita, V.V.; Poelman, H.; Bliznuk, V.; Detavernier, C.; Marin, G.B. CeO_2-modified Fe_2O_3 for CO_2 utilization via chemical looping. *Ind. Eng. Chem. Res.* **2013**, *52*, 8416–8426. [CrossRef]

23. Galvita, V.V.; Poelman, H.; Detavernier, C.; Marin, G.B. Catalyst-assisted chemical looping for CO_2 conversion to CO. *Appl. Catal. B* **2015**, *164*, 184–191. [CrossRef]

24. Ishida, M.; Zheng, D.; Akehata, T. Evaluation of a chemical-looping combustion power-generation system by graphic exergy analysis. *Energy* **1987**, *12*, 147–154. [CrossRef]

25. Fan, L.-S. *Chemical Looping Systems for Fossil Energy Conversions*; Wiley: Hoboken, NJ, USA, 2010.

26. Nandy, A.; Loha, C.; Gu, S.; Sarkar, P.; Karmakar, M.K.; Chatterjee, P.K. Present status and overview of chemical looping combustion technology. *Renew. Sustain. Energy Rev.* **2016**, *59*, 597–619. [CrossRef]

27. Yu, F.-C.; Phalak, N.; Sun, Z.; Fan, L.-S. Activation strategies for calcium-based sorbents for CO_2 capture: A perspective. *Ind. Eng. Chem. Res.* **2011**, *51*, 2133–2142. [CrossRef]

28. Dou, B.; Wang, C.; Song, Y.; Chen, H.; Jiang, B.; Yang, M.; Xu, Y. Solid sorbents for in-situ CO_2 removal during sorption-enhanced steam reforming process: A review. *Renew. Sustain. Energy Rev.* **2016**, *53*, 536–546. [CrossRef]

29. Fan, L.-S. *Chemical Looping Partial Oxidation: Gasification, Reforming, and Chemical Syntheses*; Cambridge University Press, Technology & Engineering: Cambridge, UK, 2017.

30. Galvita, V.V.; Poelman, H.; Marin, G.B. Hydrogen production from methane and carbon dioxide by catalyst-assisted chemical looping. *Top. Catal.* **2011**, *54*, 907–913. [CrossRef]

31. Chen, X.; Jiang, J.; Li, K.; Tian, S.; Yan, F. Energy-efficient biogas reforming process to produce syngas: The enhanced methane conversion by O_2. *Appl. Energy* **2017**, *185*, 687–697. [CrossRef]

32. Christian Enger, B.; Lødeng, R.; Holmen, A. A review of catalytic partial oxidation of methane to synthesis gas with emphasis on reaction mechanisms over transition metal catalysts. *Appl. Catal. A* **2008**, *346*, 1–27. [CrossRef]

33. Chen, W.-H.; Lin, S.-C. Characterization of catalytic partial oxidation of methane with carbon dioxide utilization and excess enthalpy recovery. *Appl. Energy* **2016**, *162*, 1141–1152. [CrossRef]

34. Chen, W.-H.; Lin, M.-R.; Lu, J.-J.; Chao, Y.; Leu, T.-S. Thermodynamic analysis of hydrogen production from methane via autothermal reforming and partial oxidation followed by water gas shift reaction. *Int. J. Hydrog. Energy* **2010**, *35*, 11787–11797. [CrossRef]

35. Kohn, M.P.; Castaldi, M.J.; Farrauto, R.J. Auto-thermal and dry reforming of landfill gas over a Rh/γ-Al$_2$O$_3$ monolith catalyst. *Appl. Catal. B* **2010**, *94*, 125–133. [CrossRef]

36. Deublein, D.; Steinhauser, A. *Biogas from Waste and Renewable Resources*; Weily-VCH: Weinheim, Germany, 2008.

37. Rao, C.N.R.; Dey, S. Solar thermochemical splitting of water to generate hydrogen. *Proc. Natl. Acad. Sci. USA* **2017**, *114*, 13385–13393. [CrossRef] [PubMed]

38. Protasova, L.; Snijkers, F. Recent developments in oxygen carrier materials for hydrogen production via chemical looping processes. *Fuel* **2016**, *181*, 75–93. [CrossRef]

39. Zhao, X.; Zhou, H.; Sikarwar, V.S.; Zhao, M.; Park, A.-H.A.; Fennell, P.S.; Shen, L.; Fan, L.-S. Biomass-based chemical looping technologies: The good, the bad and the future. *Energy Environ. Sci.* **2017**, *10*, 1885–1910. [CrossRef]

40. Hu, J.; Galvita, V.V.; Poelman, H.; Detavernier, C.; Marin, G.B. A core-shell structured Fe$_2$O$_3$/ZrO$_2$@ZrO$_2$ nanomaterial with enhanced redox activity and stability for CO_2 conversion. *J. CO2 Util.* **2017**, *17*, 20–31. [CrossRef]

41. González, B.; Blamey, J.; McBride-Wright, M.; Carter, N.; Dugwell, D.; Fennell, P.; Abanades, J.C. Calcium looping for CO_2 capture: Sorbent enhancement through doping. *Energy Procedia* **2011**, *4*, 402–409. [CrossRef]

42. Marinković, D.M.; Stanković, M.V.; Veličković, A.V.; Avramović, J.M.; Miladinović, M.R.; Stamenković, O.O.; Veljković, V.B.; Jovanović, D.M. Calcium oxide as a promising heterogeneous catalyst for biodiesel production: Current state and perspectives. *Renew. Sustain. Energy Rev.* **2016**, *56*, 1387–1408. [CrossRef]

43. Blamey, J.; Anthony, E.J.; Wang, J.; Fennell, P.S. The calcium looping cycle for large-scale CO_2 capture. *Prog. Energy Combust. Sci.* **2010**, *36*, 260–279. [CrossRef]

44. Dean, C.C.; Blamey, J.; Florin, N.H.; Al-Jeboori, M.J.; Fennell, P.S. The calcium looping cycle for CO_2 capture from power generation, cement manufacture and hydrogen production. *Chem. Eng. Res. Des.* **2011**, *89*, 836–855. [CrossRef]

45. Perejón, A.; Romeo, L.M.; Lara, Y.; Lisbona, P.; Martínez, A.; Valverde, J.M. The calcium-looping technology for CO_2 capture: On the important roles of energy integration and sorbent behavior. *Appl. Energy* **2016**, *162*, 787–807. [CrossRef]

46. Erans, M.; Manovic, V.; Anthony, E.J. Calcium looping sorbents for CO_2 capture. *Appl. Energy* **2016**, *180*, 722–742. [CrossRef]

47. Liu, F.; Zhang, Y.; Chen, L.; Qian, D.; Neathery, J.K.; Kozo, S.; Liu, K. Investigation of a canadian ilmenite as an oxygen carrier for chemical looping combustion. *Energy Fuels* **2013**, *27*, 5987–5995. [CrossRef]

48. Keller, M.; Arjmand, M.; Leion, H.; Mattisson, T. Interaction of mineral matter of coal with oxygen carriers in chemical-looping combustion (CLC). *Chem. Eng. Res. Des.* **2014**, *92*, 1753–1770. [CrossRef]

49. Lin, C.-L.; Kuo, J.-H.; Wey, M.-Y.; Chang, S.-H.; Wang, K.-S. Inhibition and promotion: The effect of earth alkali metals and operating temperature on particle agglomeration/defluidization during incineration in fluidized bed. *Powder Technol.* **2009**, *189*, 57–63. [CrossRef]

50. Gu, H.; Shen, L.; Zhong, Z.; Zhou, Y.; Liu, W.; Niu, X.; Ge, H.; Jiang, S.; Wang, L. Interaction between biomass ash and iron ore oxygen carrier during chemical looping combustion. *Chem. Eng. J.* **2015**, *277*, 70–78.

51. Theofanidis, S.A.; Galvita, V.V.; Poelman, H.; Marin, G.B. Enhanced carbon-resistant dry reforming Fe-Ni catalyst: Role of Fe. *ACS Catal.* **2015**, *5*, 3028–3039. [CrossRef]

52. Lyngfelt, A. Oxygen carriers for chemical looping combustion—4000 h of operational experience. *Oil Gas Sci. Technol.* **2011**, *66*, 161–172. [CrossRef]

53. De Diego, L.F.; García-Labiano, F.; Gayán, P.; Abad, A.; Cabello, A.; Adánez, J.; Sprachmann, G. Performance of Cu- and Fe-based oxygen carriers in a 500 W$_{th}$ CLC unit for sour gas combustion with high H$_2$S content. *Int. J. Greenh. Gas Control* **2014**, *28*, 168–179. [CrossRef]

54. Rydén, M.; Cleverstam, E.; Johansson, M.; Lyngfelt, A.; Mattisson, T. Fe$_2$O$_3$ on Ce-, Ca-, or Mg-stabilized ZrO$_2$ as oxygen carrier for chemical-looping combustion using NiO as additive. *AIChE J.* **2010**, *56*, 2211–2220.

55. De Diego, L.F.; García-Labiano, F.; Adánez, J.; Gayán, P.; Abad, A.; Corbella, B.M.; María Palacios, J. Development of Cu-based oxygen carriers for chemical-looping combustion. *Fuel* **2004**, *83*, 1749–1757. [CrossRef]

56. Tian, H.; Simonyi, T.; Poston, J.; Siriwardane, R. Effect of hydrogen sulfide on chemical looping combustion of coal-derived synthesis gas over bentonite-supported metal-oxide oxygen carriers. *Ind. Eng. Chem. Res.* **2009**, *48*, 8418–8430. [CrossRef]

57. Wang, B.; Gao, C.; Wang, W.; Zhao, H.; Zheng, C. Sulfur evolution in chemical looping combustion of coal with MnFe$_2$O$_4$ oxygen carrier. *J. Environ. Sci.* **2014**, *26*, 1062–1070. [CrossRef]

58. Adánez, J.; De Diego, L.F.; García-Labiano, F.; Gayán, P.; Abad, A. Selection of oxygen carriers for chemical-looping combusion. *Energy Fuels* **2004**, *18*, 371–377. [CrossRef]

59. Zafar, Q.; Mattisson, T.; Gevert, B. Redox investigation of some oxides of transition-state metals Ni, Cu, Fe, and Mn supported on SiO$_2$ and MgAl$_2$O$_4$. *Energy Fuels* **2006**, *20*, 34–44. [CrossRef]

60. Kelly, J.R.; Denry, I. Stabilized zirconia as a structural ceramic: An overview. *Dent. Mater.* **2008**, *24*, 289–298. [CrossRef] [PubMed]

61. Johansson, M.; Mattisson, T.; Lyngfelt, A. Investigation of Mn$_3$O$_4$ with stabilized ZrO$_2$ for chemical-looping combustion. *Chem. Eng. Res. Des.* **2006**, *84*, 807–818. [CrossRef]

62. Jin, H.; Okamoto, T.; Ishida, M. Development of a novel chemical-looping combustion synthesis of a looping material with a double metal oxide of CoO-NiO. *Energy Fuels* **1998**, *12*, 1272–1277. [CrossRef]

63. Mattisson, T.; Järdnäs, A.; Lyngfelt, A. Reactivity of some metal oxides supported on alumina with alternating methane and oxygen application for chemical-looping combustion. *Energy Fuels* **2003**, *17*, 643–651. [CrossRef]

64. Sedor, K.E.; Hossain, M.M.; De Lasa, H.I. Reactivity and stability of Ni/Al$_2$O$_3$ oxygen carrier for chemical-looping combustion (CLC). *Chem. Eng. Sci.* **2008**, *63*, 2994–3007. [CrossRef]

65. Dueso, C.; Abad, A.; García-Labiano, F.; De Diego, L.F.; Gayán, P.; Adánez, J.; Lyngfelt, A. Reactivity of a NiO/Al$_2$O$_3$ oxygen carrier prepared by impregnation for chemical-looping combustion. *Fuel* **2010**, *89*, 3399–3409. [CrossRef]

66. Quddus, M.R.; Hossain, M.M.; de Lasa, H.I. Ni based oxygen carrier over γ-Al$_2$O$_3$ for chemical looping combustion: Effect of preparation method on metal support interaction. *Catal. Today* **2013**, *210*, 124–134. [CrossRef]

67. Corbella, B.M.; De Diego, L.F.; García-Labiano, F.; Adánez, J.; Palacios, J.M. Performance in a fixed-bed reactor of titania-supported nickel oxide as oxygen carriers for the chemical-looping combustion of methane in multicycle tests. *Ind. Eng. Chem. Res.* **2006**, *45*, 157–165. [CrossRef]

68. Antzara, A.; Heracleous, E.; Silvester, L.; Bukur, D.B.; Lemonidou, A.A. Activity study of NiO-based oxygen carriers in chemical looping steam methane reforming. *Catal. Today* **2016**, *272*, 32–41. [CrossRef]

69. Silvester, L.; Antzara, A.; Boskovic, G.; Heracleous, E.; Lemonidou, A.A.; Bukur, D.B. NiO supported on Al$_2$O$_3$ and ZrO$_2$ oxygen carriers for chemical looping steam methane reforming. *Int. J. Hydrog. Energy* **2015**, *40*, 7490–7501. [CrossRef]

70. Rydén, M.; Johansson, M.; Lyngfelt, A.; Mattisson, T. NiO supported on Mg-ZrO$_2$ as oxygen carrier for chemical-looping combustion and chemical-looping reforming. *Energy Environ. Sci.* **2009**, *2*, 970–981. [CrossRef]

71. Ishida, M.; Jin, H.; Okamoto, T. A fundamental study of a new kind of medium material for chemical-looping combustion. *Energy Fuels* **1996**, *10*, 958–963. [CrossRef]

72. Readman, J.E.; Olafsen, A.; Smith, J.B.; Blom, R. Chemical looping combustion using NiO/NiAl$_2$O$_4$: Mechanisms and kinetics of reduction−oxidation (red-ox) reactions from in situ powder X-ray diffraction and thermogravimetry experiments. *Energy Fuels* **2006**, *20*, 1382–1387. [CrossRef]

73. Mattisson, T.; Jerndal, E.; Linderholm, C.; Lyngfelt, A. Reactivity of a spray-dried NiO/NiAl$_2$O$_4$ oxygen carrier for chemical-looping combustion. *Chem. Eng. Sci.* **2011**, *66*, 4636–4644. [CrossRef]

74. Ryden, M.; Lyngfelt, A.; Mattisson, T. Synthesis gas generation by chemical-looping reforming in a continuously operating laboratory reactor. *Fuel* **2006**, *85*, 1631–1641. [CrossRef]

75. Gayán, P.; De Diego, L.F.; García-Labiano, F.; Adánez, J.; Abad, A.; Dueso, C. Effect of support on reactivity and selectivity of Ni-based oxygen carriers for chemical-looping combustion. *Fuel* **2008**, *87*, 2641–2650. [CrossRef]

76. Cabello, A.; Gayán, P.; García-Labiano, F.; De Diego, L.F.; Abad, A.; Izquierdo, M.T.; Adánez, J. Relevance of the catalytic activity on the performance of a NiO/CaAl$_2$O$_4$ oxygen carrier in a CLC process. *Appl. Catal. B* **2014**, *147*, 980–987. [CrossRef]

77. Bhavsar, S.; Veser, G. Reducible supports for Ni-based oxygen carriers in chemical looping combustion. *Energy Fuels* **2013**, *27*, 2073–2084. [CrossRef]

78. He, F.; Wang, H.; Dai, Y. Application of Fe$_2$O$_3$/Al$_2$O$_3$ composite particles as oxygen carrier of chemical looping combustion. *J. Nat. Gas Chem.* **2007**, *16*, 155–161. [CrossRef]

79. Wang, B.; Yan, R.; Lee, D.H.; Zheng, Y.; Zhao, H.; Zheng, C. Characterization and evaluation of Fe$_2$O$_3$/Al$_2$O$_3$ oxygen carrier prepared by sol–gel combustion synthesis. *J. Anal. Appl. Pyrolysis* **2011**, *91*, 105–113. [CrossRef]

80. Mei, D.; Abad, A.; Zhao, H.; Adánez, J.; Zheng, C. On a highly reactive Fe$_2$O$_3$/Al$_2$O$_3$ oxygen carrier for in situ gasification chemical looping combustion. *Energy Fuels* **2014**, *28*, 7043–7052. [CrossRef]

81. Cabello, A.; Dueso, C.; García-Labiano, F.; Gayán, P.; Abad, A.; De Diego, L.F.; Adánez, J. Performance of a highly reactive impregnated Fe$_2$O$_3$/Al$_2$O$_3$ oxygen carrier with CH$_4$ and H$_2$S in a 500 W$_{th}$ CLC unit. *Fuel* **2014**, *121*, 117–125. [CrossRef]

82. Rihko-Struckmann, L.K.; Datta, P.; Wenzel, M.; Sundmacher, K.; Dharanipragada, N.V.R.A.; Poelman, H.; Galvita, V.V.; Marin, G.B. Hydrogen and carbon monoxide production by chemical looping over iron-aluminium oxides. *Energy Technol.* **2016**, *4*, 304–313. [CrossRef]

83. Qin, W.; Chen, Q.; Wang, Y.; Dong, C.; Zhang, J.; Li, W.; Yang, Y. Theoretical study of oxidation–reduction reaction of Fe$_2$O$_3$ supported on MgO during chemical looping combustion. *Appl. Surf. Sci.* **2013**, *266*, 350–354. [CrossRef]

84. Corbella, B.M.; Palacios, J.M. Titania-supported iron oxide as oxygen carrier for chemical-looping combustion of methane. *Fuel* **2007**, *86*, 113–122. [CrossRef]

85. Ksepko, E.; Sciazko, M.; Babinski, P. Studies on the redox reaction kinetics of Fe$_2$O$_3$-CuO/Al$_2$O$_3$ and Fe$_2$O$_3$/TiO$_2$ oxygen carriers. *Appl. Energy* **2014**, *115*, 374–383. [CrossRef]

86. Liu, Y.-C.; Ku, Y.; Tseng, Y.-H.; Lee, H.-Y.; Kuo, Y.-L. Fabrication of Fe$_2$O$_3$/TiO$_2$ oxygen carriers for chemical looping combustion and hydrogen generation. *Aerosol Air Qual. Res.* **2016**, *16*, 2023–2032. [CrossRef]

87. Zafar, Q.; Mattisson, T.; Gevert, B. Integrated hydrogen and power production with CO$_2$ capture using chemical-looping reforming redox reactivity of particles of CuO, Mn$_2$O$_3$, NiO, and Fe$_2$O$_3$ using SiO$_2$ as a support. *Ind. Eng. Chem. Res.* **2005**, *44*, 3485–3496. [CrossRef]

88. Johansson, M.; Mattisson, T.; Lyngfelt, A. Investigation of Fe$_2$O$_3$ with MgAl$_2$O$_4$ for chemical-looping combustion. *Ind. Eng. Chem. Res.* **2004**, *43*, 6978–6987. [CrossRef]

89. Dharanipragada, N.V.R.A.; Buelens, L.C.; Poelman, H.; De Grave, E.; Galvita, V.V.; Marin, G.B. Mg–Fe–Al–O for advanced CO$_2$ to CO conversion: Carbon monoxide yield vs. oxygen storage capacity. *J. Mater. Chem. A* **2015**, *3*, 16251–16262. [CrossRef]

90. Hafizi, A.; Rahimpour, M. Inhibiting Fe–Al spinel formation on a narrowed mesopore-sized MgAl$_2$O$_4$ support as a novel catalyst for H$_2$ production in chemical looping technology. *Catalysts* **2018**, *8*, 27. [CrossRef]

91. De Vos, Y.; Jacobs, M.; Van Driessche, I.; Van Der Voort, P.; Snijkers, F.; Verberckmoes, A. Processing and characterization of Fe-based oxygen carriers for chemical looping for hydrogen production. *Int. J. Greenh. Gas Control* **2018**, *70*, 12–21. [CrossRef]

92. Ismail, M.; Liu, W.; Scott, S.A. The performance of Fe$_2$O$_3$-CaO oxygen carriers and the interaction of iron oxides with CaO during chemical looping combustion and H$_2$ production. *Energy Procedia* **2014**, *63*, 87–97. [CrossRef]

93. Ismail, M.; Liu, W.; Dunstan, M.T.; Scott, S.A. Development and performance of iron based oxygen carriers containing calcium ferrites for chemical looping combustion and production of hydrogen. *Int. J. Hydrog. Energy* **2016**, *41*, 4073–4084. [CrossRef]

94. Liu, W.; Dennis, J.S.; Scott, S.A. The effect of addition of ZrO$_2$ to Fe$_2$O$_3$ for hydrogen production by chemical looping. *Ind. Eng. Chem. Res.* **2012**, *51*, 16597–16609. [CrossRef]

95. Tan, Q.; Qin, W.; Chen, Q.; Dong, C.; Li, W.; Yang, Y. Synergetic effect of ZrO$_2$ on the oxidation–reduction reaction of Fe$_2$O$_3$ during chemical looping combustion. *Appl. Surf. Sci.* **2012**, *258*, 10022–10027. [CrossRef]

96. Kang, K.-S.; Kim, C.-H.; Bae, K.-K.; Cho, W.-C.; Jeong, S.-U.; Lee, Y.-J.; Park, C.-S. Reduction and oxidation properties of Fe_2O_3/ZrO_2 oxygen carrier for hydrogen production. *Chem. Eng. Res. Des.* **2014**, *92*, 2584–2597. [CrossRef]

97. Zhu, X.; Li, K.; Wei, Y.; Wang, H.; Sun, L. Chemical-looping steam methane reforming over a CeO_2-Fe_2O_3 oxygen carrier: Evolution of its structure and reducibility. *Energy Fuels* **2014**, *28*, 754–760. [CrossRef]

98. Machida, M.; Kawada, T.; Fujii, H.; Hinokuma, S. The role of CeO_2 as a gateway for oxygen storage over CeO_2-grafted Fe_2O_3 composite materials. *J. Phys. Chem. C* **2015**, *119*, 24932–24941. [CrossRef]

99. Sun, S.; Zhao, M.; Cai, L.; Zhang, S.; Zeng, D.; Xiao, R. Performance of CeO_2-modified iron-based oxygen carrier in the chemical looping hydrogen generation process. *Energy Fuels* **2015**, *29*, 7612–7621. [CrossRef]

100. Dharanipragada, N.V.R.A.; Meledina, M.; Galvita, V.V.; Poelman, H.; Turner, S.; Van Tendeloo, G.; Detavernier, C.; Marin, G.B. Deactivation study of Fe_2O_3–CeO_2 during redox cycles for CO production from CO_2. *Ind. Eng. Chem. Res.* **2016**, *55*, 5911–5922. [CrossRef]

101. Jin, H.; Ishida, M. Reactivity study on natural-gas-fueled chemical-looping combustion by a fixed-bed reactor. *Ind. Eng. Chem. Res.* **2002**, *41*, 4004–4007. [CrossRef]

102. Hossain, M.M.; De Lasa, H.I. Reactivity and stability of Co-Ni/Al_2O_3 oxygen carrier in multicycle CLC. *AIChE J.* **2007**, *53*, 1817–1829. [CrossRef]

103. Wang, S.; Wang, G.; Jiang, F.; Luo, M.; Li, H. Chemical looping combustion of coke oven gas by using Fe_2O_3/CuO with $MgAl_2O_4$ as oxygen carrier. *Energy Environ. Sci.* **2010**, *3*, 1353. [CrossRef]

104. Mungse, P.; Saravanan, G.; Rayalu, S.; Labhsetwar, N. Mixed oxides of iron and manganese as potential low-cost oxygen carriers for chemical looping combustion. *Energy Technol.* **2015**, *3*, 856–865. [CrossRef]

105. Azimi, G.; Leion, H.; Rydén, M.; Mattisson, T.; Lyngfelt, A. Investigation of different Mn–Fe oxides as oxygen carrier for chemical-looping with oxygen uncoupling (CLOU). *Energy Fuels* **2012**, *27*, 367–377. [CrossRef]

106. Azimi, G.; Mattisson, T.; Leion, H.; Rydén, M.; Lyngfelt, A. Comprehensive study of Mn–Fe–Al oxygen-carriers for chemical-looping with oxygen uncoupling (CLOU). *Int. J. Greenh. Gas Control* **2015**, *34*, 12–24. [CrossRef]

107. Frick, V.; Rydén, M.; Leion, H. Investigation of Cu–Fe and Mn–Ni oxides as oxygen carriers for chemical-looping combustion. *Fuel Process. Technol.* **2016**, *150*, 30–40. [CrossRef]

108. Pans, M.A.; Gayán, P.; Abad, A.; García-Labiano, F.; De Diego, L.F.; Adánez, J. Use of chemically and physically mixed iron and nickel oxides as oxygen carriers for gas combustion in a CLC process. *Fuel Process. Technol.* **2013**, *115*, 152–163. [CrossRef]

109. Bhavsar, S.; Veser, G. Bimetallic Fe–Ni oxygen carriers for chemical looping combustion. *Ind. Eng. Chem. Res.* **2013**, *52*, 15342–15352. [CrossRef]

110. Wei, G.; He, F.; Zhao, Z.; Huang, Z.; Zheng, A.; Zhao, K.; Li, H. Performance of Fe–Ni bimetallic oxygen carriers for chemical looping gasification of biomass in a 10 kw_{th} interconnected circulating fluidized bed reactor. *Int. J. Hydrog. Energy* **2015**, *40*, 16021–16032. [CrossRef]

111. Wei, G.; He, F.; Zhao, W.; Huang, Z.; Zhao, K.; Zhao, Z.; Zheng, A.; Wu, X.; Li, H. Experimental investigation of Fe–Ni–Al oxygen carrier derived from hydrotalcite-like precursors for the chemical looping gasification of biomass char. *Energy Fuels* **2017**, *31*, 5174–5182. [CrossRef]

112. Lim, H.S.; Kang, D.; Lee, J.W. Phase transition of Fe_2O_3–NiO to $NiFe_2O_4$ in perovskite catalytic particles for enhanced methane chemical looping reforming-decomposition with CO_2 conversion. *Appl. Catal. B* **2017**, *202*, 175–183. [CrossRef]

113. Mihai, O.; Chen, D.; Holmen, A. Catalytic consequence of oxygen of lanthanum ferrite perovskite in chemical looping reforming of methane. *Ind. Eng. Chem. Res.* **2011**, *50*, 2613–2621. [CrossRef]

114. Antigoni, E.; Vassilis, Z.; Lori, N. $La_{1−x}Sr_xFeO_{3−\delta}$ perovskites as redox materials for application in a membrane reactor for simultaneous production of pure hydrogen and synthesis gas. *Fuel* **2010**, *89*, 1265–1273.

115. He, F.; Li, X.; Zhao, K.; Huang, Z.; Wei, G.; Li, H. The use of $La_{1−x}Sr_xFeO_3$ perovskite-type oxides as oxygen carriers in chemical-looping reforming of methane. *Fuel* **2013**, *108*, 465–473. [CrossRef]

116. Lori, N.; Antigoni, E.; Vassilis, Z. $La_{1−x}Sr_xM_yFe_{1−y}O_{3−\delta}$ perovskites as oxygen-carrier materials for chemical-looping reforming. *Int. J. Hydrog. Energy* **2011**, *36*, 6657–6670.

117. Li, Z.; Mo, L.; Kathiraser, Y.; Kawi, S. Yolk–satellite–shell structured Ni–yolk@Ni@SiO_2 nanocomposite: Superb catalyst toward methane CO_2 reforming reaction. *ACS Catal.* **2014**, *4*, 1526–1536. [CrossRef]

118. Baktash, E.; Littlewood, P.; Schomäcker, R.; Thomas, A.; Stair, P.C. Alumina coated nickel nanoparticles as a highly active catalyst for dry reforming of methane. *Appl. Catal. B* **2015**, *179*, 122–127. [CrossRef]

119. Park, J.N.; Zhang, P.; Hu, Y.S.; McFarland, E.W. Synthesis and characterization of sintering-resistant silica-encapsulated Fe_3O_4 magnetic nanoparticles active for oxidation and chemical looping combustion. *Nanotechnology* **2010**, *21*, 225708. [CrossRef] [PubMed]

120. Bhavsar, S.; Najera, M.; Veser, G. Chemical looping dry reforming as novel, intensified process for CO_2 activation. *Chem. Eng. Technol.* **2012**, *35*, 1281–1290. [CrossRef]

121. Shafiefarhood, A.; Galinsky, N.; Huang, Y.; Chen, Y.; Li, F. Fe_2O_3@$La_xSr_{1-x}FeO_3$ core-shell redox catalyst for methane partial oxidation. *ChemCatChem* **2014**, *6*, 790–799. [CrossRef]

122. Neal, L.; Shafiefarhood, A.; Li, F. Effect of core and shell compositions on MeO_x@$La_ySr_{1-y}FeO_3$ core–shell redox catalysts for chemical looping reforming of methane. *Appl. Energy* **2015**, *157*, 391–398. [CrossRef]

123. Sarshar, Z.; Sun, Z.; Zhao, D.; Kaliaguine, S. Development of sinter-resistant core–shell $LaMn_xFe_{1-x}O_3$@$mSiO_2$ oxygen carriers for chemical looping combustion. *Energy Fuels* **2012**, *26*, 3091–3102. [CrossRef]

124. Tian, X.; Wei, Y.; Zhao, H. Evaluation of a hierarchically-structured CuO@TiO_2-Al_2O_3 oxygen carrier for chemical looping with oxygen uncoupling. *Fuel* **2017**, *209*, 402–410. [CrossRef]

125. Dharanipragada, N.V.R.A.; Galvita, V.V.; Poelman, H.; Buelens, L.C.; Detavernier, C.; Marin, G.B. Bifunctional Co- and Ni-ferrites for catalyst-assisted chemical looping with alcohols. *Appl. Catal. B* **2018**, *222*, 59–72. [CrossRef]

126. Hu, J.; Galvita, V.V.; Poelman, H.; Detavernier, C.; Marin, G.B. Catalyst-assisted chemical looping auto-thermal dry reforming: Spatial structuring effects on process efficiency. *Appl. Catal. B* **2018**, *231*, 123–136. [CrossRef]

127. Vladimir, G.V.; Hilde, P.; Esteban, F.; Mark, S.; Marin, G.B. Development and performance of iron based oxygen carriers for chemical looping, in nanotechnology in catalysis. In *Applications in the Chemical Industry, Energy Development, and Environment Protection*; Voorde, M.V.d., Sels, B., Eds.; Wiley-VCH GmbH & Co. KGaA: Weinheim, Germany, 2017.

128. Zhu, T.; Flytzani-Stephanopoulos, M. Catalytic partial oxidation of methane to syngas over Ni-CeO_2. *Appl. Catal. A* **2001**, *208*, 403–417. [CrossRef]

129. Pengpanich, S.; Meeyoo, V.; Rirksomboon, T.; Bunyakiat, K. Catalytic oxidation of methane over CeO_2-ZrO_2 mixed oxide solid solution catalysts prepared via urea hydrolysis. *Appl. Catal. A* **2002**, *234*, 221–233. [CrossRef]

130. Tobias, M.; Anders, L.; Paul, C. The use of iron oxide as an oxygen carrier in chemical-looping combustion of methane with inherent separation of CO_2. *Fuel* **2001**, *80*, 1953–1962.

131. Rifflart, S.; Lambert, A.; Delebarre, A.; Salmi, J.; Durand, B.; Carpentier, S. Climate project-material development for chemical looping combustion. In Proceedings of the 1st International Conference on Chemical Looping, Lyon, France, 17–19 March 2010.

132. Tian, Q.; Che, L.; Ding, B.; Wang, Q.; Su, Q. Performance of a Cu-Fe-based oxygen carrier combined with a Ni-based oxygen carrier in a chemical-looping combustion process based on fixed-bed reactors. *Greenh. Gas Sci. Technol.* **2018**, *8*, 1–15. [CrossRef]

133. Johansson, M.; Mattisson, T.; Lyngfelt, A. Creating a synergy effect by using mixed oxides of iron- and nickel oxides in the combustion of methane in a chemical-looping combustion reactor. *Energy Fuels* **2006**, *20*, 2399–2407. [CrossRef]

134. Pishahang, M.; Larring, Y.; Sunding, M.; Jacobs, M.; Snijkers, F. Performance of perovskite-type oxides as oxygen-carrier materials for chemical looping combustion in the presence of H_2S. *Energy Technol.* **2016**, *4*, 1305–1316. [CrossRef]

135. Dai, X.P.; Li, R.J.; Yu, C.C.; Hao, Z.P. Unsteady-state direct partial oxidation of methane to synthesis gas in a fixed-bed reactor using $AFeO_3$ (A = La, Nd, Eu) perovskite-type oxides as oxygen storage. *J. Phys. Chem. B* **2006**, *110*, 22525–22531. [CrossRef] [PubMed]

136. Li, R.J.; Yu, C.C.; Ji, W.J.; Shen, S.K. Methane oxidation to synthesis gas using lattice oxygen. In *Studies in Surface Science and Catalysis*; Bao, X., Xu, Y., Eds.; Elsevier: Amsterdam, The Netherlands, 2004; Volume 147, pp. 199–204.

137. Solunke, R.D.; Veser, G.t. Hydrogen production via chemical looping steam reforming in a periodically operated fixed-bed reactor. *Ind. Eng. Chem. Res.* **2010**, *49*, 11037–11044. [CrossRef]

138. Zhang, J.; Li, F. Coke-resistant Ni@SiO$_2$ catalyst for dry reforming of methane. *Appl. Catal. B* **2015**, *176–177*, 513–521. [CrossRef]

139. Park, J.C.; Bang, J.U.; Lee, J.; Ko, C.H.; Song, H. Ni@SiO$_2$ yolk-shell nanoreactor catalysts: High temperature stability and recyclability. *J. Mater. Chem.* **2010**, *20*, 1239–1246. [CrossRef]

140. Li, L.; He, S.; Song, Y.; Zhao, J.; Ji, W.; Au, C.-T. Fine-tunable Ni@porous silica core–shell nanocatalysts: Synthesis, characterization, and catalytic properties in partial oxidation of methane to syngas. *J. Catal.* **2012**, *288*, 54–64. [CrossRef]

141. Xu, Z.; Zhao, H.; Wei, Y.; Zheng, C. Self-assembly template combustion synthesis of a core–shell CuO@TiO$_2$-Al$_2$O$_3$ hierarchical structure as an oxygen carrier for the chemical-looping processes. *Combust. Flame* **2015**, *162*, 3030–3045. [CrossRef]

142. Li, S.; Zhang, C.; Huang, Z.; Wu, G.; Gong, J. A Ni@ZrO$_2$ nanocomposite for ethanol steam reforming: Enhanced stability via strong metal-oxide interaction. *Chem. Commun. (Camb.)* **2013**, *49*, 4226–4228. [CrossRef] [PubMed]

143. Ushakov, S.V.; Navrotsky, A.; Yang, Y.; Stemmer, S.; Kukli, K.; Ritala, M.; Leskelä, M.A.; Fejes, P.; Demkov, A.; Wang, C.; et al. Crystallization in hafnia- and zirconia-based systems. *Phys. Status Solidi* **2004**, *241*, 2268–2278. [CrossRef]

144. Zhao, F.; Gao, Y.; Li, W.; Luo, H. Nanocoating Fe$_2$O$_3$ powders with a homogeneous ultrathin ZrO$_2$ shell. *J. Ceram. Soc. Jpn.* **2008**, *116*, 1164–1166. [CrossRef]

145. Gao, Y.; Zhao, F.; Liu, Y.; Luo, H. Synthesis and characterization of ZrO$_2$ capsules and crystalline ZrO$_2$ thin layers on Fe$_2$O$_3$ powders. *CrystEngComm* **2011**, *13*, 3511–3514. [CrossRef]

146. Arnal, P.M.; Weidenthaler, C.; Schüth, F. Highly monodisperse zirconia-coated silica spheres and zirconia-silica hollow spheres with remarkable textural properties. *Chem. Mater.* **2006**, *18*, 2733–2739. [CrossRef]

147. Feyen, M.; Weidenthaler, C.; Guttel, R.; Schlichte, K.; Holle, U.; Lu, A.H.; Schuth, F. High-temperature stable, iron-based core-shell catalysts for ammonia decomposition. *Chemistry* **2011**, *17*, 598–605. [CrossRef] [PubMed]

148. O'Neill, B.J.; Jackson, D.H.K.; Lee, J.; Canlas, C.; Stair, P.C.; Marshall, C.L.; Elam, J.W.; Kuech, T.F.; Dumesic, J.A.; Huber, G.W. Catalyst design with atomic layer deposition. *ACS Catal.* **2015**, *5*, 1804–1825. [CrossRef]

149. Galinsky, N.L.; Shafiefarhood, A.; Chen, Y.; Neal, L.; Li, F. Effect of support on redox stability of iron oxide for chemical looping conversion of methane. *Appl. Catal. B* **2015**, *164*, 371–379. [CrossRef]

150. Evdou, A.; Zaspalis, V.; Nalbandian, L. Ferrites as redox catalysts for chemical looping processes. *Fuel* **2016**, *165*, 367–378. [CrossRef]

151. Wang, X.; Chen, Z.; Hu, M.; Tian, Y.; Jin, X.; Ma, S.; Xu, T.; Hu, Z.; Liu, S.; Guo, D.; et al. Chemical looping combustion of biomass using metal ferrites as oxygen carriers. *Chem. Eng. J.* **2017**, *312*, 252–262. [CrossRef]

152. Srdic, V.; Mojic, B.; Nikolic, M.; Ognjanovic, S. Recent progress on synthesis of ceramics core/shell nanostructures. *Process. Appl. Ceram.* **2013**, *7*, 45–62. [CrossRef]

153. Yoo, J.B.; Kim, H.S.; Kang, S.H.; Lee, B.; Hur, N.H. Hollow nickel-coated silica microspheres containing rhodium nanoparticles for highly selective production of hydrogen from hydrous hydrazine. *J. Mater. Chem. A* **2014**, *2*, 18929–18937. [CrossRef]

154. Yang, Y.; Liu, J.; Li, X.; Liu, X.; Yang, Q. Organosilane-assisted transformation from core–shell to yolk–shell nanocomposites. *Chem. Mater.* **2011**, *23*, 3676–3684. [CrossRef]

155. Stöber, W.; Fink, A. Controlled growth of monodisperse silica spheres in the micron size range. *J. Colloid Interface Sci.* **1968**, *26*, 62–69. [CrossRef]

materials

MDPI

Review

Recent Advances in Transmission Electron Microscopy for Materials Science at the EMAT Lab of the University of Antwerp

Giulio Guzzinati [1], Thomas Altantzis [1], Maria Batuk [1], Annick De Backer [1], Gunnar Lumbeeck [1], Vahid Samaee [1], Dmitry Batuk [1], Hosni Idrissi [1,2], Joke Hadermann [1], Sandra Van Aert [1], Dominique Schryvers [1], Johan Verbeeck [1] and Sara Bals [1,*]

[1] EMAT, University of Antwerp, Groenenborgerlaan 171, Antwerp 2020, Belgium;
     giulio.guzzinati@uantwerpen.be (G.G.); thomas.altantzis@uantwerpen.be (T.A.);
     maria.batuk@uantwerpen.be (M.B.); annick.debacker@uantwerpen.be (A.D.B.);
     gunnar.lumbeeck@uantwerpen.be (G.L.); vahid.samaeeaghmiyoni@uantwerpen.be (V.S.);
     dmitry.batuk@gmail.com (D.B.); hosni.idrissi@uantwerpen.be (H.I.); joke.hadermann@uantwerpen.be (J.H.);
     sandra.vanaert@uantwerpen.be (S.V.A.); nick.schryvers@uantwerpen.be (D.S.);
     jo.verbeeck@uantwerpen.be (J.V.)
[2] Institute of Mechanics, Materials and Civil Engineering, Université catholique de Louvain,
     Louvain-la-Neuve 1348, Belgium
*   Correspondence: sara.bals@uantwerpen.be; Tel.: +32-3-265-3284

Received: 29 June 2018; Accepted: 26 July 2018; Published: 28 July 2018

Abstract: The rapid progress in materials science that enables the design of materials down to the nanoscale also demands characterization techniques able to analyze the materials down to the same scale, such as transmission electron microscopy. As Belgium's foremost electron microscopy group, among the largest in the world, EMAT is continuously contributing to the development of TEM techniques, such as high-resolution imaging, diffraction, electron tomography, and spectroscopies, with an emphasis on quantification and reproducibility, as well as employing TEM methodology at the highest level to solve real-world materials science problems. The lab's recent contributions are presented here together with specific case studies in order to highlight the usefulness of TEM to the advancement of materials science.

Keywords: TEM; electron diffraction tomography; STEM; atom counting; electron tomography; compressed sensing; EDS; EELS; nanomechanical testing; ACOM TEM

---

1. Introduction

Transmission electron microscopy (TEM) is a very rapidly developing field. The scope, breadth of information, and power of the various types of TEM techniques are being expanded every year. A modern TEM constitutes a complex characterization facility, capable of collecting diffraction patterns from volumes of a few cubic nanometres in size and imaging samples down to the atomic scale. When combined with tomography, a technique which derives three-dimensional (3D) information from two-dimensional (2D) images, one is able to determine the structure and shape of nanostructures in 3D, even with atomic resolution [1,2]. More than this, TEM allows performing spectroscopies that can analyse, once more down to the atomic scale, the composition of the sample, but also bonding, optical, and electronic structure properties both in 2D and 3D. Finally, a growing range of experiments can now be performed in situ in the TEM with simultaneous characterization, offering valuable insight in a variety of processes.

For this special issue on Materials Science in Belgium, we will focus on the contribution of the EMAT research group located at the University of Antwerp. EMAT is the foremost electron microscopy laboratory in Belgium, and one of the leading electron microscopy groups in the world, currently composed of more than 50 researchers and equipped with a wide range of state of the art instrumentation. Already the name, a contraction of electron microscopy for materials science, reveals the strong focus on materials, reflected also in the dense network of national and international collaboration with leading groups. Every characterization technique that is available here can be applied at the state of the art, and EMAT constantly strives to develop them further and push the boundaries of what is possible.

In this paper, we showcase a selection of examples of applied TEM as well as of instrumental developments that were contributed by EMAT to the field of materials science. We show how TEM can recover the different types of information from different materials, the methodological and instrumental advances pioneered by this lab, and case studies that showcase the determinant role of electron microscopy in solving different materials science problems.

2. TEM as a Structural Characterization Tool

In functional materials, a deep understanding of the relation between the properties and the crystal structure is the key to designing new materials and improving existing ones. TEM provides unique opportunities for crystal structure analysis at a very local scale, which, in many ways, complements powder diffraction techniques, such as X-ray or neutron powder diffraction, where the data is collected from a comparatively large volume of the material. TEM can access structural data both in reciprocal and direct space. Electron diffraction (ED) patterns contain information on the symmetry of the crystal and, in some cases, the intensities of the reflections can be used quantitatively to solve the crystal structure [3]. Coupling of scanning transmission electron microscopy (STEM) with atomic resolution spectroscopic techniques, such as energy-dispersive X-ray spectroscopy (EDS) and electron energy loss spectroscopy (EELS), allows analysis of the chemical composition, oxidation state, and coordination number of the individual atomic columns in the structure. In this section, we will give some examples of how these techniques were used to unveil structures that were out of reach for powder diffraction techniques due to different reasons.

Some types of samples can be difficult to study with powder diffraction, but are excellent for study using TEM, for example multiphase samples, samples with local defects and modulated materials [4–7]. While often such materials can still be analysed with powder diffraction, the data might be difficult for interpretation due to a variety of factors such as the presence of a large number of reflections, reflection overlap and anisotropic reflection broadening. There are clear advantages to the use of TEM: as electrons in a TEM interact much more strongly with matter than X-rays or neutrons, volumes of a few tens of nm^3 are already enough to provide a clear diffraction pattern which is then from a single crystal, while images of the structure in direct space can already be taken with only a few atoms. As an extra advantage, the electron wavelength is a few orders of magnitude smaller than that of typical X-rays or neutrons, increasing the radius of the Ewald sphere (which is equal to the wave vector) and making its curvature almost flat, thus making ED patterns (almost) two-dimensional sections of the reciprocal lattice of the crystal [8]. As a result, ED provides easily interpretable information about the symmetry of the crystal, on a scale ranging from micrometres down to nanometres. The two-dimensional sections can also be combined to reconstruct the three-dimensional reciprocal lattice, using, in principle, simple pen-and-paper [3].

Structure solution and refinement from TEM has proven especially useful for such prominent materials as cathode materials for lithium-ion batteries. Understanding structure transformations upon reversible intercalation (during discharge) and de-intercalation (during charge) of the Li^+ or Na^+ ions is crucial for the improvement of the capacity, charge density, and lifetime of the battery. By design, batteries are multi-component devices, where the cathode materials are primarily used in powder form to ensure a homogenous mixing with amorphous additives (carbon) as electric conductors and with an electrolyte. This limits the possibilities of powder diffraction for the analysis of the structures. As ED patterns can be collected from single-crystal particles mere nanometres in diameter, the powder mix is effectively a collection of single-crystal samples for ED. On lithium battery materials, EMAT's most important contributions through TEM are based on the use of electron diffraction tomography (EDT) or STEM images.

The specific advantage of EDT for lithium battery materials stems from the higher sensitivity of electron diffraction to elements with small atomic number (Z), such as lithium, compared to X-ray diffraction, as well as the single-crystal nature of the electron diffraction data [9,10]. However, direct structure solutions from ED patterns are hampered by multiple scattering of the electrons on their path through the crystal, even in samples with a thickness of only a few nanometres [11]. This makes the intensities of the reflections very dependent on the thickness and orientation of the crystals, resulting in intensities that deviate significantly from the intensities expected in the kinematic (i.e., single scattering) approximation. The detrimental effect of multiple electron scattering can be mitigated using either precession electron diffraction (PED) [12] or electron diffraction tomography. Both techniques result in off-zone patterns (although in the case of precession electron diffraction many off-zone patterns are summed and recombined into a seemingly in-zone pattern) which decreases the amount of possible multiple scattering paths and, thus, results in intensities closer to the kinematic ones (and, thus, called quasi-kinematical) [13]. Both techniques have been usefully applied for determining the structures of lithium battery cathode materials. In PED (whose applications are also explored in Section 8.1 of this paper) the crystal is tilted in the zone axis orientation, then the incident electron beam is tilted slightly (~1°) off the optical axis and rocked azimuthally around the optical axis, while keeping the tilt by using electromagnetic coils. This change in the direction of incidence produces the off-axis patterns. A second set of coils placed after the sample cancels out the beam tilt recombining the different patterns together. PED was used to solve and refine, for the first time, the structure of Li_2CoPO_4F [14]. Currently, the EDT technique (combinable with PED) is more often used because of its finer sampling of the reciprocal space [15], as simple PED only provides a few diffraction patterns and, hence, only a few slices of reciprocal space, while EDT probes a large fraction of the reciprocal space. In EDT, the sample holder is tilted inside the TEM column and ED patterns are collected using an increment of ~1°, avoiding major crystallographic zone axes. The acquired ED patterns are used as an input for a reconstruction algorithm, which produces a 3D dataset containing quasi-kinematical diffraction intensities. They can be used for crystal structure solutions by the conventional methods of single-crystal X-ray crystallography. Using this technique, many structures of cycled cathode materials were already successfully solved and refined. For example, for cycled $AVPO_4F$ (A = K, Li) the results of EDT revealed that K is not completely extracted from the charged material and that Li in the discharged material occupies new crystallographic positions. Hence, the K atoms in this material not only act as mobile species, but also as structural pillars supporting the VPO_4F framework in the charged material [16]. In the Li_2FePO_4F compound, prepared by electrochemical substitution of Na in $LiNaFePO_4F$ by Li, the EDT structure analysis unveiled the origin of capacity fading in the material. Removal of Li upon charging creates "dangling" P—O bonds resulting in a substantial bond imbalance, which is compensated through Fe migration towards the Li positions giving rise to Li/Fe anti-site disorder [17]. In $LiRhO_2$, the EDT study showed that the layered structure transforms upon charging into a tunnel structure with rutile and ramsdellite channels [18].

Another crucial technique for lithium battery cathode materials is STEM [19,20]. TEM and STEM can both be used to visualize the atomic arrangement of the structure. The development of aberration-corrected TEMs has enabled imaging the atom columns with a resolution of 50 pm [21]. The first TEM method capable of atomic resolution was high resolution (HR) TEM imaging, where elastic interaction of coherent parallel electron beam with the crystalline lattice of the material results in an interference pattern (HRTEM image) that has the same periodicity as the crystal and can be used to retrieve information on the symmetry [22]. Inelastic interaction provides, due to the chromatic aberrations in the lenses, just a featureless background that reduces the images' signal to noise ratio. The direct interpretation of these images is hindered by high sensitivity of the interference pattern to the sample thickness and the exact defocus of the microscope. Nevertheless, this method has been widely used for crystal structure visualization [23–25]. During the last two decades, atomic resolution STEM is more often used for this purpose, mainly due to the fact that it is more robust to the variations in the experimental settings and, thus, HR-STEM images are easier to interpret than HR-TEM images. In this method, a focused electron beam is scanned over the specimen and the transmitted electrons are collected by (typically) annular detectors below the sample. Depending on the acquisition settings of the microscope, there are two main STEM imaging modes useful in structure analysis. In the so-called high angle annular dark field STEM (HAADF-STEM) mode, the intensity of the acquired projection images is proportional to the average atomic number of the projected column ($I \sim Z^2$) and scales with the thickness of the specimen, hence delivering chemically-sensitive information [26]. While in some case even light elements can be detected by HAADF-STEM [27], many materials combine elements with widely different atomic numbers. As a result, the contrast in the images is excellent for the high Z elements, but very poor for the low-Z ones and, in most situations, the accurate localization of the light elements is impossible due to the finite signal to noise ratio. The information from atomic columns of light elements, such as O, or even H, can be more easily obtained using another STEM-imaging mode, called annular bright-field STEM (ABF-STEM) [28–31].

Atomic-resolution HAADF-STEM and ABF-STEM imaging, for example, revealed the crystal structure transformation of the model battery cathode materials Li_2IrO_3 and Na_2IrO_3 [32,33]. These compounds belong to a family of Li, Na-rich layered rock-salt compounds, which demonstrate capacities larger than those expected for the pure cationic redox activity. STEM imaging showed the formation of short O-O distances, confirming that the excess capacity is due to participation of both cationic and anionic sublattices (Figure 1). HAADF-STEM imaging can be used to analyse the migration of the transition metal cations into the Li positions upon cycling, as it was done for Li-rich $Li_2Ru_{1-y}Ti_yO_3$ phases. This migration is only partially reversible during the discharge, leaving a fraction of the transition metal cations trapped in the tetrahedral sites, hence resulting in a gradual voltage fade of the material [34]. In $LiRhO_2$, the HAADF-STEM images could be used to infer a transition mechanism from layered into a 3D structure, which involves local migration of Rh and O species [18]. Sometimes, the defect analysis using real space imaging can greatly facilitate interpretation of the powder diffraction data. In the $Li_3Ru_yNb_{1-y}O_4$ family of rock salt structures, which was used to test the limits of the anionic redox activity [35,36], the $(Nb,Ru)O_6$ octahedra form either zigzag, helical, jagged quasi-1D chains, or 0D clusters depending on the Nb/Ru concentration. The HAADF-STEM images unveiled a plethora of extended planar defects, which give rise to anisotropic broadening of reflections in the powder X-ray diffraction. Modelling these defects significantly improved the crystal structure refinement.

Another type of materials where TEM has proven to be indispensable for structure solution is modulated materials, whether commensurately or incommensurately modulated. Often the satellites or superstructure reflections are weaker compared to the parent cell reflections and form a very dense set of reflection in the powder diffraction patterns, which complicate the Rietveld refinement. Using ED, one can take clear two dimensional sections through the reciprocal lattice to determine the cell parameters and symmetry, with clearly separated satellite reflections in most cases. This can be complemented with a direct view of the structure in direct space using high-resolution TEM or STEM techniques. Structure models for numerous modulated materials were solved using this combination of techniques, most frequently followed by subsequent refinement using powder diffraction data. Examples range from scheelites [37] to perovskite-based structures [38]; a description of the typical solution route has already been published in a previous review paper by Batuk et al. [4].

When the materials are sufficiently stable under the electron beam (the doses for a high resolution spectroscopic map are in the order of few $pC/\text{Å}^2$), the diffraction patterns and images can be complemented with high-resolution EDS or EELS maps, allowing to support the structure models with direct knowledge on the distribution of the elements over the different solved atomic positions [39]. This was necessary, for example, to determine the structure at the interface between a perovskite substrate and fluorite film, which showed unexpected features in the STEM images. The STEM images showed where the atoms were, but only the HR-EDS and EELS studies demonstrated which elements were at those positions, revealing a redistribution of the cations over several layers at the interface [40]. The vast majority of the crystalline solid-state materials are indeed stable under the electron beam and the use of high-resolution TEM imaging provides a still image of the atomic arrangement. Some materials are too beam-sensitive, and even a short exposure to the electron beam (doses for a high-resolution image are of the order of $fC/Å^2$) decomposes them or renders them amorphous, thus requiring special procedures for their TEM analysis [41]. However, there is a narrow group of materials, in which the energy transfer from the electron beam triggers interesting structure transformations. This approach has been used to analyse the chemistry and dynamics of lone pair Bi^{3+} cations in the layered $Bi_{3n+1}Ti_7Fe_{3n-3}O_{9n+11}$ perovskite-anatase intergrowth materials with variable thickness of the perovskite layers [42,43] (Video S1). The lone pair cations are prone to an asymmetric coordination environment, which can induce electric polarization of the material. The HAADF-STEM and ABF-STEM data revealed that in the $Bi_{3n+1}Ti_7Fe_{3n-3}O_{9n+11}$ family, the Bi^{3+} cations trigger off-centre displacements of the transition metal cations. Although the materials are antiferroelectric, the exact pattern of polar atomic displacements depends on the thickness of the perovskite blocks. The dynamic changes in the local configuration of the structure, e.g., upon absorbing energy from the electron beam, occur cooperatively, so that a large number of atoms change their position at the same time, which could be recorded in real-time using a fast HAADF-STEM acquisition.

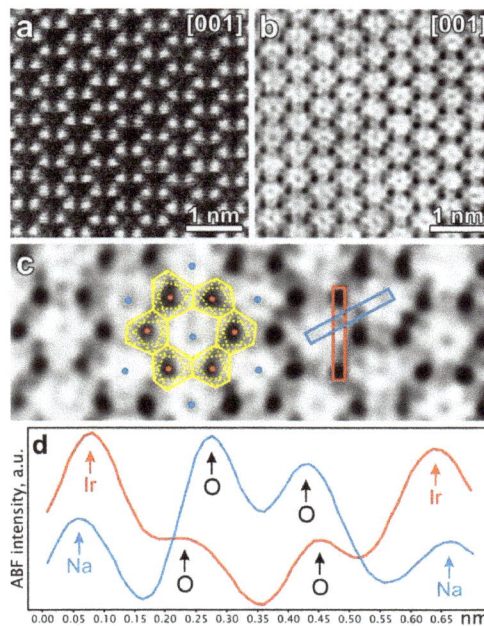

Figure 1. Complementary [001] HAADF-STEM (**a**) and ABF-STEM (**b**) images of fully-charged Na2-xIrO3 with an O1-type structure. Magnified fragment of the ABF-STEM image (**c**) with marked projections of the IrO6 octahedra (red dots: Ir columns; blue: Na; and yellow lines: O octahedral projections). Intensity profiles measured in the areas marked in blue and red highlighting short and long projected O−O distances (**d**). Adapted with permission from [33]. Copyright 2016 American Chemical Society.

3. Quantitative Imaging in Transmission Electron Microscopy

Although high-resolution HAADF-STEM imaging is considered nowadays as a standard technique for the structural and compositional characterization of different nanomaterials at the atomic scale, a quantitative analysis is often very challenging. Since the beginning of the decade, a great deal of effort has been put on the development of approaches to reliably quantify ((S)TEM) data [44]. The main objective is to extract precise and accurate numbers for unknown structure parameters including atomic positions, chemical concentrations, and atomic numbers. In order to extract these quantitative measurements from atomic resolution (S)TEM images, statistical analysis methods are needed. For this purpose, statistical parameter estimation theory has been shown to provide reliable results [45]. In this framework, images are purely considered as data planes [46], from which structure parameters have to be determined using a parametric model describing the images. Atomically-resolved HAADF-STEM images are described with a parametric model in which the projection of an atomic column corresponds to a Gaussian peaked at the column's position. The parameters of this model, including the atom positions, the height and the width of the Gaussian peaks, are determined using the least squares estimator [44,47,48]. As such, the positions of atom columns can be measured with a precision of the order of a few picometres [49–55], even though the resolution of the electron microscope is still one or two orders of magnitude larger. Moreover, small differences in average atomic number, which cannot be distinguished visually, can be quantified using HAADF-STEM images [44]. In addition, this theory allows one to measure compositional changes at interfaces [56–58], to count atoms with single atom sensitivity [59–61], and as we will see in a next part also to reconstruct atomic structures in three dimensions (3D) [1,59,62–65]. Making this well-established quantification method easier to apply for all scientists was the motivation behind the development of StatSTEM, a user-friendly software for the quantification of high-resolution STEM images (Figure 2a) [47]. Here, two recent applications of advanced quantification methods in atomic resolution STEM will be briefly discussed.

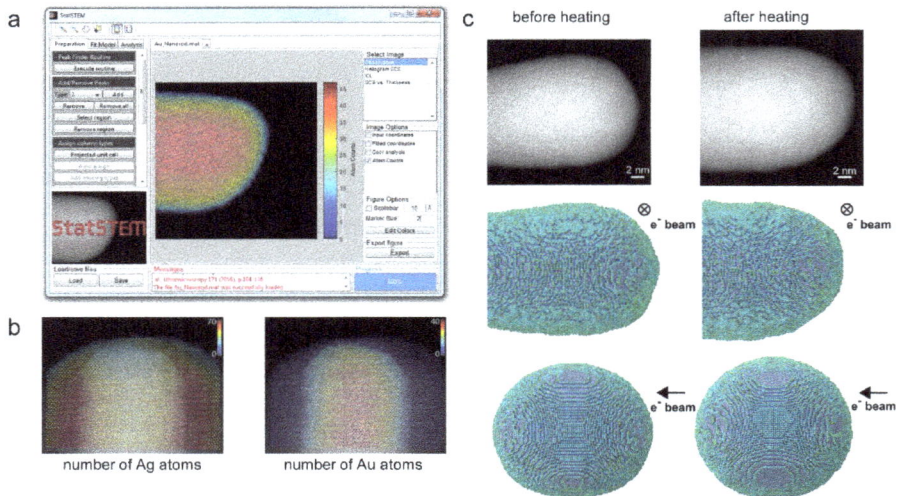

Figure 2. (**a**) Screen shot of the StatSTEM software, showing the atom-counting results on an experimental image of a Au nanorod; (**b**) Counting the number of atoms from a HAADF-STEM image of a Ag-coated Au nanorod; and (**c**) 3D atomic scale characterization of a Au nanodumbbell before and after heating.

In HAADF-STEM, the image intensities scale with the thickness of the sample and with the atomic number ($I \sim Z^2$). For homogeneous materials the scattering cross-sections, i.e., the quantified integrated intensities at the atomic column positions, only depend on the thickness and can, therefore, be used to count the number of atoms in each atomic column by either employing advanced statistical methods [60,61] or by comparison with simulations [66]. For heterogeneous materials, all types of elements will contribute differently to the scattering cross-sections, thus, significantly complicating atom-counting with respect to monotype nanostructures. In addition, small changes in the atom ordering in the column have an effect on the scattering cross-sections. In order to unscramble this information, an atomic lensing model was introduced which allows to predict scattering cross-sections as a function of composition, configuration, and thickness [67]. When using this model the number of both Ag and Au atoms could be counted from an experimental HAADF-STEM image of a Ag-coated Au nanorod, as demonstrated in Figure 2b. This type of atom-counting results can also be used to gain access to the 3D atomic structure, as will be extensively shown in part 5.

4. Investigation of Beam Sensitive Materials in a TEM

Compressed Sensing

For all the wealth of information that TEM can provide about a sample, its usefulness in some fields is still limited by the beam-sensitivity of the samples, and a great deal of work in the field is devoted into obtaining the most information for the electron dose. Examples of this are the application of rigorous statistical methods to the quantification of noisy high-resolution images, and the use of compressed sensing (CS) approaches in the development of electron tomographic algorithms [68].

Compressed sensing is a signal processing approach which consists of expressing a signal into a mathematical basis where it appears sparse and, hence, can be described with less information. As such, this approach allows reconstructing a dataset from a limited amount of samples. Since experimental images have a high degree of natural sparsity, it is possible to apply these methods to STEM imaging with the aim of reconstructing an image by measuring only a fraction of the pixels, randomly distributed within the image frame.

By using a specially developed fast electromagnetic beam shutter capable of operating at microsecond speeds, and synchronizing it with the microscope's scan engine, it becomes possible to selectively illuminate only a selected number of pixels in the image [69]. Early experimental tests, where this technique was applied to the reconstruction of high-resolution images (Figure 3) or to the imaging of highly beam-sensitive samples [70], have been successful. While theoretical doubts have been raised on whether CS can beat conventional denoising algorithms on a purely statistical basis [71], the experimental results suggest that this method of reducing beam damage is effective beyond expectations, and that beam-sensitive samples can undergo healing processes that substantially limit beam damage when exposure is not continuous [70].

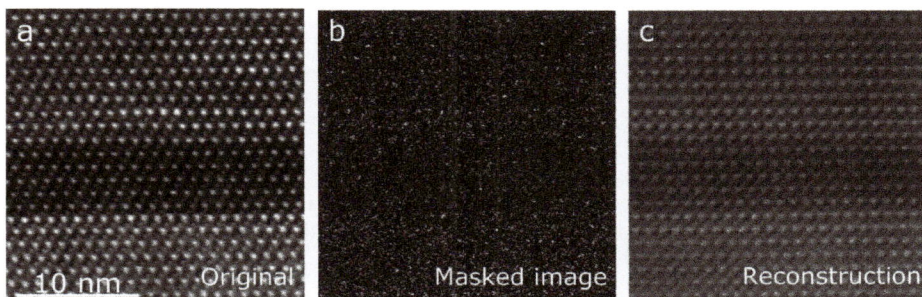

Figure 3. Experimental realization of compressed sensing for STEM imaging. (**a**) Conventional high resolution STEM image; (**b**) image over the same region of the same sample, where 20% of the pixels are acquired and 80% are blocked; and (**c**) the reconstructed image.

5. Atomic Resolution Tomography

While TEM and STEM are ideal techniques to investigate materials at both the nanometre and the atomic scale and have therefore been widely used in the study of nanomaterials, the obtained images only correspond to 2D projections of 3D objects [72]. Therefore, electron tomography, a technique which derives 3D information from 2D projections, can be used in order to obtain the necessary 3D information [73]. Most tomography results have addressed nanometre scale characterization, but recent advances in material sciences raised questions that could only be answered by the atomic scale 3D structures and have fostered the extension of the technique to the atomic scale.

The first work in atomic resolution tomography, which is also considered as a breakthrough in the field, was published almost seven years ago by Van Aert et al. By combining atom-counting results (Figure 4b,d) obtained from images of a 3 nm Ag nanoparticle embedded in an Al matrix under different viewing directions (Figure 4a,c), the 3D space could be reconstructed for the first time at an atomic level using so-called discrete tomography (Figure 4e) [59]. Here, it was assumed that all atoms are positioned on a fixed face-cantered-cubic grid without any vacancies being present. The 3D reconstruction was compared to projection images that were acquired along additional zone axes and an excellent match was found, thereby validating the quality of the reconstruction. This technique, however, assumes that atoms are located on a fixed regular lattice, while deviations often occur because of defects, strain, or lattice relaxation and are of utmost importance as they determine the physical properties of nanomaterials.

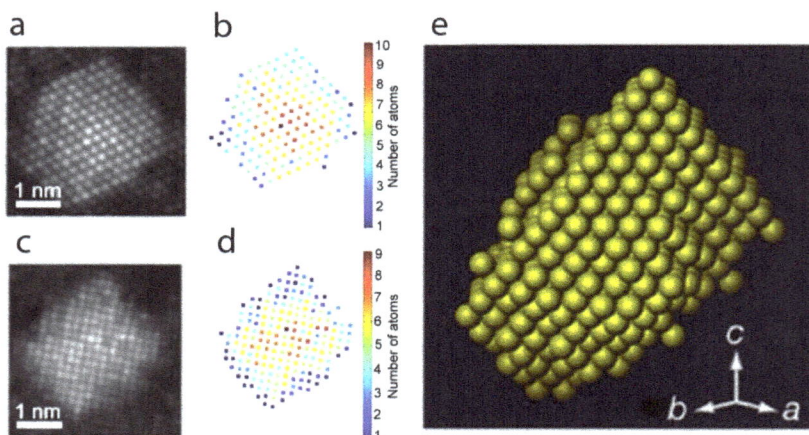

Figure 4. (**a,c**) Refined models for two HAADF-STEM projection images of a Ag atomic cluster embedded in an Al matrix; (**b,d**) Number of Ag atoms per projected atomic column for both images; (**e**) 3D representation of the reconstructed volume of the Ag nanoparticle, based on discrete tomography [59].

One year later, a different approach was proposed by Goris et al. [2], which was applied to Au nanorods in order to visualize the atom positions in 3D. Hereby, four atomic resolution projection images were used as an input for an advanced reconstruction algorithm based on compressive sensing that exploits the natural sparsity deriving from the discrete atomic structure of the material, and does not require assumptions on the crystal structure [68]. Indeed, for high-resolution electron tomography, only a limited number of voxels are expected to contain an atomic core, and most voxels will correspond to vacuum. By exploiting this in the algorithm it is possible to create a reliable reconstruction of the atomic lattice from only a few projections [2]. The methodology enabled a precise determination of the surface facets of the investigated Au nanorod, as illustrated in Figure 5a.

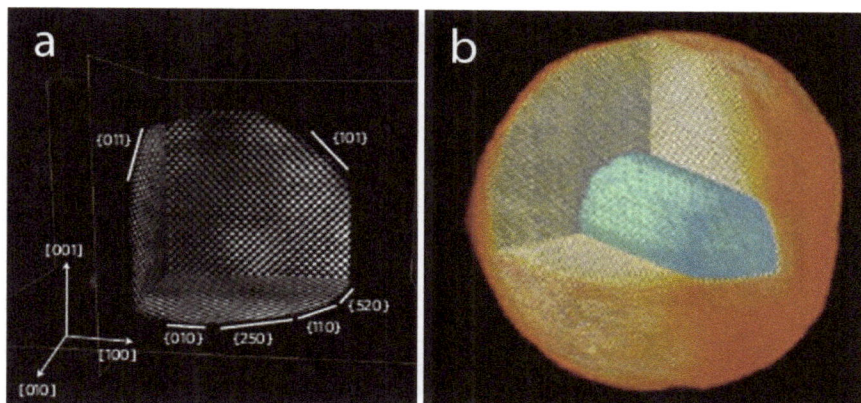

Figure 5. (**a**) Slices through the atomic resolution 3D reconstruction of a Au nanorod, revealing the atomic lattice and the surface facets present; (**b**) 3D visualization of a core—shell Au@Ag nanorod, where the Au core is rendered green, and the Ag atoms are visualized in orange [2,74].

In a subsequent work, the same approach was used by Goris et al. for the characterization of core–shell bimetallic Au@Ag nanorods, where five atomic resolution HAADF-STEM images were acquired (Figure 5b) [74]. In this case, since the intensities depend on the atomic number in the individual HAADF-STEM projections, a careful analysis of the intensities in the reconstructed volume allowed to distinguish between Ag and Au atoms. Therefore, the atomic structure at the core/shell interface could be reliably determined.

While these approaches allowed to study the 3D atomic structure, they do not allow to determine the atomic coordinates with sufficient precision to three dimensionally study, e.g., strain. Recently, Goris et al., conducted a study to compare strain investigations relying on 2D projections with 3D measurements based on high-resolution electron tomographic reconstructions [75]. The 3D measurements displayed an outward relaxation of the crystal lattice which could not be seen from a single 2D projection image. This was done by acquiring a continuous tilt series of HAADF-STEM projections with a tilt increment of $2°$, and applying a dedicated alignment procedure and reconstruction algorithm. Typically, during conventional alignment, the tilt angles are considered a fixed parameter, set to the nominal value used during acquisition. Here, the angles were optimized during the reconstruction in an iterative manner. As previously discussed, the outcome of a 3D reconstruction at the atomic scale is usually a continuous 3D volume of intensity values, from which the centre of each atom can only be determined by additional successive analysis. These datasets are very large, and in the absence of objective and automatic segmentation procedures it becomes difficult to obtain atom coordinates. Goris et al., could overcome this limitation by assuming that the 3D atomic potential can be modelled by 3D Gaussian functions. This assumption significantly simplifies the reconstruction problem, yielding the coordinates of the individual atoms as a direct outcome of the reconstruction. This highlights how using shape models can be a powerful tool to solve many problems in the field of electron tomography. The approach was applied to an Au nanodecahedron containing more than 90,000 atoms [75] and, since the atomic coordinates are a direct output of the reconstruction, it becomes straightforward to calculate the 3D displacement. The displacements were calculated with respect to a reference region in the middle of the segment. The εxx and εzz strain field components were obtained by computing the derivatives of the 3D displacement. Slices through the resulting εxx and εzz volumes extracted through the middle of the segment and the results are presented in Figure 6a,b.

From the previous section it is clear that tomography typically requires several images demanding a substantial electron dose. To circumvent this problem, atom-counting results from just a single projections can be used as an input to retrieve the 3D atomic structure [62,64–66]. In combination with prior knowledge about a material's crystal structure, an initial 3D model is generated. Next, an energy minimization is performed to relax the nanoparticle's 3D structure. This new approach opens up the possibility for the study of beam-sensitive materials, 2D self-assembled structures [64], and in conjunction with in situ experiments where the dedicated hardware makes tomography impossible. As an example this was recently applied by De Backer et al., as shown in Figure 2c, to retrieve the 3D atomic structure of a nanodumbbell on an in situ heating holder allowing only a limited tilt range [65]. From the reconstructions, the surface facets can be clearly observed for the entire tip of the nanostructure and a significant increase in low index facets is observed after heating which can be expected to be of critical importance for the study of catalysis.

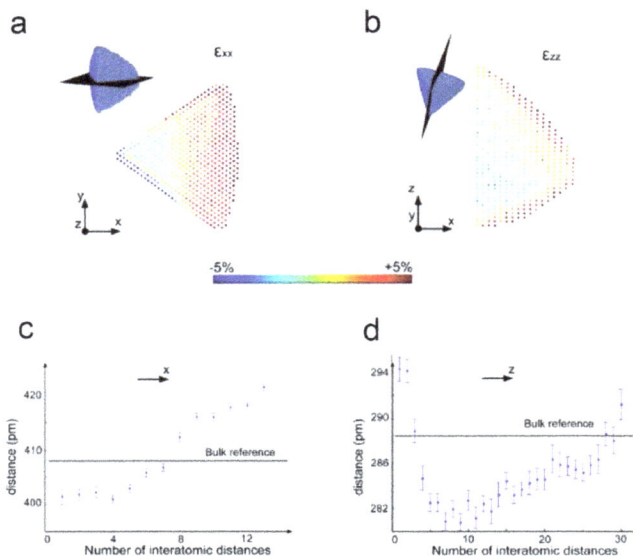

Figure 6. (**a**,**b**) εxx and εzz strain field showing surface relaxation in both directions. The colour scale indicates the expansion of the lattice parameter with respect to a reference; (**c**,**d**) The surface relaxation was confirmed by measuring the lattice parameter on slices through the reconstructions at the positions indicated in (**a**,**b**) [75].

6. Spectroscopic Techniques in a TEM

While as illustrated in the previous sections, (S)TEM can provide quantitative structural information in 3D for systems with one or two atomic species as long as the atomic number differs sufficiently, relevant samples for materials science contain often a much wider variety. The question is then how can we retrieve the distribution of different elements in the specimen and can we even extract bonding information. In order to answer these questions, spectroscopic techniques are an essential part of materials characterization methods.

EDS and EELS are spectroscopic techniques which are almost always present in modern TEMs. By using them in combination with STEM, a complete 2D mapping of the chemical elements in the specimen can be obtained. Recent advances in instrumentation, such as the implementation of electron monochromators, allow EELS edges to be acquired at energy resolutions close to that of synchrotron-based spectroscopies providing extra information which cannot be obtained by EDS. For example, investigating the core-loss region of a spectrum, can reveal information concerning the valency of cations, oxygen coordination, and bond elongation, with spatial resolutions going down to the atomic scale. Analysing electrons which have lost few eV, belonging to the so-called low-loss part of the spectrum, allows to study inter-band transitions and optical properties of nanomaterials.

High-Resolution STEM-EELS

The combination of high-resolution imaging and high energy resolution spectroscopies has proven to be a formidable tool to solve materials science problems. While the former offers, with sub-angstrom resolution, structural information, including properties such as strain, the latter can provide high-resolution chemical composition, as well as oxidation state [76], and even bonding information, forming a compelling picture of the material's properties, which can be particularly valuable in the case of localized systems such as interfaces.

In particular, interfaces between oxides have been under intense study for the variety of novel emerging quantum phenomena they exhibit, depending on a variety of factors (strain, oxygen stoichiometry, cation intermixing, etc.). As an example, we can consider how the magnetic properties of epitaxial manganite films depend heavily on the chosen substrate. In the bulk case $La_2/3Sr_1/3MnO_3$ (LSMO) is ferromagnetic with an easy magnetization axis along the b crystallographic direction and a Curie temperature of 350 K. However, when a thin film (of less than 20 unit cells) of LSMO is grown on an $NdGaO_3$ (NGO) substrate, its Curie temperature drops dramatically (180 K for 10 unit cells). In such a very thin film the easy magnetization axis is oriented along the direction [77–79]. The main driving mechanism for this behaviour is oxygen octahedral coupling. Indeed, the oxygen octahedra in the film have to rotate in order to retain corner connectivity across the interface (Figure 7a), thus deviating from the natural tilt system of LSMO. In a state of the art microscope, these effects can be visualized and quantified by high-resolution STEM imaging (Figure 7b), determining the new tilt angle unit cell by unit cell and, thus, the extent of the perturbation (Figure 7c). This deviation has a dramatic effect on the properties of the material, as it affects the overlap between the atomic orbitals and therefore tuning the oxygen-mediated superexchange interaction between the B cations (Mn) which determine the magnetic properties of the material [79,80].

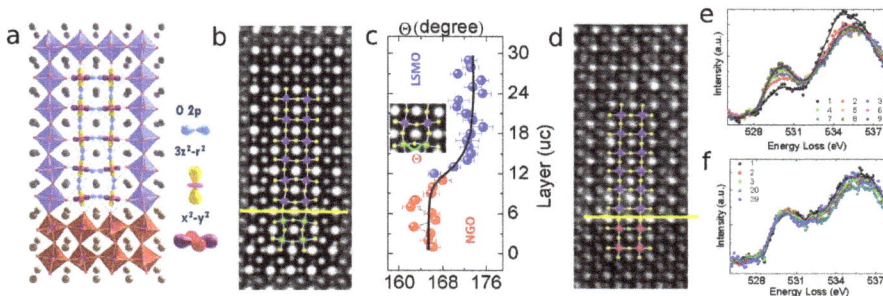

Figure 7. Octahedral tilt and orbital hybridization in a perovskite thin film. (**a**) At the interface between perovskites with different octahedral tilt systems, the oxygen octahedra have to rotate to maintain corner connectivity across the interface; (**b**) inverted-contrast ABF-STEM image showing the octahedral tilt in a LSMO on NGO film. In the inset, the image simulated from the obtained atomic structure (at the centre); (**c**) B-O-B angle between the oxygen atoms and the B cations as a function of the atomic plane; (**d**) inverted ABF-image of a similar film grown on top of a STO buffer layer, displaying no tilt; (**e**) the fine structure of the EELS oxygen peak measured in the buffered film. Each spectrum is measured from a region with a one unit cell thickness. The numbers in the inset indicate the distance of the chosen layer from the interface, expressed in unit cells. There is an obvious dependence of the fine structure from the distance; and (**f**) the fine structure of the EELS oxygen peak measured in the non-buffered film. There does not appear to be any dependence of the shape of the fine structure on the distance from the interface.

This can be effectively confirmed by observing the fine structure of the oxygen edge in high spatial and energy resolution EELS data (Figure 7f), which displays a pre-peak associated to the hybridization of the p-orbitals of the oxygen with the d-orbitals of the transition metal cations. The intensity of this prepeak proportional to the orbital overlap, is clearly reduced closer to the interface due to the octahedral tilt that reduces the overlap orbital, in agreement with the explanation. Understanding these effects allows to effectively tune and engineer them [61]. A thin (eight unit cells) buffer of SrTiO₃ between the NGO substrate and the LSMO film effectively suppresses the octahedral tilt in the first LSMO unit cells, restoring the original structure of the LSMO layer (Figure 7d), whose properties are then much closer to the bulk material, as well as the fine structure of the oxygen peak (Figure 7e). Indeed, even a one unit cell buffer has a major impact [78,79].

7. Spectroscopic Tomography

As demonstrated earlier (Figure 5, atomic resolution tomography), the HAADF-STEM intensity, depending on both sample thickness and atomic number Z, allows studying of the chemical composition in 3D. However, it becomes very challenging to use HAADF-STEM tomography for samples where different elements are mixed, or the differences in atomic number Z are small. In such cases, tomographic methods need to be combined with spectroscopic investigations to determine in 3D not only the structure, but also chemical composition, and even oxidation.

7.1. EDS Tomography

EDS studies the characteristic X-rays which are emitted during the beam-sample interaction, and whose energies correspond to the difference between specific atomic energy levels. Since the number of generated X-rays scales with the thickness of the sample, a 2D elemental map can, in principle, serve as a projection image for electron tomography. With previous generation instruments, attempts to obtain 3D information by EDS were severely hindered by the directionality and inefficiency of the sample-detector geometry, as the EDS detector is typically placed at a specific angle from the specimen. Therefore, the optimal signal can only be collected when the sample is tilted towards the detector while at different tilt angles the sample holder can cause significant shadowing. New-generation detection systems where four X-ray detectors are placed symmetrically around the sample reduce the blocking of the generated X-rays [81]. Using this type of system allows to combine EDS with tomography, therefore, enabling the determination of the composition of complex nanostructures in 3D.

An early demonstration of the combination of EDS with tomography has been realized by Goris et al., in 2014. In the first work, compositional changes in nanoparticles containing Au and Ag could be determined, for different steps of a galvanic replacement reaction, Figure 8a–c [82]. Typically, during such a reaction, solid metal nanoparticles become hollow through oxidation of one metal by another with a larger reduction potential, so that the metal with the larger reduction potential gets deposited in the form of so-called nanocages. In the second work, the technique was used for the determination of the distribution of Fe (Z = 26) and Co (Z = 27) in Fe-Co hetero-nanostructures in 3D, Figure 8d [83].

Further steps are needed to move from analysing the elemental distribution, to actual quantitative data. A recent proposal is based on combining ratio maps, obtained from the so-called ζ-factor (zeta-factor) method [84], with thickness information extracted from a HAADF-STEM reconstruction of the same nano-object [85]. Conventional EDS quantification uses the Cliff-Lorimer method, originally developed for the investigation of thin foils [86]. However, tilting the sample varies the amount of X-rays which are absorbed by the sample itself, which is an important parameter for accurate EDS quantification, in turn hampering a straightforward quantitative 3D reconstruction. The ζ-factor method presents major advantages, since reliable ratio maps are obtained, not affected by shadowing effects.

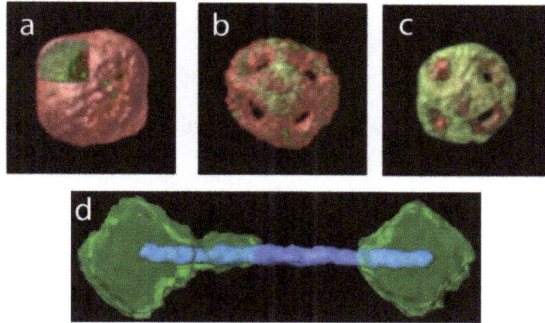

Figure 8. (**a–c**) 3D visualization of the EDS reconstructions of three different nanoparticles at different stages of a galvanic process reaction. Au and Ag are represented by red and green, respectively; (**d**) 3D visualization of the EDS reconstruction of a Fe-Co nanodumbbell, revealing the distribution of the elements in 3D. Co and Fe are represented by blue and green, respectively [82,83].

7.2. EELS Tomography

Recently, in the work of Goris et al., it was shown that it is possible to visualize in 3D the distribution of Ce^{3+}/Ce^{4+} in ceria nanoparticles, by combining monochromated EELS with electron tomography [87]. Initially, 2D maps of high energy resolution EELS data of the Ce edge were acquired from CeO_{2-x} nanoparticles along different tilt angles. Instead of analysing these maps and using the resulting valency maps as an input for a reconstruction algorithm, a different approach was followed, where complete EELS data cubes were used, leading to a 4D dataset from which an energy loss spectrum could be extracted for each reconstructed voxel. Although such experiments are challenging and computationally demanding, 3D quantitative results can be obtained in a more straightforward and reliable manner, since it prevents the accumulation of the errors introduced during the extraction of the 2D valency maps. The technique was applied at two different nanoparticles with different morphology, enabling the determination of the amount of Ce^{3+} at different crystallographic facets, which, in turn, provides a direct explanation of the observed lower catalytic activity between different surfaces.

8. TEM for Functional Characterization

While the TEM is the ideal tool to characterize the structure and chemical composition of a sample down to the atomic level, it can provide a wealth of information well beyond that, and characterize a variety of functional properties, such as optical, electronic, mechanical, magnetic, and more [88]. The techniques used to access these different properties can vary widely, from careful applications of more conventional electron spectroscopies, to in situ experiments realized through dedicated equipment. Here we will present recent advances at EMAT in nanomechanical testing, performed with a specialized holder with pN force sensitivity, as well as in the study of electronic and optical properties, performed with EELS.

8.1. Nanomechanical Testing in the TEM

Nanostructured materials—including thin films, nanocrystalline materials, nanocomposites and nanoporous foams and small-scale materials have shown remarkable variations in several mechanical properties, such as improved strength and toughness of nanocrystalline materials and the high strength of small-scale materials. These different properties have steered the transition from macro-scale engineering to micro/nano-scale technologies, e.g., now thin films are the basic components of micro/nanoelectronics and various industrial sensors are gradually being replaced by the cheaper micro-electro-mechanical systems (MEMS)-based devices and microsystems. However, such small components are often very delicate and, thus, their mechanical properties need to be properly investigated. Indeed, measuring the mechanical properties, ideally under in situ conditions, and understanding the governing deformation and failure mechanisms of such structures are the key issues to guide the design of more reliable nanocomponents.

While the excellent resolution of TEM allows to investigate defects in materials even down to the atomic scale, TEM investigations typically focus on very small volumes or individual defects, such as dislocations, interfaces, and so on [89–93], without being able to get good statistical data, essential to completely understand and/or model the behaviour of the nanomaterials. In order to bridge the gap between large area scanning electron microscopy (SEM) and atomic scale TEM, an automated crystal orientation mapping (ACOM) tool became recently available in EMAT [94]. This system is used to scan a region of a few μm² and automatically indexes grains with a 1 nm resolution by template matching nanobeam diffraction patterns with calculated ones (Figure 9a). Furthermore, PED can be applied to obtain quasi-kinematical diffraction conditions to facilitate crystal orientation indexation yielding quantitative analysis on the granular level. Although similar to electron backscatter diffraction (EBSD) the spatial resolution is higher, the electron dose is significantly lower in comparison to the backscattering technique, making ACOM-TEM applicable to beam-sensitive materials as well. In the past, ACOM-TEM was applied mostly to investigate grain size distribution in nanocrystalline materials [95–98]. For example, Wang et al. [99] investigated grain boundary processes in nanocrystalline Pd after lab-on-chip tensile deformation. Nowadays, ACOM-TEM is used for in situ experiments, such as monitoring local phase changes during heating and/or grain growth during deformation using dedicated holders [100,101]. Furthermore, applying the accompanying TOPSPIN software, the recorded nano-diffraction patterns can be used to visualize the local strain field by measuring the change in the reciprocal lattice vectors [102]. For example, Figure 9b,c show the long range strain field due to the presence of dislocation walls in single-crystal Ni micropillar after fatigue tests.

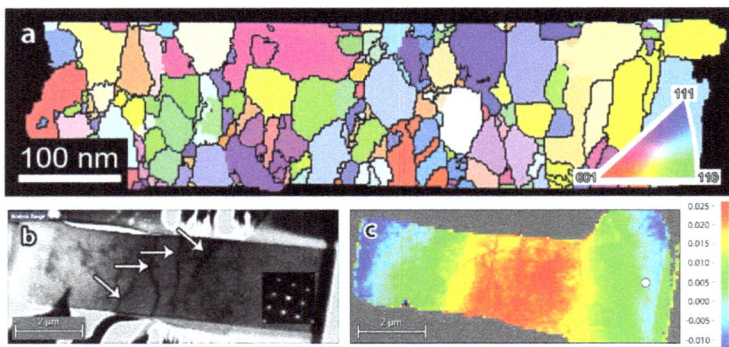

Figure 9. (**a**) Example of ACOM-TEM on a nanocrystalline Pd thin film. The colour code inverse pole figure is shown in the inset; (**b**,**c**) a virtual BF image and the corresponding local strain field map of a fatigued single-crystal Ni micropillar with dislocation walls seen as black curved lines running between opposite sides and are highlighted with white arrows.

New in situ TEM nanomechanical testing holders allow for a one-to-one relationship between loading-displacement data and direct TEM observation of the elementary plasticity and failure mechanisms. Conventional holders induce strain on a partially electro-polished metallic dog-bone sample with the size of a few millimetres [103]. An uncontrolled location of the hole as well as thickness variations in the sample make the loading conditions very complex. Therefore, such experiments are designed only for qualitative real-time investigation of microstructure evolution.

For example, Tirry and Schryvers [104] used this holder to investigate stress-induced transformation of polycrystalline and single crystal sheets of Ni-Ti and observed the role of twinning in the irreversibility of a martensitic transformation.

Recent MEMS-based developments have introduced various quantitative nanomechanical testing instruments. The PI 95 Picoindenter holder (Brucker Inc., Billerica, MA, USA), Figure 10a, is a MEMS-based in situ TEM holder which not only provides quantitative mechanical data, but also enables performing different modes of mechanical testing, e.g., tension, compression, bending, and indentation. Such experiments require well-defined micron level sized samples which are often prepared by focused ion beam (FIB) milling. For in situ tensile experiments a so-called push-to-pull (PTP) device, Figure 10b,c is used. The device is designed to convert the compressive loading of the indenter on the semi-circular part, Figure 10b, to a uniaxial tensile loading on the sample mounted in the middle gap, Figure 10c. Tensile testing experiments are of high interest since the uniaxial loading condition is satisfied uniformly during the experiment and interpretation of the mechanical results is straightforward. For instance, Idrissi et al. [105], could investigate the low-temperature rheology of olivine by quantitative in situ TEM tensile tests performed on FIB-prepared micron sized olivine samples, a common mineral in the lithospheric mantle. Heidari et al. [106] used this technique to investigate the mechanical properties of melamine–formaldehyde microcapsules. Idrissi et al. [107] also combined ACOM-TEM and the PTP device to investigate in situ the role of grain rotation in deformed ultrafine-grained Al thin films.

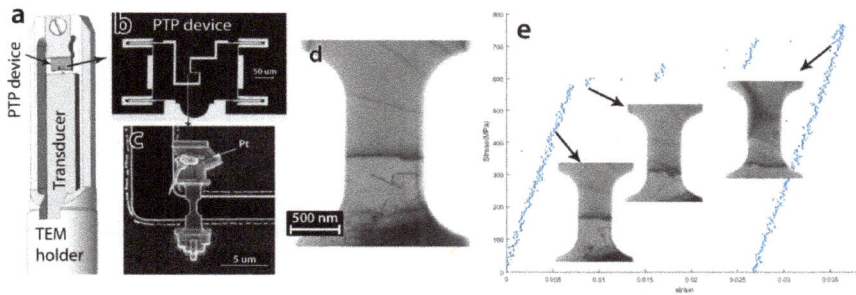

Figure 10. (**a**) Schematic of the PicoIndenter holder; (**b**) TEM BF image of a PTP device; (**c**) SEM image of a tensile sample mounted on a PTP device; (**d**) TEM BF image of a FIB near-defect-free single crystal Ni sample containing only a few dislocations, and (**e**) engineering stress-strain curve of the single crystal Ni sample, the images in the insets are snapshots from the recorded deformation movie, corresponding to the points at the arrows' origin. Note the strain bursts induced by the operation of the individual SAS (single arm source).

In spite of FIB versatility in small-scale sample preparation, FIB can induce surface defects in metallic or other samples which can not only change the mechanical response of the material [108] but also block the sight to observe the active mechanisms. Samaeeaghmiyoni et al. [109] have developed a new sample preparation method, allowing preparing near defect-free sub-micron metallic single crystals for quantitative in situ TEM tensile testing, Figure 10d. The method also enables designing the sample in terms of crystallographic orientations and imaging conditions as well as pre-selected defects, such as structural boundaries, so that interactions between various defects can be studied. In situ tensile tests on samples prepared by this method revealed new phenomena and demonstrated a direct correlation between the intrinsic mechanical properties and their governing mechanisms. For instance, in situ TEM tensile testing on a single-crystal Ni sample with a very low dislocation density revealed for the first time a hardening mechanism based on the operation of one single arm source (SAS),

Figure 10e. The results also shed light on the origin of the intermittent plastic flow characteristics of small-sized single crystals.

8.2. LowLoss EELS for Electrical and Optical Properties

EELS also allows to measure and evaluate a number of material properties beyond just the chemistry of a material. Analysing the so-called low-loss range of energies, allows to measure a range of optical and electrical properties of a material, such as the band gap [110], localized plasmon resonances [111,112], phononic excitations, and many others [113,114].

The wealth of information that can be acquired is expanding, and boundaries are constantly pushed.

The measurement of band gap with EELS is not as commonplace as one might expect. Despite the attractiveness of measuring such properties at the nanoscale, experimental spectra are affected by the presence of "parasitic" losses, mainly the emission of Cherenkov radiation and the excitation of surface guided modes, which blur the onset of the interband transition signal [115]. However, recent advances have presented a practical solution to these problems [116–118]. As first suggested by Stöger-Pollach et al. [118] and then developed in detail by Korneychuk et al. [117], the large difference in scattering angles between the interband and "parasitic" signals can be exploited by collecting electrons scattered away from the forward direction within a carefully chosen angular range that allows the faithful recovery of band gap while avoiding the disturbances (Figure 11). The further addition of a custom designed annular aperture in the illumination system of the TEM allows to realize this scattering geometry while avoiding any asymmetry that could be introduced by going off-axis [117,118].

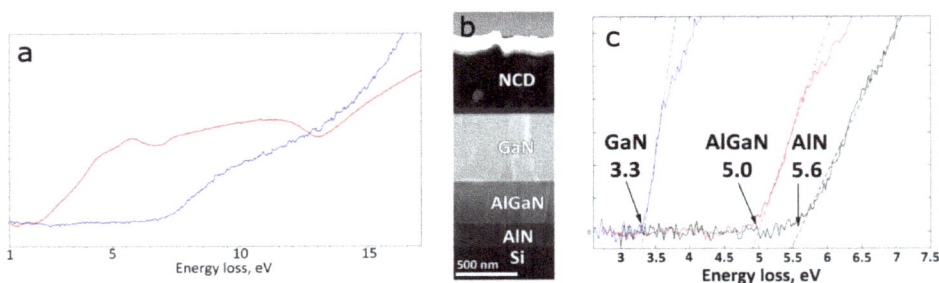

Figure 11. Improvement of band gap measurement. (**a**) Low loss spectrum acquired with the improved setup (blue) contrasted with one acquired in a conventional experiment (red). The distortion of the data caused by the parasitic losses is clearly visible; (**b**) ADF-STEM image of a multilayer sample composed of several wide-band gap semiconductors, and (**c**) recorded spectra clearly showing the onset of the inter-band transitions and, hence, the band gaps, for the different portions of the sample.

Another field where recent advances have greatly broadened the information that can be obtained from EELS is the mapping of localized surface plasmon resonances (SPRs). While TEM already has an important role in the analysis of plasmonic resonances, a great deal of information is hidden by the conventional approach. Only the squared field intensity is mapped, and then only in the component parallel to the optical axis [119,120]. This constitutes a strong limit to perform, e.g., reliable and direct plasmon field tomography, or effectively retrieve the surface charge distribution, the real field direction, or the dynamical behaviour. A wealth of recent developments however have now broadened the scope significantly. Pulsed electron sources have allowed to observe the dynamical behaviour of reproducible phenomena with femtosecond resolution [121]. Even more recently, by analysing the angular distribution of inelastically scattered electrons, effectively joining EELS and momentum-resolved STEM, Krehl et al. have measured the thus-far elusive transverse field produced by localized surface plasmon resonances [122].

Finally, it has also been shown that the dynamic electrical potential of the SPRs has an effect on the electron beam. Just like an electrostatic potential causes a phase shift on an electron beam, the dynamical one causes a phase shift in the inelastically scattered part of the beam. This has been exploited by Guzzinati et al., manipulating the complex wave function of the impinging electron beam, to directly test the charge symmetry of plasmonic modes [123]. While such experiment are cumbersome with current methods, development of more flexible techniques to manipulate the electron beam's phase is progressing rapidly and a first prototype of a programmable phase plate has been presented by Verbeeck et al. [124]. Such developments are expected to help make phase-based experiments more common [123,124].

Supplementary Materials: The following is available online at http://www.mdpi.com/1996-1944/11/8/1304/s1. Video S1 showing the time series of the [010] HAADF-STEM images for $Bi_{10}Ti_7Fe_6O_{38}$ and unveiling the cooperative displacements of Bi atomic columns (A cations) in the structure. Adjacent A2 columns displace in opposite directions that is visible as a rotation of the A2 atomic column pairs about the b axis. The rotation of the A2 pairs induces displacements in the neighboring A1 columns. The displacements in the A3 atomic columns are very subtle and difficult to distinguish because of scan distortion of the HAADF-STEM images. Reprinted with permission from [42], copyright 2016 American Chemical Society.

Funding: G. Guzzinati, T. Altantzis and A. De Backer have been supported by postdoctoral fellowship grants from the Research Foundation Flanders (FWO). Funding was also received from the European Research Council (starting grant no. COLOURATOM 335078), the European Research Council (ERC) under the European Union's Horizon 2020 research and innovation programme (grant agreement no. 770887), the Research Foundation Flanders (FWO, Belgium) through project fundings (G.0502.18N, G.0267.18N, G.0120.12N, G.0365.15N, G.0934.17N, S.0100.18N, G.0401.16N) and from the University of Antwerp through GOA project Solarpaint. Funding for the TopSPIN precession system under grant AUHA13009, as well as for the Qu-Ant-EM microscope, is acknowledged from the HERCULES Foundation. H. Idrissi is mandated by the Belgian National Fund for Scientific Research (F.R.S.-FNRS).

Acknowledgments: The authors thank G. Van Tendeloo for bringing the group to its present level of excellence and G. Van Tendeloo and Dirk Van Dyck for contributing to several of the topics discussed in this paper.

Conflicts of Interest: The authors declare no conflict of interest.

References

1. Bals, S.; Casavola, M.; Van Huis, M.A.; Van Aert, S.; Batenburg, K.J.; Van Tendeloo, G.; Vanmaekelbergh, D. Three-dimensional atomic imaging of colloidal core-shell nanocrystals. *Nano Lett.* **2011**, *11*, 3420–3424. [CrossRef] [PubMed]

2. Goris, B.; Bals, S.; Van Den Broek, W.; Carbó-Argibay, E.; Gómez-Graña, S.; Liz-Marzán, L.M.; Van Tendeloo, G. Atomic-scale determination of surface facets in gold nanorods. *Nat. Mater.* **2012**, *11*, 930–935. [CrossRef] [PubMed]

3. Zou, X.; Hovmöller, S.; Oleynikov, P. *Electron Crystallography. Electron Microscopy and Electron Diffraction*, 2nd ed.; Oxford Science Publications-Oxford University Press: Oxford, UK, 2011; ISBN -13 978-0-19-958020-0.

4. Batuk, D.; Batuk, M.; Abakumov, A.M.; Hadermann, J. Synergy between transmission electron microscopy and powder diffraction: Application to modulated structures. *Acta Crystallogr. Sect. B Struct. Sci. Cryst. Eng. Mater.* **2015**, *71*, 127–143. [CrossRef] [PubMed]

5. Batuk, D.; Batuk, M.; Tsirlin, A.A.; Hadermann, J.; Abakumov, A.M. Trapping of Oxygen Vacancies at Crystallographic Shear Planes in Acceptor-Doped Pb-Based Ferroelectrics. *Angew. Chem. Int. Ed.* **2015**, *54*, 14787–14790. [CrossRef] [PubMed]

6. Batuk, M.; Turner, S.; Abakumov, A.M.; Batuk, D.; Hadermann, J.; Van Tendeloo, G. Atomic Structure of Defects in Anion-Deficient Perovskite-Based Ferrites with a Crystallographic Shear Structure. *Inorg. Chem.* **2014**, *53*, 2171–2180. [CrossRef] [PubMed]

7. Cassidy, S.J.; Batuk, M.; Batuk, D.; Hadermann, J.; Woodruff, D.N.; Thompson, A.L.; Clarke, S.J. Complex Microstructure and Magnetism in Polymorphic CaFeSeO. *Inorg. Chem.* **2016**, *55*, 10714–10726. [CrossRef] [PubMed]

8. Pecharsky, V.; Zavalij, P. *Fundamentals of Powder Diffraction and Structural Characterization of Materials*, 2nd ed.; Springer: New York, NY, USA, 2005; ISBN 0387241477.

9. Yun, Y.; Zou, X.; Hovmöller, S.; Wan, W. Three-dimensional electron diffraction as a complementary technique to powder X-ray diffraction for phase identification and structure solution of powders. *IUCrJ* **2015**, *2*, 267–282. [CrossRef] [PubMed]

10. Vainshtein, B.K.; Zvyagin, B.B.; Avilov, A.S. Electron Diffraction Structure Analysis. In *Electron Diffraction Techniques*; Cowley, J.M., Ed.; Oxford University Press: New York, NY, USA, 1992.

11. Blackman, M. On the Intensities of Electron Diffraction Rings. *Proc. R. Soc. A Math. Phys. Eng. Sci.* **1939**, *173*, 68–82. [CrossRef]

12. Vincent, R.; Midgley, P.A. Double conical beam-rocking system for measurement of integrated electron diffraction intensities. *Ultramicroscopy* **1994**, *53*, 271–282. [CrossRef]

13. Gorelik, T.E.; Stewart, A.A.; Kolb, U. Structure solution with automated electron diffraction tomography data: Different instrumental approaches. *J. Microsc.* **2011**, *244*, 325–331. [CrossRef] [PubMed]

14. Hadermann, J.; Abakumov, A.M.; Turner, S.; Hafideddine, Z.; Khasanova, N.R.; Antipov, E.V.; Van Tendeloo, G. Solving the Structure of Li Ion Battery Materials with Precession Electron Diffraction: Application to Li_2CoPO_4F. *Chem. Mater.* **2011**, *23*, 3540–3545. [CrossRef]

15. Kolb, U.; Mugnaioli, E.; Gorelik, T.E. Automated electron diffraction tomography—A new tool for nano crystal structure analysis. *Cryst. Res. Technol.* **2011**, *46*, 542–554. [CrossRef]

16. Fedotov, S.S.; Khasanova, N.R.; Samarin, A.S.; Drozhzhin, O.A.; Batuk, D.; Karakulina, O.M.; Hadermann, J.; Abakumov, A.M.; Antipov, E.V. AVPO4F (A = Li, K): A 4 V Cathode Material for High-Power Rechargeable Batteries. *Chem. Mater.* **2016**, *28*, 411–415. [CrossRef]

17. Karakulina, O.M.; Khasanova, N.R.; Drozhzhin, O.A.; Tsirlin, A.A.; Hadermann, J.; Antipov, E.V.; Abakumov, A.M. Antisite Disorder and Bond Valence Compensation in Li_2FePO_4F Cathode for Li-Ion Batteries. *Chem. Mater.* **2016**, *28*, 7578–7581. [CrossRef]

18. Mikhailova, D.; Karakulina, O.M.; Batuk, D.; Hadermann, J.; Abakumov, A.M.; Herklotz, M.; Tsirlin, A.A.; Oswald, S.; Giebeler, L.; Schmidt, M.; et al. Layered-to-Tunnel Structure Transformation and Oxygen Redox Chemistry in $LiRhO_2$ upon Li Extraction and Insertion. *Inorg. Chem.* **2016**, *55*, 7079–7089. [CrossRef] [PubMed]

19. Batson, P.E.; Dellby, N.; Krivanek, O.L. Sub-ångstrom resolution using aberration corrected electron optics. *Nature* **2002**, *418*, 617–620. [CrossRef] [PubMed]

20. Haider, M.; Rose, H.; Uhlemann, S.; Schwan, E.; Kabius, B.; Urban, K. A spherical-aberration-corrected 200 kV transmission electron microscope. *Ultramicroscopy* **1998**, *75*, 53–60. [CrossRef]

21. Erni, R.; Rossell, M.D.; Kisielowski, C.; Dahmen, U. Atomic-resolution imaging with a sub-50-pm electron probe. *Phys. Rev. Lett.* **2009**, *102*, 096101. [CrossRef] [PubMed]

22. Lee, J.S.; Bodnarchuk, M.I.; Shevchenko, E.V.; Talapin, D.V. "Magnet-in-the-Semiconductor" FePt−PbS and FePt−PbSe Nanostructures: Magnetic Properties, Charge Transport, and Magnetoresistance. *J. Am. Chem. Soc.* **2010**, *132*, 6382–6391. [CrossRef] [PubMed]

23. Abakumov, A.M.; Hadermann, J.; Van Tendeloo, G.; Shpanchenko, R.V.; Oleinikov, P.N.; Antipov, E.V. Anion Ordering in Fluorinated La_2CuO_4. *J. Solid State Chem.* **1999**, *142*, 440–450. [CrossRef]

24. Hadermann, J.; Abakumov, A.M.; Gillie, L.J.; Martin, C.; Hervieu, M. Coupled cation and charge ordering in the $CaMn_3O_6$ tunnel structure. *Chem. Mater.* **2006**, *18*, 5530–5536. [CrossRef]

25. D'Hondt, H.; Abakumov, A.M.; Hadermann, J.; Kalyuzhnaya, A.S.; Rozova, M.G.; Antipov, E.V.; Van Tendeloo, G. Tetrahedral Chain Order in the $Sr_2Fe_2O_5$ Brownmillerite. *Chem. Mater.* **2008**, *20*, 7188–7194. [CrossRef]

26. Hartel, P.; Rose, H.; Dinges, C. Conditions and reasons for incoherent imaging in STEM. *Ultramicroscopy* **1996**, *63*, 93–114. [CrossRef]

27. Krivanek, O.L.; Chisholm, M.F.; Nicolosi, V.; Pennycook, T.J.; Corbin, G.J.; Dellby, N.; Murfitt, M.F.; Own, C.S.; Szilagyi, Z.S.; Oxley, M.P.; et al. Atom-by-atom structural and chemical analysis by annular dark-field electron microscopy. *Nature* **2010**, *464*, 571–574. [CrossRef] [PubMed]

28. Findlay, S.D.; Shibata, N.; Sawada, H.; Okunishi, E.; Kondo, Y.; Ikuhara, Y. Dynamics of annular bright field imaging in scanning transmission electron microscopy. *Ultramicroscopy* **2010**, *110*, 903–923. [CrossRef] [PubMed]

29. Ishikawa, R.; Okunishi, E.; Sawada, H.; Kondo, Y.; Hosokawa, F.; Abe, E. Direct imaging of hydrogen-atom columns in a crystal by annular bright-field electron microscopy. *Nat. Mater.* **2011**, *10*, 278–281. [CrossRef] [PubMed]

30. Findlay, S.D.; Shibata, N.; Sawada, H.; Okunishi, E.; Kondo, Y.; Yamamoto, T.; Ikuhara, Y. Robust atomic resolution imaging of light elements using scanning transmission electron microscopy. *Appl. Phys. Lett.* **2009**, *95*, 191913. [CrossRef]

31. Nellist, P.D. The Principles of STEM Imaging. In *Scanning Transmission Electron Microscopy*; Springer: New York, NY, USA, 2011; pp. 91–115, ISBN 978-1-4419-7199-9.

32. McCalla, E.; Abakumov, A.M.; Saubanère, M.; Foix, D.; Berg, E.J.; Rousse, G.; Doublet, M.L.; Gonbeau, D.; Novák, P.; Van Tendeloo, G.; et al. Visualization of O-O peroxo-like dimers in high-capacity layered oxides for Li-ion batteries. *Science* **2015**, *350*, 1516–1521. [CrossRef] [PubMed]

33. Perez, A.J.; Batuk, D.; Saubanère, M.; Rousse, G.; Foix, D.; McCalla, E.; Berg, E.J.; Dugas, R.; Van Den Bos, K.H.W.; Doublet, M.L.; et al. Strong oxygen participation in the redox governing the structural and electrochemical properties of Na-rich layered oxide Na_2IrO_3. *Chem. Mater.* **2016**, *28*, 8278–8288. [CrossRef]

34. Sathiya, M.; Abakumov, A.M.; Foix, D.; Rousse, G.; Ramesha, K.; Saubanère, M.; Doublet, M.L.; Vezin, H.; Laisa, C.P.; Prakash, A.S.; et al. Origin of voltage decay in high-capacity layered oxide electrodes. *Nat. Mater.* **2015**, *14*, 230–238. [CrossRef] [PubMed]

35. Jacquet, Q.; Perez, A.; Batuk, D.; Van Tendeloo, G.; Rousse, G.; Tarascon, J.M. The $Li_3Ru_yNb_{1-y}O_4$ ($0 \leq y \leq 1$) System: Structural Diversity and Li Insertion and Extraction Capabilities. *Chem. Mater.* **2017**, *29*, 5331–5343. [CrossRef]

36. Perez, A.J.; Jacquet, Q.; Batuk, D.; Iadecola, A.; Saubanère, M.; Rousse, G.; Larcher, D.; Vezin, H.; Doublet, M.L.; Tarascon, J.M. Approaching the limits of cationic and anionic electrochemical activity with the Li-rich layered rocksalt Li_3IrO_4. *Nat. Energy* **2017**, *2*, 954–962. [CrossRef]

37. Morozov, V.A.; Bertha, A.; Meert, K.W.; Van Rompaey, S.; Batuk, D.; Martinez, G.T.; Van Aert, S.; Smet, P.F.; Raskina, M.V.; Poelman, D.; et al. Incommensurate modulation and luminescence in the $CaGd_{2(1-x)}Eu_{2x}(MoO_4)_{4(1-y)}(WO_4)_{4y}$ ($0 \leq x \leq 1, 0 \leq y \leq 1$) red phosphors. *Chem. Mater.* **2013**, *25*, 4387–4395. [CrossRef]

38. Abakumov, A.M.; Hadermann, J.; Bals, S.; Nikolaev, I.V.; Antipov, E.V.; Van Tendeloo, G. Crystallographic shear structures as a route to anion-deficient perovskites. *Angew. Chem. Int. Ed.* **2006**, *45*, 6697–6700. [CrossRef] [PubMed]

39. Pennycook, S.J.; Nellist, P.D. (Eds.) *Scanning Transmission Electron Microscopy*; Springer: New York, NY, USA, 2011; ISBN 978-1-4419-7199-9.

40. O'Sullivan, M.; Hadermann, J.; Dyer, M.S.; Turner, S.; Alaria, J.; Manning, T.D.; Abakumov, A.M.; Claridge, J.B.; Rosseinsky, M.J. Interface control by chemical and dimensional matching in an oxide heterostructure. *Nat. Chem.* **2016**, *8*, 347–353. [CrossRef] [PubMed]

41. Egerton, R.F. Control of radiation damage in the TEM. *Ultramicroscopy* **2013**, *127*, 100–108. [CrossRef] [PubMed]

42. Batuk, D.; Tsirlin, A.A.; Filimonov, D.S.; Zakharov, K.V.; Volkova, O.S.; Vasiliev, A.; Hadermann, J.; Abakumov, A.M. $Bi_{3n+1}Ti_7Fe_{3n-3}O_{9n+11}$ Homologous Series: Slicing Perovskite Structure with Planar Interfaces Containing Anatase-like Chains. *Inorg. Chem.* **2016**, *55*, 1245–1257. [CrossRef] [PubMed]

43. Batuk, D.; Batuk, M.; Filimonov, D.S.; Zakharov, K.V.; Volkova, O.S.; Vasiliev, A.N.; Tyablikov, O.A.; Hadermann, J.; Abakumov, A.M. Crystal Structure, Defects, Magnetic and Dielectric Properties of the Layered $Bi_{3n+1}Ti_7Fe_{3n-3}O_{9n+11}$ Perovskite-Anatase Intergrowths. *Inorg. Chem.* **2017**, *56*, 931–942. [CrossRef] [PubMed]

44. Van Aert, S.; Verbeeck, J.; Erni, R.; Bals, S.; Luysberg, M.; Van Dyck, D.; Van Tendeloo, G. Quantitative atomic resolution mapping using high-angle annular dark field scanning transmission electron microscopy. *Ultramicroscopy* **2009**, *109*, 1236–1244. [CrossRef] [PubMed]

45. Van Aert, S.; De Backer, A.; Martinez, G.T.; Den Dekker, A.J.; Van Dyck, D.; Bals, S.; Van Tendeloo, G. Advanced electron crystallography through model-based imaging. *IUCrJ* **2016**, *3*, 71–83. [CrossRef] [PubMed]

46. Van den Bos, A. *Parameter Estimation for Scientists and Engineers*; John Wiley & Sons: Hoboken, NJ, USA, 2007; ISBN 0470173858.

47. De Backer, A.; van den Bos, K.H.W.; Van den Broek, W.; Sijbers, J.; Van Aert, S. StatSTEM: An efficient approach for accurate and precise model-based quantification of atomic resolution electron microscopy images. *Ultramicroscopy* **2016**, *171*, 104–116. [CrossRef] [PubMed]

48. Den Dekker, A.J.; Van Aert, S.; Van Den Bos, A.; Van Dyck, D. Maximum likelihood estimation of structure parameters from high resolution electron microscopy images. Part I: A theoretical framework. *Ultramicroscopy* **2005**, *104*, 83–106. [CrossRef] [PubMed]

49. Gonnissen, J.; Batuk, D.; Nataf, G.F.; Jones, L.; Abakumov, A.M.; Van Aert, S.; Schryvers, D.; Salje, E.K.H. Direct Observation of Ferroelectric Domain Walls in LiNbO$_3$: Wall-Meanders, Kinks, and Local Electric Charges. *Adv. Funct. Mater.* **2016**, *26*, 7599–7604. [CrossRef]

50. Gauquelin, N.; van den Bos, K.H.W.; Béché, A.; Krause, F.F.; Lobato, I.; Lazar, S.; Rosenauer, A.; Van Aert, S.; Verbeeck, J. Determining oxygen relaxations at an interface: A comparative study between transmission electron microscopy techniques. *Ultramicroscopy* **2017**, *181*, 178–190. [CrossRef] [PubMed]

51. Van der Stam, W.; Geuchies, J.J.; Altantzis, T.; Van Den Bos, K.H.W.; Meeldijk, J.D.; Van Aert, S.; Bals, S.; Vanmaekelbergh, D.; De Mello Donega, C. Highly Emissive Divalent-Ion-Doped Colloidal CsPb$_{1-x}$M$_x$Br$_3$ Perovskite Nanocrystals through Cation Exchange. *J. Am. Chem. Soc.* **2017**, *139*, 4087–4097. [CrossRef] [PubMed]

52. De Backer, A.; van Aert, S.; van Dyck, D. High precision measurements of atom column positions using model-based exit wave reconstruction. *Ultramicroscopy* **2011**, *111*, 1475–1482. [CrossRef] [PubMed]

53. Van den Bos, K.H.W.; Krause, F.F.; Béché, A.; Verbeeck, J.; Rosenauer, A.; Van Aert, S. Locating light and heavy atomic column positions with picometer precision using ISTEM. *Ultramicroscopy* **2017**, *172*, 75–81. [CrossRef] [PubMed]

54. Van Aert, S.; Turner, S.; Delville, R.; Schryvers, D.; Van Tendeloo, G.; Salje, E.K.H. Direct observation of ferrielectricity at ferroelastic domain boundaries in CaTiO$_3$ by electron microscopy. *Adv. Mater.* **2012**, *24*, 523–527. [CrossRef] [PubMed]

55. Bals, S.; Van Aert, S.; Van Tendeloo, G.; Ávila-Brande, D. Statistical estimation of atomic positions from exit wave reconstruction with a precision in the picometer range. *Phys. Rev. Lett.* **2006**, *96*, 096106. [CrossRef] [PubMed]

56. Martinez, G.T.; Rosenauer, A.; De Backer, A.; Verbeeck, J.; Van Aert, S. Quantitative composition determination at the atomic level using model-based high-angle annular dark field scanning transmission electron microscopy. *Ultramicroscopy* **2014**, *137*, 12–19. [CrossRef] [PubMed]

57. Akamine, H.; Van Den Bos, K.H.W.; Gauquelin, N.; Farjami, S.; Van Aert, S.; Schryvers, D.; Nishida, M. Determination of the atomic width of an APB in ordered CoPt using quantified HAADF-STEM. *J. Alloys Compd.* **2015**, *644*, 570–574. [CrossRef]

58. Jany, B.R.; Gauquelin, N.; Willhammar, T.; Nikiel, M.; Van Den Bos, K.H.W.; Janas, A.; Szajna, K.; Verbeeck, J.; Van Aert, S.; Van Tendeloo, G.; et al. Controlled growth of hexagonal gold nanostructures during thermally induced self-assembling on Ge(001) surface. *Sci. Rep.* **2017**, *7*, 42420. [CrossRef] [PubMed]

59. Van Aert, S.; Batenburg, K.J.; Rossell, M.D.; Erni, R.; Van Tendeloo, G. Three-dimensional atomic imaging of crystalline nanoparticles. *Nature* **2011**, *470*, 374–377. [CrossRef] [PubMed]

60. De Backer, A.; Martinez, G.T.; Rosenauer, A.; Van Aert, S. Atom counting in HAADF STEM using a statistical model-based approach: Methodology, possibilities, and inherent limitations. *Ultramicroscopy* **2013**, *134*, 23–33. [CrossRef] [PubMed]

61. Van Aert, S.; De Backer, A.; Martinez, G.T.; Goris, B.; Bals, S.; Van Tendeloo, G.; Rosenauer, A. Procedure to count atoms with trustworthy single-atom sensitivity. *Phys. Rev. B Condens. Matter Mater. Phys.* **2013**, *87*, 064107. [CrossRef]

62. Bals, S.; Van Aert, S.; Romero, C.P.; Lauwaet, K.; Van Bael, M.J.; Schoeters, B.; Partoens, B.; Yücelen, E.; Lievens, P.; Van Tendeloo, G. Atomic scale dynamics of ultrasmall germanium clusters. *Nat. Commun.* **2012**, *3*, 897. [CrossRef] [PubMed]

63. Peters, J.L.; Van Den Bos, K.H.W.; Van Aert, S.; Goris, B.; Bals, S.; Vanmaekelbergh, D. Ligand-Induced Shape Transformation of PbSe Nanocrystals. *Chem. Mater.* **2017**, *29*, 4122–4128. [CrossRef] [PubMed]

64. Geuchies, J.J.; Van Overbeek, C.; Evers, W.H.; Goris, B.; De Backer, A.; Gantapara, A.P.; Rabouw, F.T.; Hilhorst, J.; Peters, J.L.; Konovalov, O.; et al. In situ study of the formation mechanism of two-dimensional superlattices from PbSe nanocrystals. *Nat. Mater.* **2016**, *15*, 1248–1254. [CrossRef] [PubMed]

65. De Backer, A.; Jones, L.; Lobato, I.; Altantzis, T.; Goris, B.; Nellist, P.D.; Bals, S.; Van Aert, S. Three-dimensional atomic models from a single projection using Z-contrast imaging: Verification by electron tomography and opportunities. *Nanoscale* **2017**, *9*, 8791–8798. [CrossRef] [PubMed]

66. Jones, L.; Macarthur, K.E.; Fauske, V.T.; Van Helvoort, A.T.J.; Nellist, P.D. Rapid estimation of catalyst nanoparticle morphology and atomic-coordination by high-resolution Z-contrast electron microscopy. *Nano Lett.* **2014**, *14*, 6336–6341. [CrossRef] [PubMed]

67. Van Den Bos, K.H.W.; De Backer, A.; Martinez, G.T.; Winckelmans, N.; Bals, S.; Nellist, P.D.; Van Aert, S. Unscrambling Mixed Elements using High Angle Annular Dark Field Scanning Transmission Electron Microscopy. *Phys. Rev. Lett.* **2016**, *116*, 246101. [CrossRef] [PubMed]

68. Goris, B.; Van den Broek, W.; Batenburg, K.J.; Heidari Mezerji, H.; Bals, S. Electron tomography based on a total variation minimization reconstruction technique. *Ultramicroscopy* **2012**, *113*, 120–130. [CrossRef]

69. Béché, A.; Goris, B.; Freitag, B.; Verbeeck, J. Development of a fast electromagnetic beam blanker for compressed sensing in scanning transmission electron microscopy. *Appl. Phys. Lett.* **2016**, *108*, 093103. [CrossRef]

70. Béché, A.; Al, E. Compressed sensing in (S)TEM—Imaging materials with reduced electron dose. In Proceedings of the Microscopy Conference 2017 (MC 2017), Lausanne, Switzerland, 21–25 August 2017.

71. Van den Broek, W. Statistical Experimental Design in Compressed Sensing Set-ups for Optical and Transmission Electron Microscopy. *arXiv* **2018**, arXiv:1801.02388.

72. Zanaga, D.; Bleichrodt, F.; Altantzis, T.; Winckelmans, N.; Palenstijn, W.J.; Sijbers, J.; De Nijs, B.; Van Huis, M.A.; Sánchez-Iglesias, A.; Liz-Marzán, L.M.; et al. Quantitative 3D analysis of huge nanoparticle assemblies. *Nanoscale* **2016**, *8*, 292–299. [CrossRef] [PubMed]

73. Midgley, P.A.; Weyland, M. 3D electron microscopy in the physical sciences: The development of Z-contrast and EFTEM tomography. *Ultramicroscopy* **2003**, *96*, 413–431. [CrossRef]

74. Goris, B.; De Backer, A.; Van Aert, S.; Gómez-Graña, S.; Liz-Marzán, L.M.; Van Tendeloo, G.; Bals, S. Three-dimensional elemental mapping at the atomic scale in bimetallic nanocrystals. *Nano Lett.* **2013**, *13*, 4236–4241. [CrossRef] [PubMed]

75. Goris, B.; De Beenhouwer, J.; De Backer, A.; Zanaga, D.; Batenburg, K.J.; Sánchez-Iglesias, A.; Liz-Marzán, L.M.; Van Aert, S.; Bals, S.; Sijbers, J.; et al. Measuring Lattice Strain in Three Dimensions through Electron Microscopy. *Nano Lett.* **2015**, *15*, 6996–7001. [CrossRef] [PubMed]

76. Tan, H.; Turner, S.; Yücelen, E.; Verbeeck, J.; Van Tendeloo, G. 2D Atomic Mapping of Oxidation States in Transition Metal Oxides by Scanning Transmission Electron Microscopy and Electron Energy-Loss Spectroscopy. *Phys. Rev. Lett.* **2011**, *107*, 107602. [CrossRef] [PubMed]

77. Liao, Z.; Huijben, M.; Zhong, Z.; Gauquelin, N.; Macke, S.; Green, R.J.; Van Aert, S.; Verbeeck, J.; Van Tendeloo, G.; Held, K.; et al. Controlled lateral anisotropy in correlated manganite heterostructures by interface-engineered oxygen octahedral coupling. *Nat. Mater.* **2016**, *15*, 425–431. [CrossRef] [PubMed]

78. Liao, Z.; Green, R.J.; Gauquelin, N.; Macke, S.; Li, L.; Gonnissen, J.; Sutarto, R.; Houwman, E.P.; Zhong, Z.; Van Aert, S.; et al. Long-Range Domain Structure and Symmetry Engineering by Interfacial Oxygen Octahedral Coupling at Heterostructure Interface. *Adv. Funct. Mater.* **2016**, *26*, 6627–6634. [CrossRef]

79. Liao, Z.; Gauquelin, N.; Green, R.J.; Macke, S.; Gonnissen, J.; Thomas, S.; Zhong, Z.; Li, L.; Si, L.; Van Aert, S.; et al. Thickness Dependent Properties in Oxide Heterostructures Driven by Structurally Induced Metal-Oxygen Hybridization Variations. *Adv. Funct. Mater.* **2017**, *27*, 1606717. [CrossRef]

80. Blundell, S.; Thouless, D. Magnetism in Condensed Matter. *Am. J. Phys.* **2003**, *71*, 94. [CrossRef]

81. Schlossmacher, P.; Klenov, D.O.; Freitag, B.; von Harrach, H.S. Enhanced Detection Sensitivity with a New Windowless XEDS System for AEM Based on Silicon Drift Detector Technology. *Microsc. Today* **2010**, *18*, 14–20. [CrossRef]

82. Goris, B.; Polavarapu, L.; Bals, S.; Van Tendeloo, G.; Liz-Marzán, L.M. Monitoring galvanic replacement through three-dimensional morphological and chemical mapping. *Nano Lett.* **2014**, *14*, 3220–3226. [CrossRef] [PubMed]

83. Liakakos, N.; Gatel, C.; Blon, T.; Altantzis, T.; Lentijo-Mozo, S.; Garcia-Marcelot, C.; Lacroix, L.M.; Respaud, M.; Bals, S.; Van Tendeloo, G.; et al. Co-Fe nanodumbbells: Synthesis, structure, and magnetic properties. *Nano Lett.* **2014**, *14*, 2747–2754. [CrossRef] [PubMed]

84. Watanabe, M.; Williams, D.B. The quantitative analysis of thin specimens: A review of progress from the Cliff-Lorimer to the new ζ-factor methods. *J. Microsc.* **2006**, *221*, 89–109. [CrossRef] [PubMed]

85. Zanaga, D.; Altantzis, T.; Polavarapu, L.; Liz-Marzán, L.M.; Freitag, B.; Bals, S. A New Method for Quantitative XEDS Tomography of Complex Heteronanostructures. *Part. Part. Syst. Charact.* **2016**, *33*, 396–403. [CrossRef]

86. Cliff, G.; Lorimer, G.W. The quantitative analysis of thin specimens. *J. Microsc.* **1975**, *103*, 203–207. [CrossRef]

87. Goris, B.; Turner, S.; Bals, S.; Van Tendeloo, G. Three-dimensional valency mapping in ceria nanocrystals. *ACS Nano* **2014**, *8*, 10878–10884. [CrossRef] [PubMed]

88. Van Tendeloo, G.; Van Dyck, D.; Pennycook, S.J. *Handbook of Nanoscopy*; Wiley-VCH Verlag GmbH & Co. KGaA: Weinheim, Germany, 2012; ISBN 9783527641864.

89. Li, K.; Idrissi, H.; Sha, G.; Song, M.; Lu, J.; Shi, H.; Wang, W.; Ringer, S.P.; Du, Y.; Schryvers, D. Quantitative measurement for the microstructural parameters of nano-precipitates in Al-Mg-Si-Cu alloys. *Mater. Charact.* **2016**, *118*, 352–362. [CrossRef]

90. Berg, L.K.; Gjønnes, J.; Hansen, V.; Li, X.Z.; Knutson-Wedel, M.; Waterloo, G.; Schryvers, D.; Wallenberg, L.R. GP-zones in Al-Zn-Mg alloys and their role in artificial aging. *Acta Mater.* **2001**, *49*, 3443–3451. [CrossRef]

91. Amin-Ahmadi, B.; Idrissi, H.; Delmelle, R.; Pardoen, T.; Proost, J.; Schryvers, D. High resolution transmission electron microscopy characterization of fcc → 9R transformation in nanocrystalline palladium films due to hydriding. *Appl. Phys. Lett.* **2013**, *102*, 071911. [CrossRef]

92. Idrissi, H.; Amin-Ahmadi, B.; Wang, B.; Schryvers, D. Review on TEM analysis of growth twins in nanocrystalline palladium thin films: Toward better understanding of twin-related mechanisms in high stacking fault energy metals. *Phys. Status Solidi Basic Res.* **2014**, *251*, 1111–1124. [CrossRef]

93. Schouteden, K.; Amin-Ahmadi, B.; Li, Z.; Muzychenko, D.; Schryvers, D.; Van Haesendonck, C. Electronically decoupled stacking fault tetrahedra embedded in Au(111) films. *Nat. Commun.* **2016**, *7*, 14001. [CrossRef] [PubMed]

94. Rauch, E.F.; Véron, M. Automated crystal orientation and phase mapping in TEM. *Mater. Charact.* **2014**, *98*, 1–9. [CrossRef]

95. Gong, X.; Marmy, P.; Volodin, A.; Amin-Ahmadi, B.; Qin, L.; Schryvers, D.; Gavrilov, S.; Stergar, E.; Verlinden, B.; Wevers, M.; et al. Multiscale investigation of quasi-brittle fracture characteristics in a 9Cr-1Mo ferritic-martensitic steel embrittled by liquid lead-bismuth under low cycle fatigue. *Corros. Sci.* **2016**, *102*, 137–152. [CrossRef]

96. Yao, X.; Amin-Ahmadi, B.; Li, Y.; Cao, S.; Ma, X.; Zhang, X.-P.; Schryvers, D. Optimization of Automated Crystal Orientation Mapping in a TEM for Ni4Ti3 Precipitation in All-Round SMA. *Shape Mem. Superelast.* **2016**, *2*, 286–297. [CrossRef]

97. Amin-Ahmadi, B.; Connétable, D.; Fivel, M.; Tanguy, D.; Delmelle, R.; Turner, S.; Malet, L.; Godet, S.; Pardoen, T.; Proost, J.; et al. Dislocation/hydrogen interaction mechanisms in hydrided nanocrystalline palladium films. *Acta Mater.* **2016**, *111*, 253–261. [CrossRef]

98. Amin-Ahmadi, B.; Idrissi, H.; Galceran, M.; Colla, M.S.; Raskin, J.P.; Pardoen, T.; Godet, S.; Schryvers, D. Effect of deposition rate on the microstructure of electron beam evaporated nanocrystalline palladium thin films. *Thin Solid Films* **2013**, *539*, 145–150. [CrossRef]

99. Wang, B.; Idrissi, H.; Galceran, M.; Colla, M.S.; Turner, S.; Hui, S.; Raskin, J.P.; Pardoen, T.; Godet, S.; Schryvers, D. Advanced TEM investigation of the plasticity mechanisms in nanocrystalline freestanding palladium films with nanoscale twins. *Int. J. Plast.* **2012**, *37*, 140–156. [CrossRef]

100. Kobler, A.; Kashiwar, A.; Hahn, H.; Kübel, C. Combination of in situ straining and ACOM TEM: A novel method for analysis of plastic deformation of nanocrystalline metals. *Ultramicroscopy* **2013**, *128*, 68–81. [CrossRef] [PubMed]

101. Mompiou, F.; Legros, M. Quantitative grain growth and rotation probed by in-situ TEM straining and orientation mapping in small grained Al thin films. *Scr. Mater.* **2015**, *99*, 5–8. [CrossRef]

102. Gammer, C.; Kacher, J.; Czarnik, C.; Warren, O.L.; Ciston, J.; Minor, A.M. Local and transient nanoscale strain mapping during in situ deformation. *Appl. Phys. Lett.* **2016**, *109*, 081906. [CrossRef]

103. Legros, M. In situ mechanical TEM: Seeing and measuring under stress with electrons. *Comptes Rendus Phys.* **2014**, *15*, 224–240. [CrossRef]

104. Tirry, W.; Schryvers, D. In situ transmission electron microscopy of stress-induced martensite with focus on martensite twinning. *Mater. Sci. Eng. A* **2008**, *481–482*, 420–425. [CrossRef]

105. Idrissi, H.; Bollinger, C.; Boioli, F.; Schryvers, D.; Cordier, P. Low-temperature plasticity of olivine revisited with in situ TEM nanomechanical testing. *Sci. Adv.* **2016**, *2*, e1501671. [CrossRef] [PubMed]

106. Heidari, H.; Rivero, G.; Idrissi, H.; Ramachandran, D.; Cakir, S.; Egoavil, R.; Kurttepeli, M.; Crabbé, A.C.; Hauffman, T.; Terryn, H.; et al. Melamine-formaldehyde microcapsules: Micro- and nanostructural characterization with electron microscopy. *Microsc. Microanal.* **2016**, *22*, 1222–1232. [CrossRef] [PubMed]

107. Idrissi, H.; Kobler, A.; Amin-Ahmadi, B.; Coulombier, M.; Galceran, M.; Raskin, J.P.; Godet, S.; Kübel, C.; Pardoen, T.; Schryvers, D. Plasticity mechanisms in ultrafine grained freestanding aluminum thin films revealed by in-situ transmission electron microscopy nanomechanical testing. *Appl. Phys. Lett.* **2014**, *104*, 101903. [CrossRef]

108. Shim, S.; Bei, H.; Miller, M.K.; Pharr, G.M.; George, E.P. Effects of focused ion beam milling on the compressive behavior of directionally solidified micropillars and the nanoindentation response of an electropolished surface. *Acta Mater.* **2009**, *57*, 503–510. [CrossRef]

109. Samaeeaghmiyoni, V.; Idrissi, H.; Groten, J.; Schwaiger, R.; Schryvers, D. Quantitative in-situ TEM nanotensile testing of single crystal Ni facilitated by a new sample preparation approach. *Micron* **2017**, *94*, 66–73. [CrossRef] [PubMed]

110. Stöger-Pollach, M. Optical properties and bandgaps from low loss EELS: Pitfalls and solutions. *Micron* **2008**, *39*, 1092–1110. [CrossRef] [PubMed]

111. Nelayah, J.; Kociak, M.; Stéphan, O.; García de Abajo, F.J.; Tencé, M.; Henrard, L.; Taverna, D.; Pastoriza-Santos, I.; Liz-Marzán, L.M.; Colliex, C. Mapping surface plasmons on a single metallic nanoparticle. *Nat. Phys.* **2007**, *3*, 348–353. [CrossRef]

112. Bosman, M.; Keast, V.J.; Watanabe, M.; Maaroof, A.I.; Cortie, M.B. Mapping surface plasmons at the nanometre scale with an electron beam. *Nanotechnology* **2007**, *18*, 165505. [CrossRef]

113. Egoavil, R.; Gauquelin, N.; Martinez, G.T.; Van Aert, S.; Van Tendeloo, G.; Verbeeck, J. Atomic resolution mapping of phonon excitations in STEM-EELS experiments. *Ultramicroscopy* **2014**, *147*, 1–7. [CrossRef] [PubMed]

114. Krivanek, O.L.; Lovejoy, T.C.; Dellby, N.; Aoki, T.; Carpenter, R.W.; Rez, P.; Soignard, E.; Zhu, J.; Batson, P.E.; Lagos, M.J.; et al. Vibrational spectroscopy in the electron microscope. *Nature* **2014**, *514*, 209–212. [CrossRef] [PubMed]

115. Stöger-Pollach, M.; Franco, H.; Schattschneider, P.; Lazar, S.; Schaffer, B.; Grogger, W.; Zandbergen, H.W. Čerenkov losses: A limit for bandgap determination and Kramers–Kronig analysis. *Micron* **2006**, *37*, 396–402. [CrossRef] [PubMed]

116. Vatanparast, M.; Egoavil, R.; Reenaas, T.W.; Verbeeck, J.; Holmestad, R.; Vullum, P.E. Bandgap measurement of high refractive index materials by off-axis EELS. *Ultramicroscopy* **2017**, *182*, 92–98. [CrossRef] [PubMed]

117. Korneychuk, S.; Partoens, B.; Guzzinati, G.; Ramaneti, R.; Derluyn, J.; Haenen, K.; Verbeeck, J. Exploring possibilities of band gap measurement with off-axis EELS in TEM. *Ultramicroscopy* **2018**, *189*, 76–84. [CrossRef] [PubMed]

118. Stöger-Pollach, M.; Schachinger, T.; Biedermann, K.; Beyer, V. Valence EELS below the limit of inelastic delocalization using conical dark field EFTEM or Bessel beams. *Ultramicroscopy* **2017**, *173*, 24–30. [CrossRef] [PubMed]

119. García de Abajo, F.J. Optical excitations in electron microscopy. *Rev. Mod. Phys.* **2010**, *82*, 209–275. [CrossRef]

120. Boudarham, G.; Kociak, M. Modal decompositions of the local electromagnetic density of states and spatially resolved electron energy loss probability in terms of geometric modes. *Phys. Rev. B* **2012**, *85*, 245447. [CrossRef]

121. Piazza, L.; Lummen, T.T.A.; Quiñonez, E.; Murooka, Y.; Reed, B.W.; Barwick, B.; Carbone, F. Simultaneous observation of the quantization and the interference pattern of a plasmonic near-field. *Nat. Commun.* **2015**, *6*, 6407. [CrossRef] [PubMed]

122. Krehl, J.; Guzzinati, G.; Schultz, J.; Potapov, P.; Pohl, D.; Martin, J.; Verbeeck, J.; Fery, A.; Büchner, B.; Lubk, A. Spectral Field Mapping in Plasmonic Nanostructures with Nanometer Resolution. *arXiv* **2018**. Available online: https://arxiv.org/abs/1803.04399 (accessed on 12 March 2018).

123. Guzzinati, G.; Béché, A.; Lourenço-Martins, H.; Martin, J.; Kociak, M.; Verbeeck, J. Probing the symmetry of the potential of localized surface plasmon resonances with phase-shaped electron beams. *Nat. Commun.* **2017**, *8*, 14999. [CrossRef] [PubMed]
124. Verbeeck, J.; Béché, A.; Müller-Caspary, K.; Guzzinati, G.; Luong, M.A.; Den Hertog, M. Demonstration of a 2 × 2 programmable phase plate for electrons. *Ultramicroscopy* **2018**, *190*, 58–65. [CrossRef] [PubMed]

materials

MDPI

Article

LaAlO$_3$:Mn^{4+} as Near-Infrared Emitting Persistent Luminescence Phosphor for Medical Imaging: A Charge Compensation Study

Jiaren Du, Olivier Q. De Clercq, Katleen Korthout and Dirk Poelman *

LumiLab, Department of Solid State Sciences, Ghent University, Krijgslaan 281-S1, 9000 Ghent, Belgium; Jiaren.Du@ugent.be (J.D.); Olivier.DeClercq@ugent.be (O.Q.D.C.); Katleen.Korthout@ugent.be (K.K.)
* Correspondence: Dirk.Poelman@ugent.be; Tel.: +32-9264-4367

Received: 10 November 2017; Accepted: 9 December 2017; Published: 12 December 2017

Abstract: Mn^{4+}-activated phosphors are emerging as a novel class of deep red/near-infrared emitting persistent luminescence materials for medical imaging as a promising alternative to Cr^{3+}-doped nanomaterials. Currently, it remains a challenge to improve the afterglow and photoluminescence properties of these phosphors through a traditional high-temperature solid-state reaction method in air. Herein we propose a charge compensation strategy for enhancing the photoluminescence and afterglow performance of Mn^{4+}-activated LaAlO$_3$ phosphors. LaAlO$_3$:Mn^{4+} (LAO:Mn^{4+}) was synthesized by high-temperature solid-state reaction in air. The charge compensation strategies for LaAlO$_3$:Mn^{4+} phosphors were systematically discussed. Interestingly, Cl$^-$/Na$^+$/Ca^{2+}/Sr^{2+}/Ba^{2+}/Ge^{4+} co-dopants were all found to be beneficial for enhancing LaAlO$_3$:Mn^{4+} luminescence and afterglow intensity. This strategy shows great promise and opens up new avenues for the exploration of more promising near-infrared emitting long persistent phosphors for medical imaging.

Keywords: persistent luminescence; Mn^{4+}-activated phosphors; charge compensation; LaAlO$_3$:Mn^{4+}; solid-state reaction

1. Introduction

Persistent luminescence materials relate to a particular optical phenomenon whereby the light emission can last for several hours after the excitation has stopped [1–3]. The basic principles and physics behind persistent luminescence materials are related to two kinds of active centers: traps and emitters. Emitters release light in the wavelength range of interest and traps contribute to the duration time of the long persistent phosphor. The latter are believed to originate from lattice defects, co-dopants or impurities. The persistent luminescence phenomenon can be related to an optical battery as discussed in other papers [4].

The past two decades have witnessed rapid development and enormous advances to establish persistent luminescence materials for various applications including night-vision, emergency route illumination, security signs, traffic night signage, dials, decorative objects and toys [5]. Representative examples are SrAl$_2$O$_4$:Eu^{2+},Dy^{3+} (green emission) [1] and CaAl$_2$O$_4$:Eu^{2+},Nd^{3+} (blue emission) [6]. A pioneer proof-of-concept work published in 2007 first realized the application of in vivo imaging by using the near-infrared persistent luminescent material Ca$_{0.2}$Zn$_{0.9}$Mg$_{0.9}$Si$_2$O$_6$:Eu^{2+},Dy^{3+},Mn^{2+} as biomarker, hereby opening up new avenues for the widespread uses of persistent luminescent phosphors [7]. As for the use of in vivo imaging, the emitting wavelength of the phosphor is required to be located in the biological optical window (i.e., the first near-infrared window in the wavelength between 650 nm and 950 nm or the second near-infrared window in the wavelength between 1000 nm and 1350 nm). In these wavelength ranges, scattering, absorption and auto-fluorescence are limited and biological tissue is partly transparent [8]. Thus, the development of deep red/near-infrared emitting persistent phosphors has attracted much attention for application to in vivo bio-imaging systems or medical imaging.

Near-infrared emitting persistent phosphors have several advantages in comparison with other optically active particles, such as quantum dots and upconversion nanoparticles. Near-infrared quantum dots are limited by their potential toxicity and upconversion luminescent particles are hindered by the need for high energy lasers, which can lead to tissue damage [9,10]. Autofluorescence, one of the main drawbacks encountered with classic imaging probes, can be greatly reduced by using the near-infrared emitting persistent probes. Before its injection, the probe is well pre-excited outside the body of small animals to avoid autofluorescence. The signal to background ratio can be enhanced and local heating effects coming from high-power laser excitation can be avoided. Toxic effects can also be reduced when choosing appropriate phosphors, although long-term cytotoxicity studies still need to be undertaken [11]. Efforts to work on persistent luminescent nanoprobes for in vivo bioimaging applications were made, providing highly sensitive optical detection from living tissues and showing promising prospects in future practical use [12].

Traditionally, Cr^{3+}-doped deep red/near-infrared emitting persistent luminescence nanoparticles (PLNPs) are used as the key phosphors for in vivo imaging in small animals. Many Cr^{3+}-doped phosphors are widely investigated, such as $ZnGa_2O_4:Cr^{3+}$ [13], $Zn_3Ga_2Ge_2O_{10}:Cr^{3+}$ [14], $Zn_3Ga_2SnO_8:Cr^{3+}$ [15] and $LiGa_5O_8:Cr^{3+}$ [16–18]. Currently, manganese doped compounds, especially Mn^{4+}-activated phosphors, are considered as a promising alternative to Cr^{3+}-doped nanomaterials [19]. The tetravalent manganese ions (Mn^{4+}) can be doped in both fluorides and oxides [20–23]. Great effort has been made to the development of Mn^{4+}-activated oxide compounds. However, it remains a challenge to improve Mn^{4+}-activated phosphors with long afterglow and strong luminescence intensity compared with traditional Cr^{3+}-doped nanocrystals. This problem can be caused by the difficulty of stabilizing the Mn ions in the correct oxidation state due to the charge imbalance.

In the present work, manganese doped perovskite lanthanide aluminates ($LaAlO_3$) were synthesized by a high-temperature solid-state reaction method, similar to other reported work [19,24,25] and the persistent luminescence behavior was optimized. In order to improve the near-infrared persistent luminescence intensity of $LaAlO_3:Mn^{4+}$ phosphors, a series of samples were prepared using solid state reaction in air through co-doping with a variety of ions with different valence states. The doping effects and their possible mechanisms were investigated. The afterglow and luminescence intensity of this material can be strongly improved by this strategy of co-doping, leading to charge compensation or the introduction of new trap levels.

2. Results and Discussion

2.1. Crystal Structure

Figure 1a shows the crystal structure of $LaAlO_3$ drawn on the basis of the Inorganic Crystal Structure Database (ICSD No. 153821). $LaAlO_3$ is described in the trigonal crystal system with space group R-3cH (space group number 167) and lattice parameters a = 5.3598 Å, b = 5.3598 Å, c = 13.086 Å, Volume = 325.56 Å^3 and z = 6 [26]. As is shown in Figure 1a, the crystal structure corresponds to the rhombohedral, nearly cubic perovskite structure, which involves a rotation of the AlO_6 octahedra with respect to cubic perovskite as reported elsewhere [27,28]. There are two types of units in the crystal structure: AlO_6 octahedra and LaO_{12} polyhedra. The central Al^{3+} cation is in 6-fold oxygen coordination and forms AlO_6 octahedral units (blue unit in Figure 1). The La^{3+} cations are located in a polyhedral unit with 12-fold oxygen coordination (green unit in Figure 1). It is reported that the La-sites have D_3 point symmetry [29] and Cr^{3+}-doped LAO confirm a C_{3i} symmetry for the Al site [30]. Both cation sites thus have reduced symmetry from pure O_h symmetry, which results from a contraction along and a small rotation of AlO_6 octahedra around the c-axis of the LAO host. However, an inversion center is maintained for the Al-site. The reduction from pure octahedral symmetry is expected to result in a small splitting of the ^2E level and thus the zero-phonon transition of Mn^{4+}. The ionic radius of the Mn^{4+} ion, Al^{3+} ion and La^{3+} ion is 53 pm, 53.5 pm and 136 pm in 6-fold octahedral coordination (Mn^{4+} ion, Al^{3+} ion) and 12-fold coordination (La^{3+} ion), respectively [31]. It is well known that the Mn^{4+} ion usually stabilizes in an octahedral site with 6-fold coordination [21], thus, Mn^{4+} ions will supposedly occupy the Al^{3+} ion sites in the $LaAlO_3$ host as shown in Figure 1b. The similar ionic radius helps the substitution between dopant Mn^{4+} ion and central Al^{3+} ion in the AlO_6 octahedra.

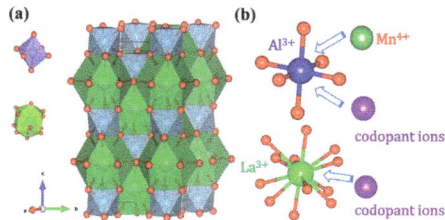

Figure 1. (**a**) Crystal structure of $LaAlO_3$ (blue unit is AlO_6 octahedron and green unit is LaO_{12} polyhedron); (**b**) Doping positions for Mn^{4+} and other dopant ions, the Mn^{4+} ion can occupy Al^{3+} ion site in the AlO_6 octahedral unit.

XRD (X-ray diffraction) patterns of un-doped $LaAlO_3$ sintered at 1350 °C, 1400 °C, 1450 °C, 1500 °C, 1550 °C, 1600 °C and 1650 °C are shown in Figure S1. At higher sintering temperatures (above 1550 °C), the XRD patterns of these samples match well with the standard XRD data of $LaAlO_3$ (No. 00-031-0022). At lower sintering temperatures, especially at 1350 °C, a different pattern is displayed. XRD patterns of samples synthesized at 1350 °C and 1550 °C are compared in Figure 2. The impurity phase at 1350 °C is found to be La_2O_3 (as shown in Figure 2) and the optimized temperature for synthesizing $LaAlO_3$ was chosen at 1550 °C. The standard XRD data of $LaAlO_3$ (No. 00-031-0022) and La_2O_3 (No. 00-005-0602) are illustrated in red and green bars respectively in Figure 2. Detailed XRD patterns of the obtained $LaAlO_3$ phosphor with different concentrations of Mn^{4+} and Mn^{4+},R sintered at 1550 °C demonstrate that doping of Mn^{4+} ions or Mn^{4+},R (R = Ge^{4+}, Si^{4+}, Ti^{4+}, Zr^{4+}, Ba^{2+}, Ca^{2+}, Mg^{2+}, Sr^{2+}, Cl^-, Li^+, Na^+) co-dopants does not cause any significant structural changes of the $LaAlO_3$ host for any of the dopants (XRD patterns not shown). Energy-dispersive X-ray (EDS) mapping indicates the homogeneous distribution of manganese in the host (not shown). In agreement with the previous reports, doping of manganese ions is perfectly feasible in the $LaAlO_3$ host [19,32].

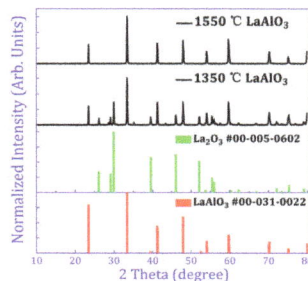

Figure 2. A comparison of XRD (X-ray diffraction) patterns between 1350 °C and 1550 °C. The impurity phase in the XRD pattern at 1350 °C is assigned to La_2O_3 and the optimized temperature for synthesizing is 1550 °C. The standard XRD data of $LaAlO_3$ and La_2O_3 are illustrated in red and green bars respectively. The intensities of XRD patterns are normalized to arbitrary units [0,1].

2.2. Luminescence Properties

The room-temperature photoluminescence (PL) spectrum of LaAlO$_3$:0.5%Mn^{4+} phosphor upon excitation at 335 nm exhibits narrow emission bands in the range 650–800 nm due to the $^2E_g\rightarrow^4A_{2g}$ spin-forbidden transitions in Mn^{4+} ions, with a maximum located at 731 nm as shown in Figure 3a. The spectrum consists of several sharp features, peaking at (from left to right) 697.5 nm, 704.5 nm, 710.5 nm, 718 nm, 724.5 nm and 731 nm, corresponding to the spin-forbidden $^2E_g\rightarrow^4A_{2g}$ transition and the vibrational sidebands of zero-phonon line (ZPL) with phonon assistance. Due to its high effective positive charge, Mn^{4+} experiences a large crystal field and hence no transitions from the $^4T_2$ level are expected, in contrast to Cr^{3+}-based phosphors [17]. The zero-phonon line (ZPL) is located at 718 nm in Figure 3. The ZPL is surrounded by both anti-Stokes and Stokes phonon side bands. The ZPL here (718 nm, ~13927 cm^{-1}) has a larger wavelength than the value of 712 nm (~14045 cm^{-1}), measured at 300 K and reported by Van Ipenburg et al. [33] but corresponds to the value reported by Li et al. [19]. An assignment of the sidebands to the type of vibration was done by Van Ipenburg et al. [33] and the anti-Stokes sidebands at 697.5 nm, 704.5 nm and 710.5 nm appear in our sample. The $^2E_g\rightarrow^4A_{2g}$ transition has a small electron-phonon coupling and the excited ions usually relax non-radiatively to ^2E_g followed by the spin-forbidden $^2E_g\rightarrow^4A_{2g}$ transition, thus resulting in narrow-band emission lines. This is in contrast with the broad bands in the excitation spectrum that correspond to spin-allowed $^4A_2\rightarrow^4T_1$ and $^4A_2\rightarrow^4T_2$ transitions, with larger electron-phonon coupling [23,34]. It has been widely reported that the spectra of Mn^{4+} ions exhibit a combination of broadband excitation bands and sharp emission lines [34,35]. Usually, the excitation and emission peaks of Mn^{4+} ions in many other oxide hosts are observed around 300 nm and above 650 nm, respectively [36–38]. This PL behavior proves that Mn is indeed incorporated in the LaAlO$_3$ lattice and is incorporated in a 4+ oxidation state, since Mn^{2+} is expected to show an entirely different and broad emission spectrum. Mostly, the excitation spectrum of Mn^{2+} is very characteristic for the d^5 electron configuration. The photoluminescence excitation (PLE) spectrum of the phosphor at room temperature is also shown in Figure 3a. The PLE spectrum (λ_{em} = 731 nm) shows a broad band with the main peak located at 335 nm and ranging from 250 nm to 450 nm. The broad PLE band is mainly attributed to $^4A_2\rightarrow^4T_1$ and $^4A_2\rightarrow^4T_2$ transitions of Mn^{4+} ion as illustrated in the Tanabe-Sugano energy level diagram for 3d^3 ions [39]. When Mn^{4+} ions are situated in the LaAlO$_3$ host with octahedral coordination, the dependence of energy levels of Mn^{4+} on crystal field strength can be clearly illustrated by the Tanabe-Sugano energy level diagram in Figure 3b. A comparison of PL spectra of LaAlO$_3$:0.1%Mn^{4+}, LaAlO$_3$:0.2%Mn^4, LaAlO$_3$:0.5%Mn^{4+}, LaAlO$_3$:1%Mn^{4+}, LaAlO$_3$:2%Mn^{4+} and LaAlO$_3$:5%Mn^{4+} is illustrated in Figure 4. The PL intensity of LaAlO$_3$:Mn^{4+} phosphors increases with increasing Mn^{4+} ion concentration within the range 0.1% to 0.5% and decreases upon further increasing Mn^{4+} ion concentration from 0.5% to 5%. The former effects are presumably due to the effective Mn^{4+} concentration, which dominatingly determines PL intensity in this host. When the content of the doping Mn^{4+} ions is relatively low, the effective Mn^{4+} concentration is approximately proportional to the content of the doping Mn^{4+} ions. Thus, increasing the content of Mn^{4+} ions from 0.1% to 0.5%, the PL intensity of LaAlO$_3$:Mn^{4+} phosphors increases synchronously. The latter observation could be attributed to the concentration quenching phenomenon of Mn^{4+} ions. It can be seen that the optimal doping concentration of Mn^{4+} ions is 0.5%.

Figure 5 shows the persistent luminescence decay curves of LaAlO$_3$:0.1%Mn^{4+}, LaAlO$_3$:0.2%Mn^{4+}, LaAlO$_3$:0.5%Mn^{4+}, LaAlO$_3$:1%Mn^{4+}, LaAlO$_3$:2%Mn^{4+}, LaAlO$_3$:5%Mn^{4+} phosphors after 5 min of irradiation with a Xenon arc lamp. It was also found that 0.5% Mn^{4+} is also the optimal doping concentration for the afterglow intensity and duration.

(a)

(b)

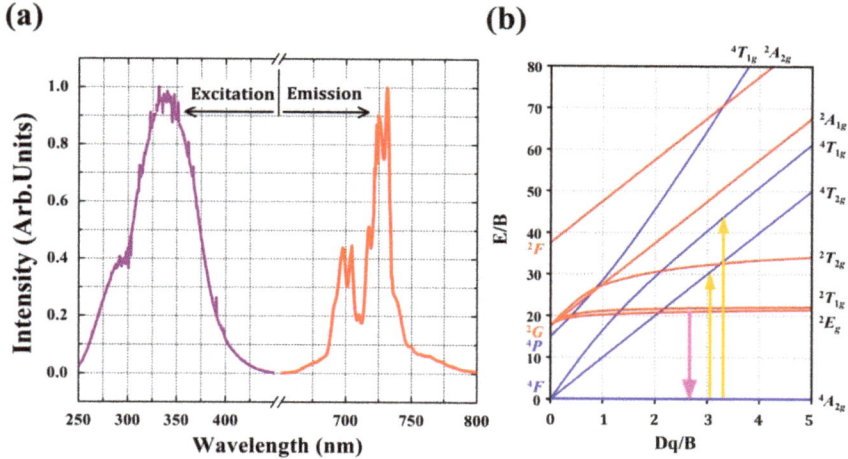

Figure 3. (**a**) Photoluminescence (PL) and photoluminescence excitation (PLE) spectra of LaAlO$_3$:0.5%Mn^{4+} phosphor. PL and PLE spectra (λ_{ex} = 335 nm and λ_{em} = 731 nm) are in the range 250–450 nm and 650–800 nm, respectively; (**b**)Tanabe-Sugano energy level diagram of a d^3 configuration (Mn^{4+}ion in the octahedron).

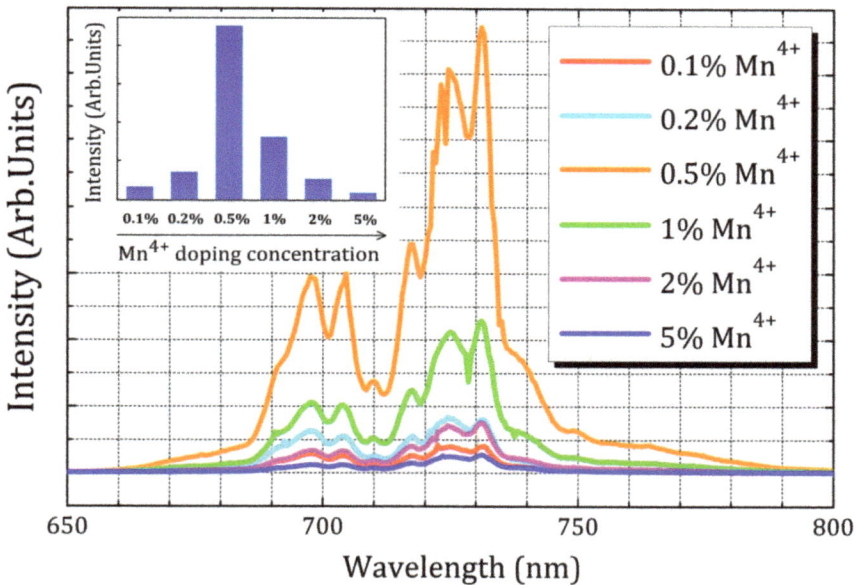

Figure 4. A comparison of photoluminescence (PL) spectra of LaAlO$_3$:0.1%Mn^{4+}, LaAlO$_3$:0.2%Mn^{4+}, LaAlO$_3$:0.5%Mn^{4+}, LaAlO$_3$:1%Mn^{4+}, LaAlO$_3$:2%Mn^{4+} and LaAlO$_3$:5%Mn^{4+} phosphors. The emission spectrum is acquired under 335 nm excitation. The inset gives the relative emission intensity while increasing Mn^{4+} doping concentration from 0.1% to 5%.

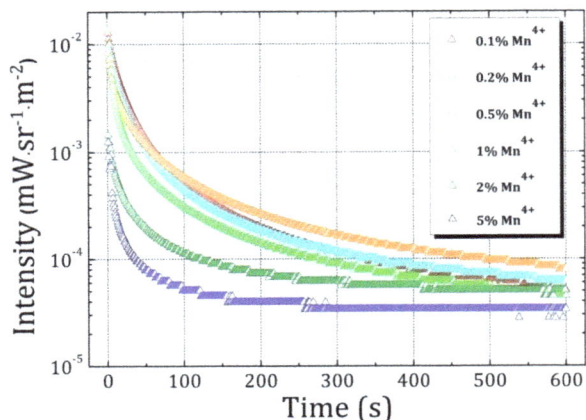

Figure 5. Persistent luminescence decay curves of Mn^{4+}-doped LaAlO$_3$ phosphors after 5 min of irradiation with a Xenon arc lamp. The concentrations of Mn^{4+}ions in LaAlO$_3$host are 0.1%, 0.2%, 0.5%, 1%, 2%, 5% as shown in different colors.

2.3. Charge Compensation Strategy

To enhance the photoluminescence and afterglow performance of Mn^{4+}-activated LaAlO$_3$ phosphor, a charge compensation strategy was proposed. As mentioned before, Mn^{4+} ions are supposed to substitute the Al^{3+} ions in the LaAlO$_3$ host. Thus, a charge imbalance occurs when the substitution happens. With the aid of lower valence state ions to balance the superfluous positive charge of tetravalent manganese, charge compensation is expected. A schematic of charge compensation strategies is illustrated in Figure 6 for LaAlO$_3$:Mn^{4+} phosphor. For the divalent cations such as Ca^{2+} in Figure 6a, the unit of Mn^{4+}-Ca^{2+} can be selected as a substitutable alternative to the unit of Al^{3+}-La^{3+} to the charge compensation. As for the monovalent cations (for example, Na$^+$ in Figure 6b), the units of Mn^{4+}-Na$^+$-Mn^{4+} and Al^{3+}-La^{3+}-Al^{3+} are equivalent in number of charges. We expect Mn^{4+} to substitute for Al^{3+} and Ca^{2+} to substitute for La^{3+} based on the following considerations: Ionic radii, coordinated environment and chemical stability. For chemical stability, it is well known that the Mn^{4+} ion usually stabilizes in an octahedral site with 6-fold coordination (in Al^{3+} site). Mn^{4+} ion could hardly be situated in 12-fold coordinated environment (La^{3+} site). In an octahedral environment (in Al^{3+} site), the Mn-3d states are split into three- and two-fold degenerate t$_{2g}$ and e$_g$ states, respectively. The three Mn-3d electrons of Mn^{4+} exactly fill the majority-spin t$_{2g}$ states. The crystal field splitting creates a large gap between the t$_{2g}$ and e$_g$ states, stabilizing the 4+ oxidation state [21,40,41]. Therefore, Mn^{4+} ions are usually found on octahedral sites of solids [40,41]. For Ge^{4+} and Ti^{4+}, they also have a stable chemical environment like the Mn^{4+} ion, so we adopted these ions as co-dopants.

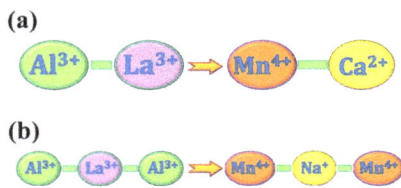

(a)

(b)

Figure 6. A schematic of charge compensation strategies for LaAlO$_3$:Mn^{4+} phosphor. (**a**) The charge compensation for the unit of Al^{3+}-La^{3+} and Mn^{4+}-Ca^{2+}; (**b**) The charge compensation for the unit of Al^{3+}-La^{3+}-Al^{3+} and Mn^{4+}-Na$^+$-Mn^{4+}.

Suitable charge-compensating co-dopants should fulfill certain requirements. The difference in ionic radius between doping ions and central ions plays a critical role in possible substitution and consummate incorporation. The substitution of ions in a crystal lattice has been discussed in depth [42]. It is possible to replace an ion in a specific lattice position with a dopant and not disturb the crystal structure when both ions differ in size by no more than a certain radius ratio for a given coordination number, according to the work of Linus Pauling [42]. It is expected that a smaller radius difference (no more than 15% or 20%) leads to a better substitution. 6-fold coordinated Al^{3+} ion has an ionic radius of 53.5 pm and 12-fold coordinated La^{3+} ion has an ionic radius of 136 pm. Figure 7 shows the radius difference between possible doping ions and central ions [31]. Ions in blue stars are supposed to substitute on the Al^{3+} site with coordination number VI and ions while half-filled pink hexagons are supposed to substitute on the La^{3+} site with coordination number XII (details in Tables S1 and S2). Both tetravalent cations and lower valence state ions are selected for comparison. Stars and hexagons located in the green wireframe are feasible candidates for doping, considering the similarity in ionic radius and coordinated environment in the host. In the case of $LaAlO_3$, Ge^{4+} ions and Na^+, Ca^{2+}, Sr^{2+} ions are located in the two green wireframes. Ge^{4+} ions are supposed to replace Al^{3+} ions, while Na^+, Ca^{2+} and Sr^{2+} ions are substitutable for La^{3+} ions (shown in Figure 6).

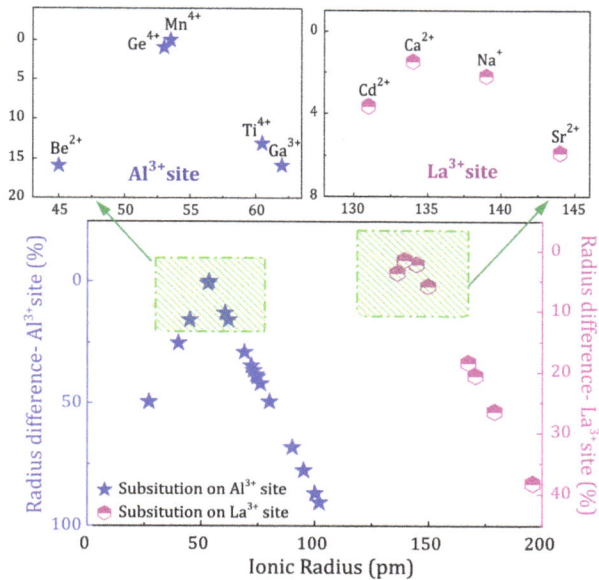

Figure 7. Radius difference between doping ions and substitutable central ions. Blue stars are supposed to substitute on Al^{3+} site with coordination number VI and half-filled pink hexagons are supposed to substitute on the La^{3+} site with coordination number XII. Stars and hexagons located in the green wireframe are feasible candidates for doping.

2.4. Influence of Various Co-Dopants

To obtain systematic information on the influence of other dopants, various kinds of ions can be taken into account. Some other research groups also found that tetravalent cations (such as Ge^{4+}) or negative charge ions (such as Cl^-) could be added as co-dopants in $CaAl_{12}O_{19}$ and $SrMgAl_{10}O_{17}$ hosts [43–45]. Thus, tetravalent cations (Ge^{4+}, Si^{4+}, Ti^{4+}, Zr^{4+}), divalent cations (Ba^{2+}, Ca^{2+}, Mg^{2+}, Sr^{2+}), monovalent cations (Li^+, Na^+) and negative charge ions (Cl^-) were all selected as discussed above. A series of $2\%Mn^{4+},2\%R$ (R = Ge^{4+}, Si^{4+}, Ti^{4+}, Zr^{4+}, Ba^{2+}, Ca^{2+}, Mg^{2+}, Sr^{2+}, Li^+, Na^+, Cl^-) co-doped phosphors were synthesized at 1550 °C and PL measurements were performed at room temperature. PL spectra of 2%Mn^{4+},2%R (R = Ge^{4+}, Si^{4+}, Ti^{4+}, Zr^{4+}, Ba^{2+}, Ca^{2+}, Mg^{2+}, Sr^{2+}, Li^{+}, Na^{+}, Cl^{-}) phosphors are shown in Figures S2–S4. Some co-dopants such as Li^{+}, Mg^{2+} and Si^{4+} ions did not contribute much to the enhancement of PL intensity. Interestingly, it was found that co-doping with R (R = Na^{+}, Cl^{-}, Ge^{4+}, Ca^{2+}, Sr^{2+} and Ba^{2+}) increases the PL intensity of LaAlO$_3$:2%Mn^{4+} phosphor. In particular, the PL emission of LaAlO$_3$:2%Mn^{4+},2%Na^{+}, LaAlO$_3$:2%Mn^{4+},2%Ca^{2+}, LaAlO$_3$:2%Mn^{4+},2%Ba^{2+}, and LaAlO$_3$:2%Mn^{4+},2%Sr^{2+} was several times stronger than that of LaAlO$_3$:2%Mn^{4+} phosphor. It indicates that employing the charge compensation strategy with appropriate co-dopants such as Na^{+}, Cl^{-}, Ge^{4+}, Ca^{2+}, Sr^{2+} and Ba^{2+} ions helps to achieve the improvement of luminescence. Mn^{4+} ions and these R co-dopants were successfully incorporated into the LaAlO$_3$ host compound.

In order to optimize the properties of LaAlO$_3$:Mn^{4+} phosphors with the charge compensation strategy, six groups of LaAlO$_3$:Mn^{4+}, R (R = Na^{+}, Cl^{-}, Ge^{4+}, Ca^{2+}, Sr^{2+}, Ba^{2+}) phosphors with different co-dopant concentrations were synthesized using the same conditions as discussed above, now using the optimum Mn^{4+} concentration of 0.5%. The concentration of R (R = Na^{+}, Cl^{-}, Ge^{4+}, Ca^{2+}, Sr^{2+}, Ba^{2+}) was 0.5%, 1%, 2%, 3% and 5%. The persistent luminescence decay curves were measured after 5 min of irradiation with a Xenon arc lamp and the detailed persistent luminescence decay curves of LaAlO$_3$:0.5%Mn^{4+},yR (R = Na^{+}, Cl^{-}, Ge^{4+}, Ca^{2+}, Sr^{2+}, Ba^{2+}; y = 0.5%, 1%, 2%, 3% and 5%) phosphors are shown respectively in Figures S5–S10. A comparison of the afterglow duration time of the six groups is illustrated in Figure 8, showing the time after excitation when the intensity of the afterglow luminescence drops to 5 × 10^{-4} mW/sr/m^2. This benchmark intensity roughly corresponds to the same intensity as the 0.3 mcd/m^2 value, used as a benchmark for visible persistent luminescence [3]. The first red column corresponds to LaAlO$_3$:0.5%Mn^{4+} as an intensity and duration reference of persistent luminescence. The concentration of R (R = Ca^{2+}, Sr^{2+}, Ba^{2+}, Cl^{-}, Na^{+}, Ge^{4+}) is increasing from left to right in each doping group in Figure 8. The afterglow duration time is prolonged when increasing the concentration of co-dopants and decreases upon further increasing concentrations in the case of Ca^{2+}, Sr^{2+}, Cl^{-} co-dopants, similar to the Mn^{4+} doping behavior as discussed above in Figure 5. For other co-dopants, the trend turns abnormal due to the complex interactions between the traps and defects. All the co-dopants chosen in this research can help to improve the persistent luminescence and strengthen the afterglow duration time to different extent with various doping concentrations. Steady state photoluminescence spectra of the six groups of LaAlO$_3$:Mn^{4+},R (R = Na^{+}, Cl^{-}, Ge^{4+}, Ca^{2+}, Sr^{2+}, Ba^{2+}) phosphors were measured at room temperature. For each group of LaAlO$_3$:Mn^{4+},R phosphors with different co-dopant concentrations, the intensities of photoluminescence were enhanced to a different extent for certain doping concentrations and PL spectra are shown in Figures S11–S16. This also proves the feasibility of the charge compensation strategy for LaAlO$_3$:Mn^{4+} phosphors. Figure 9 exhibits a map of the integrated intensity of photoluminescence from Mn^{4+} with R (R = Ca^{2+}, Sr^{2+}, Ba^{2+}, Cl^{-}, Na^{+}, Ge^{4+}) co-dopants. The integral intensity was calculated in the wavelength region from 600 nm to 800 nm. The concentration of Mn^{4+} ion was 0.5% in each LaAlO$_3$ host and the concentration of R is 0.5%, 1%, 2%, 3% and 5% respectively from left to right in each doping group as illustrated in Figure 9. In order to compare the influence among the different co-dopants with the different R concentrations, the integral intensity of LaAlO$_3$:0.5%Mn^{4+} emission spectrum was normalized to 1 as a benchmark of emission intensity. It indicates that co-doping with Ge^{4+}, Na^{+}, Cl^{-}, Ba^{2+}, Sr^{2+} and Ca^{2+} is beneficial for the enhancement of the PL intensity. Upon doping only one type of dopant, like Mn^{4+} ions, the PL and afterglow properties of LaAlO$_3$:Mn^{4+} phosphors have a regular trend when increasing the concentrations of Mn^{4+} ions from 0.1% to 5% as shown in Figures 4 and 5. However, upon doping two kinds of dopants with the phosphor formula 'LaAlO$_3$:Mn^{4+},R', the possible mechanism and interactions between the traps and defects turns complicated, resulting in an anomalous performance with different concentrations of dopants and co-dopants. The duration and PL intensity of the Mn^{4+} emission may not simply change monotonously when co-doping another R ions as it is shown in Figures 8 and 9. When co-doping with different ions, even isovalent Ge^{4+}, it is possible that the optimum Mn dopant concentration is changed, again leading to a more complex relation between co-dopant concentration and performance. In view of the improved performance of Mn^{4+}-activated $LaAlO_3$ phosphor, co-dopants chosen in this research can lead to a 2-fold to 4-fold increase of afterglow time and PL intensity with different doping concentrations. Based on this preliminary screening of the performance of different co-dopants, a more in-depth investigation of the effects of co-doping on afterglow performance will be conducted on selected co-dopants, using temperature dependent charging and afterglow experiments and thermoluminescence measurements.

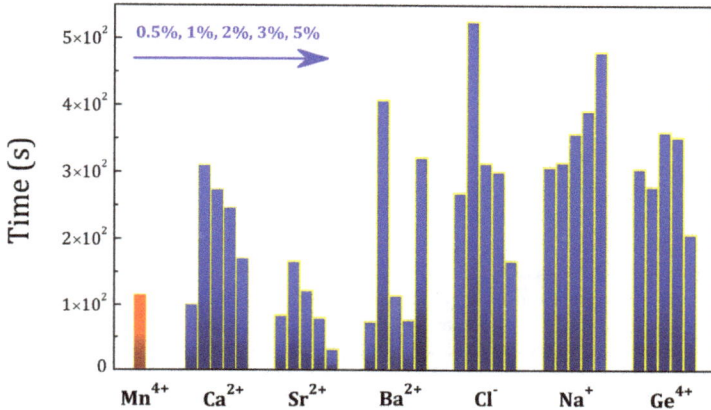

Figure 8. A comparison of the time until the intensity of afterglow luminescence drops to 5×10^{-4} mW/sr/m^2. The concentration of Mn^{4+} ion is 0.5% in each case and the concentration of R is 0.5%, 1%, 2%, 3% and 5% respectively from left to right in each doping group (R = Ca^{2+}, Sr^{2+}, Ba^{2+}, Cl^{-}, Na^{+}, Ge^{4+}). The first column corresponds to $LaAlO_3$:0.5%Mn^{4+} as an intensity and duration benchmark of persistent luminescence.

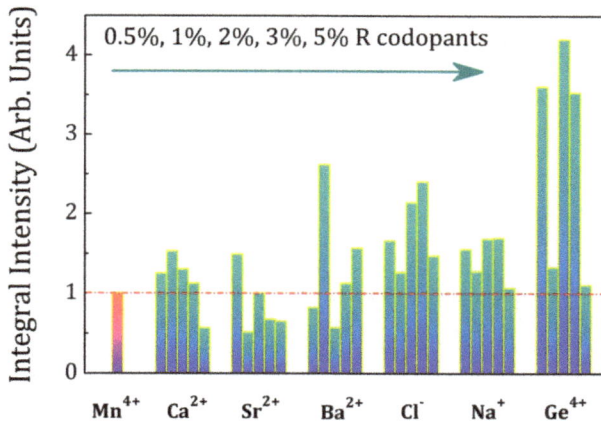

Figure 9. A map of integral intensity of photoluminescence spectra from Mn^{4+}/R (R = Ca^{2+}, Sr^{2+}, Ba^{2+}, Cl^{-}, Na^{+}, Ge^{4+}) co-dopants in the wavelength from 600 nm to 800 nm. The concentration of Mn^{4+} ions is 0.5% and the concentration of R is 0.5%, 1%, 2%, 3% and 5% respectively from left to right in each doping group.

PL intensities and afterglow duration of the LaAlO$_3$:Mn^{4+} phosphors are determined by the competition between the quantity of the effective Mn^{4+} ions (defect density of Mn^{4+} ions) in the phosphors and the interaction among the Mn^{4+} ions (Mn^{4+}-Mn^{4+} pairs). We believe that more isolated Mn^{4+} ions in the phosphors enhance the light emission and more traps and defects strengthen the afterglow duration. However, the formation and interaction from Mn^{4+}-Mn^{4+} pairs, which are inevitably formed at high annealing temperature as reported in some other papers [43,46], will decrease both defect density and the number of effective Mn^{4+} ions, resulting in quenching the emission and shorten afterglow duration. From the Figures 8 and 9, both the photoluminescence and persistent luminescence can be enhanced to a certain extent when employing the appropriate co-dopants such as Na$^+$, Cl$^-$, Ca^{2+}, Ba^{2+}, Sr^{2+} and Ge^{4+} ions. As for the improvement of persistent luminescence, two kinds of active centers are involved in persistent luminescence, namely traps and emitters. Traps originate from lattice defects or co-dopants in the phosphor and emitters release light in the region of interest. It is known that Mn^{4+} ion can act as both the trapping center and the emitting center in the perovskite LaAlO$_3$ host [19]. The afterglow duration thus relies on the effective defect density of Mn^{4+} ions as trapping centers. In addition, Mn^{4+} trapping centers have a complex dependency on the concentration of incorporated manganese ions and Mn^{4+}-Mn^{4+} pairs [43]. Mn^{4+}-Mn^{4+} pairs decrease the effective defect density of Mn^{4+} ions and weaken the persistent luminescence. The afterglow of this material is constrained by the formation of Mn^{4+}-Mn^{4+} pairs in the host, resulting in a lower effective defect density. The challenge in the perovskite LaAlO$_3$ host is to incorporate more Mn^{4+} ions as traps and avoid Mn^{4+}-Mn^{4+} pairs. Charge-compensating co-dopants help to impede the formation of Mn^{4+}-Mn^{4+} pairs. In this material, the effective incorporation and increasing defect density are strongly improved by co-doping with Ge^{4+} or ions with a lower valence state for charge compensation, such as divalent cations (Ca^{2+}, Ba^{2+} or Sr^{2+} ions), monovalent cations (Na$^+$ ions) or ions in their negative valence (Cl$^-$ ions). Thus, the enrichment of the effective defect density prolongs the persistent luminescence. As for the photoluminescence enhancement, the origin of this phenomenon is widely understood and explained by the charge compensation mechanism. The photoluminescence intensity is determined by the complex competition between the effective amount of Mn^{4+} ions and quantity of the Mn^{4+}-Mn^{4+} pairs in LaAlO$_3$:Mn^{4+} phosphors. It is believed that more effective Mn^{4+} ions increase the PL intensity while the Mn^{4+}-Mn^{4+} pairs quench the emission. The interaction of Mn^{4+}-Mn^{4+} pairs is related to the Mn^{4+}-Mn^{4+} distance [43]. Substitution of 6-fold coordinated octahedral Al^{3+} ion or 12-fold coordinated central La^{3+} ion with a lower charge state is supposed to create a negative local charge, which improves the efficiency of Mn^{4+} ions to replace Al^{3+} sites, thus increasing the effective amount of Mn^{4+} ions. Furthermore, the incorporation with lower valence state ions guarantees the charge equilibrium, which restrains the formation of Mn^{4+}-Mn^{4+} pairs leading to the enhancement of PL intensity [44]. That is the case for divalent cations (Ca^{2+}, Ba^{2+} or Sr^{2+} ions), monovalent cations (Na$^+$ ions) and negative valence ions (Cl$^-$ ions). Tetravalent cations such as Ge^{4+} ions are competitive with the substitution by Mn^{4+} due to the same valence state. However, the co-dopant Ge^{4+} ions may enrich the defect density and play an effective role in decreasing the formation and interaction of Mn^{4+}-Mn^{4+} pairs, resulting as a compensator to Mn^{4+}-Mn^{4+} pairs and enhancing the PL intensity [19,43]. In the case of the perovskite LaAlO$_3$ host, the compensation effect of co-doping Mn^{4+}/Ge^{4+} ions is found to occupy a predominant position.

3. Materials and Methods

All the raw chemicals were analytical grade, used without further purification. The precursors were La$_2$O$_3$ (Sigma Aldrich, Saint Louis, MO, USA, 99.99%), Al$_2$O$_3$ (Fluka, Schwerte, Germany, 99.5%), MnO$_2$ (Alfa Aesar, Karlsruhe, Germany, 99.997%), NH$_4$Cl (Alfa Aesar, 99.999%), Li$_2$CO$_3$ (Alfa Aesar, 99.998%), Na$_2$CO$_3$ (Alfa Aesar, 99.95%), MgO (Alfa Aesar, 99.95%), CaCO$_3$ (Alfa Aesar, 99.95%), SrCO$_3$ (Alfa Aesar, 99.99%), BaCO$_3$ (Alfa Aesar, 99.95%), SiO$_2$ (Alfa Aesar, 99.5%), GeO$_2$ (Alfa Aesar, 99.999%), TiO$_2$ (Alfa Aesar, 99.995%), ZrO$_2$ (Alfa Aesar, 99.978%). The concentrations of dopants were chosen as follows: LaAlO$_3$:xMn^{4+} (x = 0.1%, 0.2%, 0.5%, 1%, 2% and 5%); LaAlO$_3$:2%Mn^{4+},2%R;

$LaAlO_3$:0.5%Mn^{4+},yR (y = 0.5%, 1%, 2%, 3% and 5%). The molar % is defined with respect to one mole of a host phosphor chemical formula. The appropriate stoichiometric number of precursors were weighed and manually ground in an agate mortar. Subsequently, the starting materials were mixed with ethanol and put in a ZrO_2 grinding jar. Grinding was performed in a Retsch PM 100 Planetary ball mill for 6 h to reduce the particle size. After evaporating the remaining ethanol, repeated grinding was performed in an agate mortar to improve the mixing homogeneity of the precursors. The phosphors of $LaAlO_3$:Mn^{4+} and a series of $LaAlO_3$:Mn^{4+},R (R = Cl^-, Li^+, Na^+, Mg^{2+}, Ca^{2+}, Sr^{2+}, Ba^{2+}, Si^{4+}, Ge^{4+}, Ti^{4+}, Zr^{4+}) were synthesized through a traditional high-temperature solid-state reaction method, testing various temperatures between 1350 °C and 1650 °C, for 6 h in air. The employed heating rate was 300 °C/h using a tube furnace (ETF30-50/18-S furnace, ENTECH, Ängelholm, Sweden). All samples were allowed to cool naturally inside the tube furnace. The sintered samples were well ground again. To compare the luminescence properties and afterglow intensity, all the phosphors were synthesized under the same experimental conditions.

The crystal structures of $LaAlO_3$:Mn^{4+} and $LaAlO_3$:Mn^{4+}, R were characterized by Powder X-ray diffraction (XRD) (Bruker, Leiderdorp, The Nederlands). Crystallographic phases of the obtained powders were verified on a Siemens D5000 diffractometer (40 kV, 40 mA, Bruker) using Cu Kα1 radiation (λ = 0.154 nm). The XRD data were collected in the range 2θ from 10° to 80° at room temperature. A comparison of the obtained XRD patterns with the reference pattern (No. 00-031-0022) was made to check the phase purity.

Steady state photoluminescence excitation and emission spectra were measured using an Edinburgh FS920 (Edinburgh Instruments Ltd., Livingston, UK) fluorescence spectrometer with a monochromated 450 W Xe-arc lamp as the excitation source.

The powder was put into a metal disc with a diameter of 12.5 mm and persistent luminescence was measured with a photosensor amplifier (C9329, Hamamatsu, Japan) and a Centronics OSD100-5T (Centronic Ltd., Croydon, UK) silicon photodiode. The afterglow decay profiles were then calibrated to the absolute radiance (in unit of $mW/sr/m^2$), since the usual units of luminance, cd/m^2, are not relevant for near-infrared emitting phosphors [3]. All persistent luminescent decay curves of $LaAlO_3$ samples were recorded at room temperature after excitation for 5 min by the light of an unfiltered Xenon arc lamp at an intensity of 1000 lux.

4. Conclusions

In summary, a series of novel near-infrared emitting persistent luminescent phosphors $LaAlO_3$:Mn^{4+} (LAO:Mn^{4+}) were synthesized by a convenient high-temperature solid-state reaction in air. Various kinds of co-dopants with Mn^{4+} ions were optimized and incorporated in different concentrations. Impressively, co-dopants such as Cl^-, Na^+, Ca^{2+}, Sr^{2+}, Ba^{2+} and Ge^{4+} ions were all found to be beneficial for improving the $LaAlO_3$:Mn^{4+} luminescence and afterglow intensity. The charge compensation strategies for $LaAlO_3$:Mn^{4+} phosphors were systematically discussed. Employing this charge compensation strategy is believed to open up new avenues for the exploration of more promising near-infrared emitting long persistent phosphors for medical imaging.

Supplementary Materials: The following are available online at www.mdpi.com/1996-1944/10/12/1422/s1, Figure S1: XRD pattern of $LaAlO_3$ synthesized through a solid-state reaction method, Figure S2: Photoluminescence (PL) spectra of $LaAlO_3$:2%Mn^{4+},2%Li^+, $LaAlO_3$:2%Mn^{4+},2%Na^+, and $LaAlO_3$:2%Mn^{4+},2%Cl^- phosphors, Figure S3: Photoluminescence (PL) spectra of $LaAlO_3$:2%Mn^{4+},2%Ge^{4+}, $LaAlO_3$:2%Mn^{4+},2%Si^{4+}, $LaAlO_3$:2%Mn^{4+},2%Ti^{4+}, and $LaAlO_3$:2%Mn^{4+},2%Zr^{4+} phosphors, Figure S4: Photoluminescence (PL) spectra of $LaAlO_3$:2%Mn^{4+},2%Ba^{2+}, $LaAlO_3$:2%Mn^{4+},2%Ca^{2+}, $LaAlO_3$:2%Mn^{4+},2%Mg^{2+},and $LaAlO_3$:2%Mn^{4+},2%Sr^{2+} phosphors, Figure S5: Persistent luminescence decay curves of $LaAlO_3$:0.5%Mn^{4+},ySr^{2+} (y = 0.5%, 1%, 2%, 3%, and 5%) phosphors after 5 min of irradiation with a Xenon arc lamp, Figure S6: Persistent luminescence decay curves of $LaAlO_3$:0.5%Mn^{4+},yGe^{4+} (y = 0.5%, 1%, 2%, 3%, and 5%) phosphors after 5 min of irradiation with a Xenon arc lamp, Figure S7: Persistent luminescence decay curves of $LaAlO_3$:0.5%Mn^{4+},yCa^{2+} (y = 0.5%, 1%, 2%, 3%, and 5%) phosphors after 5 min of irradiation with a Xenon arc lamp, Figure S8: Persistent luminescence decay curves of $LaAlO_3$:0.5%Mn^{4+},yBa^{2+} (y = 0.5%, 1%, 2%, 3%, and 5%) phosphors after 5 min of irradiation with a Xenon arc lamp, Figure S9: Persistent luminescence decay curves of LaAlO$_3$:0.5%Mn^{4+},yCl$^-$ (y = 0.5%, 1%, 2%, 3%, and 5%) phosphors after 5 min of irradiation with a Xenon arc lamp, Figure S10: Persistent luminescence decay curves of LaAlO$_3$:0.5%Mn^{4+},yNa$^+$ (y = 0.5%, 1%, 2%, 3%, and 5%) phosphors after 5 min of irradiation with a Xenon arc lamp, Figure S11: Photoluminescence (PL) spectra of LaAlO$_3$:0.5%Mn^{4+},0.5%Ge^{4+}, LaAlO$_3$:0.5%Mn^{4+},1%Ge^{4+}, LaAlO$_3$:0.5%Mn^{4+},2%Ge^{4+}, LaAlO$_3$:0.5%Mn^{4+},3%Ge^{4+} and LaAlO$_3$:0.5%Mn^{4+},5%Ge^{4+} phosphors, Figure S12: Photoluminescence (PL) spectra of LaAlO$_3$:0.5%Mn^{4+},0.5%Ba^{2+}, LaAlO$_3$:0.5%Mn^{4+},1%Ba^{2+}, LaAlO$_3$:0.5%Mn^{4+},2%Ba^{2+}, LaAlO$_3$:0.5%Mn^{4+},3%Ba^{2+} and LaAlO$_3$:0.5%Mn^{4+},5%Ba^{2+} phosphors, Figure S13: Photoluminescence (PL) spectra of LaAlO$_3$:0.5%Mn^{4+},0.5%Sr^{2+}, LaAlO$_3$:0.5%Mn^{4+},1%Sr^{2+}, LaAlO$_3$:0.5%Mn^{4+},2%Sr^{2+}, LaAlO$_3$:0.5%Mn^{4+},3%Sr^{2+} and LaAlO$_3$:0.5%Mn^{4+},5%Sr^{2+} phosphors, Figure S14: Photoluminescence (PL) spectra of LaAlO$_3$:0.5%Mn^{4+},0.5%Ca^{2+}, LaAlO$_3$:0.5%Mn^{4+},1%Ca^{2+}, LaAlO$_3$:0.5%Mn^{4+},2%Ca^{2+}, LaAlO$_3$:0.5%Mn^{4+},3%Ca^{2+} and LaAlO$_3$:0.5%Mn^{4+},5%Ca^{2+} phosphors, Figure S15: Photoluminescence (PL) spectra of LaAlO$_3$:0.5%Mn^{4+},0.5%Na$^+$, LaAlO$_3$:0.5%Mn^{4+},1%Na$^+$, LaAlO$_3$:0.5%Mn^{4+},2%Na$^+$, LaAlO$_3$:0.5%Mn^{4+},3%Na$^+$ and LaAlO$_3$:0.5%Mn^{4+},5%Na$^+$ phosphors, Figure S16: Photoluminescence (PL) spectra of LaAlO$_3$:0.5%Mn^{4+},0.5%Cl$^-$, LaAlO$_3$:0.5%Mn^{4+},1%Cl$^-$, LaAlO$_3$:0.5%Mn^{4+},2%Cl$^-$, LaAlO$_3$:0.5%Mn^{4+},3%Cl$^-$ and LaAlO$_3$:0.5%Mn^{4+},5%Cl$^-$ phosphors, Table S1: Ionic radius of some common dopant cations for the substitution on Al^{3+} site, Table S2: Ionic radius of some common dopant cations for the substitution on La^{3+} site.

Acknowledgments: Jiaren Du gratefully acknowledges the China Scholarship Council (Grant number 201606170077). Olivier Q. De Clercq acknowledges the financial support of the Ghent University's Special Research Fund (BOF).

Author Contributions: Dirk Poelman and Jiaren Du conceived and designed the experiments; Jiaren Du performed the experiments; Dirk Poelman, Jiaren Du, Olivier Q. De Clercq and Katleen Korthout the authors analyzed the data and wrote the paper.

Conflicts of Interest: The authors declare no conflict of interest.

References

1. Matsuzawa, T.; Aoki, Y.; Takeuchi, N.; Murayama, Y. A new long phosphorescent phosphor with high brightness, SrAl$_2$O$_4$:Eu^{2+}, Dy^{3+}. *J. Electrochem. Soc.* **1996**, *143*, 2670–2673. [CrossRef]

2. Aitasalo, T.; Dereń, P.; Hölsä, J.; Jungner, H.; Krupa, J.-C.; Lastusaari, M.; Legendziewicz, J.; Niittykoski, J.; Stręk, W. Persistent luminescence phenomena in materials doped with rare earth ions. *J. Solid State Chem.* **2003**, *171*, 114–122. [CrossRef]

3. Smet, P.F.; Van den Eeckhout, K.; De Clercq, O.Q.; Poelman, D. Persistent phosphors. In *Handbook on the Physics and Chemistry of Rare Earths*; Elsevier: Amsterdam, The Netherlands, 2015; Volume 48, pp. 1–108.

4. Viana, B.; Sharma, S.; Gourier, D.; Maldiney, T.; Teston, E.; Scherman, D.; Richard, C. Long term in vivo imaging with Cr^{3+} doped spinel nanoparticles exhibiting persistent luminescence. *J. Lumin.* **2016**, *170*, 879–887. [CrossRef]

5. Li, Y.; Gecevicius, M.; Qiu, J. Long persistent phosphors- from fundamentals to applications. *Chem. Soc. Rev.* **2016**, *45*, 2090–2136. [CrossRef] [PubMed]

6. Yamamoto, H.; Matsuzawa, T. Mechanism of long phosphorescence of SrAl$_2$O$_4$:Eu^{2+}, Dy^{3+} and CaAl$_2$O$_4$:Eu^{2+}, Nd^{3+}. *J. Lumin.* **1997**, *72*, 287–289. [CrossRef]

7. De Chermont, Q.L.M.; Chanéac, C.; Seguin, J.; Pellé, F.; Maîtrejean, S.; Jolivet, J.-P.; Gourier, D.; Bessodes, M.; Scherman, D. Nanoprobes with near-infrared persistent luminescence for in vivo imaging. *Proc. Natl. Acad. Sci. USA* **2007**, *104*, 9266–9271. [CrossRef] [PubMed]

8. Smith, A.M.; Mancini, M.C.; Nie, S. Bioimaging: Second window for in vivo imaging. *Nat. Nanotechnol.* **2009**, *4*, 710–711. [CrossRef] [PubMed]

9. Duan, H.; Nie, S. Cell-penetrating quantum dots based on multivalent and endosome-disrupting surface coatings. *J. Am. Chem. Soc.* **2007**, *129*, 3333–3338. [CrossRef] [PubMed]

10. Chen, G.; Ohulchanskyy, T.Y.; Kumar, R.; Ågren, H.; Prasad, P.N. Ultrasmall monodisperse NaYF$_4$:Yb^{3+}/ Tm^{3+} nanocrystals with enhanced near-infrared to near-infrared upconversion photoluminescence. *ACS Nano* **2010**, *4*, 3163–3168. [CrossRef] [PubMed]

11. Singh, S. Red and near infrared persistent luminescence nano-probes for bioimaging and targeting applications. *RSC Adv.* **2014**, *4*, 58674–58698. [CrossRef]

12. Maldiney, T.; Viana, B.; Bessière, A.; Gourier, D.; Bessodes, M.; Scherman, D.; Richard, C. In vivo imaging with persistent luminescence silicate-based nanoparticles. *Opt. Mater.* **2013**, *35*, 1852–1858. [CrossRef]

13. Bessière, A.; Jacquart, S.; Priolkar, K.; Lecointre, A.; Viana, B.; Gourier, D. ZnGa$_2$O$_4$:Cr^{3+}: A new red long-lasting phosphor with high brightness. *Opt. Express* **2011**, *19*, 10131–10137. [CrossRef] [PubMed]

14. Pan, Z.; Lu, Y.-Y.; Liu, F. Sunlight-activated long-persistent luminescence in the near-infrared from Cr^{3+}-doped zinc gallogermanates. *Nat. Mater.* **2012**, *11*, 58. [CrossRef] [PubMed]

15. Li, Y.; Zhou, S.; Li, Y.; Sharafudeen, K.; Ma, Z.; Dong, G.; Peng, M.; Qiu, J. Long persistent and photo-stimulated luminescence in Cr^{3+}-doped Zn–Ga–Sn–O phosphors for deep and reproducible tissue imaging. *J. Mater. Chem. C* **2014**, *2*, 2657–2663. [CrossRef]

16. Liu, F.; Yan, W.; Chuang, Y.-J.; Zhen, Z.; Xie, J.; Pan, Z. Photostimulated near-infrared persistent luminescence as a new optical read-out from Cr^{3+}-doped LiGa$_5$O$_8$. *Sci. Rep.* **2013**, *3*, 1554. [CrossRef] [PubMed]

17. De Clercq, O.Q.; Martin, L.I.; Korthout, K.; Kusakovskij, J.; Vrielinck, H.; Poelman, D. Probing the local structure of the near-infrared emitting persistent phosphor LiGa$_5$O$_8$:Cr^{3+}. *J. Mater. Chem. C* **2017**, *5*, 10861. [CrossRef]

18. De Clercq, O.Q.; Poelman, D. Local, temperature-dependent trapping and detrapping in the LiGa$_5$O$_8$:Cr infrared emitting persistent phosphor. *ECS J. Solid State Sci. Technol.* **2017**, *7*, R3171–R3175. [CrossRef]

19. Li, Y.; Li, Y.-Y.; Sharafudeen, K.; Dong, G.-P.; Zhou, S.-F.; Ma, Z.-J.; Peng, M.-Y.; Qiu, J.-R. A strategy for developing near infrared long-persistent phosphors: Taking MAlO$_3$:Mn^{4+}, Ge^{4+} (M = La, Gd) as an example. *J. Mater. Chem. C* **2014**, *2*, 2019–2027. [CrossRef]

20. Sijbom, H.F.; Joos, J.J.; Martin, L.I.; Van den Eeckhout, K.; Poelman, D.; Smet, P.F. Luminescent behavior of the K$_2$SiF$_6$:Mn^{4+} red phosphor at high fluxes and at the microscopic level. *ECS J. Solid State Sci. Technol.* **2016**, *5*, R3040–R3048. [CrossRef]

21. Zhu, H.; Lin, C.C.; Luo, W.; Shu, S.; Liu, Z.; Liu, Y.; Kong, J.; Ma, E.; Cao, Y.; Liu, R.-S. Highly efficient non-rare-earth red emitting phosphor for warm white light-emitting diodes. *Nat. Commun.* **2014**, *5*, 4312. [CrossRef] [PubMed]

22. Liu, J.-M.; Liu, Y.-Y.; Zhang, D.-D.; Fang, G.-Z.; Wang, S. Synthesis of GdAlO$_3$:Mn^{4+}, Ge^{4+}@Au core-shell nanoprobes with plasmon-enhanced near-infrared persistent luminescence for in vivo trimodality bioimaging. *ACS Appl. Mater. Interfaces* **2016**, *8*, 29939–29949. [CrossRef] [PubMed]

23. Sijbom, H.F.; Verstraete, R.; Joos, J.J.; Poelman, D.; Smet, P.F. K$_2$SiF$_6$:Mn^{4+} as a red phosphor for displays and warm-white LEDs: A review of properties and perspectives. *Opt. Mater. Express* **2017**, *7*, 3332–3365. [CrossRef]

24. Katayama, Y.; Kobayashi, H.; Tanabe, S. Deep-red persistent luminescence in Cr^{3+}-doped LaAlO$_3$ perovskite phosphor for in vivo imaging. *Appl. Phys. Express* **2014**, *8*, 012102. [CrossRef]

25. Xu, J.; Murata, D.; Katayama, Y.; Ueda, J.; Tanabe, S. Cr^{3+}/Er^{3+} co-doped LaAlO$_3$ perovskite phosphor: A near-infrared persistent luminescence probe covering the first and third biological windows. *J. Mater. Chem. B* **2017**, *5*, 6385–6393. [CrossRef]

26. Hayward, S.; Morrison, F.; Redfern, S.; Salje, E.; Scott, J.; Knight, K.; Tarantino, S.; Glazer, A.; Shuvaeva, V.; Daniel, P. Transformation processes in LaAlO$_3$: Neutron diffraction, dielectric, thermal, optical and Raman studies. *Phys. Rev. B* **2005**, *72*, 054110. [CrossRef]

27. Sanz-Ortiz, M.N.; Rodríguez, F.; Rodríguez, J.; Demazeau, G. Optical and magnetic characterisation of Co^{3+} and Ni^{3+} in LaAlO$_3$: Interplay between the spin state and Jahn—Teller effect. *J. Phys. Condens. Matter* **2011**, *23*, 415501. [CrossRef] [PubMed]

28. Srivastava, A.; Brik, M. Crystal field studies of the Mn^{4+} energy levels in the perovskite, LaAlO$_3$. *Opt. Mater.* **2013**, *35*, 1544–1548. [CrossRef]

29. Faucher, M.; Caro, P. Optical study of LaAlO$_3$:Eu at temperatures approaching the rhombohedric→cubic transition. *J. Chem. Phys.* **1975**, *63*, 446–454. [CrossRef]

30. Wen-Chen, Z. Theoretical studies of the electron paramagnetic resonance and optical spectra of Cr^{3+} ions in the rhombohedral phase of a LaAlO$_3$ crystal. *J. Phys. Condens. Matter* **1995**, *7*, 4499. [CrossRef]

31. Shannon, R.D. Revised effective ionic radii and systematic studies of interatomic distances in halides and chalcogenides. *Acta Crystallogr. A* **1976**, *32*, 751–767. [CrossRef]

32. Cao, R.; Ceng, D.; Liu, P.; Yu, X.; Guo, S.; Zheng, G. Synthesis and photoluminescence properties of LaAlO$_3$:Mn^{4+}, Na$^+$ deep red-emitting phosphor. *Appl. Phys. A* **2016**, *122*, 299. [CrossRef]

33. Van Ipenburg, M.; Dirksen, G.; Blasse, G. Charge-transfer excitation of transition-metal-ion luminescence. *Mater. Chem. Phys.* **1995**, *39*, 236–238. [CrossRef]

34. Brik, M.; Srivastava, A. On the optical properties of the Mn^{4+} ion in solids. *J. Lumin.* **2013**, *133*, 69–72. [CrossRef]

35. Brik, M.; Camardello, S.; Srivastava, A. Influence of covalency on the $Mn^{4+}\,{}^2E_g{\rightarrow}^4A_{2g}$ emission energy in crystals. *ECS J. Solid State Sci. Technol.* **2015**, *4*, R39–R43. [CrossRef]

36. Du, M.-H. Chemical trends of Mn^{4+} emission in solids. *J. Mater. Chem. C* **2014**, *2*, 2475–2481. [CrossRef]

37. Cao, R.; Zhang, F.; Cao, C.; Yu, X.; Liang, A.; Guo, S.; Xue, H. Synthesis and luminescence properties of $CaAl_2O_4$:Mn^{4+} phosphor. *Opt. Mater.* **2014**, *38*, 53–56. [CrossRef]

38. Seki, K.; Uematsu, K.; Toda, K.; Sato, M. Novel deep red emitting phosphors $Ca_{14}Zn_6M_{10}O_{35}$:Mn^{4+} (M = Al^{3+} and Ga^{3+}). *Chem. Lett.* **2014**, *43*, 1213–1215. [CrossRef]

39. Tanabe, Y.; Sugano, S. On the absorption spectra of complex ions II. *J. Phys. Soc. Jpn.* **1954**, *9*, 766–779. [CrossRef]

40. Srivastava, A.M.; Brik, M.G.; Camardello, S.J.; Comanzo, H.A.; Garcia-Santamaria, F. Optical spectroscopy and crystal field studies of the Mn^{4+} ion ($3d^3$) in the double perovskite $NaLaMgTeO_6$. *Z. Naturforschung B* **2014**, *69*, 141–149. [CrossRef]

41. Zhou, Z.; Zhou, N.; Xia, M.; Yokoyama, M.; Hintzen, H.B. Research progress and application prospects of transition metal Mn^{4+}-activated luminescent materials. *J. Mater. Chem. C* **2016**, *4*, 9143–9161. [CrossRef]

42. Pauling, L. The principles determining the structure of complex ionic crystals. *J. Am. Chem. Soc.* **1929**, *51*, 1010–1026. [CrossRef]

43. Shu, W.; Jiang, L.; Xiao, S.; Yang, X.; Ding, J. GeO_2 dopant induced enhancement of red emission in $CaAl_{12}O_{19}$:Mn^{4+} phosphor. *Mater. Sci. Eng. B* **2012**, *177*, 274–277. [CrossRef]

44. Xu, Y.; Zhang, Y.; Wang, L.; Shi, M.; Liu, L.; Chen, Y. Red emission enhancement for $CaAl_{12}O_{19}$:Cr^{3+} and $CaAl_{12}O_{19}$:Mn^{4+} phosphors. *J. Mater. Sci. Mater. Electron.* **2017**, *28*, 12032–12038. [CrossRef]

45. Meng, L.; Liang, L.; Wen, Y. Deep red phosphors $SrMgAl_{10}O_{17}$:Mn^{4+}, M (M = Li^+, Na^+, K^+, Cl^-) for warm white light emitting diodes. *J. Mater. Sci. Mater. Electron.* **2014**, *25*, 2676–2681. [CrossRef]

46. Pan, Y.; Liu, G. Enhancement of phosphor efficiency via composition modification. *Opt. Lett.* **2008**, *33*, 1816–1818. [CrossRef] [PubMed]

materials

MDPI

Article

Highly Efficient Low-Temperature N-Doped TiO$_2$ Catalysts for Visible Light Photocatalytic Applications

Julien G. Mahy [1,*], Vincent Cerfontaine [1], Dirk Poelman [2], François Devred [3], Eric M. Gaigneaux [3], Benoît Heinrichs [1] and Stéphanie D. Lambert [1]

[1] Department of Chemical Engineering—Nanomaterials, Catalysis & Electrochemistry, University of Liège, B6a, Quartier Agora, Allée du six Août 11, 4000 Liège, Belgium; vincent_cerfontaine@hotmail.com (V.C.); b.heinrichs@uliege.be (B.H.); stephanie.lambert@uliege.be (S.D.L.)
[2] LumiLab, Department of Solid State Sciences, Ghent University, 9000 Gent, Belgium; Dirk.Poelman@UGent.be
[3] Institute of Condensed Matter and Nanosciences—MOlecules, Solids and ReactiviTy (IMCN/MOST), Université catholique de Louvain, Place Louis Pasteur 1, Box L4.01.09, 1348 Louvain-La-Neuve, Belgium; francois.devred@uclouvain.be (F.D.); eric.gaigneaux@uclouvain.be (E.M.G.)
* Correspondence: julien.mahy@uliege.be; Tel.: +32-4-366-35-63

Received: 22 February 2018; Accepted: 6 April 2018; Published: 10 April 2018

Abstract: In this paper, TiO$_2$ prepared with an aqueous sol-gel synthesis by peptization process is doped with nitrogen precursor to extend its activity towards the visible region. Three N-precursors are used: urea, ethylenediamine and triethylamine. Different molar N/Ti ratios are tested and the synthesis is adapted for each dopant. For urea- and trimethylamine-doped samples, anatase-brookite TiO$_2$ nanoparticles of 6–8 nm are formed, with a specific surface area between 200 and 275 m$^2 \cdot$g^{-1}. In ethylenediamine-doped samples, the formation of rutile phase is observed, and TiO$_2$ nanoparticles of 6–8 nm with a specific surface area between 185 and 240 m$^2 \cdot$g^{-1} are obtained. X-ray photoelectron spectroscopy (XPS) and diffuse reflectance measurements show the incorporation of nitrogen in TiO$_2$ materials through Ti–O–N bonds allowing light absorption in the visible region. Photocatalytic tests on the remediation of water polluted with *p*-nitrophenol show a marked improvement for all doped catalysts under visible light. The optimum doping, taking into account cost, activity and ease of synthesis, is up-scaled to a volume of 5 L and compared to commercial Degussa P25 material. This up-scaled sample shows similar properties compared to the lab-scale sample, i.e., a photoactivity 4 times higher than commercial P25.

Keywords: N-doped TiO$_2$; aqueous sol-gel process; photocatalysis; *p*-nitrophenol degradation; multiple crystalline phase catalyst

1. Introduction

Nowadays, environmental issues are a main problem in our lives, and a lot of research is conducted to reduce them. Pollutants can be very diverse molecules, occurring as aromatic compounds, pesticides, chlorinated compounds, heavy metals or petroleum hydrocarbons [1]. Different methods exist to reduce them, such as photocatalysis, which is an efficient process for degrading organic pollution [2].

This technique consists of redox reactions between organic pollutants and radical species produced by the illumination of a photocatalyst [3,4].

The most commonly used photocatalyst is TiO$_2$ [1,5], which is a semiconductor sensitive to UV radiation; it is non-toxic and cheap [6]. The photon energy required to activate the TiO$_2$ is 3.2 eV (band gap width); this value corresponds to radiation with a wavelength lower than 388 nm (for anatase phase). However, in the case of illumination by solar light, only 5–8% of the spectrum will be used for activation (only the UV part) [7]. When energetic light illuminates TiO$_2$, electrons (e-) are promoted from the valence band to the conduction band, leading to the formation of positive holes (h+) in the valence band. When these photoactive species reach the surface of the material, they react with water and oxygen to produce radicals, such as superoxide and hydroxyl radicals, and then these radicals are able to degrade adsorbed organic molecules [4,5,8,9].

In recent years, sol-gel synthesis has been used widely for the preparation of TiO$_2$ films or powders allowing a control of the nanostructure and surface properties [6,10–13]. This process is a soft chemistry method, using low temperature and low pressure, where titanium precursors sustain two main reactions: hydrolysis and condensation [14]. By adapting the rate of these reactions, a colloidal suspension or a solid gel can be obtained.

Sol-gel TiO$_2$ materials can be synthesized via several paths depending on the solvent used, which can include organic solvents or water. Usually, when the sol-gel synthesis is conducted in an organic solvent, it is for the purpose of controlling its reactivity by complexation of titanium alkoxide (Ti-(OR)$_4$), where R is an alkyl group (–CH$_3$, –C$_2$H$_5$, . . .). Only a small amount of water is added to avoid precipitation [10,15,16]. Then the material sustains drying and calcination to remove residual organic molecules and to crystallize amorphous TiO$_2$ into anatase or rutile phase [17].

When water is used as solvent, peptizing agents are used to form small crystalline TiO$_2$ nanoparticles at low temperature [7,18,19]. With this method, crystalline TiO$_2$ materials are synthesized at low temperature without organic solvent. This process is thus well adapted for industrialization [8,9].

Indeed, in previous works [8,9], a global process has been developed to produce pure TiO$_2$ films deposited on steel at large scale in order to obtain an easy-to-clean surface [8] and doped TiO$_2$ with enhanced photocatalytic properties [9]. In these works, pure and doped TiO$_2$ aqueous syntheses, using titanium tetraisopropoxide in acidified water, were simplified from [19] to facilitate extrapolation towards an industrial scale, and then pure and doped TiO$_2$ sols were scaled up to volumes of 5 L [8,9].

As mentioned above, TiO$_2$ needs energetic light to be photoactive, and so, when only the solar spectrum is available for an application, pure TiO$_2$ is not very effective. In order to improve the photocatalytic properties of TiO$_2$ in the visible range, three main routes can be followed: (i) the introduction of metallic nanoparticles [20] or ions [10,21,22], (ii) the introduction of non-metallic elements like N [23–28], B [25,29], F [24,25], P [30,31] or S [24,27], and (iii) doping with dye photosensitizers such as porphyrins [32,33].

With regard to the introduction of non-metallic elements, N-doping is considered to be an ideal candidate, because N 2*p* states could effectively mix with O 2*p* states [34]. Indeed, nitrogen can be easily introduced into the TiO$_2$ structure, due to its atomic size, which is comparable with that of oxygen, its low ionization energy and high stability [24]. The visible absorption of the N-doped samples is due to the reduction of the band gap. Indeed, substitutional N atoms in TiO$_2$ lattice narrow the band gap by creating an energy level above the valence band maximum or creating an intermediate band for the electron below the conduction band [24,26,34].

The aim of the present work is to N-dope the previously developed synthesis [8,9]. This synthesis makes it possible to produce a crystalline TiO$_2$ colloid at low temperature in aqueous media. This material can be produced at large scale, showing a high photocatalytic efficiency, but only under UV radiation. Synthesis at pilot scale for N-doped TiO$_2$ catalyst by sol-gel process is not reported in the literature. In this work, three N-precursors are used to extend the photoactivity towards visible light: urea, ethylenediamine and triethylamine. The resulting materials are characterized by X-ray diffraction (XRD), transmission electron microscopy (TEM), nitrogen adsorption–desorption measurements, XPS and diffuse reflectance spectroscopy measurements in order to study the crystalline phase formation, the doping influence and the texture of the catalysts. In the second part of this study, the photocatalytic activity of the samples is tested in the degradation of *p*-nitrophenol (PNP, C$_6$H$_5$NO$_3$) in polluted water under visible light to show the influence of N-doping on the photoactivity and identify the best N-precursor for low-temperature N-doping of TiO$_2$. For large-scale application, a large volume

(5 L) synthesis of the best N-doped TiO_2 is performed and compared to laboratory-scale synthesis. Although TiO_2 N-doping has been widely studied, the novelty of this research is, from an easy and environmentally friendly synthesis of photoactive TiO_2 colloid, to try to N-dope this material in an easy way without heating and to try to up-scale this synthesis to a larger volume for further visible applications. A comparison with commercial Degussa P25 will be made.

2. Materials and Methods

2.1. Samples Preparation

2.1.1. Pure TiO_2 Powder Synthesis

Titanium (IV) tetraisopropoxide (TTIP > 97%, Sigma-Aldrich, Saint Louis, MO, USA), nitric acid (HNO_3, 65%, Merck, Kenneth Fort Worth, NJ, USA), isopropanol (IsoP, 99.5%, Acros, Hull, Belgium) and distilled water are used as starting materials. Pure TiO_2 is synthesized according to Mahy et al. [8] using water as solvent and nitric acid for TTIP peptization. After a reaction time of 4 h, a light blue transparent liquid sol is obtained. Then, the sol is dried under an ambient air flow to obtain a xerogel, which is crushed in a white-yellow powder [8]. Pure TiO_2 sample is denoted as "TiO_2 pure A".

An alternative synthesis of pure TiO_2 is also tested. In this case, a mixture of TTIP and isopropanol is precipitated in water. Then this precipitate is washed three times with water. This slurry is then mixed in water with HNO_3 (pH = 1) at 50 °C for 12 h. After this reaction time, a light blue transparent liquid sol is obtained. Then, the sol is dried under an ambient air flow to obtain a xerogel, which is crushed in a white-yellow powder [8]. This pure TiO_2 sample is denoted as "TiO_2 pure B".

2.1.2. N-Doped TiO_2 Powder Synthesis

TiO_2 catalyst is doped by nitrogen with three different precursors: urea ((NH_2)$_2$CO, Sigma-Aldrich, 98%), ethylenediamine ($NH_2CH_2CH_2NH_2$, Sigma-Aldrich, puriss. p.a., absolute, ≥99.5% (GC)), and triethylamine ((C_2H_5)$_3$N, Sigma-Aldrich, ≥99%).

For urea doping, distilled water is mixed with urea then acidified by HNO_3 to a pH equal to 1. Then, TTIP is added to IsoP, and the mixture is stirred at room temperature for 30 min. The TTIP/IsoP mixture is added to acidified urea/water solution under vigorous stirring. The solution stays under stirring for 4 h at 80 °C. After this time, a light white-blue transparent liquid sol is obtained and kept in ambient atmosphere. The molar ratio between urea and TTIP is equal to either 1, 2, 4 or 10 leading to four urea-doped samples. The samples are called "TiO_2/UX", with X corresponding to the molar ratio between urea and TTIP.

For ethylenediamine doping, TTIP/IsoP solution is precipitated in water containing ethylenediamine. The precipitate is washed three times with water. Then it is mixed in water with HNO_3 (pH = 1) at 50 °C for 12 h. After reaction, a light blue transparent liquid sol is obtained. The molar ratio between ethylenediamine and TTIP is equal to either 1, 2, 4 or 10 leading to four ethylenediamine-doped samples. The samples are called "TiO_2/EtDNX", with X corresponding to the molar ratio between ethylenediamine and TTIP.

For triethylamine doping, a pure TiO_2 sol is synthesized like TiO_2 pure B sample. Then this colloid is mixed with an excess of triethylamine for 12 h. A yellowish opaque liquid sol is obtained. The molar ratio between triethylamine and TTIP is equal to 28 or 42, leading to two trimethylamine-doped samples. The samples are called "TiO_2/Et$_3$NX", with X corresponding to the molar ratio between triethylamine and TTIP.

All N-doped samples are dried under ambient air at room temperature, then dried at 100 °C for 1 h. The resulting powders are washed three times with water, then finally dried at 100 °C for 12 h.

2.1.3. Urea-Doped TiO$_2$ Powder Synthesis at a Large Scale

The synthesis of the TiO$_2$/U2 sample is scaled up to a volume of 5 L in a glass batch reactor with a water recirculation cooling system (jacketed reactor) [9]. 3.6 L of a solution urea/distilled water is acidified with HNO$_3$ to a pH equal to 1. Then, 480 g of TTIP is added to 168.8 g of isopropanol (IsoP). The mixture is stirred at room temperature for 30 min. The TTIP-IsoP mixture is added to the acidified urea water and stirred by a propeller at 300 rpm. The liquid stays under stirring for 4 h at a temperature of 80 °C. After the reaction, a light white-blue transparent liquid sol is obtained, similar to the laboratory-scale sol. As for the laboratory-scale TiO$_2$ sample, the large-scale urea-doped TiO$_2$ catalyst is dried under ambient air at room temperature, then dried at 100 °C for 1 h. The resulting powder is washed three times with water, then finally dried at 100 °C for 12 h.

The large-scale urea-doped TiO$_2$ sample is denoted as "TiO$_2$/U2-LS".

2.2. Sample Characterization

The crystallographic properties are studied through the X-Ray Diffraction (XRD) patterns recorded with a Bruker D8 Twin-Twin powder diffractometer using Cu-K$_\alpha$ radiation. The Scherrer formula (Equtaion (1)) is used to determine the size of the TiO$_2$ crystallites, d_{XRD} [35]:

$$d_{XRD} = 0.9 \frac{\lambda}{B \cos(\theta)} \tag{1}$$

where d_{XRD} is the crystallite size (nm), B the peak full-width at half maximum after correction of the instrumental broadening (rad), λ the X-ray wavelength (0.154 nm), and θ the Bragg angle (rad).

The repartition of the crystallographic phases is estimated with the Rietveld method using "Profex" software (Profex 3.12.1, Nicola Döbelin, Solothurn, Switzerland) [36]. The amount of crystalline phase is estimated with CaF$_2$ internal standard (calcium fluoride, Sigma-Aldrich, anhydrous powder, 99.99% trace metal basis), also using "Profex" software [37].

The TiO$_2$ textural properties are characterized by nitrogen adsorption-desorption isotherms in an ASAP 2420 multi-sampler adsorption-desorption volumetric device from Micromeritics. From these isotherms, the microporous volume is calculated from the Dubinin-Radushkevich theory (V_{DR}). The surface area is evaluated using the Brunauer, Emmett and Teller theory (S_{BET}) [38]. An average particle size, d_{BET}, can be calculated from S_{BET} values by assuming non-porous TiO$_2$ anatase nanoparticles using the following formula [7]:

$$\frac{d_{BET}}{6} = \frac{\frac{1}{\rho_{anatase}}}{S_{BET}} \tag{2}$$

where $\rho_{anatase}$ is the apparent density of TiO$_2$-anatase considered equal to 3.89×10^6 g·m^{-3} [8,9].

The sizes of TiO$_2$ nanoparticles are estimated by transmission electron microscopy (TEM) by averaging the measurement of approximately 100 particles on TEM micrographs obtained with a Phillips CM 100 device (accelerating voltage 200 Kv, Amsterdam, The Netherlands). First, samples are dispersed in distilled water using an ultrasonic treatment. Then a drop of the dispersion is placed on a copper grid (Formvar/Carbon 200 Mesh Cu from Agar Scientific, Essex, UK).

The sample's optical properties are evaluated by using diffuse reflectance spectroscopy measurements in the region 300–800 nm with a Varian Cary 500 UV–Vis-NIR spectrophotometer, equipped with an integrating sphere (Varian External DRA-2500, Palo Alto, CA, USA) and using BaSO$_4$ as reference. The UV–Vis spectra recorded in diffuse reflectance (R_{sample}) mode are transformed by using the Kubelka–Munk function:

$$F(R_\infty) = \frac{(1 - R_\infty)^2}{2 R_\infty} \tag{3}$$

where R_∞ is defined as $R_\infty = R_{sample}/R_{reference}$ [8,39,40] with $R_{reference}$, the diffuse reflectance measured for the BaSO$_4$ reference. To compare them, all spectra are normalized to 1.0 by dividing each spectrum by its maximum intensity [7,41]. Using the well-known equation:

$$(F(R_\infty)h\nu)^{1/m} = C\left(h\nu - E_g\right) \tag{4}$$

where C is a constant and m is a constant that depends on the optical transition mode, the direct and indirect optical band-gap values, $E_{g,direct}$ (eV) and $E_{g,indirect}$ (eV) are obtained by plotting, respectively, $(F(R_\infty)h\nu)^2$ and $(F(R_\infty)h\nu)^{1/2}$ as functions of the photon energy $h\nu$ and by determining the intersection of the linear part of the curve and the *x*-axis [8,42].

X-ray photoelectron spectra are obtained with a SSI-X-probe (SSX-100/206) spectrometer equipped with a monochromatized microfocused Al X-ray source (1486.6 eV), operating at 10 kV and 20 mA. Samples are placed in the analysis chamber, where the residual pressure was of about 10^{-6} Pa. The charging effect is adjusted using flood gun energy at 8 eV and a fine-meshed nickel grid placed 3 mm above the sample surface [43]. The pass energy is 150 eV, and the spot size is 1.4 mm^2. Angle between the normal to the sample surface and the direction of electron collection is 55°. Under these conditions, the mid-height width (FWHM) of the Au $4f_{7/2}$ peak photo-peak measured on a standard sample of cleaned gold is about 1.6 eV. The following sequence of spectra is recorded: general spectrum, C 1*s*, O 1*s*, N 1*s* and Ti 2*p* and again C 1*s* to check the stability of charge compensation with time and absence of degradation of the samples.

The C–(C,H) component of the carbon C 1*s* peak is fixed at 284.8 eV to calibrate the scale in binding energy. Two other components of the carbon peak (C–(O,N), C=O or O–C–O) are resolved, notably to determine the amount of oxygen due to contamination. The O–C=O component that would show up at slightly higher binding energy is not observed in our samples. Data processing is carried out with the CasaXPS program (Casa Software Ltd., Teignmouth, UK). Some spectra are decomposed using the Gaussian and Lorentzian function product model (least squares fitting) after subtraction of a nonlinear baseline [44]. The molar fractions are calculated using the normalized peak areas based on acquisition parameters and sensitivity factors supplied by the manufacturer.

2.3. Photocatalytic Tests

The photocatalytic activity of the samples in the form of powders is evaluated by following the degradation of *p*-nitrophenol (PNP) after 0, 8 and 24 h, in triplicate, in water medium. For each test, the degradation percentage of PNP, D_{PNPi}, is given by Equation (5) [9]:

$$D_{PNPi}(\%) = \left(1 - \frac{[PNP]_i}{[PNP]_0}\right) \times 100 \tag{5}$$

where $[PNP]_i$ represents the residual concentration of PNP at time $t = i$ h and $[PNP]_0$ represents the initial concentration of PNP at time $t = 0$ h.

The photocatalytic activity of the samples is estimated under halogen light (UV/visible light) and under the same lamp covered with a UV-filter that removes wavelengths lower than 390 nm, this condition will be called low-energy light [9]. The experimental setup is shown in [8,45]. The test is the same as the procedure described in [8,9]. The concentration of PNP is estimated by UV/Vis spectroscopy (GENESYS 10S UV–Vis from Thermo Scientific, Waltham, MA, USA) at 318 nm. For each catalyst tested, three flasks are exposed to light to calculate the PNP degradation and one is kept in the dark (dark test) to evaluate PNP adsorption on the samples [8,9]. Additionally, a flask with only PNP without any catalyst is exposed to the light for 24 h (blank test), to show that no natural PNP concentration occurs under halogen illumination. In each flask, the initial concentration of catalyst and PNP are 1 g·L^{-1} and 10^{-4} M, respectively [8,9]; the initial pH is 4. Experiments are conducted in test tubes fitted with a sealing cap. These tubes are placed in a cylindrical glass reactor with the halogen lamp in the center. The halogen lamp has a continuous spectrum from 300 nm (or 390 nm with UV filter) to 800 nm (300 W, 220 V), measured with a Mini-Spectrometer TM-UV/vis C10082MD from Hamamatsu (Hamamatsu, Japan) [8,9]. The reactor is maintained at constant temperature (20 °C) by a cooling system that functions by recirculating water; the lamp is also cooled by a similar system [8,9]. Aluminum foil is used to cover the outer wall of the reactor to prevent any interactions with the room lighting. The volume of each flask is 10 mL, agitated by a magnetic stirrer. The PNP degradation due to photocatalysis is equal to the total PNP degradation minus the PNP adsorption estimated with the dark test [8,9].

3. Results

3.1. X-ray Diffraction (XRD) of TiO$_2$ Samples

Figure 1 represents the XRD patterns for TiO$_2$ pure B sample and ethylenediamine-doped samples.

For TiO$_2$ pure B sample, mainly anatase peaks are observed (reference pattern (A)), along with a peak corresponding to brookite phase at around 31° (reference pattern (B)). The phase distribution calculated with "Profex" [36] is given in Table 1. Indeed, the anatase phase is the main crystalline phase of this sample, corresponding to 75% of the sample, the brookite phase amounts to around 10%, and the amorphous fraction corresponds to 15%.

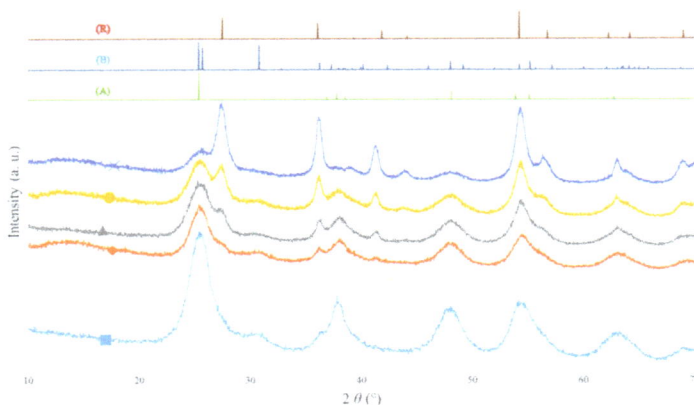

Figure 1. X-ray diffraction (XRD) patterns of ethylenediamine-doped samples: (■) TiO$_2$ pure B, (♦) TiO$_2$/EtDN1, (▲) TiO$_2$/EtDN2, (●) TiO$_2$/EtDN4 and (×) TiO$_2$/EtDN10. (A) reference pattern of anatase, (B) reference pattern of brookite and (R) reference pattern of rutile.

When the amount of ethylenediamine increases, the intensity of rutile peaks increases and that of anatase peaks decreases. This observation is confirmed by the phase distribution in Table 1; for example, for the TiO$_2$/EtDN1 sample, the repartition is 65% anatase, 10% rutile and 5% brookite, while for the TiO$_2$/EtDN10 sample, the repartition is modified, with 20% anatase, 40% rutile and 5% brookite. The TiO$_2$/EtDN10 sample has the biggest amorphous fraction, reaching 35%.

For the TiO$_2$ pure A sample, the urea-doped and trimethylamine-doped samples (not shown), the same XRD patterns are observed as for the TiO$_2$ pure B sample (mainly anatase + a small fraction of brookite phase). For these samples, the phase distributions are given in Table 1. Similar repartitions are obtained with mainly anatase (65–75%), a small amount of brookite (5–10%) and an amorphous fraction (20–30%).

The crystallite size can be estimated from XRD patterns using the Scherrer formula (Equation (1)). For all samples, similar sizes are obtained between 4 and 8 nm (Table 1). When several phases are present, the size is estimated on specific peaks corresponding to each of the different phases. If different values are obtained, it is mentioned in Table 1 (as for TiO$_2$/EtDN4 sample, where anatase crystallites are around 5 nm and rutile around 8 nm). The formation of rutile phase in ethylenediamine-doped samples seems to produce larger crystallites (Table 1).

3.2. TEM Micrographs

TEM micrographs for some samples at different magnifications are presented in Figure 2. For all samples, TiO$_2$ aggregates are observed; these aggregates are composed of TiO$_2$ nanoparticles with spherical shapes. The particles are not perfectly separated from each other because the material was first dried and then deposited on the TEM grid for measurement; additionally, the TiO$_2$ nanoparticles are not very clearly observable by TEM because of their relatively low contrast in bright-field TEM conditions [9]. TiO$_2$ nanoparticles have a size of about 5–7 nm. The TiO$_2$ average size estimated from TEM (d_{TEM}) is similar to that estimated from XRD (d_{XRD}, Table 1).

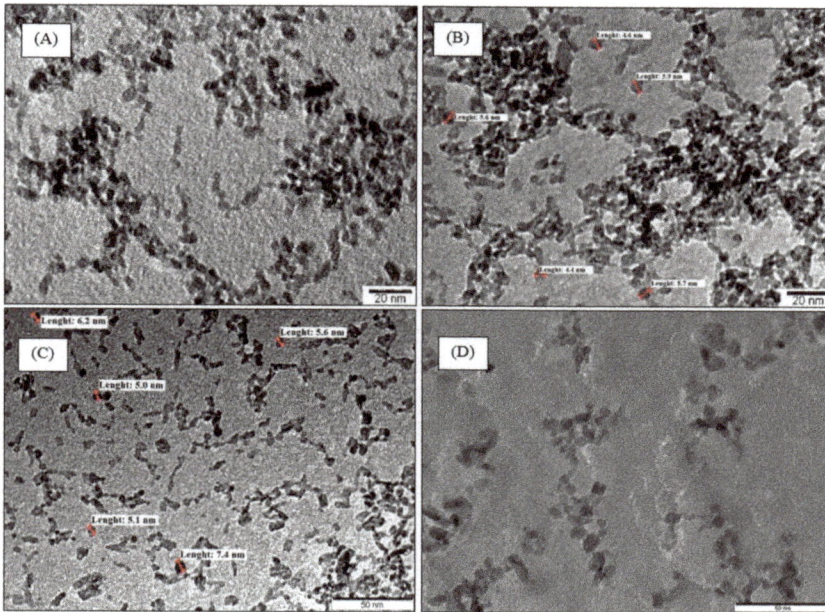

Figure 2. TEM micrographs of samples: (**A**) TiO$_2$ pur, (**B**) TiO$_2$/U2, (**C**) TiO$_2$/EtDN4 and (**D**) TiO$_2$/Et$_3$N42. Primary nanoparticles are highlighted by red lines.

Table 1. Textural and optical properties of TiO$_2$-based photocatalysts.

Sample	Phase Distribution	d_{XRD}	S_{BET}	V_{DR}	d_{BET}	d_{TEM}	$E_{g,direct}$	$E_{g,indirect}$
	(%)	(nm)	(m^{2}g^{-1})	(cm^{3}g^{-1})	(nm)	(nm)	(eV)	(eV)
	±5	±1	±5	±0.01	±1	±1	±0.01	±0.01
TiO$_2$ pure A	Am[25%] + A[65%] + B[10%]	4	205	0.11	8	5	3.35	3.06
TiO$_2$/U1	Am[30%] + A[65%] + B[5%]	5	235	0.15	7	6	3.25	2.97
TiO$_2$/U2	Am[30%] + A[65%] + B[5%]	4	255	0.16	6	6	3.24	2.97
TiO$_2$/U4	Am[30%] + A[65%] + B[5%]	4	260	0.16	6	7	3.27	3.04

Table 1. *Cont.*

Sample	Phase Distribution (%) ±5	d_{XRD} (nm) ±1	S_{BET} (m²g⁻¹) ±5	V_{DR} (cm³g⁻¹) ±0.01	d_{BET} (nm) ±1	d_{TEM} (nm) ±1	$E_{g,direct}$ (eV) ±0.01	$E_{g,indirect}$ (eV) ±0.01
TiO₂/U10	Am[25%] + A[70%] + B[5%]	4	270	0.16	6	5	3.36	3.05
TiO₂ pure B	Am[15%] + A[75%] + B[10%]	4	195	0.10	8	5	3.42	3.12
TiO₂/EtDN1	Am[20%] + A[65%] + B[5%] + R[10%]	7	225	0.13	7	6	3.30	2.94
TiO₂/EtDN2	Am[15%] + A[60%] + B[5%] + R[20%]	4	240	0.13	6	6	3.39	2.99
TiO₂/EtDN4	Am[15%] + A[45%] + B[5%] + R[35%]	5 [a]–8 [b]	195	0.11	8	7	3.30	2.94
TiO₂/EtDN10	Am[35%] + A[20%] + B[5%] + R[40%]	8	185	0.11	8	5	3.43	2.96
TiO₂/Et₃N28	Am[25%] + A[70%] + B[5%]	4	230	0.13	7	5	-[c]	-[c]
TiO₂/Et₃N42	Am[20%] + A[75%] + B[5%]	4	275	0.16	6	6	-[c]	-[c]
TiO₂/U2-LS	Am[30%] + A[65%] + B[5%]	6	245	0.15	6	7	-[d]	-[d]

Am: amorphous TiO₂ phase; A: anatase TiO₂ phase; B: Brookite TiO₂ phase; R: rutile TiO₂ phase; d_{XRD}: mean diameter of TiO₂ crystallites measured by the Scherrer method; [a] measured from anatase peak; [b] measured from rutile peak; S_{BET}: specific surface area determined by the BET method; V_{DR}: specific micropore volume determined by Dubinin–Raduskevitch theory; d_{BET}: mean diameter of TiO₂ nanoparticles calculated from S_{BET} values; d_{TEM}: mean diameter of TiO₂ nanoparticles measured by TEM; $E_{g,direct}$: direct optical band-gap values calculated using the transformed Kubelka–Munk function; $E_{g,indirect}$: indirect optical band-gap values calculated using the transformed Kubelka–Munk function; -[c] not applicable; -[d] not measured.

3.3. Sample Textural Properties

The textural properties of the samples are summarized in Table 1. Figures 3 and 4 represent the raw nitrogen adsorption-desorption isotherms for the urea-doped and ethylenediamine-doped samples, respectively, with the corresponding pure TiO₂ sample as reference.

For both pure TiO₂ samples, the isotherms are characterized by a sharp increase at low relative pressure followed by a plateau, corresponding to type I isotherms (microporous materials) according to the BDDT classification [38].

For the urea-doped samples (Figure 3), the same shape of isotherms is obtained for the 4 samples, presenting a sharp increase at low pressure, and a triangular hysteresis followed by a plateau. This type of isotherm corresponds to a mix of type I (microporous materials) and IV (mesoporous materials) isotherms according to the BDDT classification [38]. The specific surface area, S_{BET}, and the microporous volume, V_{DR}, increase with the amount of urea introduced in the synthesis (Table 1).

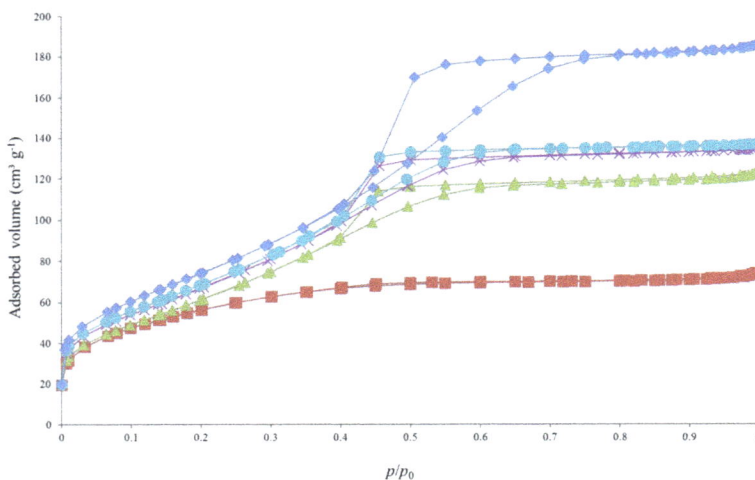

Figure 3. Nitrogen adsorption-desorption isotherms of urea doped samples: (■) TiO₂ pure A, (▲) TiO₂/U1, (×) TiO₂/U2, (●) TiO₂/U4 and (◆) TiO₂/U10.

For the ethylenediamine-doped samples (Figure 4), the same shape of isotherms is obtained for the 3 first samples, presenting a sharp increase at low pressure, and a triangular hysteresis followed by a plateau. This type of isotherm corresponds to a mix of type I (microporous materials) and IV (mesoporous materials) isotherms according to the BDDT classification [38]. The $TiO_2/EtDN10$ sample presents a different shape, with a less marked nitrogen uptake at low p/p_0 values and a hysteresis extending from $p/p_0 = 0.5$ to $p/p_0 = 1$, which corresponds to larger mesopores [46]. This type of isotherm corresponds to a mix of type I (microporous materials) and IV (mesoporous materials) isotherms according to the BDDT classification [38]. The specific surface area, S_{BET}, and the microporous volume, V_{DR}, first increases with the doping ratio then decreases (Table 1).

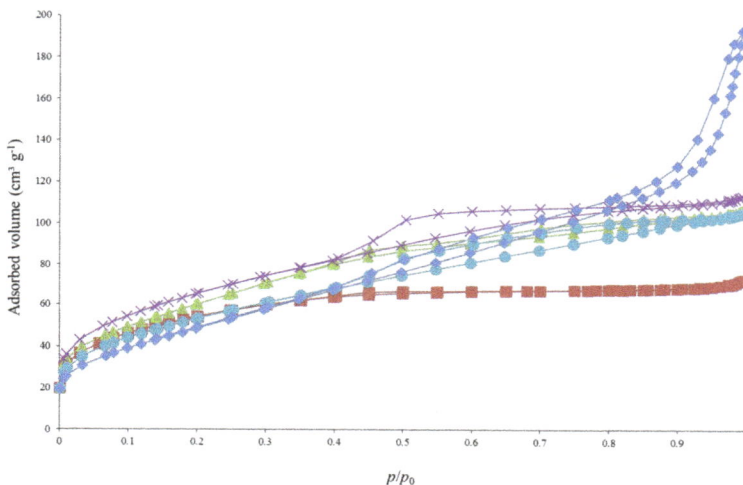

Figure 4. Nitrogen adsorption-desorption isotherms of ethylenediamine doped samples: (■) TiO_2 pure B, (▲) $TiO_2/EtDN1$, (×) $TiO_2/EtDN2$, (●) $TiO_2/EtDN4$ and (♦) $TiO_2/EtDN10$.

For the trimethylamine-doped samples, similar isotherms compared to isotherms for the urea-doped samples are obtained (as Figure 3), corresponding to a mix of type I (microporous materials) and IV (mesoporous materials) isotherms according to the BDDT classification [38]. The specific surface area, S_{BET}, and the microporous volume, V_{DR}, increase with the amount of triethylamine (Table 1).

For all samples, an estimation of TiO_2 nanoparticles sizes (d_{BET}) can be calculated from Equation (2), and in all cases, the TiO_2 sizes are between 6–8 nm (Table 1), corresponding to both the TEM and XRD estimates.

3.4. Optical Properties

The evolution of the normalized Kubelka–Munk function $F(R_\infty)$ with wavelength (λ) is presented for TiO_2 pure A, $TiO_2/U2$, TiO_2 pure B and $TiO_2/EtDN4$ samples in Figure 5. The TiO_2 pure A and B samples present absorption at 360 nm, while for both N-doped samples, the absorption is shifted towards higher wavelengths. Both samples present the largest shift in their respective series. For the other doped samples, although smaller, shifts towards the visible domain are also observed. The direct and indirect band gaps are calculated for all samples (Table 1); a decrease in the band gap value is observed for the N-doped samples compared to the corresponding pure TiO_2 samples. For the triethylamine doped samples, the absorption spectrum is not exploitable because of the dark color of the sample; therefore, the band gap values cannot be properly calculated.

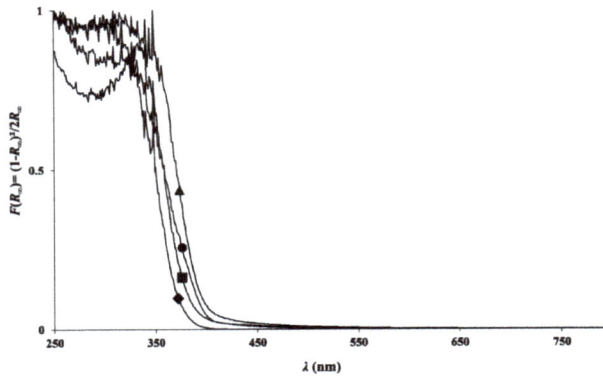

Figure 5. Normalized Kubelka–Munk function F(R∞) calculated from DR-UV–Vis spectra for samples: (■) TiO_2 pure A, (▲)TiO_2/U2 samples, (♦) TiO_2 pure B and (●)TiO_2/EtDN4 samples.

3.5. XPS Measurements

The general XPS spectrum for the TiO_2 pure B sample is presented in Figure 6. The different peaks for carbon, oxygen and titanium are indexed. It is important to note that nitrogen is present in the TiO_2 reference sample. N 1*s*, Ti 2*p*, O 1*s* and C 1*s* spectra are presented on Figure 7. For all samples, similar spectra are obtained. On the Ti 2*p* spectrum (Figure 7c), the Ti 2*p*1/2 and Ti 2*p*3/2 are observed at 464.1 and 458.5 eV, respectively, and are attributed to Ti^{4+} species, and thus to TiO_2. No differences between samples are noticeable in the Ti 2*p* area. Indeed, the shift to a slightly lower binding energy induced by a change in the titanium chemical environment (as in the case of Ti–O–N bonding) would be too small to be distinguished from the main peak [47]. On the other hand, the systematic absence of a peak around 455–456 eV excludes the presence of Ti–N species.

The C 1*s* contribution is divided into three components (Figure 7a). The C–(C,H) contribution at 248.8 eV is a classical aliphatic carbon contamination used to calibrate the measurements. In our standard routine procedure for decomposing the C 1*s* contribution, we define the contribution of carbon involved in simple bond with O or N at 1.5 eV higher (286.3 eV). The signal at a binding energy of around 289 eV is attributed to the contribution of C doubly bonded to O. The pure TiO_2 samples contain 17 to 19% (C–H) component. The other samples show comparable decomposition of the C signal. This suggests that there are no remaining organic compounds present.

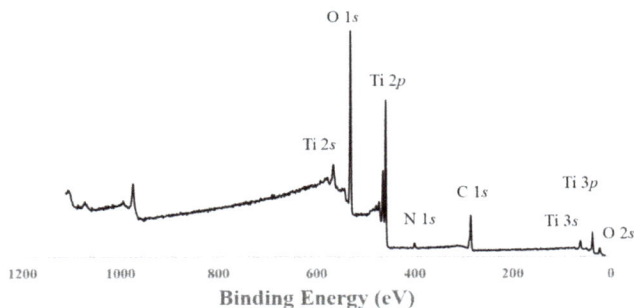

Figure 6. XPS general spectrum of TiO_2 pure B sample.

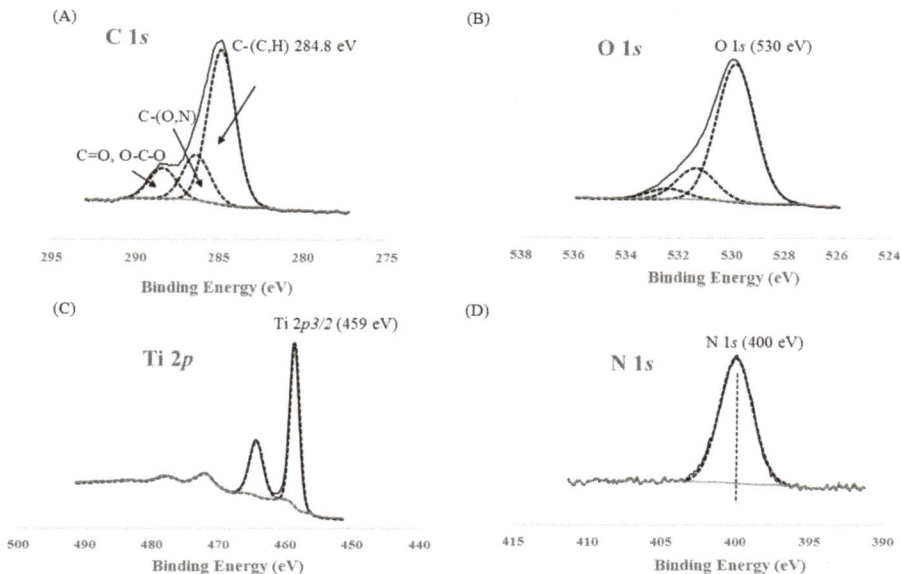

Figure 7. XPS spectra of TiO$_2$ pure B sample: (**A**) C 1s region, (**B**) O 1s region, (**C**) Ti 2p region and (**D**) N 1s region.

The O 1s contribution (Figure 7b) is decomposed into 3 peaks. The main one at 530 eV corresponds to Ti–O in TiO$_2$. For each sample, the O 1s (530 eV)/Ti 2p ratio is close to 2, corresponding to stoichiometric TiO$_2$. The two other components at higher binding energy are complicated to attribute, as they correspond to oxygen bonded to carbon due to carbon contamination and/or nitrogen bonding (N–O–Ti, N=O) [47].

For the N 1s spectrum (Figure 7d), a peak centered on 400 eV is observed. For 3 samples (TiO$_2$ pure A, TiO$_2$/EtDN2 and TiO$_2$/EtDN4), a small peak around 407 eV was also observed. This peak has been attributed to residual nitrate due to the residual nitric acid from the synthesis. This contribution was not taken into account in the latter quantification. In the literature, a N 1s peak around 400 eV may correspond to many contributions such as NH$_x$, NO$_x$, NHOH potentially due to impurities but also interstitial Ti–O–N [47,48]. Substitutional Ti–N–O can be excluded in our case, since the specific corresponding peak around 395 eV is never observed. At this stage, the complexity of the decomposition of the N 1s peak does not allow to claim the presence of Ti–O–N species only. An estimation of the amount of nitrogen present in the samples (N/Ti molar ratio) is presented in Figure 8 and Table 2 for each series in front of the corresponding pure TiO$_2$ sample. An increase in the amount of nitrogen is observed with the three precursors. For each series, a sample presents a maximum in nitrogen concentration: TiO$_2$/U2, TiO$_2$/EtDN2 and TiO$_2$/Et$_3$N28 show the three maxima for the urea, the ethylenediamine and the triethylamine doping, respectively. The highest nitrogen amount is obtained with TiO$_2$/Et$_3$N28 sample with a N/Ti molar ratio of 0.32. The results are presented in Figure 8 and Table 2, and are illustrated in Figure 9 for the triethylamine series.

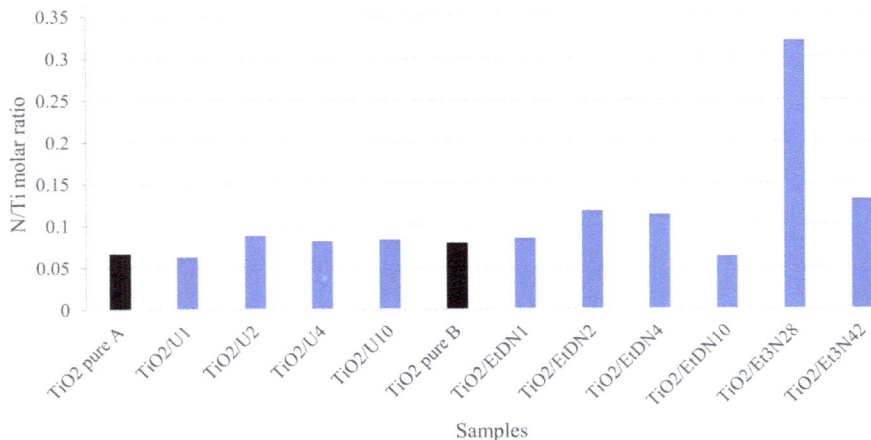

Figure 8. Estimated N/Ti molar ratio calculated from XPS measurements.

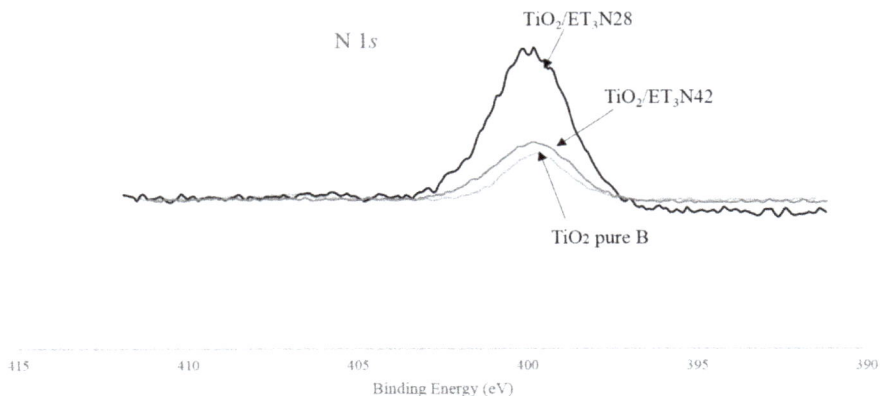

Figure 9. Comparison of normalized N 1s region XPS spectra of TiO_2 pure B, TiO_2/Et3N28 and TiO_2/Et3N42 samples.

3.6. Photocatalytic Activity

3.6.1. Under UV/Visible Light

For each sample, the test in the dark shows no PNP adsorption on the catalyst. Moreover, no spontaneous PNP degradation appears under UV/visible light in the absence of the catalyst. The catalytic test is performed over 24 h with an estimation of PNP degradation (D_{PNPi},%) after $i = 0$, 8 and 24 h. D_{PNP8} is used to compare the different catalytic activities of the samples because the differences between catalysts are the most noticeable after 8 hours of activitiy [9].

D_{PNP8} is presented in Figure 10 (dark grey) and in Table 2 for all samples. For both TiO_2 pure A and B samples, the same activity is obtained at around 50%. Concerning the urea- and trimethylamine-doped samples, the photoactivity stays nearly constant for each sample at around 45–50% of PNP degradation. For the ethylenediamine-doped samples, the activity decreases for all samples when the amount of ethylenediamine increases except for the TiO_2/EtDN4 sample, which presents 45% of PNP degradation.

If the photocatalytic activity is divided by the specific surface area, S_{BET}, the same observations can be made (Figure 11 in dark grey).

Table 2. Photocatalytic properties and XPS results of TiO_2-based samples.

Sample	D_{PNP8} under UV/Visible	D_{PNP8} under UV/Visible	D_{PNP24} under Visible	D_{PNP24} under Visible	N/Ti
	(%)	(% g·m^{-2})	(%)	(% g·m^{-2})	(mol/mol)
	±3	±0.02	±3	±0.02	
TiO_2 pure A	47	0.23	28	0.14	0.07
TiO_2/U1	43	0.18	42	0.18	0.06
TiO_2/U2	44	0.17	41	0.16	0.09
TiO_2/U4	45	0.17	41	0.16	0.08
TiO_2/U10	44	0.16	38	0.14	0.08
TiO_2 pure B	50	0.26	20	0.10	0.08
TiO_2/EtDN1	38	0.17	31	0.14	0.08
TiO_2/EtDN2	39	0.16	31	0.13	0.12
TiO_2/EtDN4	45	0.23	33	0.17	0.11
TiO_2/EtDN10	25	0.14	28	0.15	0.06
TiO_2/Et$_3$N28	43	0.19	43	0.19	0.32
TiO_2/Et$_3$N42	49	0.18	69	0.25	0.13
TiO_2/U2-LS	46	0.19	43	0.18	-[a]

D_{PNP8}: the degradation percentage of PNP after 8 h of illumination; D_{PNP24}: the degradation percentage of PNP after 24 h of illumination, values in% g·m^{-2} are obtained by dividing values in% by S_{BET}; N/Ti: molar ration between N and Ti measured by XPS; -[a] not measured.

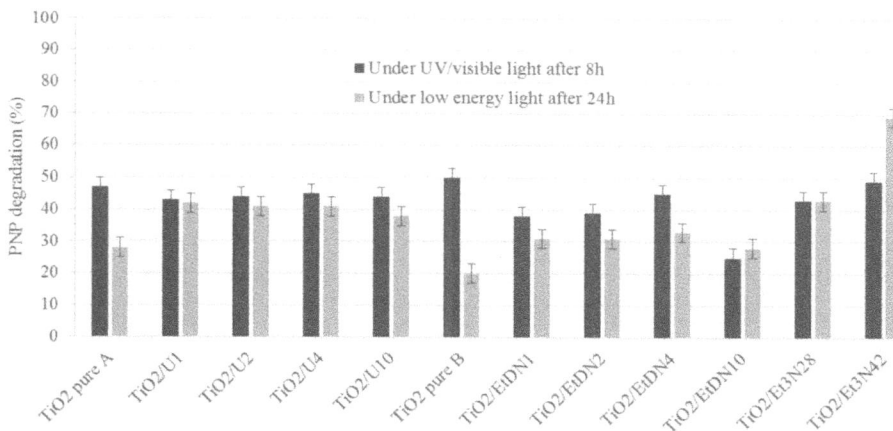

Figure 10. PNP degradation (%) for all samples under UV/visible light after 8 h of irradiation (dark grey) and under low-energy light (with filter to remove λ lower than 390 nm) after 24 h of irradiation (light grey).

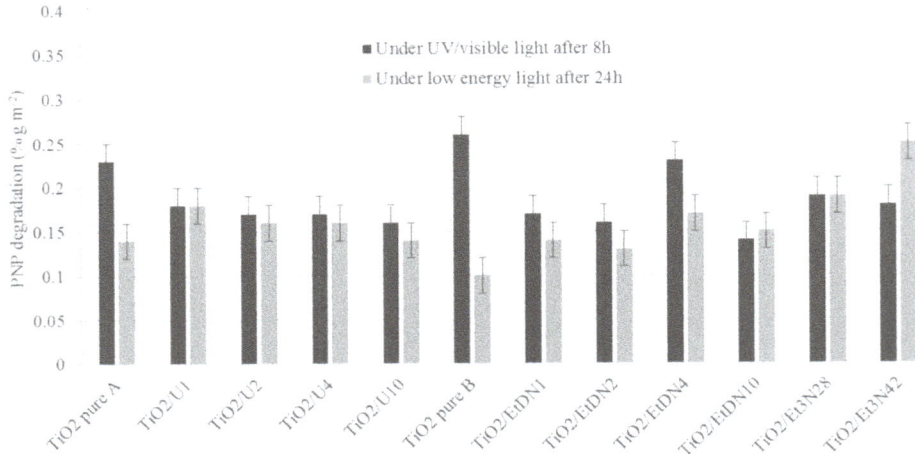

Figure 11. PNP degradation (% g·m^{-2}) calculated by dividing the PNP degradation in% by S_{BET} for all the samples under UV/visible after 8 h of irradiation (dark grey) and under low-energy light (with filter to remove λ lower than 390 nm) after 24 h of irradiation (light grey).

3.6.2. Under Low-Energy Light (λ > 390 nm)

For low-energy light conditions, D_{PNP24} is used to compare the different catalytic activities of samples because under visible light, the reactions are slower [9].

D_{PNP24} is presented in Figure 10 (light grey) and Table 2 for all samples. For TiO$_2$ pure A and B samples, the activity is slightly different between both syntheses with values of 28 and 20% for the PNP degradation respectively. For the urea-doped samples, the photoactivity increases compared to the corresponding TiO$_2$ pure A sample, exhibiting between 39% and 42% PNP degradation. For the ethylenediamine-doped samples, the photoactivity also increases compared to the corresponding TiO$_2$ pure B sample, with the optimal value being obtained for the TiO$_2$/EtDN4 sample, with 33% PNP degradation. For the trimethylamine-doped samples, the photoactivity also increases compared to the corresponding TiO$_2$ pure B sample with the optimal value obtained for the TiO$_2$/Et$_3$N42 sample, with 69% PNP degradation.

If the activity is divided by the specific surface area, S_{BET}, the same observations can be made (Figure 11, in light grey).

Concerning the use of UV/visible spectroscopy to measure the PNP degradation, it has been reported in the literature [49,50] that the presence of intermediate species associated with the partial degradation of PNP can be detected by the presence of peaks corresponding to the intermediates (4-nitrocatechol, 1,2,4-benzenetriol, hydroquinone) in the UV/Vis spectrum measured between 200 and 500 nm after several hours under illumination [9]. In the present study, no supplementary peaks are measured in the UV/Vis spectra between 200 and 500 nm, which is consistent with the complete mineralization of the pollutant, and therefore it is concluded that the photocatalysts developed in this study promote the complete mineralization of PNP [9]. Furthermore, total mineralization of PNP during homologous photocatalytic tests on a similar installation has been shown in a previous work [9,10].

3.7. Characterization of the Large-Scale Urea-Doped TiO$_2$ Photocatalyst

A urea-doped TiO$_2$ sample (TiO$_2$/U2-LS) is successfully synthesized at a large scale (5 L). It is characterized by XRD and BET measurements, and its photocatalytic activity is tested. All the results are shown in Tables 1 and 2. Concerning the physico-chemical properties, analogous characteristics are obtained compared to the laboratory scale sample (TiO$_2$/U2), with mainly anatase TiO$_2$ nanoparticles with a size of 7 nm and an S_{BET} value of 245 m^2·g^{-1} (Table 1). For the photocatalytic activity, similar values are obtained as for the laboratory scale sample under UV/visible and low-energy light, namely 46% and 43%, respectively (Table 2).

4. Discussion

All these syntheses are based on the precipitation-acidic peptization of a titanium precursor, where the acid allows dissolving amorphous TiO$_2$ precipitate by hydrolysis followed by the crystallization of small TiO$_2$ nanoparticles. This leads to the formation of stable crystalline TiO$_2$ colloids [51].

4.1. Crystallinity and Texture of TiO$_2$ Based Samples

Urea and triethylamine doping does not influence the crystallinity of the samples. Indeed, the TiO$_2$ phase distribution stays nearly constant between the pure and doped samples (Table 1). For these dopings, the synthesis conditions have not been modified and so same crystalline phases are obtained (anatase and brookite, Table 1). On the opposite, ethylenediamine doping modifies clearly the TiO$_2$ phase distribution leading to the formation of rutile TiO$_2$ when the amount of dopant increases (Table 1). For these syntheses, the amount of nitric acid has to be increased to obtain a peptization process of titania precursor. Indeed, ethylenediamine has a strong basic character [52], which requires more acid to reach peptization. In this case, if the amount of acid is increased, more oxolation bonds between titanium atoms are broken and more OH groups are produced around a single titanium atom [53], which facilitates the crystallization of TiO$_2$ in its most thermodynamically stable phase, i.e., the rutile phase [51,53].

Concerning the shape of the samples, TEM pictures, BET measurements and XRD crystallite size (Table 1), similar values between 4 and 8 nm are obtained. It is assumed that the samples are made of TiO$_2$ nanoparticles where a nanoparticle corresponds to a crystallite [7,54].

For the influence of doping on the texture, a modification of the isotherms compared to pure TiO$_2$ samples is observed. Indeed, doping leads to the formation of mesopores. For urea and triethylamine doping, the addition of dopant modifies the electrostatic interactions between the TiO$_2$ particles due to the incorporation of nitrogen, leading to a different particle stacking compared to pure TiO$_2$ samples when the samples are dried [54]. The difference of stacking between pure and doped TiO$_2$ nanoparticles has been previously described in [54]; it becomes less compact with doping which creates the mesoporosity [54]. Concerning ethylenediamine doping, two effects modify the isotherms: (i) the modification of the electrostatic interactions between the TiO$_2$ particles due to the incorporation of nitrogen as for the other dopings, and (ii) the modification of the crystalline phase distribution. In this second case, the apparition of rutile induces a change in the shape of the N$_2$ physisorption isotherm as described in [18]. Indeed, when the amount of ethylenediamine increases, mesopores appear (hysteresis of TiO$_2$/EtDN10 sample on Figure 4) due to the effect of N-doping and the amount of rutile remains <20%. However, when the amount of rutile exceeds 50%, the hysteresis is modified and reveals larger pores due to the rutile [18], which seems to produce larger rutile crystallite of 8 nm compared to anatase crystallite of 5 nm (d_{XRD} in Table 1). Concerning the two pure TiO$_2$ syntheses, these lead to similar materials as their specific surface area, crystalline phase and particle shapes are analogous (Table 1).

4.2. Photoactivity and N-Doping

For pure TiO$_2$ samples, the degradation of PNP (D_{PNPi}) is 47 and 50% after 8 h under UV/visible light for TiO$_2$ pure A and B, respectively, while it is 28 and 20% after 24 h under low-energy light. The activity under low-energy light is the characteristic of interest in this research, as an efficient visible photocatalyst is aimed for. This low-energy light is mainly "visible light", as the UV filter removes $\lambda < 390$ nm.

If both syntheses are compared, similar activities are obtained, which is consistent with the same type of materials that are observed through the characterizations (Table 1). An activity under low-energy light is observed for both catalysts. At the same time, XPS measurements (Figures 7d and 8) show that nitrogen is still present in both pure materials, with the peak at 400 eV seeming to correspond to Ti–O–N bonds [55,56]. So, a small nitrogen doping has occurred due to nitric acid used during the synthesis. A comparison is made with the commercial Degussa P25 catalyst, which has no trace of nitrogen [57] and is poorly active under visible light [57,58]. In this case, the PNP degradation after 24 h is merely 10%, which could be induced by the remaining UV radiation, which is not removed by the UV filter as explained above. Therefore, the TiO_2 pure A and B samples have an activity under low energy (and better than P25) not only due to the remaining UV light, but likely due to a trace of N-doping.

Concerning the N-doped samples, an increase in the activity is observed for all samples compared to the corresponding pure TiO_2 materials. All samples present a peak at 400 eV, which may correspond to Ti–O–N bonds [47,48,55,56].

For the urea-doped samples, XPS measurements (Figure 8) and diffuse reflectance spectroscopy (Figure 5) show an N-doping higher than the corresponding TiO_2 pure A sample and a reduction of the band gap value (Table 1). In this case, the nitrogen doping has been effective. As explained in [30,31,33], nitrogen doping can reduce the band gap by creating an intermediate band for the electron below the conduction band or above the valence band (Figure 12). It seems that in the four samples, the amount of nitrogen introduced in the samples is quite similar, even if different amounts were added during the synthesis, leading to materials with similar photocatalytic activity (D_{PNP24} between 37 and 42%) and optical properties (similar band gaps, Table 1). Urea seems to be a N precursor effective for doping, leading to visible active photocatalyst as described in [23].

Concerning the ethylenediamine-doped samples, the doping seems to also produce effective visible photocatalysts. Indeed, the activity increases for the four samples (Table 2), along with the band gap decreases (Table 1), and the amount of nitrogen increases for the TiO_2/EtDN2 and TiO_2/EtDN4 samples (Figure 8). The activity under low-energy light evolves, depending on the amount of nitrogen; when the amount increases, the activity increases (from samples TiO_2/EtDN1 to TiO_2/EtDN4, see Table 2 and Figure 8), but for the TiO_2/EtDN10 sample when the amount decreases, the activity decreases. The bad gap values follow the same trend (Table 1). Another important aspect for this series is that rutile TiO_2 is present in different percentages for the different samples, which can impact the photocatalytic activity. The rutile phase increases from TiO_2/EtDN1 sample to TiO_2/EtDN10 sample, but rutile is a less active phase than the anatase phase [6] and is also poorly active under visible light [6,24]. Therefore, the activity would decrease through the samples, but it is not observed under UV/visible light. Hence, the N-doping is successful and clearly linked to the activity of samples. On the other hand, a synergetic effect between anatase and rutile can also exist [24], leading to a better activity for the TiO_2/EtDN4 sample with the composition of 45% anatase and 35% rutile. This synergetic effect is also observed for Degussa P25, which has a composition of 80% anatase and 20% rutile [59]. A same mechanism for nitrogen doping as for urea doping can thus be considered (Figure 12).

For the trimethylamine-doped samples, the best activity is observed especially for the TiO_2/Et$_3$N42 sample, which reaches 69% PNP degradation. For this series of samples, the band gap values are not available as explained in Section 3.4. Only XPS measurements can therefore be used and show an increased nitrogen amount compared to the TiO_2 pure B sample (Figure 8), and that the N amount is the highest, when compared to the other series of samples. The very good activity shows that the N-doping is effective as described in [23,55] and a same mechanism for nitrogen doping as for urea and ethylenediamine doping can thus be considered (Figure 12).

The three precursors show that visible N-doped catalyst is obtained at low temperature, and that triethylamine seems to be the best nitrogenous agent, but requires the largest amount of precursors for doping.

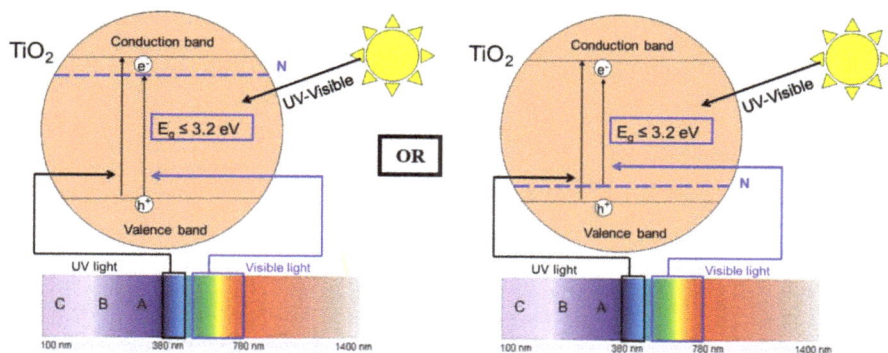

Figure 12. Effect of N-doping on TiO$_2$ photocatalysis.

4.3. Large-Scale Synthesis

For a visible light application, large amounts of catalyst will be needed, this is why a large-scale synthesis of N-doped TiO$_2$ of 5 L has been performed. The urea precursor is chosen because of its easy availability, low cost and low toxicity [60]. The urea/Ti precursor ratio of 2 has been chosen according to the good results obtained for the TiO$_2$/U2 sample at laboratory scale, and is thereby denoted TiO$_2$/U2-LS.

The results show that the large-scale synthesis does not influence the physico-chemical and photocatalytic properties of urea-doped TiO$_2$ samples (Tables 1 and 2). The PNP degradation of this large-scale sample is also higher than P25 activity under low-energy light (Table 2).

If this sample is compared to the commercial catalyst Degussa P25, the main advantages are the following [9]: (i) crystalline urea-doped TiO$_2$ samples are obtained without a calcination step; (ii) the synthesis is better for the environment, as water is the solvent (low use of organic compounds); (iii) the synthesis protocol is easy; (iv) the only cost is the TiO$_2$ precursor (titanium tetraisopropoxide) and urea, while the synthesis of commercial Degussa P25 involves the use of an aerosol process, which is known to be expensive, environmentally "unfriendly", and involves high-temperature treatments (1000–1300 °C) [7,61]; and (v) the sample is 4 times more active than P25 under visible light.

5. Conclusions

In this chapter, the previously developed aqueous titania sol-gel synthesis is doped with nitrogen precursors to extend its activity towards visible region. Three N-precursors are used: urea, ethylenediamine and triethylamine. Different molar ratios have been tested for each dopant. Results showed the formation of anatase-brookite TiO$_2$ nanoparticles of 6–8 nm with a specific surface area of between 200 and 275 m$^2 \cdot$g^{-1} for the urea and triethylamine series. Using ethylenediamine, the formation of rutile phase is observed when the amount of ethylenediamine increases due to the addition of nitric acid in order to maintain the peptization process. In this series, TiO$_2$ nanoparticles of 6–8 nm are also obtained with a specific surface area between 185 and 240 m$^2 \cdot$g^{-1}.

The combination of XPS and diffuse reflectance measurements suggests the incorporation of nitrogen in the TiO$_2$ materials through Ti–O–N bonds, allowing absorption in the visible region. Catalytic tests show a marked improvement of performance under visible radiation for all doped catalysts in the remediation of polluted water with *p*-nitrophenol. In this case, nitrogen doping can reduce the band gap by creating an intermediate band for the electrons below the conduction band or above the valence band, allowing activity in the visible range.

The best doping, taking into account cost, activity and ease of synthesis, is up-scaled to a volume of 5 L and compared to the commercial Degussa P25 material. This urea-doped large-scale catalyst shows analogous properties as the lab-scale corresponding synthesis and photoactivity 4 times higher than the commercial catalysts Degussa P25.

Acknowledgments: Stéphanie D. Lambert thanks the Belgian National Funds for Scientific Research (F.R.S.-F.N.R.S.) for her Associate Researcher position. The authors also thank the Ministère de la Région Wallonne Direction Générale des Technologies, de la Recherche et de l'Energie (DG06).

Author Contributions: Julien G. Mahy and Vincent Cerfontaine conceived and performed the photocatalysts syntheses, their morphological characterizations and their photocatalytic activity tests ; François Devred and Éric M. Gaigneaux performed and analyzed the XPS data; Dirk Poelman performed and analyzed the diffuse reflectance data; Julien G. Mahy, Benoît Heinrichs and Stéphanie D. Lambert wrote the paper and analyzed all the data; all the authors corrected the paper before submission and after the reviewer comments.

Conflicts of Interest: The authors declare no conflict of interest.

References

1. Hoffmann, M.R.; Martin, S.T.; Choi, W.; Bahnemann, D.W. Environmental Applications of Semiconductor Photocatalysis. *Chem. Rev.* **1995**, *95*, 69–96. [CrossRef]
2. Mills, A.; Le Hunte, S. An overview of semiconductor photocatalysis. *J. Photochem. Photobiol. A Chem.* **1997**, *108*, 1–35. [CrossRef]
3. Di Paola, A.; García-López, E.; Marcì, G.; Palmisano, L. A survey of photocatalytic materials for environmental remediation. *J. Hazard. Mater.* **2012**, *211–212*, 3–29. [CrossRef] [PubMed]
4. Rauf, M.A.; Ashraf, S.S. Fundamental principles and application of heterogeneous photocatalytic degradation of dyes in solution. *Chem. Eng. J.* **2009**, *151*, 10–18. [CrossRef]
5. Fujishima, A.; Hashimoto, K.; Watanabe, T. *TiO₂ Photocatalysis: Fundamentals and Applications*; BKC, Inc.: Tokyo, Japan, 1999.
6. Carp, O. Photoinduced reactivity of titanium dioxide. *Prog. Solid State Chem.* **2004**, *32*, 33–177. [CrossRef]
7. Malengreaux, C.M.; Douven, S.; Poelman, D.; Heinrichs, B.; Bartlett, J.R. An ambient temperature aqueous sol–gel processing of efficient nanocrystalline doped TiO₂-based photocatalysts for the degradation of organic pollutants. *J. Sol-Gel Sci. Technol.* **2014**, *71*, 557–570. [CrossRef]
8. Mahy, J.G.; Leonard, G.L.-M.; Pirard, S.; Wicky, D.; Daniel, A.; Archambeau, C.; Liquet, D.; Heinrichs, B. Aqueous sol-gel synthesis and film deposition methods for the large-scale manufacture of coated steel with self-cleaning properties. *J. Sol-Gel Sci. Technol.* **2017**, *81*, 27–35. [CrossRef]
9. Mahy, J.G.; Lambert, S.D.; Leonard, G.L.-M.; Zubiaur, A.; Olu, P.-Y.; Mahmoud, A.; Boschini, F.; Heinrichs, B. Towards a large scale aqueous sol-gel synthesis of doped TiO₂: Study of various metallic dopings for the photocatalytic degradation of p-nitrophenol. *J. Photochem. Photobiol. A Chem.* **2016**, *329*, 189–202. [CrossRef]
10. Malengreaux, C.M.; Léonard, G.M.-L.; Pirard, S.L.; Cimieri, I.; Lambert, S.D.; Bartlett, J.R.; Heinrichs, B. How to modify the photocatalytic activity of TiO₂ thin films through their roughness by using additives. A relation between kinetics, morphology and synthesis. *Chem. Eng. J.* **2014**, *243*, 537–548. [CrossRef]
11. Tasseroul, L.; Pirard, S.L.; Lambert, S.D.; Páez, C.A.; Poelman, D.; Pirard, J.-P.; Heinrichs, B. Kinetic study of p-nitrophenol photodegradation with modified TiO₂ xerogels. *Chem. Eng. J.* **2012**, *191*, 441–450. [CrossRef]
12. Anderson, C.; Bard, A.J. An Improved Photocatalyst of TiO₂/SiO₂ Prepared by a Sol-Gel Synthesis. *J. Phys. Chem.* **1995**, *99*, 9882–9885. [CrossRef]
13. Gratzel, M. Sol-gel processed TiO₂ films for photovoltaic applications. *J. Sol-Gel Sci. Technol.* **2001**, *22*, 7–13. [CrossRef]
14. Brinker, C.J.; Scherer, G.W. *Sol-Gel Science: The Physics and Chemistry of Sol-Gel Processing*; Academic Press: San Diego, CA, USA, 1990.
15. Kapusuz, D.; Park, J.; Ozturk, A. Effect of initial water content and calcination temperature on photocatalytic properties of TiO₂ nanopowders synthesized by the sol-gel process. *Ceram. Int.* **2015**, *41*, 12788–12797. [CrossRef]
16. Malengreaux, C.M.; Timmermans, A.; Pirard, S.L.; Lambert, S.D.; Pirard, J.-P.; Poelman, D.; Heinrichs, B. Optimized deposition of TiO₂ thin films produced by a non-aqueous sol–gel method and quantification of their photocatalytic activity. *Chem. Eng. J.* **2012**, *195–196*, 347–358. [CrossRef]
17. Schubert, U. Chemical modification of titanium alkoxides for sol–gel processing. *J. Mater. Chem.* **2005**, *15*, 3701–3715. [CrossRef]

18. Khalil, K.M.S.; El-Khatib, R.M.; Ali, T.T.; Mahmoud, H.A.; Elsamahy, A.A. Titania nanoparticles by acidic peptization of xerogel formed by hydrolysis of titanium(IV) isopropoxide under atmospheric humidity conditions. *Powder Technol.* **2013**, *245*, 156–162. [CrossRef]

19. Mahshid, S.; Askari, M.; Ghamsari, M.S. Synthesis of TiO$_2$ nanoparticles by hydrolysis and peptization of titanium isopropoxide solution. *J. Mater. Process. Technol.* **2007**, *189*, 296–300. [CrossRef]

20. Braconnier, B.; Páez, C.A.; Lambert, S.D.; Alié, C.; Henrist, C.; Poelman, D.; Pirard, J.-P.; Cloots, R.; Heinrichs, B. Ag- and SiO$_2$-doped porous TiO$_2$ with enhanced thermal stability. *Microporous Mesoporous Mater.* **2009**, *122*, 247–254. [CrossRef]

21. Di Paola, A.; Marcì, G.; Palmisano, L.; Schiavello, M.; Uosaki, K.; Ikeda, S.; Ohtani, B. Preparation of polycrystalline Tio$_2$ photocatalysts impregnated with various transition metal ions: Characterization and photocatalytic activity for the degradation of 4-nitrophenol. *J. Phys. Chem. B* **2002**, *106*, 637–645. [CrossRef]

22. Ruggieri, F.; Di, D. Electrospun Cu-, W- and Fe-doped TiO$_2$ nanofibres for photocatalytic degradation of rhodamine 6G. *J. Nanoparticle Res.* **2013**, *15*, 1982. [CrossRef]

23. Qiu, X.; Burda, C. Chemically synthesized nitrogen-doped metal oxide nanoparticles. *Chem. Phys.* **2007**, *339*, 1–10. [CrossRef]

24. Pelaez, M.; Nolan, N.T.; Pillai, S.C.; Seery, M.K.; Falaras, P.; Kontos, A.G.; Dunlop, P.S.M.; Hamilton, J.W.J.; Byrne, J.A.; O'Shea, K.; et al. A review on the visible light active titanium dioxide photocatalysts for environmental applications. *Appl. Catal. B Environ.* **2012**, *125*, 331–349. [CrossRef]

25. Di Valentin, C.; Pacchioni, G. Trends in non-metal doping of anatase TiO$_2$: B, C, N and F. *Catal. Today* **2013**, *206*, 12–18. [CrossRef]

26. Di Valentin, C.; Pacchioni, G.; Selloni, A.; Livraghi, S.; Cozzi, V.R. Characterization of Paramagnetic Species in N-Doped TiO$_2$ Powders by EPR Spectroscopy and DFT Calculations. *J. Phys. Chem. B Lett.* **2005**, *109*, 11414–11419. [CrossRef] [PubMed]

27. Cheng, X.; Liu, H.; Chen, Q.; Li, J.; Wang, P. Electrochimica Acta Construction of N, S codoped TiO$_2$ NCs decorated TiO$_2$ nano-tube array photoelectrode and its enhanced visible light photocatalytic mechanism. *Electrochim. Acta* **2013**, *103*, 134–142. [CrossRef]

28. Selvaraj, A.; Sivakumar, S.; Ramasamy, A.K.; Balasubramanian, V. Photocatalytic degradation of triazine dyes over N-doped TiO$_2$ in solar radiation. *Res. Chem. Intermed.* **2013**, *39*, 2287–2302. [CrossRef]

29. Patel, N.; Dashora, A.; Jaiswal, R.; Fernandes, R.; Yadav, M.; Kothari, D.C.; Ahuja, B.L.; Miotello, A. Experimental and Theoretical Investigations on the Activity and Stability of Substitutional and Interstitial Boron in TiO$_2$ Photocatalyst. *J. Phys. Chem. C* **2015**, *119*, 18581–18590. [CrossRef]

30. Iwase, M.; Yamada, K.; Kurisaki, T.; Prieto-mahaney, O.O.; Ohtani, B.; Wakita, H. Visible-light photocatalysis with phosphorus-doped titanium (IV) oxide particles prepared using a phosphide compound. *Appl. Catal. B Environ.* **2013**, *132–133*, 39–44. [CrossRef]

31. Bodson, C.J.; Heinrichs, B.; Tasseroul, L.; Bied, C.; Mahy, J.G.; Wong Chi Man, M.; Lambert, S.D. Efficient P- and Ag-doped titania for the photocatalytic degradation of waste water organic pollutants. *J. Alloys Compd.* **2016**, *682*, 144–153. [CrossRef]

32. Lee, C.; Hyeon, T.; Lee, H. Visible Light-Induced Degradation of Carbon Tetrachloride on. *Environ. Sci. Technol.* **2001**, *35*, 966–970.

33. Granados, G.O.; Páez, C.A.M.; Martínez, F.O.; Páez-Mozo, E.A. Photocatalytic degradation of phenol on TiO$_2$ and TiO$_2$/Pt sensitized with metallophthalocyanines. *Catal. Today* **2005**, *107–108*, 589–594. [CrossRef]

34. Hu, L.; Wang, J.; Zhang, J.; Zhang, Q.; Liu, Z. An N-doped anatase/rutile TiO$_2$ hybrid from low-temperature direct nitridization: Enhanced photoactivity under UV-/visible-light. *RSC Adv.* **2014**, *4*, 420–427. [CrossRef]

35. Sing, K.S.W.; Rouquerol, J.; Bergeret, H.J.G.; Gallezot, P.; Vaarkamp, M.; Koningsberger, D.C.; Datye, A.K.; Niemantsverdriet, J.W.; Butz, T.; Engelhardt, G.; et al. *Handbook of Heterogenous Catalysis*; Weitkamp, J., Ed.; Wiley-VCH Verlag GmbH: Weinheim, Germany, 1997; pp. 428–582.

36. Doebelin, N.; Kleeberg, R. Profex: A graphical user interface for the Rietveld refinement program BGMN. *J. Appl. Crystallogr.* **2015**, *48*, 1573–1580. [CrossRef] [PubMed]

37. Madsen, I.C.; Finney, R.J.; Flann, R.C.A.; Frost, M.T.; Wilson, B.W. Quantitative Analysis of High-alumina Refractories Using X-ray Powder Diffraction Data and the Rietveld Method. *J. Am. Ceram. Soc.* **1991**, *74*, 619–624. [CrossRef]

38. Lecloux, A. Exploitation des isothermes d'adsorption et de désorption d'azote pour l'étude de la texture des solides poreux. *Mém. Soc. R. Sci. Liège* **1971**, *4*, 169–209.

39. KUBELKA, P. Ein Beitrag zur Optik der Farban striche. *Z. Tech. Phys.* **1931**, *12*, 593–603.

40. Kubelka, P. New contributions to the optics of intensely light-scattering materials. *J. Opt. Soc. Am.* **1948**, *38*, 448–457. [CrossRef] [PubMed]

41. Páez, C.A.; Poelman, D.; Pirard, J.P.; Heinrichs, B. Unpredictable photocatalytic ability of H_2-reduced rutile-TiO_2 xerogel in the degradation of dye-pollutants under UV and visible light irradiation. *Appl. Catal. B Environ.* **2010**, *94*, 263–271. [CrossRef]

42. Escobedo Morales, A.; Sánchez Mora, E.; Pal, U. Use of diffuse reflectance spectroscopy for optical characterization of un-supported nanostructures. *Rev. Mex. Fis.* **2007**, *53*, 18–22.

43. Bryson, C.E. Surface potential control in XPS. *Surf. Sci.* **1987**, *189/190*, 50–58. [CrossRef]

44. Shirley, D.A. High-Resolution X-Ray Photoemission Spectrum of the Valence Bands of Gold. *Phys. Rev. B* **1972**, *5*, 4709–4714. [CrossRef]

45. Páez, C.A.; Liquet, D.Y.; Calberg, C.; Lambert, S.D.; Willems, I.; Germeau, A.; Pirard, J.P.; Heinrichs, B. Study of photocatalytic decomposition of hydrogen peroxide over ramsdellite-MnO_2 by O_2-pressure monitoring. *Catal. Commun.* **2011**, *15*, 132–136. [CrossRef]

46. Lambert, S.D.; Alie, C.; Pirard, J.P.; Heinrichs, B. Study of textural properties and nucleation phenomenon in Pd/SiO_2, Ag/SiO_2 and Cu/SiO_2 cogelled xerogel catalysts. *J. Non-Cryst. Solids* **2004**, *342*, 70–81. [CrossRef]

47. Azouani, R.; Tieng, S.; Chhor, K.; Bocquet, J.F.; Eloy, P.; Gaigneaux, E.M.; Klementiev, K.; Kanaev, A.V. TiO_2 doping by hydroxyurea at the nucleation stage: Towards a new photocatalyst in the visible spectral range. *Phys. Chem. Chem. Phys.* **2010**, *12*, 1–10. [CrossRef] [PubMed]

48. Bittencourt, C.; Rutar, M.; Umek, P.; Mrzel, A.; Vozel, K.; Arcon, D.; Henzler, K.; Krüger, P.; Guttmann, P. Molecular nitrogen in N-doped TiO_2 nanoribbons. *RSC Adv.* **2015**, *5*, 23350–23356. [CrossRef]

49. Di Paola, A.; Augugliaro, V.; Palmisano, L.; Pantaleo, G.; Savinov, E. Heterogeneous photocatalytic degradation of nitrophenols. *J. Photochem. Photobiol. A Chem.* **2003**, *155*, 207–214. [CrossRef]

50. Augugliaro, V.; Palmisano, L.; Schiavello, M.; Sclafani, A.; Marchese, L.; Martra, G.; Miano, F. Photocatalytic degradation of nitrophenols in aqueous titanium dioxide dispersion. *Appl. Catal.* **1991**, *69*, 323–340. [CrossRef]

51. Bischoff, B.; Anderson, M. Peptization Process in the Sol-Gel Preparation of Porous Anatase (TiO_2). *Chem. Mater.* **1995**, *7*, 1772–1778. [CrossRef]

52. Hawley, G.G.; Lewis, R.J. *Hawley Scondensed Chemcal Dictionary*; Wiley: New York, NY, USA, 1993.

53. Yang, J.; Mei, S.; Ferreira, J.M.F.; Norby, P.; Quaresmâ, S. Fabrication of rutile rod-like particle by hydrothermal method: An insight into HNO_3 peptization. *J. Colloid Interface Sci.* **2005**, *283*, 102–106. [CrossRef] [PubMed]

54. Malengreaux, C.M.; Pirard, S.L.; Léonard, G.; Mahy, J.G.; Klobes, B.; Herlitschke, M.; Heinrichs, B.; Bartlett, J.R. Study of the photocatalytic activity of Fe^{3+}, Cr^{3+}, La^{3+} and Eu^{3+} single- doped and co-doped TiO_2 catalysts produced by aqueous sol-gel processing. *J. Alloys Compd.* **2017**, *691*, 726–738. [CrossRef]

55. Qiu, B.X.; Zhao, Y.; Burda, C. Synthesis and Characterization of Nitrogen-Doped Group IVB Visible-Light-Photoactive Metal Oxide Nanoparticles. *Adv. Mater.* **2007**, *19*, 3995–3999. [CrossRef]

56. Zhao, Z.; Qiu, X.; Burda, C. The Effects of Sintering on the Photocatalytic Activity of N-Doped TiO_2 Nanoparticles. *Chem. Mater.* **2008**, *20*, 2629–2636. [CrossRef]

57. Gole, J.L.; Stout, J.D.; Burda, C.; Lou, Y.; Chen, X. Highly Efficient Formation of Visible Light Tunable TiO_2 - x N x Photocatalysts and Their Transformation at the Nanoscale. *J. Phys. Chem. B* **2004**, *108*, 1230–1240. [CrossRef]

58. Liu, Y.; Chen, X.; Li, J.; Burda, C. Photocatalytic degradation of azo dyes by nitrogen-doped TiO_2 nanocatalysts. *Chemosphere* **2005**, *61*, 11–18. [CrossRef]

59. Ohtani, B.; Prieto-Mahaney, O.O.; Li, D.; Abe, R. What is Degussa (Evonik) P25? Crystalline composition analysis, reconstruction from isolated pure particles and photocatalytic activity test. *J. Photochem. Photobiol. A Chem.* **2010**, *216*, 179–182. [CrossRef]

60. Yamarik, T.A.; Elmore, A.R. Final report of the safety assessment of Urea. *Int. J. Toxicil.* **2004**, *24*, 1–56.

61. Koirala, R.; Pratsinis, S.E.; Baiker, A. Synthesis of catalytic materials in flames: Opportunities and challenges. *Chem. Soc. Rev.* **2016**, *45*, 3053–3068. [CrossRef] [PubMed]

![materials logo] *materials*

MDPI

Article

Understanding the Interaction between a Steel Microstructure and Hydrogen

Tom Depover *, Aurélie Laureys, Diana Pérez Escobar, Emilie Van den Eeckhout, Elien Wallaert and Kim Verbeken

Department of Materials, Textiles and Chemical Engineering, Ghent University (UGent), Technologiepark 903, B-9052 Ghent, Belgium; aurelie.laureys@ugent.be (A.L.); diana.perezescobar@gmail.com (D.P.E.); emilie.vandeneeckhout@ugent.be (E.V.d.E.); elien.wallaert@ugent.be (E.W.); kim.verbeken@ugent.be (K.V.)
* Correspondence: Tom.Depover@ugent.be; Tel.: +32-9-331-0433

Received: 2 April 2018; Accepted: 24 April 2018; Published: 28 April 2018

Abstract: The present work provides an overview of the work on the interaction between hydrogen (H) and the steel's microstructure. Different techniques are used to evaluate the H-induced damage phenomena. The impact of H charging on multiphase high-strength steels, i.e., high-strength low-alloy (HSLA), transformation-induced plasticity (TRIP) and dual phase (DP) is first studied. The highest hydrogen embrittlement resistance is obtained for HSLA steel due to the presence of Ti- and Nb-based precipitates. Generic Fe-C lab-cast alloys consisting of a single phase, i.e., ferrite, bainite, pearlite or martensite, and with carbon contents of approximately 0, 0.2 and 0.4 wt %, are further considered to simplify the microstructure. Finally, the addition of carbides is investigated in lab-cast Fe-C-X alloys by adding a ternary carbide forming element to the Fe-C alloys. To understand the H/material interaction, a comparison of the available H trapping sites, the H pick-up level and the H diffusivity with the H-induced mechanical degradation or H-induced cracking is correlated with a thorough microstructural analysis.

Keywords: hydrogen embrittlement; hydrogen-induced cracking; thermal desorption spectroscopy; in situ tensile testing; permeation

1. Introduction

The limited fossil fuel resources, the concerns about nuclear power, and especially global warming issues imply energy related challenges that are triggering scientists to find ecological solutions for an environmentally friendly future. Hydrogen (H) gas is an attractive replacement for fossil fuels since it is the most abundant element on earth and its combustion only generates water. Hence, no greenhouse gasses are emitted in the atmosphere [1]. Despite these advantages, H contains a negative connotation due to multiple incidents that happened in the past indicating the potential danger of its usage [2]. Another approach to make transportation more ecological is to lower the fuel consumption of vehicles. In automotive industry, the use of high-strength steels (HSS) has, for instance, been promoted since it can both guarantee an increased safety together with weight reduction, which is required to meet the stringent CO_2 emission guidelines. Unfortunately, these steels are prone to H-induced mechanical degradation [3–5], which impedes further alloy development, according to BMW [6]. Steels used in offshore industry are often protected against corrosion by cathodic protection which causes a conversion of all the active anodic sites of the metal to inactive cathodic sites. However, when cathodic overprotection occurs, atomic H absorbs into the steel and a H concentration is built up. Consequently, the material integrity degrades, and H-induced failure is promoted [7].

Johnson [8] was the first to describe the potential effect of H on the mechanical properties of iron; i.e., a temporary loss of ductility was observed. This work has inspired a lot of researchers to continue investigating this subject resulting in many reference works and international conferences [9–18].

Recently, the attention got reinforced by the development of numerous applications of high-strength steels involving H interaction, such as oil and gas pipelines, vehicles, storage tanks, offshore structures, welds, etc. [19–22]. The phenomenon of temporary ductility loss implies so called hydrogen embrittlement (HE), causing unpredictable failure. Absorbed and dissolved H diffuses through the metal, introducing extra stresses at potential crack initiation sites and/or facilitating crack propagation. HSS are very susceptible to HE because of their high stress level as well as their large number of potential fracture initiation sites [3,23]. Consequently, the potential and safe application of H still offers major challenges to the materials engineer and clearly a renewed interest in this research field occurred during the last decade in both industrial and scientific communities.

Although the detrimental H effect was already discussed in 1875 [8], no full understanding of the phenomenon governing the observed H-induced ductility loss has been achieved up to now. Several mechanisms have been proposed to describe HE, but none has been fully accepted since indications accounting for different mechanisms have been observed experimentally. The three most cited mechanisms in non-hydride forming materials are the H Enhanced DEcohesion (HEDE) [24], the H Enhanced Localized Plasticity (HELP) [25] and Adsorption-Induced Dislocation Emission (AIDE) theory [26,27]. HEDE proposes a decrease in the cohesive bond strength between the metal atoms in the presence of H causing brittle crack propagation under tensile load. HELP considers an increase in dislocation mobility due to the presence of H, resulting into highly localized plastic deformation and accelerated failure. This mechanism has attained substantial support [28–30]. AIDE combines both HEDE and HELP since it also involves localized plasticity, but it proposes that the localized plasticity occurs close to the surface at regions of stress concentrations, for instance at crack tips, including nanovoids in front of it.

Multiphase HSS, such as transformation-induced plasticity (TRIP), dual phase (DP) and high-strength low-alloyed (HSLA) steels are commonly used in the automotive industry and were already subject of several H related studies since they are prone to HE [31–33]. These HSS show, however, a multiphase microstructure, complicating the interpretation of H related data. Therefore, the interaction between H and single-phase alloys is also of interest to get an improved understanding of the damaging mechanisms [34–37]. Additionally, the presence of carbides is generally considered to be beneficial to improve the HE resistance since they are assumed to trap H efficiently removing the detrimental mobile H from the microstructure [38,39]. The size and coherency of the specific carbides were crucial in terms of H trapping ability [40,41]. However, the applied H charging conditions might affect their beneficial role, as will be evaluated further in Section 5.

This work provides an overview of our work on the interaction of H with a steel microstructure and is built up in a systematic approach. At first, the effect of H charging on HSS, i.e., TRIP, DP and HSLA, will be considered. To simplify the microstructure, generic Fe-C lab-cast alloys consisting of a single phase, i.e., ferrite, bainite, pearlite or martensite and with carbon contents of approximately 0, 0.2 and 0.4 wt %, will be studied. Finally, the addition of carbides will be investigated in Fe-C-X alloys by adding a ternary carbide forming element to the Fe-C alloys. A comparison of the H absorption level, the available trapping sites and the H diffusivity with the H-induced mechanical degradation or H-induced cracking was correlated with a detailed microstructural analysis to understand the H/material interaction.

2. Experimental Procedure

2.1. Materials Characterization

2.1.1. Advanced High-Strength Steels

The investigated multiphase, advanced HSS such as TRIP, DP and HSLA, contained different constituents, such as martensite, bainite, ferrite and retained austenite. The thickness of the TRIP steel sheets was 0.7 mm, while DP and HSLA steels had a thickness of 1.1 mm. These thicknesses were reached after hot and cold rolling, followed by subsequent annealing via industrial annealing parameters required to obtain the desired microstructure. TRIP steel contained ferrite, martensite, bainite and retained austenite. The amount of retained austenite, quantified by XRD, was 9.6%. The carbon content of the retained austenite was determined to be 1.2 wt % [42]. The DP steel consisted of ferrite and 23.6% martensite [43]. In the HSLA steel, cementite (Fe_3C) was observed in the ferrite matrix, while this steel also contained Ti- and Nb-based carbonitrides [44]. Chemical compositions are summarized in Table 1. Notched tensile samples, with stress concentration factor of 4.2, were made by spark erosion, the tensile axis being parallel to the rolling direction [45].

Table 1. Chemical compositions in wt %. Fe content in balance.

Material/Element	C	Mn	Si	Other
TRIP	0.17	1.60	0.40	1–2% Al, 0.04–0.1% P
DP	0.07	1.50	0.25	0.4–0.8% Cr + Mo
HSLA	0.07	0.95	0.00	0.08–0.12% Ti + Nb

2.1.2. Pure Iron, Generic Fe-C and Fe-C-X Alloys

Generic steels were studied to avoid the effect of complex microstructural characteristics. The lab-cast generic alloys were "pure iron", ultra-low carbon (ULC) steel and two Fe-C-based materials containing 0.2 and 0.4% of carbon. This difference in carbon content will allow evaluating the impact of the carbon content on HE. A commercial Armco pure iron was used as a reference as well. The chemical compositions are given in Table 2. The alloys were produced in a Pfeiffer VSG100 vacuum (Pfeiffer, Asslar, Hessen, Germany) melting and casting unit, operating under an argon gas protective atmosphere. The materials were hot and cold rolled resulting in sheet material with final thicknesses ranging between 1 and 2 mm while different heat treatments were applied to induce a ferritic, bainitic, martensitic or pearlitic structure, for which the authors refer to the corresponding works [46–49]. The effect of deformation-induced defects on the H behavior in Armco pure iron and ULC steel was investigated by applying varying degrees of cold deformation and heat treatments [49,50].

Table 2. Chemical composition of the used materials in wt %. Fe content in balance.

Material/Element	C	Mn	Si	Other
"Pure Iron"	0.0015	0.0003	0.00	<0.02% Al, P
Ultra-low C	0.0214	0.2500	0.00	<0.05% Al, P, N
Armco iron	0.001	0.050	0.003	<0.005% Al, P, N
0.2% C	0.199	0.004	<0.0002	<0.0008 P, N
0.4% C	0.374	0.002	<0.0001	<0.0007 P, N

Additionally, the H/carbide interaction was studied extensively. Fe-C-X grades with a stoichiometric amount of a ternary alloying element X, i.e., Ti, Cr, Mo, W or V, were processed. Each grade was incrementally cast into three alloys with increasing carbon content (cf. Table 3). The carbon increase allows a reliable assessment of the effect of the carbides with varying strength level of the alloys and gives an opportunity to evaluate their role in different Fe-C-X alloys. The materials were again processed in a Pfeiffer VSG100 incremental vacuum melting and casting unit under an argon gas atmosphere. Subsequently, hot rolling was performed till 1.5 mm and an appropriate heat treatment was applied to obtain two main conditions: a martensitic as-quenched (as-Q) and subsequent quenched and tempered (Q&T) state, with carbides generated during tempering. A variation in the tempering time was further applied to change the carbide characteristics in the Q&T matrix allowing to thoroughly analyze the trapping ability of the carbides. Tensile specimens without a notch were prepared for these alloys with the tensile axis parallel to the rolling direction. The carbide characteristics were also studied in a ferritic microstructure since this facilitates the microstructural characterization. NbC and TiC were selected for this purpose and the experimental considered steels contained 0.013 wt % C–0.1% Nb and 0.025 wt % C–0.09 % Ti.

Table 3. Chemical compositions of the Fe-C-X materials in wt %. Fe content in balance.

Alloy Fe-C-X		C	X	Other Elements
Fe-C-Ti	Alloy A	0.099	0.380	
	Alloy B	0.202	0.740	
	Alloy C	0.313	1.340	
Fe-C-Cr	Alloy A	0.097	1.300	
	Alloy B	0.143	1.800	
	Alloy C	0.184	2.200	
Fe-C-Mo	Alloy A	0.100	1.700	Al: 200–300 wt. ppm
	Alloy B	0.142	2.380	Other elements: traces
	Alloy C	0.177	2.990	
Fe-C-W	Alloy A	0.096	2.670	
	Alloy B	0.186	6.130	
	Alloy C	0.277	8.700	
Fe-C-V	Alloy A	0.100	0.570	
	Alloy B	0.190	1.090	
	Alloy C	0.286	1.670	

2.1.3. Microstructural Characterization

Vickers hardness measurements were performed to determine the hardness level and the tempering temperature at which secondary hardening was most effective for the Fe-C-X alloys. Hardness measurements were performed with a drop weight of 2 kg and a pyramidal diamond indenter tip. The microstructures were investigated by light optical microscopy (LOM) and scanning electron microscopy (SEM) (FEI Quanta FEG 450, ThermoFisher Scientific, Hillsboro, OR, USA), transmission electron microscopy (TEM) (JEOL JEM-2200FS, JEOL, Tokyo, Japan), scanning transmission electron microscopy (STEM) analysis, as well as energy dispersive X-ray (EDX) spectroscopy allowed to characterize specific microstructural features such as carbides in terms of size, size distribution and morphology. Therefore, carbon replicas and thin foils were prepared. Carbon replication was made by sputtering carbon on top of a polished (1 μm) sample, followed by carbide extraction from the sample by putting the sample in a Nital 4% solution. Thin foil samples were prepared by grinding and polishing the samples to a thickness below 100 μm. Subsequently, the thin foils were electropolished, using a TenuPol-5 electropolishing unit in a 10% perchloric acid and 90% acetic acid solution. Fracture surfaces were analyzed by SEM. H assisted cracks were investigated by SEM and electron backscatter diffraction (EBSD) to characterize their morphology and crystallographic characteristics. The detection of blisters was done by LOM surface imaging permitting determination of their morphology, distribution, size, and areal density. Cross sections were analyzed to obtain information on the morphology and depth of internal cracks in H charged samples.

2.2. Determination of the Degree of Hydrogen-Induced Mechanical Degradation

By comparing tensile tests performed in air with tests done on in situ charged, H-saturated samples, the impact of H on the mechanical properties was determined. H was introduced in the alloys by electrochemical pre-charging using a 1 g/L thiourea 0.5 M H_2SO_4 solution at a current density of 0.8 mA/cm^2 for 1 or 2 h, depending on the H saturation time. In situ charging continued during the tensile test. The conditions were chosen in such a way that they did not create blisters or any internal damage [51], which has been demonstrated to be important in earlier work [47]. The tensile tests were done at a cross-head displacement speed of 5 or 0.05 mm/min, which corresponds with a strain rate of

1.11×10^{-3} or 0.0111×10^{-3} s^{-1}, respectively. The %*HE* is calculated to compare the sensitivity to H-induced mechanical degradation and defined as [45]:

$$\%HE = 100 \left(1 - \frac{\varepsilon_{ch}}{\varepsilon_{un}} \right) \tag{1}$$

with ε_{ch} and ε_{un} being the elongation of the charged and uncharged sample, respectively. Hence, the %*HE* varies between 0 and 1, with 0 meaning that there is no ductility loss and the material is insensitive to *HE*. When an index of 1 is obtained, the ductility drop is 100% and *HE* is maximal. HAC was investigated by performing similar tensile tests. However, instead of performing the tensile test until fracture, tests were interrupted when reaching the tensile strength. As such, H assisted cracks could be investigated more efficiently. Blisters were obtained by electrochemical H charging in a 1 g/L thiourea 0.5 M H$_2$SO$_4$ solution with varying charging conditions, i.e., charging time and current density. No external load is applied during these tests.

2.3. Determination of Diffusible and Total Hydrogen Content

Melt/hot extraction was used to determine the H saturation level, for which the samples were charged electrochemically as described above. During melt extraction, the sample is kept at 1600 °C in a pulse furnace which allows measurement of the total amount of H present in the sample. The amount of diffusible hydrogen, defined as proposed by Akiyama et al. [52,53] is measured by holding the sample at 300 °C in an infrared furnace. The metallic sample releases its H as gaseous H$_2$ which is dragged along by a nitrogen flow. This mixture (N$_2$–H$_2$) is directed to a thermal conductivity measuring cell. Essentially, the thermal conductivity of the mixture depends on the H$_2$ concentration since the conductivity of H$_2$ and N$_2$ differs significantly. Hence, the software can calculate the H concentration based on the variation in thermal conductivity. An average of 10 samples was taken to determine both hydrogen contents.

2.4. Determination of Hydrogen Trapping Capacity

The H trapping sites and their activation energy were determined by performing thermal desorption spectroscopy analysis, for which circular discs with a diameter of 20 mm and a thickness of 1 mm were used. The samples were H charged, similarly as described above, and three different heating rates of 200, 600 and 1200 °C/h were used for TDS analysis. However, the applied experimental procedure required one hour between the end of H charging and the start of the TDS measurement as an appropriate vacuum level needs to be created in the analysis chamber before the TDS measurement could start. The method based on the work of Lee et al. [54–56] was used to determine the E_a of H traps related to the peaks observed in the TDS spectra. Equation (2) is a simplification of the original formula of Kissinger [57]. Hydrogen re-trapping and diffusion is however not considered in this simplified equation.

$$\frac{d\left(ln\frac{\Phi}{T_{max}^2} \right)}{d\left(\frac{1}{T_{max}} \right)} = -\frac{E_a}{R} \tag{2}$$

where Φ is the heating rate (K/min), T_{max} (K) the TDS peak temperature, E_a (J/mol) the detrapping activation energy for specific H trapping sites associated with T_{max} and R (J·K^{-1}·mol^{-1}) the universal gas constant. After TDS measurements using different heating rates, deconvolution of the results and determining the corresponding peak temperatures for the different traps, plotting $ln(\Phi/T_{max}^2)$ vs. $(1/T_{max})$ allows to obtain the corresponding E_a of the specific trapping sites.

2.5. Determination of Hydrogen Diffusion Coefficient

The H diffusion coefficient is determined by electrochemical permeation tests using a set-up based on the Devanathan and Stachurski method [58]. Two compartments were filled with a 0.1 M NaOH solution and a polished circular sample of 20 mm diameter and 1 mm thickness was clamped in between. Temperature was maintained constant at 25 °C and the oxygen content was limited by stirring the electrolyte using nitrogen gas bubbles. The entry side acted as the cathode by applying a current density of 3 mA/cm^2, while the exit side (anode) was potentiostatically kept at -500 mV according to a Hg/Hg_2SO_4 reference electrode. The apparent H diffusion coefficient was calculated from the permeation transient using the following formula [59]:

$$D_{app} = \frac{L^2}{7.7\,t} \left(m^2/s \right) \tag{3}$$

where t is the time (s) when the normalized steady state value has reached 0.1 and L is the sample thickness (m). The determination of the apparent diffusion coefficient includes the assumption of one dimensional diffusion and both initial and exit site hydrogen concentration is assumed to be zero. Although palladium coating is well known to avoid corrosion and to enhance H oxidation at the exit side, defects and/or oxides at the metal/Pd interface are easily introduced during the plating process. These heterogeneities will hence affect the H diffusion in an uncontrolled manner, which is not desirable. Since identical experimental conditions were used, the relative variations in the permeation transients are reliable. Hence, the permeation tests were done without Pd coating.

3. The Interaction of Hydrogen with Advanced High-Strength Steels

The impact of H on the mechanical properties of TRIP and HSLA steels was evaluated by tensile tests with in situ H charging at a deformation rate of 5 mm/min on notched tensile specimens [45]. The corresponding results on both H charged and uncharged samples are presented in Figure 1. The curves of the samples measured under the same conditions were nearly identical, confirming the good reproducibility of the test. A high HE degree of 60% was obtained for TRIP steel, while only 8% of ductility loss was found for HSLA steel. This was correlated to the presence of Nb- and Ti-based precipitates, providing effective trapping sites for H as such decreasing the H diffusivity. Generally, carbide addition has been stated to improve the HE resistance [38,39,60]. The obtained degree of HE, based on the elongation, was in good agreement with the one based on the reduction of area determined by the fracture surfaces [45]. Moreover, the fracture surface of the uncharged samples showed a considerable ductile necking zone, while the H charged samples showed a brittle transgranular cleavage failure near the edges with transition zone to some ductile features in the center of the specimen.

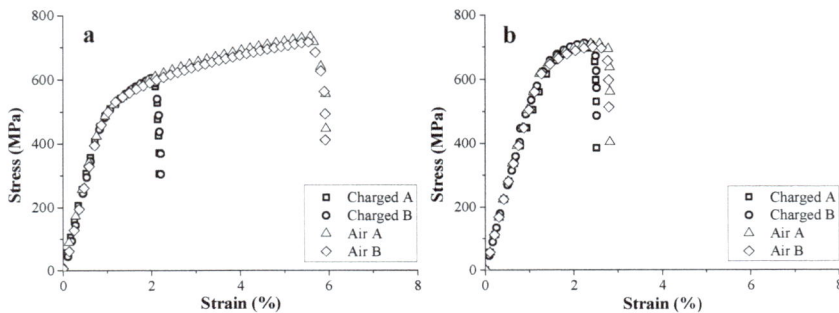

Figure 1. Stress-strain curves for TRIP (**a**) and HSLA (**b**). Similar tests indicated with A and B.

The tensile specimens contained a notch to control fracture initiation. However, the presence of this notch induces a specific stress distribution in the sample. To evaluate whether this would influence the failure mechanism in the presence of H, notched tensile samples were compared with unnotched samples for both TRIP steel and pure iron was used as a reference [61]. The HAC behavior of both materials was also compared to assess the effect of the microstructure on the general HAC behavior of a material. The materials exhibited a slightly different HAC behavior, since H assisted secondary crack formation occurred to a much lower extent in pure iron than in TRIP steel. HAC analysis showed that crack initiation is more stress-controlled for TRIP steel, while being strain-controlled for pure iron as a much smaller number of secondary cracks was present in pure iron which were generally wider compared to those in TRIP steel. More deformation was present on the surface of the pure iron samples as well. Both effect result from the ductile nature of pure iron, allowing more crack blunting and sample deformation prior to fracture. Sofronis and McMeeking [62] showed that the hydrostatic stress effect on the H distribution at a notch or crack is more significant for high-strength steels. Softer materials will accommodate the arising stresses more easily during plastic deformation. Therefore, a larger amount of lattice defects will be present in these regions which will trap H and hence impede H diffusion to critical high stress regions. Therefore, a strain-induced crack propagation of the principal crack is promoted, rather than stress-induced as observed for TRIP steel.

Further crack propagation of the secondary cracks was for both materials mainly stress-controlled. Hence, H related cracks exhibited a similar S-shape, independently of the microstructure type (cf. Figure 2). The observations in [61] indicate that the cracks propagate in two stages. First, an initiated crack propagates perpendicular to the tensile direction, which is a direct result of mode I crack tip opening in a uniaxial load situation. At a certain point (stage two), the crack starts interacting with the stress field of the approaching principal crack, which started at the notch root, causing a deviation of its crack tip by shearing in a direction determined by the local stress field. Nevertheless, the kinetics of crack initiation and propagation were presumably affected by the present microstructure and will hence occur faster for TRIP steel. Therefore, a higher number of H related cracks were formed in the TRIP steel allowing easier crack propagation. The presence of a notch similarly accelerates the HAC phenomenon, without affecting other HAC characteristics. Therefore, independently of the material's microstructure or the presence of a notch, H assisted cracks showed a characteristic S-shape and stepwise coalescence occurred. Therefore, it was concluded that the stress state surrounding the crack tip has a major impact on the HAC features, while the stress concentration induced by the notch had little effect.

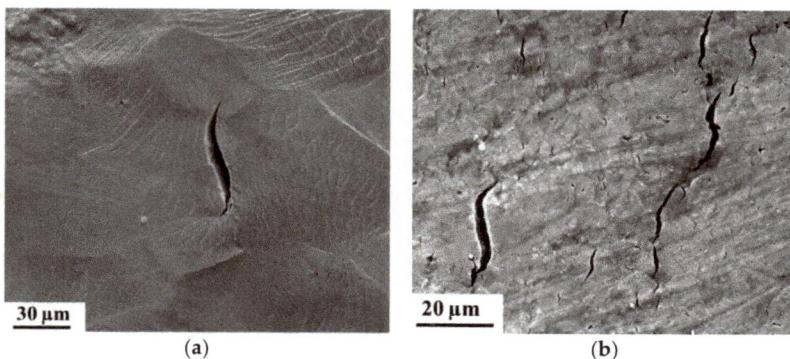

(a) (b)

Figure 2. SEM images showing similar S-shape of crack for pure iron (**a**) and TRIP steel (**b**).

The same HSS as well as DP and again pure iron, as a reference, were studied further by TDS and hot extraction experiments to evaluate the H interaction with their specific microstructural constituents [44]. The corresponding TDS spectra are presented in Figure 3a. The HSLA steel showed the highest TDS peak, while it contained less diffusible H, determined by hot extraction, compared to TRIP for instance. This was again correlated to the presence of Ti- and Nb-based precipitates, lowering the H diffusivity and thus increasing the HE resistance of HSLA (cf. Figure 1). Furthermore, TRIP showed a high temperature peak, which was not observed for the other materials. This peak is enlarged in Figure 3b. To correlate it to a microstructural feature, further analysis was performed.

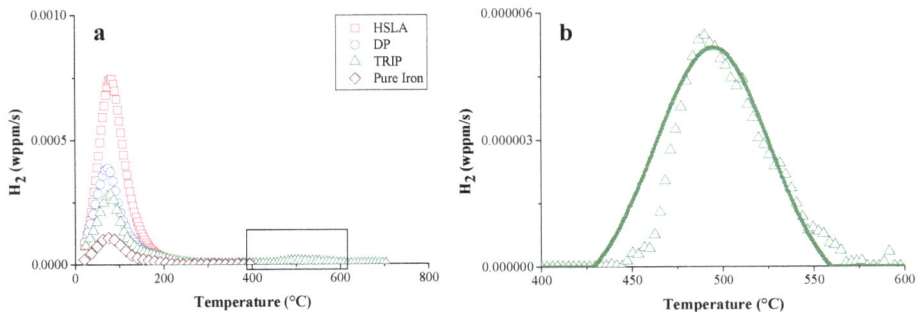

Figure 3. TDS spectra of HSLA, DP, TRIP steel and pure iron (**a**) with enlarged the high temperature peak of the TRIP (green) steel (**b**) (heating rate: 400 °C/h).

TDS was also done on TRIP steel and pure iron (as a reference material), in both undeformed and cold deformed (0–5–10–15%) conditions [42]. The corresponding TDS spectra are presented in Figure 4. The TDS spectra of TRIP consist of a single low temperature peak with significantly increasing intensity with increasing deformation and a small high temperature peak, which was also affected by the amount of cold deformation. Additional characterization techniques, such as X-ray diffraction, differential scanning calorimetry and scanning electron microscopy, were used to correlate the different peaks with the microstructural evolution induced by deformation. The high temperature peak was associated with the presence of retained austenite in TRIP steel. Indeed, when the deformation % in TRIP samples increases, the amount of retained austenite decreases since it transforms into martensite which explains the decreasing peak height with increasing amount of deformation. It was demonstrated [42] that this H was introduced in the material during processing. A comparison with pure iron was done to evaluate the low temperature peak. Based on the hot extraction results (cf. Figure 5), the increase of cold deformation in pure iron, i.e., the increase in dislocation density in the material, was detectable immediately after H charging, although no difference in H content was observed after 1 hour of vacuum, which is the experimental requirement to be able to start the TDS analysis. Hence, it could be concluded that H trapped at dislocations is not detectable by TDS when operating under those specific requirements. Moreover, a comparison with pure iron and the diffusivity measurements allowed us to correlate the low temperature peak mainly with martensite formation formed by deformation-induced transformation in TRIP steel.

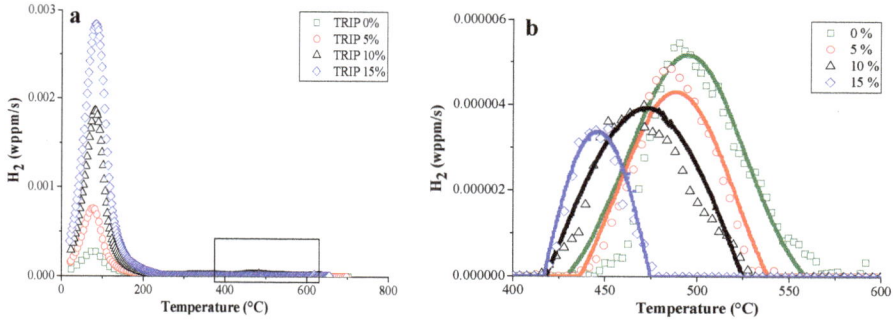

Figure 4. Effect of cold deformation on the TDS spectrum of the TRIP steel (**a**); with enlarged the high temperature peaks (**b**) (heating rate: 400 °C/h).

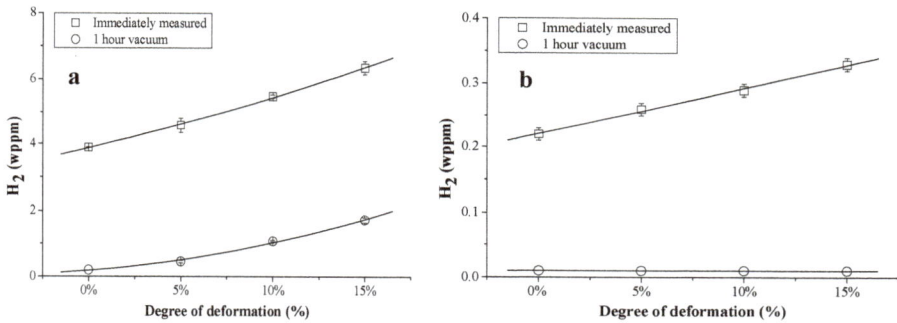

Figure 5. Hot extraction results immediately after H charging and after 1 h under vacuum as a function of the degree of cold deformation for TRIP (**a**) and pure iron (**b**).

In another work, the HAC initiation and propagation was extensively investigated for TRIP steel [63,64]. 85% of crack initiations occurred in or along martensitic regions (cf. Figure 6). The other 15% of the initiating cracks were in the ferrite and bainite grain centers. No cracks were observed along non-martensitic grain boundaries. More H is present around and in the martensitic islands in comparison to neighboring ferrite or bainite. This is linked to (i) freshly formed transformation-induced martensite which is supersaturated with H, since it originates from H containing retained austenite, which has a higher solubility (cf. Figure 4); (ii) interfaces in martensite and (iii) dislocations surrounding the martensite, which are formed due to the volume expansion during phase transformation, acting as H traps. Moreover, the high HE susceptibility of martensite makes this phase very prone to HAC after application of stress when it is H-enriched. The dominating mechanism for crack initiation was found to be martensite-martensite interface decohesion, implying a H Enhanced Interface Decohesion (HEIDE) mechanism, which is a variant of the often cited HEDE mechanism [24,65,66]. In HEIDE, H enhances specifically the decohesion across the martensitic interface. Furthermore, the crack propagation mode was found to be both inter- and transgranular, although the latter was the main one (cf. Figure 6). Although the crack propagates mainly transgranularly, it is not a cleavage fracture given that the orientation of the crack does not change when going from one grain to another. No relation was found between grain orientation and crack initiation and nucleation.

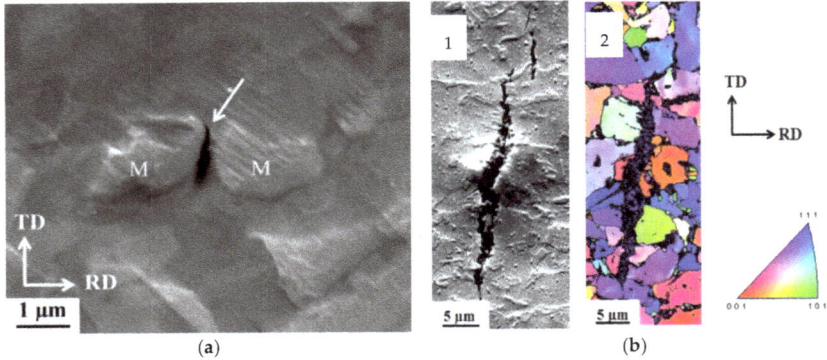

(a)   (b)

Figure 6. SEM image of initiating crack in TRIP between two martensitic (M) regions (**a**). SEM image of propagating crack (**b-1**) and [001] IPF map (**b-2**).

The H-induced mechanical degradation of DP was studied by tensile tests under variable H charging conditions, as presented in Figure 7 [43]. The ductility loss increased with H charging time until a maximal embrittlement of 50% was found, i.e., for a H-saturated sample, after which additional charging did not result in more embrittlement [43]. The influence of the cross-head deformation speed on H-saturated samples allowed evaluating the effect of H diffusion during the test. HE increased when lowering the test speed, as H was enabled to diffuse to critical regions ahead of the crack tip during the test.

Different cross-head deformation speeds were also applied on uncharged samples. Comparison between the outcomes of the charged and uncharged tests allowed to visualize the effective H diffusion distance into the sample during the test. The H distance x (cm) can be calculated by taking the square root of the product of the diffusion coefficient D (cm^2/s) and the test time t (s), i.e., $x = (D \times t)^{1/2}$ [33]. The distances H can diffuse at 5, 0.5 and 0.05 mm/min were about 65, 168 and 410 μm, respectively, which confirms the increased importance of H diffusion at lower deformation speeds. A fractography study indeed proved that the brittle H-induced features were present over a distance equivalent to the distance H can diffuse during the test (cf. Figure 8). Tests performed at the lowest cross-head speed, revealed brittle cleavage appearances in the central region which was correlated to the MnS segregation line that is prone to H-induced fracture.

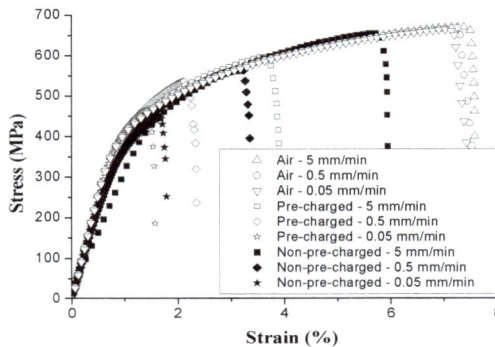

Figure 7. Stress-strain curves of H pre- and non-pre-charged DP at varying deformation speeds.

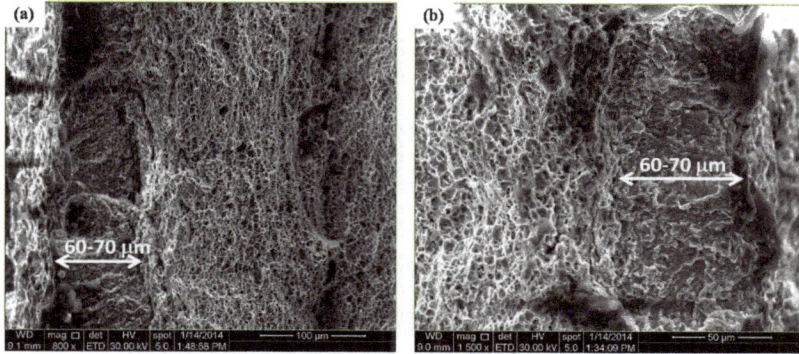

Figure 8. Detailed SEM image of both left (**a**) and right (**b**) side of the fracture surface of the sample tested at cross-head deformation speed of 5 mm/min without H pre-charging. The brittle zone is shown.

These HSS and pure iron, as a reference, were also charged with H without external load application using various severe charging conditions to evaluate both surface damage (H-induced blisters) and internal damage (H-induced cracks). To do so, different electrolytes, charging current densities and charging times were used [67]. Internal damage in DP samples was observed by the presence of cracks situated in the middle of the sample and propagating along the rolling direction (cf. Figure 9). HSLA did not show cracks for the same charging conditions. Cracks were identified to be both inter- and intragranular. Mn segregations were demonstrated for the DP and TRIP steel and elongated MnS inclusions were even clearly identified in the middle of the crack for the DP steel. This suggested that these second phase particles might play a role in the initiation of H-induced cracks. Similar results were obtained in [68,69], where inclusions which were hard, brittle, and incoherent with the matrix, such as manganese sulfides were recognized to be harmful to H-induced cracking. Similar blister formation and H-induced crack analysis was performed on pure iron as well. Pure iron showed faster blister formation compared to the HSS and was found to be more susceptible to H surface damage due to its softer matrix, as shown in Figure 10.

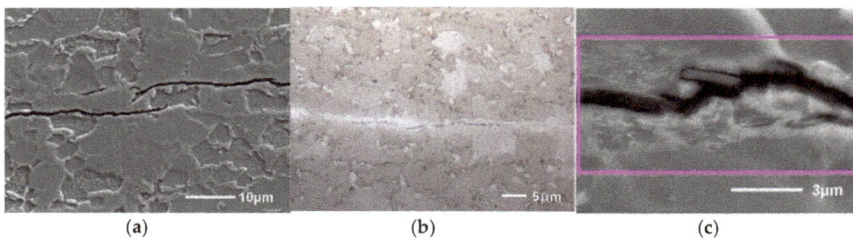

Figure 9. SEM (**a**) and LOM (**b**) images of DP revealing HIC together with SEM image (**c**) indicating the MnS containing segregation line in the center of the sample.

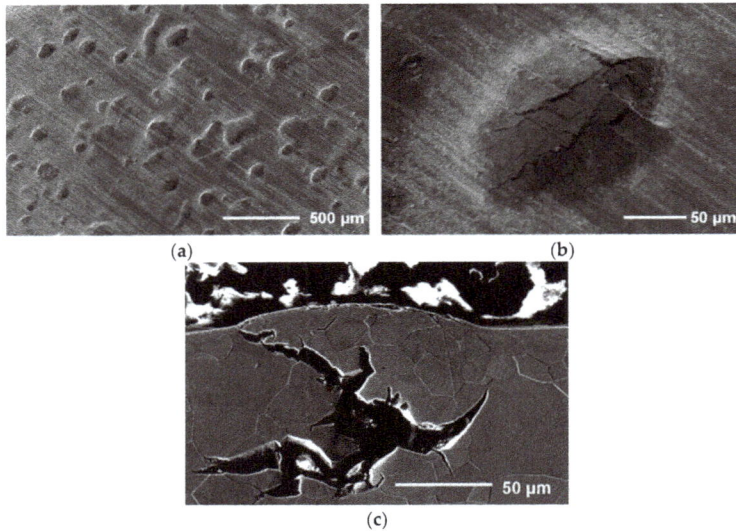

Figure 10. SEM images of surface (**a**, **b**) and cross section (**c**) of a crack propagation in pure iron charged for 1 h at 50 mA/cm^2 in arsenic-poisoned electrolyte.

4. The Interaction of Hydrogen with Lab-Cast Fe-C Alloys

The H/material interaction was further studied in generic alloys to avoid the effect of complex microstructural characteristics. A H-induced crack analysis was also done in an ultra-low carbon steel by exposing it to electrochemical H charging [49]. Three material conditions, i.e., cold deformed, recovered, and recrystallized, were considered to assess the impact of deformation-induced defects, such as dislocations and microvoids [70], on the HIC sensitivity. Generally, H diffusion is hindered by such defects, which stimulates H accumulation and thus causes easier crack initiation at these sites. Therefore, the H-induced blister formation is promoted by these defects, as shown in Figure 11. However, a certain critical concentration of H needs to be reached before initiation of blisters occur. If this critical concentration is reached, H recombination will occur, and blister formation takes place according to the internal pressure theory [66], which was confirmed by EBSD in [49]. The H charging time at a specific charging current density required to achieve this concentration is decreased for deformed material compared to heat treated material since a higher amount of crack initiation sites are present in deformed material due to the presence of a higher density of deformation-induced defects. Initiation of HICs in recrystallized material was related to debonding of inclusions which were found in the material, as illustrated in Figure 11. The same observations were made by Tiegel et al. [71]. They stated that H accumulation occurred at the interfaces of incoherent particles, causing an interface failure. Moreover, the high H concentration created vacancy stabilization providing the required space to form molecular H and the consequential internal pressure build-up resulted into cracks. Additionally, the cracking behavior was also largely affected by the applied current density (10 vs. 20 mA/cm^2), which was linked to a change in the internal crack morphology. Cracks were dominantly transgranular at low current density. The favorable crack propagation paths were associated to high dislocation density slip planes. While intergranular crack propagation along high angle grain boundaries occurred at higher current density.

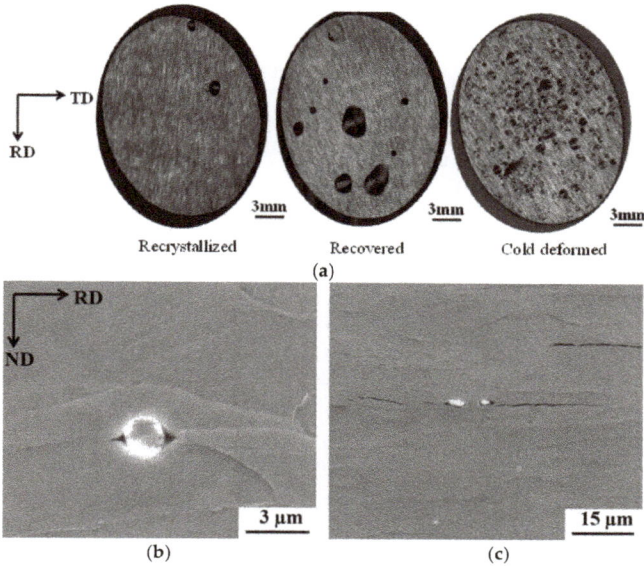

Figure 11. Blister surface appearance (**a**) of recrystallized, recovered, and cold deformed ULC steel charged at 5 mA/cm² for 2 days together with SEM images revealing crack initiation at alumina particle (**b**, **c**).

The influence of deformation-induced defects was studied further by permeation experiments. Since inclusions were present in the ULC steel, Armco pure iron was used for this purpose [50]. The material was reduced in thickness by 50 and 70% of cold rolling. Figure 12 shows how the permeation transient is affected by the microstructural changes, i.e., a H diffusivity decrease was linked to the presence of deformation-induced defects, such as dislocations and microvoids [70]. When the material was annealed to release the stresses of the 70% cold rolled grade, a higher diffusivity was observed, implying less H traps present in the material. This was more linked to rearrangements of dislocations and internal stress relief than to a considerable decrease of dislocation density [72].

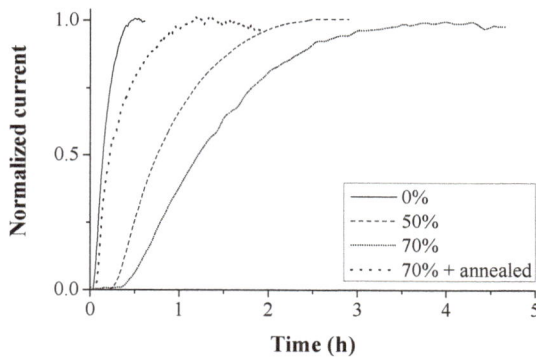

Figure 12. Permeation transients of Armco iron in specific cold deformed conditions.

The effect of H charging on the mechanical properties of Fe-C alloys was studied in [46–48]. Different phases, such as pearlite, bainite and martensite were generated in a 0.2% C Fe-C alloy by an appropriate heat treatment, referred to as P2, B2 and M2, respectively. An increase in the carbon content up to 0.4% was established for the bainitic grade, i.e., B4. Tensile tests were done at 5 mm/min on notched specimens and a significantly different HE degree was observed. The obtained degrees of HE were correlated with the amount of H present in the materials (cf. Table 4). A distinction between the amount of diffusible H and the total H was made, since diffusible H has been argued to play a crucial role in H-induced failure [5,73,74]. For the 0.2% C Fe-C alloys, the pearlitic microstructure was more susceptible to HE compared to the bainitic and martensitic grade, which exhibited a similar HE response. This was linked to the higher diffusible H content and H diffusivity for P2. Increasing the C% to 0.4% for the bainitic microstructure resulted in an increased HE due to the larger amount of diffusible H. The yield strength was also affected by H charging due to solid solution strengthening caused by the interstitial H as the increase in yield strength was correlated with the total amount of H determined by melt extraction, as observed as well in [5]. Tensile tests performed at a lower speed of 0.05 mm/min allowed to further elaborate the synergetic effect of both the amount of diffusible H and the H diffusion distance during the test. An additional ductility loss was observed at a slower cross-head speed, which was the lowest for the martensitic grade (22→30%), as this material showed the lowest H diffusivity. Although adding carbon resulted in a higher amount of diffusible H, a lower HE increase was obtained for B4 (40→63%) compared to B2 (21→50%) when lowering the cross-head deformation speed. This was related to the lower H diffusion coefficient of B4 (3.40×10^{-10} m^2/s) compared to B2 (6.71×10^{-10} m^2/s), as demonstrated in [48].

Table 4. Melt (1600 °C) and hot (300 °C) extraction results for H-saturated samples in wppm (weighted Parts per million).

	Total H (wppm)	Diffusible H (wppm)	HE% (5 mm/min)	HE% (0.05 mm/min)
B2	8.71 ± 0.12	3.82 ± 0.06	21	50
M2	10.53 ± 0.15	4.72 ± 0.07	22	30
P2	9.69 ± 0.13	5.10 ± 0.09	52	-
B4	14.65 ± 0.16	8.73 ± 0.11	40	63

5. The Interaction of Hydrogen with Lab-Cast Fe-C-X Alloys

The effect of carbide addition was evaluated in several types of Fe-C-X alloys [51,75–79], with X = Ti, V, Mo, Cr and W, respectively. At first, a martensitic microstructure was considered where the as-Q and Q&T condition, containing tempered induced carbides, were compared. Tensile tests were done in air and on H-saturated unnotched specimen at 5 mm/min. The stress-strain curves for the Fe-C-Ti and Fe-C-V alloy B (medium C content) in both conditions are depicted in Figure 13. Tempering caused an increase in strength level and ductility for Fe-C-Ti whereas the opposite effect was observed for Fe-C-V. Only a limited amount of plastic deformation was present for the Fe-C-V alloy since it includes a fast dissolution of carbides during austenitizing, resulting in a dense C rich martensitic microstructure and hence a more brittle material after quenching. For the Q&T condition, the secondary hardening effect was pronounced for Fe-C-Ti, while a decrease in dislocation density overtook the strengthening effect of the precipitates for Fe-C-V. The HE degree increased upon tempering when comparing the as-Q with the Q&T condition for Fe-C-Ti (21→60%) and Fe-C-V (28→32%).

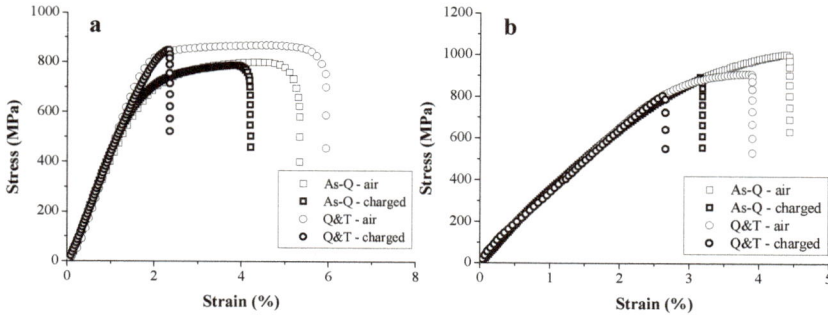

Figure 13. Stress-strain curves for (**a**) Fe-C-Ti; (**b**) and Fe-C-V alloy B at a cross-head deformation speed of 5 mm/min of uncharged (air) and H-saturated (charged) samples.

The H/material interaction was studied further to comprehend the underlying reasoning of the obtained HE%. TDS was done on all Fe-C-X materials for both the as-Q and Q&T condition. The TDS spectra of Fe-C-Ti and Fe-C-V alloy B in both conditions, together with their deconvoluted peaks, are presented in Figure 14. For the as-Q condition, only one peak was observed showing activation energies of about 27–30 kJ/mol. This first peak was therefore attributed to H trapped by martensitic lath boundaries [80]. Although H trapped at dislocations shows an E_a in the same range [9,70,81,82], this trapping site is assumed to be largely undetectable due to the experimental requirements to perform the TDS analysis (cf. Figures 4 and 5) [42,83]. Although incoherent large TiC particles were present in the as-Q condition of Fe-C-Ti [51], a single peak was present indicating that no H was trapped from electrochemical charging at their interface. This confirmed that gaseous H charging at elevated temperature was required to charge these large particles [40] and H from the gaseous charging being presumably trapped inside the carbide [37] instead of at its interface, which will be further analyzed.

When considering the Q&T condition, a significantly different trapping behavior was revealed. The additional peaks in the spectra were attributed to the presence of small TiC and V_4C_3 carbides which can trap a high amount of H, resulting in three peaks with activation energies in the range of 50–71 kJ/mol. The first deconvoluted peak was slightly increased for the Q&T condition compared to the as-Q condition, which was attributed to the slower H diffusivity due to the presence of the precipitates, as determined by permeation [51,75–78]. Since the trapping ability was found to be dependent on the carbide size, the tempering time was increased (10 and 20 h) to verify the size dependent trapping ability of TiC, as illustrated in Figure 15. Representative TEM bright field images are shown for each condition, together with the corresponding carbide size distribution map to evaluate the increase in carbide size with tempering time. The related TDS spectra is given as well to correlate the deconvoluted peaks to the microstructural characteristics. The carbide related TDS peaks decreased with tempering time, which was linked to the decreased total interfacial area between the carbides and the matrix. As such, it was confirmed that H was trapped at the carbide/matrix interface. Moreover, it was shown that no H was trapped by TiC larger than 70 nm by tempering the material for 20 h, as shown in Figure 15d. The peak shift obtained when comparing Figure 15a with Figure 15d is linked to the increased elastic strain fields surrounding the large incoherent TiC [51].

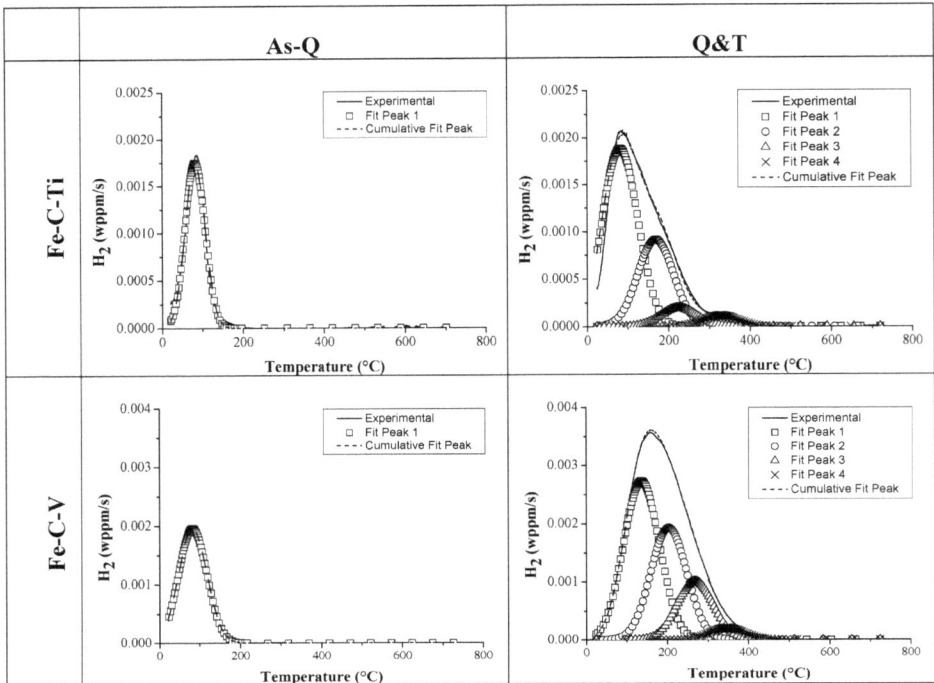

Figure 14. TDS curves of Fe-C-X grades (alloy B) in the as-Q and Q&T condition (heating rate: 600 °C/h).

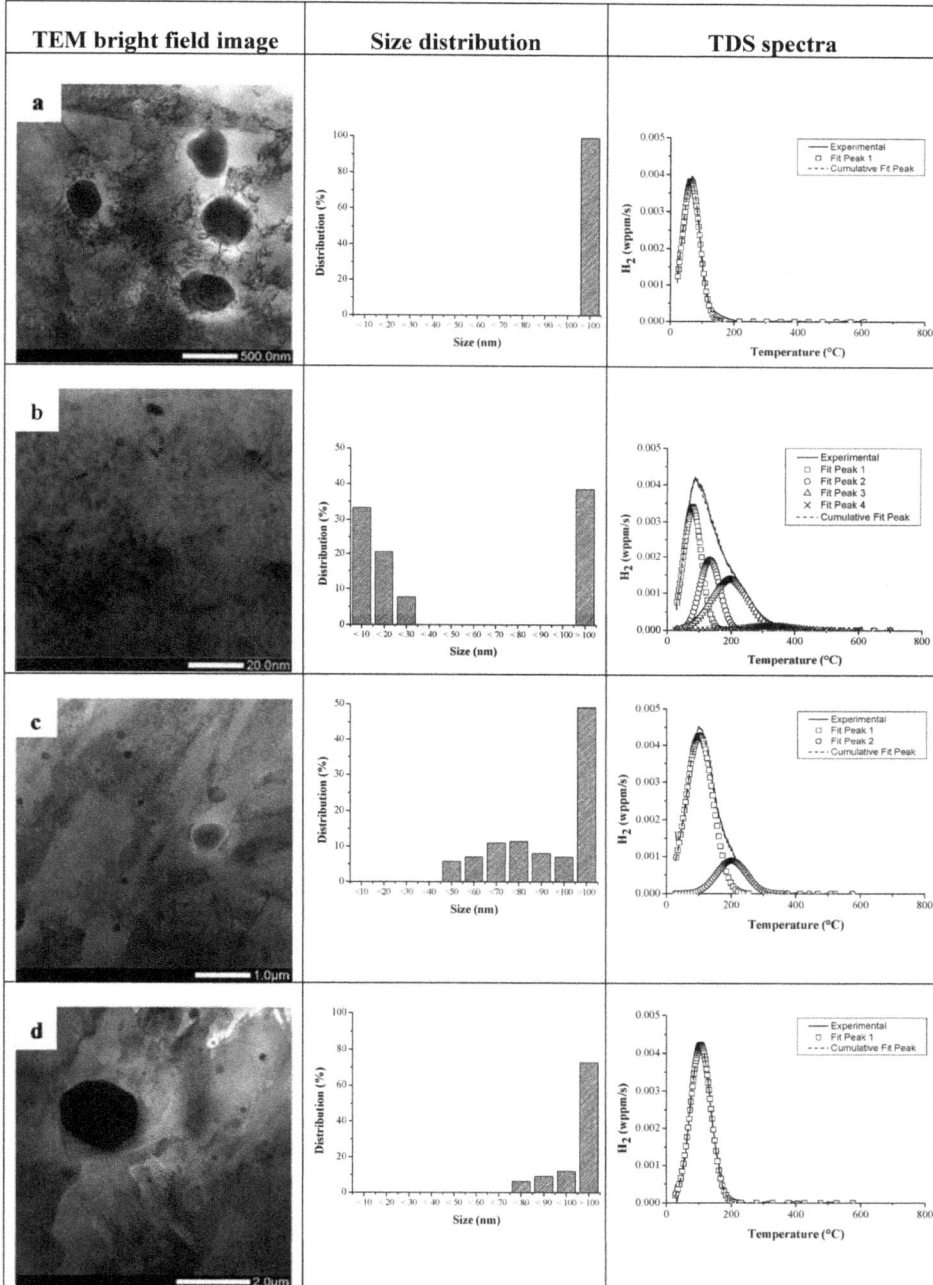

Figure 15. Representative TEM bright field image, carbides size distribution and TDS spectrum for alloy C in the as-Q (**a**); Q&T 1 h (**b**); Q&T 10 h (**c**) and Q&T 20 h (**d**) condition (heating rate: 600 °C/h).

Three different types of H were distinguished in the detailed study on Fe-C-X alloys, i.e., the total H amount measured by melt extraction, the diffusible H content measured by hot extraction and the mobile H content, which was defined as the amount of H which was released before the TDS measurement due to the abovementioned experimental requirements to achieve a vacuum level in the analysis chamber. This mobile H amount was calculated by subtracting the amount determined by TDS from the diffusible H content. This mobile H is correlated to H trapped at dislocations which is undetectable by that specific TDS set-up [83]. To analyze the correlation between the HE degree and the type and amount of H, a linear fitting of the different types of H (i.e., total, diffusible, and mobile) and HE% is made (cf. Figure 16a). For this purpose, all alloys for each Fe-C-X grade (cf. Table 3) in the as-Q and Q&T condition were included (30 different materials conditions). The correlation improved for total over diffusible to mobile H. Moreover, when the Fe-C-V materials were excluded, which is a valid assumption since they did not show any significant plastic deformation, (cf. Figure 16b), a R^2 of 95% between the degree of HE and the amount of mobile H was achieved. This clearly indicated the crucial role of H trapped by dislocations on the HE during a uniaxial tensile test which is linked to the enhancement of the dislocation mobility by the presence of H as proposed by the HELP mechanism [83]. Therefore, these experimental results present a very nice confirmation of the HELP theory.

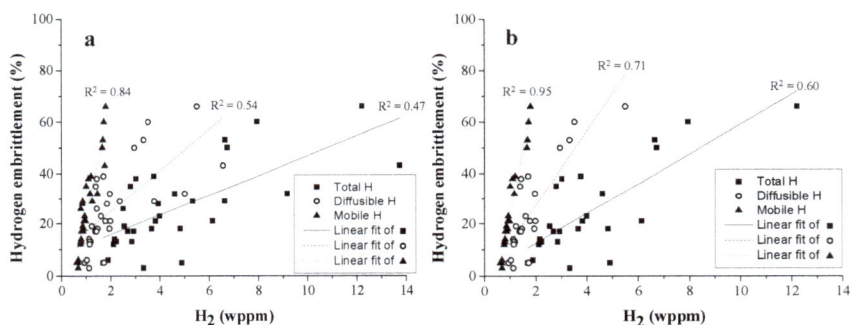

Figure 16. Degree of HE vs. the total, diffusible and mobile H content for all alloys of the Fe-C-X grades in the as-Q and Q&T condition (**a**); The Fe-C-V alloys were excluded in (**b**) to illustrate the correlation between mobile H and HE for alloys which failed past the macroscopic yield strength.

One remarkable result was the increased HE susceptibility for Q&T samples despite the presence of H trapping carbides (cf. Figures 13 and 14). Evidently, the question arose whether the presence of these precipitates is a good strategy to enhance the resistance to HE. Since a clear correlation between the amount of H and the HE degree was found (cf. Figure 16), an alternative charging procedure was applied. So far, the materials were charged electrochemically till H saturation, which implies that the amount of H present in the as-Q and Q&T condition upon testing was not equal. Indeed, for example a H-saturated Q&T material having TiC contained about double the amount of H than its as-Q variant. Consequently, a similar H content compared to the as-Q state was charged in Q&T condition of Fe-C-Ti in a second series of tensile tests. This condition will be referred to as Q&T 'charged*'. These materials were compared with the H-saturated as-Q and Q&T samples and the corresponding stress-strain curves are displayed in Figure 17a. The HE resistance increased considerably when the tempered condition was compared to the as-Q state when a similar amount of H was present. Therefore, an appropriate next step is to evaluate the H distribution in the non-saturated Q&T samples (charged*) over the different traps by TDS and to compare the resulting spectrum with the TDS spectrum of the Q&T sample which was H-saturated (charged), as shown in Figure 17b. This comparison indicated that the trapping sites with the higher activation energy, i.e., those linked with TiC carbides, were first filled. The first peak, mainly associated with H at lath boundaries, had not yet been filled in the non-saturated materials. Therefore, H was first trapped by the deeper trapping sites, which have the higher E_a. Consequently, a higher proportion of H is strongly trapped and hence the non-saturated samples have a relatively lower amount of mobile H which improves the HE resistance. This demonstrates that TiC are indeed beneficial to reduce the susceptibility to HE. Hence, the beneficial effect of carbide addition depends on the carbides characteristics, the charging conditions and effective amount of hydrogen induced in the material.

An alternative way to evaluate the available trapping sites was done by performing permeation tests, as presented in Figure 18. The H diffusivity increased from Q&T 1 h < Q&T 2 h < as-Q which was in the opposite order as the amount of available strong trapping sites. When tempering was done for 2 h, the TDS peaks linked to H at TiC, i.e., peak 2, 3 and 4 in Figure 15b, considerably decreased. Since the amount of interface between carbides and matrix decreased when the precipitates grew, and their interface became more incoherent, the present results indicated that H was trapped at this interface rather than inside the carbides. These observations are in good agreement with those obtained by Wei and Tsuzaki [40,41], Pérez Escobar et al. [84] and density functional theory calculations by Di Stefano [37].

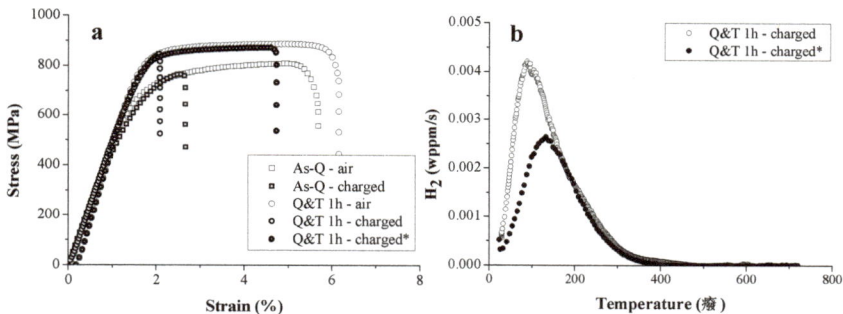

Figure 17. Stress-strain curves (**a**) and corresponding TDS spectra (**b**) for Fe-C-Ti alloy C in the as-Q and Q&T 1h condition. H charging was applied till saturation (charged) and until a similar H amount as as-Q was obtained for Q&T 1 h-charged*.

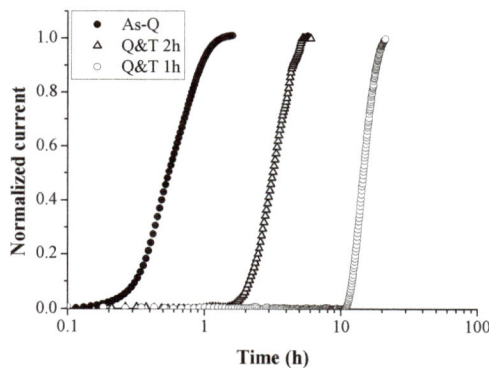

Figure 18. Permeation curves for Fe-C-Ti alloy C in the as-Q, Q&T 1 h and Q&T 2 h condition.

The H interaction with carbides was also considered in a ferritic microstructure for several carbide-forming elements. At first, the H trapping ability of NbC was considered by TDS in an experimental steel containing 0.013 wt % C–0.1 % Nb [85]. The sample was, next to gaseously charged at elevated temperature and atmospheric pressure, also electrochemically charged before the measurement. The details on the gaseous hydrogen charging procedure can be found elsewhere [85]. This resulted in a curve showing two distinct peaks, a low temperature peak and a high temperature peak (cf. Figure 19). The low temperature peak was attributed to the H that was introduced by electrochemical charging and trapped at grain boundaries, with activation energies ranging from 24–33 kJ/mol and at the interface of the smaller precipitates, with E_a of about 23–48 kJ/mol. The high temperature peak was attributed to the H that is introduced by gaseous charging and is trapped inside the vacancies of the incoherent NbC precipitates, with an activation energy ranging between 63 and 68 kJ/mol. Gaseous charging is necessary to provide the required energy to trap the H in the vacancies of the precipitates.

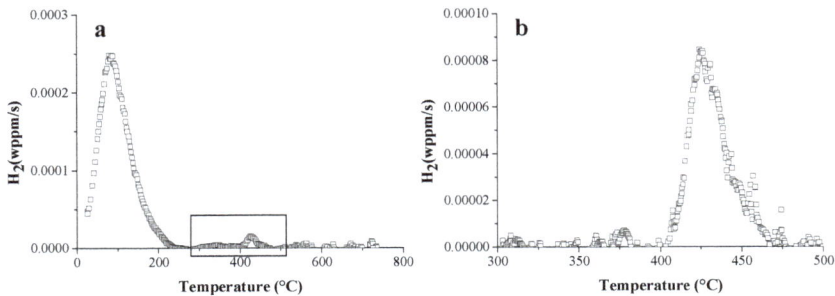

Figure 19. TDS spectrum of NbC containing ferritic steel after both electrochemical and gaseous charging (**a**); The high temperature peak linked with H inside NbC is presented in (**b**) (heating rate: 400 °C/h).

The next experimental steel, with 0.025 wt % C and 0.09 wt %Ti, showed a microstructure consisting of small ferrite grains and nanometer size TiC precipitates [84]. The steel contained some H in irreversible traps, which were TiC, and which originated from the hot rolling of the steel. After annealing in H_2, the TDS spectra contained a high temperature peak, for which both the peak area and peak temperature increased with higher annealing temperature, which was related to an increased carbide vacancy concentration at higher annealing temperature, which may act as potential H traps. Therefore, this peak was attributed to irreversible trapping by TiC particles. The TDS spectra for samples annealed in H_2 atmosphere at 800 °C and sequentially electrochemically charged for 1 h at 0.8 mA/cm^2 are presented in Figure 20. A low and high temperature peak was revealed. With increased desorption time, the low peak decreased in height, whereas the high temperature peak did not change significantly. Therefore, the low temperature peak was correlated with reversible traps, i.e., grain boundaries, while the high temperature peak was associated with irreversible trapping by the TiC particles. The amount of H in the reversible traps desorbed gradually with increased desorption time, whereas practically all H trapped by TiC remained in the material. The irreversibly trapped H by TiC in the high temperature peak showed an E_a of about 145 kJ/mol.

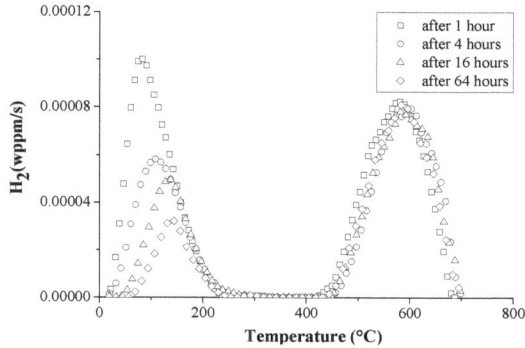

Figure 20. TDS spectra of TiC containing ferritic steel annealed at 800 °C in gaseous H_2 atmosphere, charged electrochemically and submitted to specified desorption times in vacuum (heating rate: 400 °C/h).

6. Conclusions

The H/material interaction was studied in this work. In situ tensile testing was done to evaluate the HE sensitivity, whereas the H content was determined by hot and melt extraction. TDS analysis further revealed the H trapping capacity and permeation experiments allowed the determination of the H diffusivity. Blisters or HIC were induced by severe electrochemical charging and HAC was assessed by interrupting the in situ tensile tests to understand the complete failure process in terms of crack initiation and propagation. These H characterization techniques were related with a thorough microstructural analysis of the different considered materials.

The interaction of H with multiphase high-strength steels was studied and a combined influence of both the amount of H and the diffusivity of H determined the observed HE degrees. HSLA steel showed the highest resistance to HE (8% ductility loss) which was linked to the presence of Ti- and Nb-based precipitates. TDS demonstrated that the retained austenite in TRIP showed a high activation energy trap for H, present from the processing of the material. Though, since retained austenite transformed into martensite during tensile deformation, a high HE sensitivity of 60% was obtained for this material. Further SEM and EBSD analysis showed that initiation of HAC in TRIP occurred at the martensitic islands, which is established on a H-enhanced interface decohesion mechanism. Furthermore, the HAC characteristics were not affected by the presence of a notch in the tensile sample geometry. Lastly, a fractographic SEM analysis on DP steel clearly visualized the effect of H diffusion during an in situ tensile test; the calculated H diffusion distance equals the observed transition between a H-induced brittle and ductile fracture. Moreover, HIC was demonstrated to initiate at the MnS inclusions present in this DP steel.

Nevertheless, these multiphase high-strength steels contain a complex microstructure, which motivated the fundamental H/material interaction in single phase microstructures, such as ferrite, pearlite, bainite and martensite. ULC steel, pure iron and generic Fe-C alloys were used for this purpose. Initiation of HIC in recrystallized ULC steel was correlated with debonding of inclusions present in the material. Permeation experiments allowed to evaluate the influence of deformation-induced defects in pure iron. A decrease in H diffusivity was observed and linked to the presence of deformation-induced defects, such as dislocations and microvoids. The HE susceptibility was further evaluated in Fe-C alloys and both the H amount and diffusivity determined the observed degrees of HE. Finally, the H trapping ability of carbides and their effect on the susceptibility to HE was considered in generic Fe-C-X materials. TDS revealed that the carbide trapping ability was mainly determined by their size and coherency with the matrix. In addition, gaseous H charging at elevated temperature was required to charge the incoherent carbides, showing an irreversible H peak by TDS. The amount of H present in the materials, as determined by combining hot extraction, melt extraction and TDS, was correlated with the HE sensitivity. It was demonstrated that H trapped by dislocations played a determinant role in the obtained HE degrees. This was linked to an enhanced dislocation mobility in the presence of H, which provides a clear experimental support for the HELP mechanism.

Author Contributions: T.D., A.L., D.P.E., E.V.d.E., E.W. and K.V. designed the experiments. T.D., A.L., D.P.E., E.V.d.E. and E.W. performed the experiments and analyzed the data. K.V. contributed analysis tools and T.D. wrote the paper. All authors provided feedback on the manuscript.

Funding: T.D. holds a postdoctoral fellowship via grant nr BOF01P03516. The authors wish to thank the Special Research Fund (BOF), UGent (BOF15/BAS/06), the MaDuRos program (SIM), part of the DeMoPreCI-MDT project and the Agency for Innovation by Science and Technology in Flanders (IWT) (Project nr SB141399).

Acknowledgments: The authors wish to thank Vitaliy Bliznuk for the TEM images. The authors also acknowledge the technical staff from the Department Materials, Textiles and Chemical Engineering, UGent, for their help with the experiments and/or sample preparation.

Conflicts of Interest: The authors declare no conflict of interest.

References

1. Hauch, A.; Ebbesen, S.D.; Jensen, S.H.; Mogensen, M. Highly efficient high temperature electrolysis. *J. Mater. Chem.* **2008**, *18*, 2331–2340. [CrossRef]
2. Woodtli, J.; Kieselbach, R. Damage due to hydrogen embrittlement and stress corrosion cracking. *Eng. Fail. Anal.* **2000**, *7*, 427–450. [CrossRef]
3. Hilditch, T.B.; Lee, S.B.; Speer, J.G.; Matlock, D.K. *Response to Hydrogen Charging in High Strength Automotive Sheet Steel Products*; SAE Technical Paper; SAE Int.: Warrendale, PA, USA, 2003. [CrossRef]
4. Venezuela, J.; Liu, Q.; Zhang, M.; Zhou, Q.; Atrens, A. The influence of hydrogen on the mechanical and fracture properties of some martensitic advanced high strength steels studied using the linearly increasing stress test. *Corros. Sci.* **2015**, *99*, 98–117. [CrossRef]
5. Duprez, L.; Verbeken, K.; Verhaege, M. Effect of Hydrogen on the Mechanical Properties of Multiphase High Strength Steels. In Proceedings of the 2008 International Hydrogen Conference, Jackson Lake Lodge, WY, USA, 7–10 September 2008.
6. Loidl, M.; Kolk, O. Hydrogen embrittlement in HSSs limits use in lightweight body. *Adv. Mater. Process.* **2011**, *169*, 22–25.
7. Olden, V.; Thaulow, C.; Johnsen, R.; Ostby, E.; Berstad, T. Influence of hydrogen from cathodic protection on the fracture susceptibility of 25% Cr duplex stainless steel—Constant load SENT testing and FE-modelling using hydorgen influenced cohersive zone elements. *Eng. Fract. Mech.* **2009**, *76*, 827–844. [CrossRef]
8. Johnson, W.H. On some remarkable change produced in iron and steel by the action of hydrogen and acids. *Proc. R. Soc. Lond.* **1875**, *23*, 168–179. [CrossRef]
9. Hirth, J.P. Effects of Hydrogen on the Properties of Iron and Steel. *Metall. Trans. A* **1980**, *11*, 861–890.
10. Oriani, R.A.; Hirth, J.P.; Smialowski, M. *Hydrogen Degradation of Ferrous Alloys*; Noyes Publications: Park Ridge, NJ, USA, 1985.
11. Gangloff, R.P.; Ives, M.B. (Eds.) *Environment-Induced Cracking of Metals*; NACE: Houston, TX, USA, 1990.
12. Vehoff, H. Hydrogen related material properties. *Topics Appl. Phys.* **1997**, *73*, 215–278.
13. Sofronis, P. Special issue on recent advances in engineering aspects of hydrogen embrittlement. *Eng. Fracture Mech.* **2001**, *68*. [CrossRef]
14. Pundt, A.; Kirchheim, R. Hydrogen in metals: Microstructural aspects. *Ann. Rev. Mater. Res.* **2006**, *36*, 555–608. [CrossRef]
15. Somerday, B.; Sofronis, P.; Jones, R. Effects of Hydrogen on Materials. In Proceedings of the 2008 International Hydrogen Conference, Jackson Lake Lodge, WY, USA, 7–10 September 2008.
16. Somerday, B.; Sofronis, P.; Jones, R. Hydrogen-Materials Interactions. In Proceedings of the 2012 International Hydrogen Conference, Jackson Lake Lodge, WY, USA, 9–12 September 2012.
17. Gangloff, P.; Somerday, B. *Gaseous Hydrogen Embrittlement of Materials in Energy Technologies*; Woodhead: Cambridge, UK, 2012.
18. Somerday, B.; Gangloff, P.; Jonas, R. Effects of hydrogen on materials. In Proceedings of the 2016 International Hydrogen Conference, Jackson Lake Lodge, WY, USA, 11–14 September 2016.

19. Revie, R.W.; Sastri, V.S.; Elboujdaii, M.; Ramsingh, R.R.; Lafrenière, Y. Hydrogen-induced cracking of line pipe steels used in sour service. *Corrosion* **1993**, *49*, 531–536. [CrossRef]
20. Chawla, K.K.; Rigsbee, J.M.; Woodhouse, J.B. Hydrogen-induced cracking in two linepipe steels. *J. Mater. Sci.* **1986**, *21*, 3777–3782. [CrossRef]
21. Maroef, I.; Olson, D.L.; Eberhart, M.; Edwards, G.R. Hydrogen trapping in ferritic steel weld metal. *Int. Mater. Rev.* **2002**, *47*, 191–223. [CrossRef]
22. Olden, V.; Saai, A.; Jemblie, L.; Johnsen, R. FE simulation of hydrogen diffusion in duplex stainless steel. *Int. J. Hydrog. Energy* **2014**, *39*, 1156–1163. [CrossRef]
23. Eliaz, N.; Shachar, A.; Tal, B.; Eliezer, D. Characteristics of hydrogen embrittlement, stress corrosion cracking and tempered martensite embrittlement in high-strength steels. *Eng. Fail. Anal.* **2000**, *9*, 167–184. [CrossRef]
24. Troiano, A.R. The role of Hydrogen and Other Interstitials in the Mechanical Behaviour of Metals. *Trans. ASM* **1960**, *52*, 54–80.
25. Beachem, C.D. A new Model for Hydrogen-Assisted Cracking. *Metall. Trans. A* **1972**, *3*, 437–451.
26. Lynch, S.P. Comments on A unified model of environment-assisted cracking. *Scr. Mater.* **2009**, *61*, 331–334. [CrossRef]
27. Lynch, S.P. Environmentally assisted cracking: Overview of evidence for an adsorption-induced localised-slip process. *Acta Metall.* **1988**, *36*, 2639–2661. [CrossRef]
28. Gangloff, R.P. Hydrogen assisted cracking of high strength alloys. In *Comprehensive Structural Integrity*; Elsevier: Amsterdam, The Netherlands, 2003; pp. 31–101.
29. Barnoush, A.; Vehoff, H. Electrochemical nanoindentation: A new approach to prove hydrogen/deformation interaction. *Scr. Mater.* **2006**, *55*, 195–198. [CrossRef]
30. Birnbaum, H.K.; Sofronis, P. Hydrogen-enhanced localized plasticity—A mechanism for hydrogen-related fracture. *Mater. Sci. Eng. A* **1994**, *176*, 191–202. [CrossRef]
31. Ryu, J.H.; Chun, Y.S.; Lee, C.S.; Bhadeshia, H.K.D.H.; Suh, D.W. Effect of deformation on hydrogen trapping and effusion in TRIP-assisted steel. *Acta Mater.* **2012**, *60*, 4085–4092. [CrossRef]
32. Liu, Q.; Venezuela, J.; Zhang, M.; Zhou, Q.; Atrens, A. Hydrogen trapping in some advanced high strength steels. *Corros. Sci.* **2016**, *111*, 770–785. [CrossRef]
33. Koyama, M.; Tasan, C.C.; Akiyama, E.; Tsuzaki, K.; Raabe, D. Hydrogen-assisted decohesion and localized plasticity in dual-phase steel. *Acta Mater.* **2014**, *70*, 174–187. [CrossRef]
34. Wang, S.; Martin, M.L.; Sofronis, P.; Ohnuki, S.; Hashimoto, N.; Robertson, I.M. Hydrogen-induced intergranular failure of iron. *Acta Mater.* **2014**, *69*, 275–282. [CrossRef]
35. Deng, Y.; Hajilou, T.; Wan, D.; Kheradmand, N.; Barnoush, A. In-situ micro-cantilever bending test in environmental scanning electron micrscope: Real time observation of hydrogen enhanced cracking. *Scr. Mater.* **2017**, *127*, 19–23. [CrossRef]
36. Wan, D.; Deng, Y.; Barnoush, A. Hydrogen embrittlement observed by in-situ hydrogen plasma charging on a ferritic alloy. *Scr. Mater.* **2018**, *151*, 24–27. [CrossRef]
37. Di Stefano, D.; Nazarov, R.; Hickel, T.; Neugebauer, J.; Mrovec, M. Elsässer, First principles investigation of hydrogen interaction with TiC precipitates in alpha-Fe. *Phys. Rev.* **2016**, *93*, 184108. [CrossRef]
38. Nagao, A.; Martin, M.L.; Dadfarnia, M.; Sofronis, P.; Robertson, M. The effect of nanosized (Ti,Mo)C precipitates on hydrogen embrittlement of tempered lath martensitic steel. *Acta Mater.* **2014**, *74*, 244–254. [CrossRef]
39. Asahi, H.; Hirakami, D.; Yamasaki, S. Hydrogen trapping behavior in vanadium-added steel. *ISIJ Int.* **2003**, *43*, 527–533. [CrossRef]
40. Wei, F.G.; Tsuzaki, K. Quantitative Analysis on hydrogen trapping of TiC particles in steel. *Metall. Mater. Trans. A* **2006**, *37*, 331–353. [CrossRef]
41. Wei, F.G.; Tsuzaki, K. Hydrogen Absorption of Incoherent TiC Particles in Iron from Environment at High Temperatures. *Metall. Mater. Trans. A* **2004**, *35*, 3155–3163. [CrossRef]
42. Pérez Escobar, D.; Depover, T.; Wallaert, E.; Duprez, L.; Verbeken, K.; Verhaege, M. Combined thermal desorption spectroscopy, differential scanning calorimetry, scanning electron microscopy and X-ray diffraction study of hydrogen trapping in cold deformed TRIP steel. *Acta Mater.* **2012**, *60*, 2593–2605. [CrossRef]
43. Depover, T.; Wallaert, E.; Verbeken, K. Fractographic analysis of the role of hydrogen diffusion on the hydrogen embrittlement susceptibilty of DP steel. *Mater. Sci. Eng. A* **2016**, *649*, 201–208. [CrossRef]

44. Pérez Escobar, D.; Verbeken, K.; Duprez, L.; Verhaege, M. Evaluation of hydrogen trapping in high strength steels by thermal desorption spectroscopy. *Mater. Sci. Eng. A* **2012**, *551*, 50–58. [CrossRef]

45. Depover, T.; Pérez Escobar, D.; Wallaert, E.; Zermout, Z.; Verbeken, K. Effect of in-situ hydrogen charging on the mechanical properties of advanced high strength steels. *Int. J. Hydrog. Energy* **2014**, *39*, 4647–4656. [CrossRef]

46. Depover, T.; Verbeken, K. On the synergy of diffusible hydrogen content and hydrogen diffusivity in the mechanical degradation of laboratory cast Fe-C alloys. *Mater. Sci. Eng. A* **2016**, *664*, 195–205. [CrossRef]

47. Pérez Escobar, D.; Depover, T.; Wallaert, E.; Duprez, L.; Verhaege, M.; Verbeken, K. Thermal desorption spectroscopy study of the interaction between hydrogen and different microstructural constituents in lab cast Fe-C alloys. *Corros. Sci.* **2012**, *65*, 199–208. [CrossRef]

48. Depover, T.; Van den Eeckhout, E.; Verbeken, K. The impact of hydrogen on the ductility loss of bainitic Fe-C alloys. *Mater. Sci. Technol.* **2016**, *32*, 1625–1631. [CrossRef]

49. Laureys, A.; Van den Eeckhout, E.; Petrov, R.; Verbeken, K. Effect of deformation and charging conditions on crack and blister formation during electrochemical hydrogen charging. *Acta Mater.* **2017**, *127*, 192–202. [CrossRef]

50. Van den Eeckhout, E.; Laureys, A.; Van Ingelgem, Y.; Verbeken, K. Hydrogen permeation through deformed and heat-treated Armco pure iron. *Mater. Sci. Technol.* **2017**, *33*, 1515–1523. [CrossRef]

51. Depover, T.; Verbeken, K. The effect of TiC on the hydrogen induced ductility loss and trapping behavior of Fe-C-Ti alloys. *Corros. Sci.* **2016**, *112*, 308–326. [CrossRef]

52. Wang, M.; Akiyama, E.; Tsuzaki, K. Effect of hydrogen on the fracture behavior of high strength steel during slow strain rate test. *Corros. Sci.* **2007**, *49*, 4081–4097. [CrossRef]

53. Akiyama, E.; Matsukado, K.; Wang, M.; Tsuzaki, K. Evaluation of hydrogen entry into high strength steel under atmospheric corrosion. *Corros. Sci.* **2010**, *52*, 2758–2765. [CrossRef]

54. Lee, J.L.; Lee, J.Y. Hydrogen trapping in AISI-4340 steel. *Metal Sci.* **1983**, *17*, 426–432. [CrossRef]

55. Lee, J.Y.; Lee, S.M. Hydrogen trapping phenomena in metals with bcc and fcc crystal structures by the desorption thermal-analysis technique. *Surf. Coat. Technol.* **1986**, *28*, 301–314. [CrossRef]

56. Lee, S.M.; Lee, J.Y. The trapping and transport phenomena of hydrogen in nickel. *Metall. Trans. A* **1986**, *17*, 181–187. [CrossRef]

57. Kissinger, H.E. Reaction kinetics in differential thermal analysis. *Anal. Chem.* **1957**, *29*, 1702–1706. [CrossRef]

58. Devanathan, M.A.V.; Stachurski, Z. The adsorption and diffusion of electrolytic hydrogen in palladium. *Proc. R. Soc. Lond. A* **1962**, *270*, 90–101. [CrossRef]

59. McBreen, J.; Nanis, L.; Beck, W. A method for determination of the permeation rate of hydrogen through metal membranes. *J. Electrochem. Soc.* **1966**, *113*, 1218–1222.

60. Kushida, T.; Matsumoto, H.; Kuratomi, N.; Tsumura, T.; Nakasato, F.; Kudo, T. Delayed fracture and hydrogen absorption of 1.3 GPa grade high strength bolt steel. *Testu Hagane* **1996**, *82*, 297–302. [CrossRef]

61. Laureys, A.; Depover, T.; Petrov, R.; Verbeken, K. Influence of sample geometry and microstructure on the hydrogen induced cracking characteristics under uniaxial load. *Mater. Sci. Eng. A* **2017**, *690*, 88–95. [CrossRef]

62. Sofronis, P.; McMeeking, R.M. Numerical analysis of hydrogen transport near a blunting crack tip. *J. Mech. Phys. Solids* **1989**, *37*, 317–350. [CrossRef]

63. Laureys, A.; Depover, T.; Petrov, R.; Verbeken, K. Characterization of hydrogen induced cracking in TRIP-assisted steels. *Int. J. Hydrog. Energy* **2015**, *40*, 16901–16912. [CrossRef]

64. Laureys, A.; Depover, T.; Petrov, R.; Verbeken, K. Microstructural characterization of hydrogen induced cracking in TRIP-assisted steel by EBSD. *Mater. Charact.* **2016**, *112*, 169–179. [CrossRef]

65. Oriani, R.A. A mechanistic theory on hydrogen embrittlement of steels. *Berichte Bunsenges. Phys. Chem.* **1972**, *76*, 848–857.

66. Zapffe, C.; Sims, C. Hydrogen embrittlement, internal stress and defects in steel. *Trans. AIME* **1941**, *145*, 225–232.

67. Pérez Escobar, D.; Miñambres, C.; Duprez, L.; Verbeken, K.; Verhaege, M. Internal and surface damage of multiphase steels and pure iron after electrochemical hydrogen charging. *Corros. Sci.* **2011**, *53*, 3166–3176. [CrossRef]

68. Mohtadi-Bonab, M.A.; Eskandari, M.; Karimdadashi, R.; Szpunar, J.A. Effect of Different Microstructural Parameters on Hydrogen Induced Cracking. *Met. Mater. Int.* **2017**, *23*, 726–735. [CrossRef]

69. Mohtadi-Bonab, M.A.; Eskandari, M. A focus on different factors affecting hydrogen induced cracking in oil and natural gas pipeline steel. *Eng. Fail. Anal.* **2017**, *79*, 351–360. [CrossRef]

70. Choo, W.Y.; Lee, J.Y. Thermal analysis of trapped hydrogen in pure iron. *Metall. Trans. A* **1982**, *13*, 135–140. [CrossRef]

71. Tiegel, M.C.; Martin, M.L.; Lehmberg, A.K.; Deutges, M.; Borchers, C.; Kirchheim, R. Crack and blister initiation and growth in purified iron due to hydrogen loading. *Acta Mater.* **2016**, *115*, 24–34. [CrossRef]

72. Kumnick, A.; Johson, H. Hydrogen transport through annealed and deformed Armco iron. *Metall. Trans.* **1974**, *5*, 1199–1206. [CrossRef]

73. Luppo, M.I.; Ovejero-Garcia, J. The influence of microstructure on the trapping and diffusion of hydrogen in a low carbon steel. *Corros. Sci.* **1991**, *32*, 1125–1136. [CrossRef]

74. Hadzipasic, A.B.; Malina, J.; Malina, M. The influence of microstructure on hydrogen diffusion and embrittlement of multiphase fine-grained steels with increased plasticity and strength. *Chem. Biochem. Eng. Q* **2011**, *25*, 159–169.

75. Depover, T.; Verbeken, K. Evaluation of the effect of V4C3 precipitates on the hydrogen induced mechanical degrdation in Fe-C-V alloys. *Mater. Sci. Eng. A* **2016**, *675*, 299–313. [CrossRef]

76. Depover, T.; Verbeken, K. Evaluation of the role of Mo2C in hydrogen induced ductility loss in Q&T Fe-C-Mo alloys. *Int. J. Hydrog. Energy* **2016**, *41*, 14310–14329.

77. Depover, T.; Verbeken, K. Hydrogen trapping and hydrogen induced mechanical degradation in lab cast Fe-C-Cr alloys. *Mater. Sci. Eng. A* **2016**, *669*, 134–149. [CrossRef]

78. Depover, T.; Van den Eeckhout, E.; Verbeken, K. Hydrogen induced mechanical degradation in tungsten alloyed steels. *Mater. Charact.* **2018**, *136*, 84–93. [CrossRef]

79. Depover, T.; Verbeken, K. Thermal desorption spectroscopy study of the hydrogen trapping ability of W based precipitates in a Q&T matrix. *Int. J. Hydrog. Energy* **2018**, *43*, 5760–5769.

80. Thomas, L.S.R.; Li, D.; Gangloff, R.P.; Scully, J.R. Trap-governed hydrogen diffusivity and uptake capacity in ultrahigh strength aermet 100 steel. *Metall. Mater. Trans. A* **2002**, *33*, 1991–2004. [CrossRef]

81. Wei, F.G.; Hara, T.; Tsuzaki, K. Precise determination of the activation energy for desorption of hydrogen in two Ti-added steels by a single thermal-desorption spectrum. *Metall. Mater. Trans. B* **2004**, *35*, 587–597. [CrossRef]

82. Pressouyre, G.M. Classification of hydrogen traps in steel. *Metall. Trans. A* **1979**, *10*, 1571–1573. [CrossRef]

83. Depover, T.; Verbeken, K. The detrimental effect of mobile hydrogen at dislocations on the hydrogen embrittlement susceptibility of Fe-C-X alloys: An experimental proof of the HELP mechanism. *Int. J. Hydrog. Energy* **2018**, *43*, 3050–3061. [CrossRef]

84. Pérez Escobar, D.; Wallaert, E.; Duprez, L.; Atrens, A.; Verbeken, K. Thermal Desorption Spectroscopy Study of the Interaction of Hydrogen with TiC Precipitates. *Met. Mater. Int.* **2013**, *19*, 741–748. [CrossRef]

85. Wallaert, E.; Depover, T.; Arafin, M.A.; Verbeken, K. Thermal desorption spectroscopy evaluation of the hydrogen trapping capacity of NbC and NbN precipitates. *Metall. Mater. Trans. A* **2014**, *45*, 2412–2420. [CrossRef]

Article

Pair Distribution Function Analysis of ZrO$_2$ Nanocrystals and Insights in the Formation of ZrO$_2$-YBa$_2$Cu$_3$O$_7$ Nanocomposites

Hannes Rijckaert [1], Jonathan De Roo [1], Matthias Van Zele [1], Soham Banerjee [2], Hannu Huhtinen [3], Petriina Paturi [3], Jan Bennewitz [4], Simon J. L. Billinge [2,5], Michael Bäcker [6], Klaartje De Buysser [1] and Isabel Van Driessche [1,*]

[1] Sol-gel Centre for Research on Inorganic Powders and Thin films Synthesis (SCRiPTS), Department of Chemistry, Ghent University, Krijslaan 281-S3, 9000 Ghent, Belgium; Hannes.Rijckaert@ugent.be (H.R.); Jonathan.DeRoo@ugent.be (J.D.R.); Matthias.VanZele@ugent.be (M.V.Z.); Klaartje.DeBuysser@ugent.be (K.D.B.)

[2] Department of Applied Physics and Applied Mathematics, Columbia University, 1105 S.W. Mudd, New York, NY 10027, USA; sb3519@columbia.edu (S.B.); sb2896@columbia.edu (S.J.L.B.)

[3] Wihuri Physical Laboratory, Department of Physics and Astronomy, University of Turku, 20014 Turku, Finland; Hannu.Huhtinen@utu.fi (H.H.); Petriina.Paturi@utu.fi (P.P.)

[4] BASF SE, Advanced Materials & Systems Research, Carl-Bosch-Straße 38, 67056 Ludwigshafen am Rhein, Germany; Jan.Bennewitz@basf.com

[5] Condensed Matter Physics and Materials Science Department, Brookhaven National Laboratory, Upton, NY 11973, USA

[6] Deutsche Nanoschicht GmbH, Heisenbergstraβe 16, 53359 Rheinbach, Germany; Baecker@d-nano.com

* Correspondence: Isabel.VanDriessche@ugent.be; Tel.: +32-9263-4433

Received: 24 May 2018; Accepted: 20 June 2018; Published: 23 June 2018

Abstract: The formation of superconducting nanocomposites from preformed nanocrystals is still not well understood. Here, we examine the case of ZrO$_2$ nanocrystals in a YBa$_2$Cu$_3$O$_{7-x}$ matrix. First we analyzed the preformed ZrO$_2$ nanocrystals via atomic pair distribution function analysis and found that the nanocrystals have a distorted tetragonal crystal structure. Second, we investigated the influence of various surface ligands attached to the ZrO$_2$ nanocrystals on the distribution of metal ions in the pyrolyzed matrix via secondary ion mass spectroscopy technique. The choice of stabilizing ligand is crucial in order to obtain good superconducting nanocomposite films with vortex pinning. Short, carboxylate based ligands lead to poor superconducting properties due to the inhomogeneity of metal content in the pyrolyzed matrix. Counter-intuitively, a phosphonate ligand with long chains does not disturb the growth of YBa$_2$Cu$_3$O$_{7-x}$. Even more surprisingly, bisphosphonate polymeric ligands provide good colloidal stability in solution but do not prevent coagulation in the final film, resulting in poor pinning. These results thus shed light on the various stages of the superconducting nanocomposite formation.

Keywords: chemical solution deposition; nucleation and growth; nanocomposite; thin film; YBa$_2$Cu$_3$O$_{7-\delta}$; superconductor; nanoparticles; SIMS

1. Introduction

Generators and other rotating devices used in energy conversion are important applications for the high-temperature YBa$_2$Cu$_3$O$_{7-\delta}$ (YBCO) superconductor [1,2]. However, these devices operate under magnetic fields that reduce the performance of the superconductor significantly due to vortex motion [2]. The incorporation of nano-sized structural defects as so-called 'artificial pinning centers' in the YBCO matrix can immobilize the vortices and thus create effective 'pinning' [3]. YBCO nanocomposite films, grown via pulsed laser deposition, were shown to maintain high critical currents in high magnetic fields. In these nanocomposite films, non-superconducting $BaMO_3$ nanocolumns (M = Zr, Hf and Sn) are created in the YBCO matrix by manipulating the film deposition process [4,5]. Nanocolumns generate a good in-field performance when the magnetic field (*H*) is aligned parallel to them (i.e., *H* | |*c*-axis). Such defects are correlated pinning centers along the *c*-axis and produce an enhancement of critical current densities (J_c) at *H* | |*c* with a pinning force density of more than 25 GN m^{-3} at 77 K [5]. To increase J_c at all orientations with respect to the magnetic field, non-correlated pinning centers (Y_2O_3 or $BaCeO_3$) were introduced as nanodots [6,7]. Pulsed laser deposition-based coatings thus already feature an intricate control over the size, shape, and density of pinning centers [4,8–10].

However, chemical solution deposition of YBCO films with embedded nanoparticles offers higher deposition speeds and lower processing costs. At first, nanocomposites were synthesized by a spontaneous segregation of secondary phases in the YBCO matrix, due to excess metal salts in the precursor. In this way, a variety of secondary phases were grown: Y_2O_3, $BaZrO_3$, $BaHfO_3$, $BaCeO_3$, and Ba_2YTaO_6) [11–16]. This approach enhances in-field performances compared to undoped YBCO films but the pinning force densities (77 K, *H* | |*c*) are still lower than the nanocomposite films by pulsed laser deposition. To improve performance, the nanoparticles must remain small (3–10 nm) and homogenously distributed in the YBCO matrix. In this respect, spontaneous segregation offers limited control on the formation and size distribution of the nanostructures and faces issues with reproducibility. To reproducibly gain control over the final microstructure of the nanocomposite films, we have pioneered the introduction of colloidally stable nanocrystals in the YBCO precursor solution [17,18]. A key advantage of this approach is the improved control over the composition, the particle size, and the concentration of the nanocrystals [19]. Using preformed nanocrystals, both trifluoroacetic-based and low-fluorine YBCO nanocomposite films were produced and showed a successful increase of the pinning force [17–19].

It was demonstrated that the nanocrystal surface chemistry is a crucial parameter influencing the nanocomposite formation. We determined via X-ray diffraction (XRD) and thermogravimetrical studies that the choice of stabilizing ligands affects the YBCO growth. However, it is not entirely clear how the various ligands affect the growth.

In this work, we first extensively characterize the ZrO_2 nanocrystals, synthesized in trioctylphosphine oxide. Although their native surface chemistry was already elucidated, their crystal structure was ambiguous. Here, we firmly established the nanocrystals as tetragonal and exceptionally well ordered, comparable to bulk single crystals. Second, we exchange the native hydrophobic ligands for various polar surface ligands to provide dispersibility in the methanol-based YBCO precursor solution. Subsequently, we use secondary ion mass spectroscopy (SIMS) to determine the impact of these various ligands on the distribution of metal ions in the YBCO matrix after deposition and pyrolysis of the precursor. We find that the homogeneous distribution of Ba/Y in the pyrolyzed matrix is an important requirement as it has an influence on the YBCO growth and thus also on the final superconducting properties. Moreover, the use of a phosphonate-containing copolymer ligand leads to an epitaxial YBCO structure with good superconducting properties in spite of the presence of phosphorus and thus possible degradation of superconducting properties. Even though a copolymer with a bisphosphonate group can be used as ligand and does not affect the YBCO growth, the final YBCO layer exhibits no improvement of pinning behavior as a function of the magnetic field. This is likely due to the loss of ligand stabilization from ZrO_2 nanocrystals during the pyrolysis step, leading to large coagulated $BaZrO_3$ particles in the size range of 150–200 nm. This comprehensive study provides strategies towards improving the superconducting properties of YBCO nanocomposite films and controlling pinning behavior by the careful choice of ligands.

2. Materials and Methods

2.1. Nanocrystal Synthesis and Stabilization

ZrO_2 nanocrystals in toluene were synthesized and purified according to De Keukeleere et al. [20]. In a ligand exchange step to polar solvents, 1 mL (0.3 mmol) ZrO_2 dispersion is first precipitated by addition of acetone (1:3 by volume). In a second step, the precipitate (obtained after centrifugation at 5000 rpm for 2 min) is transferred to 1 mL methanol via the addition of a 35 mg phosphonate-containing copolymer or 15 mg short carboxylate, leading to a transparent and stable nano-suspension after an ultrasonic treatment of 30–60 min, according to previous work [17].

2.2. Nanocrystal Characterization

The solvodynamic diameter of the nanocrystals was determined via Dynamic light scattering (DLS) analysis on a Malvern Nano ZS (Malvern, United Kingdom) in backscattering mode (173°). Nuclear Magnetic Resonance (NMR) measurements were recorded on a Bruker Avance II Spectrometer (Billerica, MA, USA) operating at a ^1H and ^{13}C frequency of 500.13 and 125.77 MHz respectively and featuring a ^1H, ^{13}C, ^{31}P TXI-Z probe. The sample temperature was set to 298.15 K. Diffusion measurements (2D DOSY) were performed using a double stimulated echo sequence for convection compensation and with monopolar gradient pulses. Smoothed rectangle gradient pulse shapes were used throughout. The gradient strength was varied linearly from 2–95% of the probe's maximum value (calibrated at 50.2 G cm^{-1}) in 64 steps, with the gradient pulse duration and diffusion delay optimized to ensure a final attenuation of the signal in the final increment of less than 10% relative to the first increment. The diffusion coefficients were obtained by fitting the appropriate Stejskal-Tanner equation to the signal intensity decay. Total scattering x-ray measurements were performed at the National Synchrotron Light Source II (XPD, 28-ID-2), Brookhaven National Laboratory (Upton, NY, US). Nanocrystalline powders of ZrO_2 were sealed in polyimide capillaries and diffraction patterns were collected at room temperature in a transmission geometry with an X-ray energy of 66.47 keV (λ = 0.1866 Å) using a large-area 2D PerkinElmer detector (Waltham, MA, USA). The detector was mounted with a sample-to-detector distance of 202.99 mm. The experimental geometry, 2θ range, and detector misorientations were calibrated by measuring a crystalline nickel powder directly prior to the zirconia nanocrystals, with the experimental geometry parameters refined using the Fit2D program [21]. Standardized corrections are then made to the data to obtain the total scattering structure function, $F(Q)$, which is then Fourier transformed to obtain the Pair Distribution Function (PDF), using PDFgetX3 [22] within xPDFsuite [23] The maximum range of data used in the Fourier transform (Q_{max}, where $Q = 4 \pi sin\theta / \lambda$ is the magnitude of the momentum transfer on scattering) was chosen to be 23.5 Å to give the best tradeoff between statistical noise and real-space resolution. The PDFGUI program was used to construct virtual crystal (VC) nanoparticle models from reference structures, carry out refinements, and determine the agreement between calculated PDFs and data, quantified by the residual [24]. Starting structure models for three bulk crystallographic phases of ZrO_2 were obtained from single crystal and neutron diffraction studies established in the literature [25,26]. For the tetragonal model, refined atomic coordinates are reset to their high symmetry positions prior to the first refinement (P4$_2$/nmc-I, Table 1). Refinements of the candidate phases are kept conservative. Lattice parameters and bond angles are constrained by symmetry, one isotropic atomic displacement parameters (ADPs) is applied per element (Zr, O), and a spherical particle diameter (SPD) is refined to account for the finite size of the nanocrystals [27]. Atomic positions are not refined unless specified and occupancies are kept full.

2.3. Thin Film Deposition and Processing

YBCO precursor solution is prepared by dissolving barium trifluoracetic, copper propionate, and yttrium propionate in methanol with a Y:Ba:Cu ratio of 1:2:3 and a total concentration of 1.08 M L^{-1}. Prior to spin-coating, the (100)-oriented recrystallized $LaAlO_3$ single crystal substrates were cleaned with isopropanol and heated to 400 °C to improve wettability. The substrates were spin-coated with 2000 rpm for 1 min and subsequently pyrolyzed by heating to 400 °C with a heating rate of 3–5 °C min^{-1} under a humidified O$_2$ atmosphere. The pyrolyzed YBCO films were subsequently treated to obtain the desired superconducting film with the high-temperature thermal treatment at 800 °C for 70 min in a humid 200 ppm O$_2$ in N$_2$ atmosphere which was switched to dry O$_2$ at 450 °C for 2 h during the annealing step.

2.4. Microstructural Characterization

Texture quality and phase composition of the YBCO thin films were investigated by means of X-ray diffraction (XRD) on a Bruker D4 diffractometer,(Billerica, MA, USA (Cu-K$_\alpha$). The distribution of metal ions in amorphous BYF matrix after pyrolysis was determined via Time-of-Flight (ToF) SIMS using a modified TOF-SIMS IV device from ION-TOF GmbH (Münster, Germany), equipped with a 25 kV Bi LMIG and 10 kV C60 sputter source. During the sputtering, the use of C60^{++} clusters with a sputter current of 0.5–2 nA was introduced. The raster size of the sputter beam was set at 300 × 300 µm^2. Bi$^+$-ions were used as primary ions with a pulsed target current between 0.2 and 0.5 pA. The TOF analyzer was set in positive mode.

High-resolution and high annular dark-field scanning transmission electron microscopy (HRTEM and HAADF-STEM) images were taken on a JEOL JEM-2200FS (Tokyo, Japan) TEM with a Cs corrector, operated at 200 kV. For TEM analysis, a cross-sectional TEM lamella was obtained using ion milling techniques via the FIB in-situ lift-out procedure with an Omniprobe™ (FEI, Hillsboro, OR, USA)extraction needle and top cleaning. Chemical information was obtained via the combination of HAADF-STEM with energy dispersive X-ray spectroscopy (EDX).

2.5. Electrical Characterization

The self-field critical current density J_c at 77 K was determined inductively with a 50 µV voltage criterion in a Theva Cryoscan™ system. The magnetic properties were measured with a Quantum Design Physical Property Measurement System (PPMS) with AC-measurement system.(San Diego, CA, USA). The magnetic transition temperature T_c was defined as the onset temperature of the in-phase component of the AC-magnetization at zero-field in the range of 10–100 K. The width of the magnetic transition was calculated as $\Delta T_c = T_{c,90} - T_{c,10}$. The DC-measurement was used to determine the critical current of the sample at constant temperature of 77 K as a function of the applied magnetic field perpendicular to the direction of current flow. The J_c's of all samples are calculated using the Bean critical state model from the opening of the hysteresis loop up to 8 T, obtained by DC-magnetization. The J_c was recorded with the electric field criterion of 215 µV cm^{-1}. The Bean model is widely used because of the ease of use and its accuracy. However, the obtained inductively J_c values measured via the Cryoscan™ system must be carefully compared with the magnetically J_c values obtained via PPMS system [28].

3. Results and Discussion

We chose to synthesize ZrO$_2$ nanocrystals via a heating-up synthesis with tri-*n*-octylphosphine oxide due to their high quality and already known nanocrystal surface chemistry. These nanocrystals are capped with hydrophobic phosphorus-containing ligands including di-*n*-octylphosphinic acid and P,P'-(di-*n*-octyl)pyrophosphonate which are formed upon decomposition of tri-*n*-octylphosphine oxide solvent [20]. The nanocrystals are 3.5 nm in diameter (according to TEM, Figure 1A) and stable in toluene with a solvodynamic diameter of 6.3 nm as measured by DLS (Figure 1B). The monocrystallinity and tetragonal structure of the nanocrystals was supported by atomic pair distribution function (PDF) analysis. In Figure 2 we show single phase refinements of three common ZrO$_2$ polymorphs, fitted to an experimental PDF from ∼3.5 nm ZrO$_2$ nanocrystals over the full *r*-range where structure can be resolved ($1.5 \leq r \leq 50$ Å). The refined parameters per model are provided in Table 1. The monoclinic P2$_1$/c phase is ruled out by the agreement factor (R$_w$). The cubic-fluorite (Fm-3m) and tetragonal (P4$_2$/nmc) structures are similar, but the lower symmetry tetragonal model allows for an independent refinement of the *c*-axis lattice parameter and a displacement of oxygen atoms along the same axis. These models were tested in stages as shown in Table 1. First for $P4_2/nmc$-I, where the tetragonality of the unit cell is confirmed by the significant reduction in R_w versus the cubic model, and second for $P4_2/nmc$-II, where a distortion of the eight-fold coordinated oxygen polyhedra (see Figure S1) further improves the agreement factor, and substantially reduces the average atomic displacement parameters (ADPs) for oxygen atoms. The magnitude of the oxygen displacement from the high symmetry (4*d*) site along the c-axis is ~0.2 Å. The refined oxygen position for the $P4_2/nmc$-II model is robust, and matches the position refined from neutron diffraction studies where oxygen structural parameters can be extracted even more reliably [26]. There is a small structural misfit in the low-*r* region near ~3.5 Å which may originate from ligand-nanoparticle correlations that are not included in the models. The agreement factor for the $P4_2/nmc$-II model is excellent, better than any previously reported agreement factor for a PDF refinement of nanocrystalline ZrO_2 [29–31], and decreases the likelihood of phase coexistence in these samples. Furthermore, given the agreement between the refined PDF crystallite size and supporting particle size estimates, the colloidally stable nanocrystals studied here are fully ordered and tetragonal.

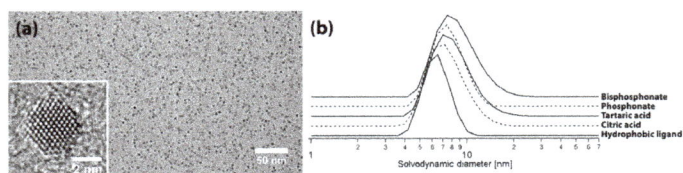

Figure 1. (**a**) Transmission electron microscopy (TEM) image of the ZrO_2 nanocrystals after the heating-up synthesis (inset shows the structure of the s crystalline grains), (**b**) Dynamic Light scattering (DLS) volume percent analysis of ZrO_2 nanocrystals before and after ligand exchange with short carboxylate and after ligand exchange with the steric dispersant.

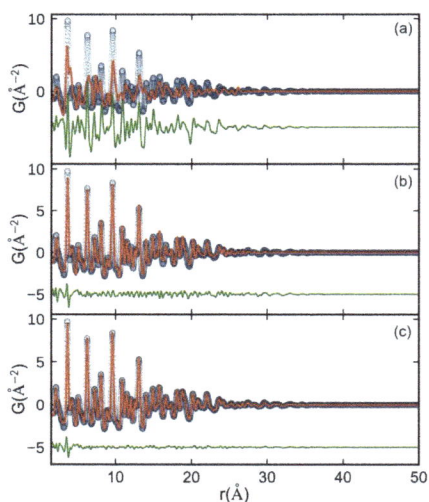

Figure 2. Measured (open circles) and calculated (red solid lines) Pair Distribution Functions (PDFs) with difference curves shown offset below (green) for three candidate ZrO_2 crystallographic phases fit to an experimental PDF from ~3.5 nm nanocrystals (**a**) monoclinic ($P2_1/c$) (**b**) cubic (Fm-3m) and (**c**) tetragonal ($P4_2/nmc$-II). Parameters used for each model are shown in Table 1 and discussed in the text.

Table 1. Refined parameters from virtual crystal nanoparticle ZrO_2 models. $P4_2/nmc$-II differs from $P4_2/nmc$-I due to an additional degree of freedom which allows oxygen atom displacement off of the $4d$ Wyckoff position as described in the text. The crystallite size is refined from a spherical shape function parameter and refers to the domain of coherent scattering in the material.

	$P2_1/c$	Fm-$3m$	$P4_2/nmc$-I	$P4_2/nmc$-II
a (Å)	5.200	5.125	3.603	3.603
b (Å)	5.231	5.125	3.603	3.603
c (Å)	5.617	5.125	5.188	5.186
β	94.8	90.0	90.0	90.0
Zr-Uiso (Å)	0.008	0.010	0.008	0.009
O-Uiso (Å)	0.046	0.072	0.072	0.041
Crystallite size (Å)	36.9	34.2	38.8	39.1
z(O1) (f.c.)	–	0.25	0.50	0.45
R_w	0.737	0.151	0.120	0.098

As the YBCO precursor solution is highly ionic and methanol based, a ligand exchange is essential to stabilize the nanocrystals in the precursor solution. According to the previous work, it is possible to stabilize ZrO_2 nanocrystals in methanol via the addition of short carboxylate (tartaric acid and citric acid) and steric dispersant (a polar copolymer with phosphonate group) [17]. Short carboxylate capped ZrO_2 nanocrystals show a polydispersity index value of around 0.90 (measured via DLS, Figure 1b, indicating more ZrO_2 agglomeration) while a copolymer with phosphonate group led to a polydispersity index value of 0.32 [17]. This indicates that phosphonates bind better to the surface than carboxylates, leading to less agglomeration, which is also confirmed in the work of De Keukeleere et al. [20]. As nanocrystals should be agglomerate-free and phosphonates bind better to nanocrystal surfaces, we introduce here a copolymer with bisphosphonate group to stabilize ZrO_2 nanocrystals even better. The use of this ligand results in a low polydispersity index value of 0.35 and solvodynamic diameter of 8.8 nm as shown in Figure 1b.

We previously characterized the surface chemistry of ZrO_2 nanocrystals stabilized with citric acid [20] or phosphonate copolymer [17] in methanol. Here, we analyze the binding of the bisphosphonate ligand to the zirconia nanocrystal surface. The ^1H NMR spectrum of bisphosphonate based ZrO_2 nanocrystals in methanol-d_4 (Figure S2) shows resonances between 0.7 and 2 ppm attributed to tri-*n*-octylphosphine oxide, di-*n*-octylphosphinic acid, and P,P′-(di-*n*-octyl) pyrophosphonate. The bisphosphonate copolymer resonances are in the range 3.3–4 ppm. In the ^{31}P spectrum, sharp resonances are detected for tri-*n*-octylphosphine oxide and di-*n*-octylphosphinic acid while the P,P′-(di-*n*-octyl)pyrophosphonate resonance is too broad to be observed (Figure S2). This indicates that the former two are displaced from the surface by the bisphosphonate while the latter remains bound to the surface. This conclusion is confirmed by the bi-exponential decay of the CH_3 resonance in Pulsed Field Gradient NMR (Figure S3). The smallest diffusion coefficient of 91 μm^2 s^{-1} can be converted to a diameter of 8.8 nm. This is somewhat larger than the original solvodynamic size in toluene of 6.3 nm due to the longer chain of the copolymer with bisphosphonate group compared to octyl chains (23 versus 8 atoms). As expected, the diffusion coefficient obtained from the CH_2 resonances of the copolymer with bisphosphonate group is exactly the same, confirming its strong binding. Thus, the stabilization of ZrO_2 nanocrystals with ligands containing phosphonate or bisphosphonate groups is more effectively compared to the short carboxylates.

However, the presence of phosphorus in ligands with phosphonate or a bisphosphonate group can lead to the degradation of the YBCO nanocomposite film [17,32] which would degrade the superconducting properties. To study the influence of phosphorus during the YBCO processing, ~1 m% copolymer (containing phosphonate or bisphosphonate) without nanocrystals was added in YBCO precursor solution. The bisphosphonate-containing YBCO has a slightly lower critical current (Table 2, average of 5 samples) compared to the undoped YBCO film while phosphonate-containing YBCO film shows a slight improvement. So, it is clear that the presence of a small amount of phosphorus has no large detrimental effect on the superconducting properties.

Table 2. Thickness and its critical current of undoped $YBa_2Cu_3O_{7-\delta}$ (YBCO) film without and with 1 m% phosphonate-containing copolymer.

Ligands	Thickness (nm)	Critical Current, I_c (A)
Undoped	275 ± 14	139 ± 25
Phosphonate	280 ± 10	144 ± 17
Bisphosphonate	282 ± 17	129 ± 14

As the thermal decomposition of the metal organic precursors is one of the critical steps in CSD-based growth, it is important that the stabilizing ligands have only a limited influence on the pyrolysis step [33]. To further unravel the behaviour of the stabilizing ligands, secondary ion mass spectroscopy (SIMS) was introduced to analyze undoped amorphous BYF matrix with CuO nanoparticles after the pyrolysis step (Figure 3A) (black line). This analysis reveals the ratio of Ba/Y (red line) and Cu/Y indicating that more CuO/Cu_2O nanoparticles are on top of the pyrolyzed matrix with a constant Ba/Y ratio throughout pyrolyzed sample. These large CuO/Cu_2O particles can lead to undesired phases during the thermal process and thus the degrading J_c of the resulting films.

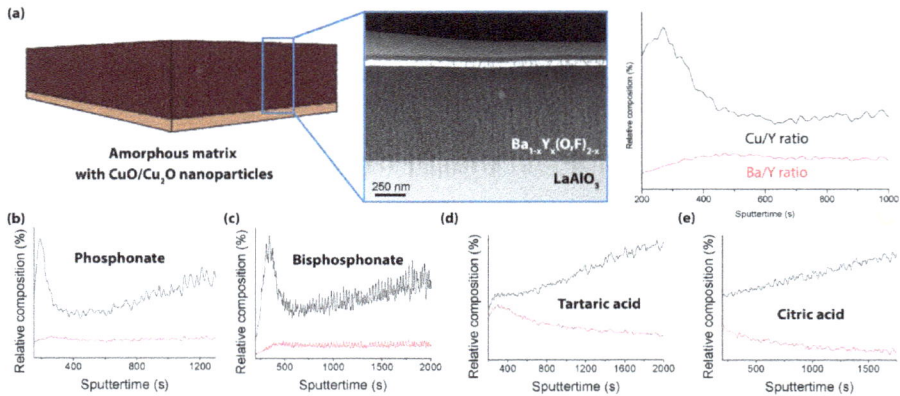

Figure 3. The relative composition of Ba/Y (red line) and Cu/Y (black line) in pyrolyzed amorphous BYF matrix of (**a**) undoped and (**b-e**) ZrO_2-doped YBCO films, determined via secondary ion mass spectroscopy (SIMS) analysis. Different sputter time steps are due to the introduction of different sputter current.

We have chosen 5 mol % ZrO_2-doped pyrolyzed YBCO samples starting from different ligands to study the metal distribution of Y, Ba, and Cu metal ions (Figure 3B–E) in the whole amorphous matrix. SIMS analysis (Figure 3B,C) shows that phosphorus-containing ligands exhibit an amorphous matrix with a constant Ba/Y ratio similar to the undoped matrix. Also on the top-surface, there is a Cu-rich zone for the copolymer with phosphonate and bisphosphonate based ZrO_2-doped sample. This metal ion distribution in the matrix yields excellent superconducting properties (e.g., high critical current density, J_c). However, short carboxylate based ZrO_2-doped films show an inhomogeneous distribution of metal ions (i.e., irregular ratio of Ba/Y) into the matrix and result in lower critical currents. It is possible that the quality of YBCO structure is sensitive to the inhomogeneity of metal content (especially Ba and Y) in the layer. To unravel this effect, all pyrolyzed samples underwent a YBCO crystallization process and were analyzed via x-ray diffraction (XRD, Figure 4A). Based on XRD patterns, short carboxylate based nanocomposite films contain more secondary phases such as $Ba_xCu_yO_z$ at $2\theta = 29.3°$ and $Y_2Cu_2O_5$—(211) at $2\theta = 31.5°$ and (204) at $2\theta = 33.4°$—while steric dispersant based films have minor secondary phases. This may be due to the disturbed nucleation mechanism of the epitaxial YBCO film because the formation of the BaF_2 phase on the $LaAlO_3$ interface is not beneficial. Also SIMS indicates there are fewer Ba^{2+} present at the bottom layer of the pyrolyzed layer. This is also confirmed via YBCO (005) reflections ($2\theta = 38.5$, Figure 4B) of crystallized samples quenched at 800 °C, indicating that the nucleation/growth mechanism is different. The XRD spectrum (Figure 3B) features crystalline BaF_2 that is in the process of reacting towards epitaxial YBCO. The difference in BaF_2 intensities is due to the competing reaction with ZrO_2 nanocrystals in combination with less availability of Ba^{2+} at $LaAlO_3$ interface during the nucleation process.

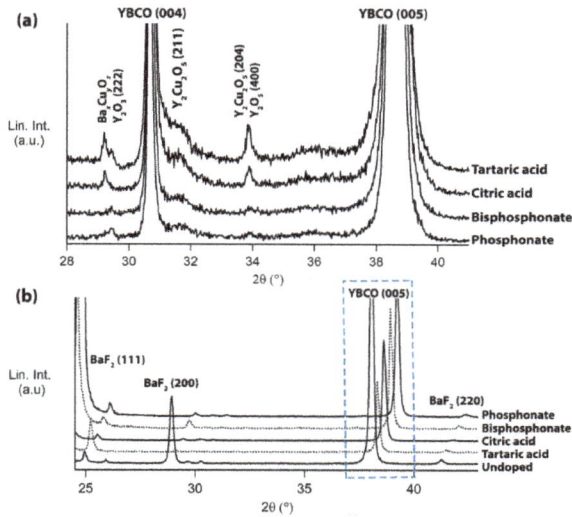

Figure 4. X-ray diffraction (XRD) analysis of (**a**) ZrO_2-doped YBCO films using different ligands after YBCO crystallization and (**b**) XRD scans of different crystallized YBCO films quenched at 800 °C, indicating the YBCO growth rate (blue rectangle marked) is different. (Reflections marked with an asterisk are related to $Ba_xCu_yO_z$ phase.).

Good biaxial YBCO texture is important but it is essential that the nanocrystals are incorporated into YBCO matrix to deliver good pinning properties. For this reason, the magnetic properties were measured with a PPMS. The onset magnetic transition (T_c) values and J_c are listed in Table 3. These magnetically measured transitions are very informative to understand the overall film quality due to the current percolation throughout the YBCO film. It can be concluded that the addition of nanocrystals hardly influence the magnetic transition. However, the tartaric acid based ZrO_2-doped YBCO film shows a slight decrease with wider width of T_c and is probably explained by a multitude of the secondary phases in YBCO matrix as shown in XRD analysis (Figure 4A). On the other hand, the magnetic field dependences of J_c's calculated at 77 K in maximum Lorentz force configuration are shown in Figure 5. It is clear that the critical current densities are in the range of 1.5–3 MA cm^{-2} except tartaric acid based ZrO_2-doped YBCO film, which only achieved a J_c of 0.74 MA cm^{-2}. The latter is due to a large amount of undesired secondary phases in the YBCO matrix as confirmed on HAADF-STEM image (Figure 6) and XRD analysis (Figure 4A).

To determine if the ZrO_2 nanocrystals in YBCO matrix act as pinning centers, we studied the shape of the $J_c(H)$ curves in Figure 5. Copolymer with phosphonate, citric acid, and tartaric acid based ZrO_2-doped YBCO films (except copolymer with bisphosphonate) show a smoother decay—round shape of $J_c(B)$—compared to undoped YBCO film. This is also confirmed by higher values of accommodation field B^* (determined by the criterion $J_c(B^*) = 0.9J_c(0)$ at 77 K) and by lower values of the slope (power-law exponent α) in log–log plot (Table 3). The low field plateau below B^* is the single vortex pinning region where each vortex is pinned to a free pinning site [34]. At B^* collective pinning effects take place and a reduction of the slope of α is seen in log–log plot. (Figure 5) The α values were estimated and are lower than 0.5, indicating that the pinning site (sizes comparable to the coherence length) is strong enough to destroy the vortex lattice and pin the vortices individually [34]. However, bisphosphonate based ZrO_2-doped YBCO film shows a straight shape of $J_c(B)$ and thus also results in a higher α value of 0.5. It is explained due to the presence of very large $BaZrO_3$ particles in the YBCO matrix (vide infra).

Table 3. Collection of magnetic transition temperature T_c and its width, critical current densities J_c at self-field and 1 T, accommodation field B^* and the power-law exponent α in undoped and ZrO_2-doped YBCO films on $LaAlO_3$ substrates.

Ligands	T_c (K)	ΔT_c (K)	$J_{c,mag}$ (0 T) (MA cm^{-2})	$J_{c,mag}$ (1 T) (kA cm^{-2})	B^* (mT)	α
Undoped	90.0	1.1	2.37	41.54	7.62	0.68
Phosphonate	90.5	1.5	2.68	237.00	17.02	0.39
Bisphosphonate	91.5	1.6	2.14	79.05	9.85	0.58
Citric acid	90.5	2.2	1.65	120.32	15.52	0.40
Tartaric acid	89.0	2.5	0.74	72.76	20.06	0.40

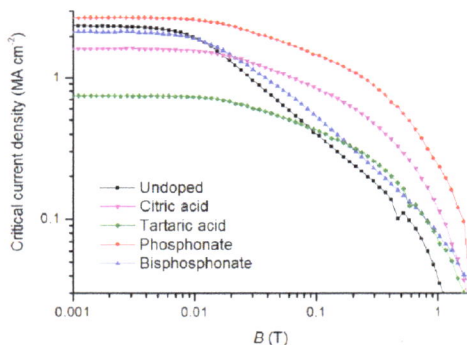

Figure 5. Double logarithmic plots of critical current density vs. magnetic field H measured at 77 K for undoped and ZrO_2-doped YBCO films on $LaAlO_3$ substrates.

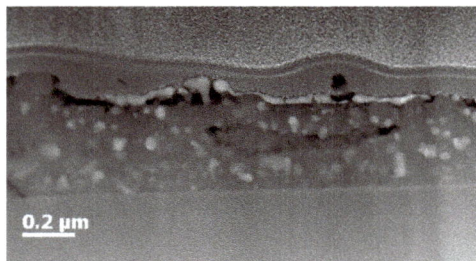

Figure 6. High annular dark-field scanning transmission electron microscopy (HAADF-STEM) cross sectional image of tartaric acid based ZrO_2-doped YBCO film, indicating lots of secondary phases in YBCO matrix.

The values of B^* and α are listed in Table 2. A clear difference between undoped and ZrO_2-doped films (except for the copolymer with bisphosphonate group) can be seen. This means that the addition of preformed nanocrystals increases pinning (leading to higher B^* values) and results in a slower decay (leading to lower α values) of critical current density in the function of the magnetic field at 77 K. However, the J_c value of tartaric acid-based ZrO_2-doped YBCO nanocomposite is still lower than any other ZrO_2-doped films and is accompanied by the presence of more undesired secondary phases in the YBCO matrix as confirmed via XRD (Figure 3A) and the HAADF-STEM image (Figure 5).

It is remarkable that the bisphosphonate based ZrO_2-doped YBCO nanocomposite did not show any improvement of pinning behaviour while the self-field J_c has an acceptable value in the range of 2–2.5 MA cm^{-2}. As shown on the HAADF-STEM image (Figure 7A) and TEM image (Figure 7B), the YBCO layer is strongly textured and grows epitaxially on the $LaAlO_3$ substrates which explains the good superconducting properties. However, some secondary phases and large $BaZrO_3$ particles in the size of 150–200 nm can be observed in the YBCO layer via EDX analysis. In order to have a better understanding of the $BaZrO_3$ particles in the YBCO matrix, HRTEM analysis was used. Figure 7C shows that this $BaZrO_3$ particle contains several aggregated small nanocrystals. So, it seems that the nanocrystals are coagulated during the YBCO growth to the size of approximately 200 nm. This particle is too large to act as a pinning center because it is not in the order of superconducting coherence length of 2–4 nm for YBCO at 77 K. It appears that the stabilizing effect of bisphosphonate copolymer is reduced or annihilated upon the initial decomposition, leading to an increased tendency to form agglomerations. However, the precise origins of this effect remain unclear.

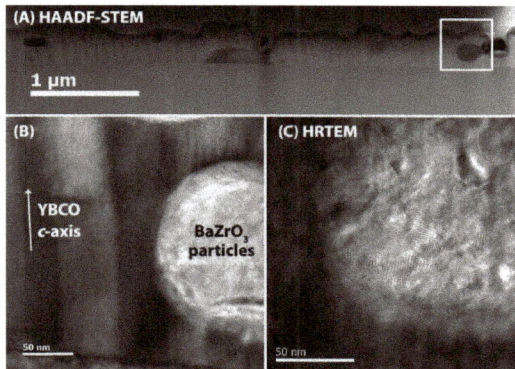

Figure 7. (a) HAADF-STEM image of copolymer with bisphosphonate capped ZrO_2-doped YBCO film grown on $LaAlO_3$ substrate, indicating big particles in the YBCO matrix, (b) TEM image showing YBCO/$BaZrO_3$ particles interface and (c) High-resolution transmission electron microscopy (HRTEM) image of $BaZrO_3$ particles, showing a coagulation of ZrO_2 nanocrystals.

4. Conclusions

ZrO_2 nanocrystals, synthesized in tri-n-octylphosphine oxide were probed with PDF analysis and found to be monocrystalline, featuring a distorted tetragonal crystal structure. They were subsequently stabilized by a steric polar ligand (Copolymer with phosphonate or bisphosphonate group) or short carboxylates (tartaric and citric acid) in low-fluorine YBCO precursor solutions. From these suspensions, we synthesized high quality superconducting ZrO_2-doped YBCO nanocomposites in a single coating step, in contrast to earlier results with amino acid stabilized nanocrystals. Interestingly, nanocrystals stabilized by short carboxylate ligands resulted in poorly superconducting nanocomposites while the phosphonate dispersant led to excellent self-field superconducting nanocomposites. This is a counter-intuitive result as one would expect that as phosphorus or/and carbon is introduced in the layer, the worse the superconductor would be. The use of copolymer with bisphosphonate group results in the coagulation of ZrO_2 nanocrystals during YBCO growth, resulting in no improved pinning properties. Given the counter-intuitive relation between the nanocrystal surface chemistry and the final nanocomposites performance, we expect that surveying a wide library of ligands will be crucial in order to obtain a good superconducting nanocomposite film with the ability to pin the vortices. The use of a copolymer with bisphosphonate group leads to coagulation of ZrO_2 nanocrystals during YBCO growth likely due to the reduction of stabilizing effects during the thermal decomposition, which do not facilitate pinning.

Supplementary Materials: The following are available online at http://www.mdpi.com/1996-1944/11/7/1066/s1, Figure S1: tetragonal and distorted tetragonal crystal structure for ZrO_2 after PDF refinement. Figure S2: ^{1}D ^{1}H spectrum and (B) ^{31}P spectrum of ZrO_2 nanocrystals stabilized with bisphosphonate (BP) in MeOD-d_4, with (C) a zoom on BP resonances. Figure S3: Bi-exponential diffusion decay fitting of the bisphosphonate ligand.

Author Contributions: I.V.D. and M.B. conceived and designed the experiments; J.D.R. synthesized ZrO_2 nanocrystals, studied the surface chemistry via NMR analysis and analyzed PDF data. S.B. and S.J.L.B. performed PDF experiments and analyzed the data; J.B. performed the XRD analysis and inductive measurements, M.V.Z. and K.D.B. performed and analyzed the SIMS data; H.H. and P.P. measured the magnetic measurements; H.R. deposited the samples and made TEM measurements. All authors reviewed the manuscript.

Funding: This work was financially supported by the European Union Horizon 2020 Marie Curie Actions under the project SynFoNY (H2020/2016-722071). J.D.R. thanks FWO Vlaanderen Fulbright, and B.A.E.F. for fellowships. P.P. and H.H. wish to thank the Jenny and Antti Wihuri Foundation for financial support. Work in the Billinge group was supported by U.S. Department of Energy, Office of Science, Office of Basic Energy Sciences (DOE-BES) under contract No. DE-SC00112704. Banerjee acknowledges support from the National Defense Science and Engineering Graduate Fellowship (DOD-NDSEG) program.

Acknowledgments: X-ray PDF measurements were conducted on beamline 28-ID-2 of the National Synchrotron Light Source II, a U.S. Department of Energy (DOE) Office of Science User Facility operated for the DOE Office of Science by Brookhaven National Laboratory under Contract No. DE-SC0012704.

Conflicts of Interest: The authors declare no conflict of interest.

References

1. Larbalestier, D.; Gurevich, A.; Feldmann, D.M.; Polyanskii, A. High-T_c superconducting materials for electric power applications. *Nature* **2001**, *414*, 368–377. [CrossRef] [PubMed]

2. Obradors, X.; Puig, T. Coated conductors for power applications: Materials challenges. *Supercond. Sci. Technol.* **2014**, *27*. [CrossRef]

3. Obradors, X.; Puig, T.; Ricart, S.; Coll, M.; Gazquez, J.; Palau, A.; Granados, X. Growth, nanostructure and vortex pinning in superconducting $YBa_2Cu_3O_7$ thin films based on trifluoroacetate solutions. *Supercond. Sci. Technol.* **2012**, *25*, 123001. [CrossRef]

4. Hänisch, J.; Cai, C.; Stehr, V.; Hühne, R.; Lyubina, J.; Nenkov, K.; Fuchs, G.; Schultz, L.; Holzapfel, B. Formation and pinning properties of growth-controlled nanoscale precipitates in $YBa_2Cu_3O_{7-\delta}$/transition metal quasi-multilayers. *Supercond. Sci. Technol.* **2006**, *19*, 534–540. [CrossRef]

5. MacManus-Driscoll, J.; Foltyn, S.; Jia, Q.; Wang, H.; Serquis, A.; Civale, L.; Maiorov, B.; Hawley, M.; Maley, M.; Peterson, D. Strongly enhanced current densities in superconducting coated conductors of $YBa_2Cu_3O_{7-x}+BaZrO_3$. *Nat. Mater.* **2004**, *3*, 439–443. [CrossRef] [PubMed]

6. Malmivirta, M.; Rijckaert, H.; Paasonen, V.; Huhtinen, H.; Hynninen, T.; Jha, R.; Awana, V.S.; Driessche, I.; Paturi, P. Enhanced flux pinning in YBCO multilayer films with BCO nanodots and segmented BZO nanorods. *Sci. Rep.* **2017**, *1*, 14682. [CrossRef] [PubMed]

7. Feighan, J.; Kursumovic, A.; MacManus-Driscoll, J. Materials design for artificial pinning centres in superconductor PLD coated conductors. *Supercond. Sci. Technol.* **2017**, *30*, 123001. [CrossRef]

8. Haugan, T.; Barnes, P.; Wheeler, R.; Meisenkothen, F.; Sumption, M. Addition of nanoparticle dispersions to enhance flux pinning of the $YBa_2Cu_3O_{7-x}$ superconductor. *Nature* **2004**, *430*, 867–870. [CrossRef] [PubMed]

9. Opherden, L.; Sieger, M.; Pahlke, P.; Hühne, R.; Schultz, L.; Meledin, A.; Van Tendeloo, G.; Nast, R.; Holzapfel, B.; Bianchetti, M. Large pinning forces and matching effects in $YBa_2Cu_3O_{7-\delta}$ thin films with $Ba_2Y(Nb/Ta)O_6$ nano-precipitates. *Sci. Rep.* **2016**, *6*, 21188. [CrossRef] [PubMed]

10. Sieger, M.; Pahlke, P.; Hänisch, J.; Sparing, M.; Bianchetti, M.; MacManus-Driscoll, J.; Lao, M.; Eisterer, M.; Meledin, A.; Van Tendeloo, G. Ba$_2$Y(Nb/Ta)O$_6$–Doped YBCO Films on Biaxially Textured Ni–5at.% W Substrates. *IEEE Trans. Appl. Supercond.* **2016**, *26*, 1–5. [CrossRef]

11. Lei, L.; Zhao, G.; Xu, H.; Wu, N.; Chen, Y. Influences of Y$_2$O$_3$ nanoparticle additions on the microstructure and superconductivity of YBCO films derived from low-fluorine solution. *Mater. Chem. Phys.* **2011**, *127*, 91–94. [CrossRef]

12. Ye, S.; Suo, H.; Wu, Z.; Liu, M.; Xu, Y.; Ma, L.; Zhou, M. Preparation of solution-based YBCO films with BaSnO$_3$ particles. *Physica C* **2011**, *471*, 265–269. [CrossRef]

13. Erbe, M.; Hänisch, J.; Hühne, R.; Freudenberg, T.; Kirchner, A.; Molina-Luna, L.; Damm, C.; Van Tendeloo, G.; Kaskel, S.; Schultz, L. BaHfO$_3$ artificial pinning centres in TFA-MOD-derived YBCO and GdBCO thin films. *Supercond. Sci. Technol.* **2015**, *28*, 114002. [CrossRef]

14. Ding, F.; Gu, H.; Zhang, T.; Wang, H.; Qu, F.; Qiu, Q.; Dai, S.; Peng, X.; Cao, J.-L. Strong enhancement flux pinning in MOD-YBa$_2$Cu$_3$O$_{7-x}$ films with self-assembled BaTiO$_3$ nanocolumns. *Appl. Surf. Sci.* **2014**, *314*, 622–627. [CrossRef]

15. Gutierrez, J.; Llordes, A.; Gazquez, J.; Gibert, M.; Roma, N.; Ricart, S.; Pomar, A.; Sandiumenge, F.; Mestres, N.; Puig, T. Strong isotropic flux pinning in solution-derived YBa$_2$Cu$_3$O$_{7-x}$ nanocomposite superconductor films. *Nat. Mater.* **2007**, *6*, 367–373. [CrossRef] [PubMed]

16. Coll, M.; Guzman, R.; Garcés, P.; Gazquez, J.; Rouco, V.; Palau, A.; Ye, S.; Magen, C.; Suo, H.; Castro, H.; et al. Size-controlled spontaneously segregated Ba$_2$YTaO$_6$ nanoparticles in YBa$_2$Cu$_3$O$_7$ nanocomposites obtained by chemical solution deposition. Supercond. *Sci. Technol.* **2014**, *27*, 044008. [CrossRef]

17. Rijckaert, H.; Pollefeyt, G.; Sieger, M.; Hänisch, J.; Bennewitz, J.; De Keukeleere, K.; De Roo, J.; Hühne, R.; Bäcker, M.; Paturi, P.; et al. Optimizing Nanocomposites through Nanocrystal Surface Chemistry: Superconducting YBa$_2$Cu$_3$O$_7$ Thin Films via Low-Fluorine Metal Organic Deposition and Preformed Metal Oxide Nanocrystals. *Chem. Mater.* **2017**, *29*, 6104–6113. [CrossRef]

18. De Keukeleere, K.; Cayado, P.; Meledin, A.; Vallès, F.; De Roo, J.; Rijckaert, H.; Pollefeyt, G.; Bruneel, E.; Palau, A.; Coll, M.; et al. Superconducting YBa$_2$Cu$_3$O$_{7-\delta}$ Nanocomposites Using Preformed ZrO$_2$ Nanocrystals: Growth Mechanisms and Vortex Pinning Properties. *Adv. Electr. Mater.* **2016**, *2*, 1600161. [CrossRef]

19. Cayado, P.; De Keukeleere, K.; Garzón, A.; Perez-Mirabet, L.; Meledin, A.; De Roo, J.; Vallés, F.; Mundet, B.; Rijckaert, H.; Pollefeyt, G.; et al. Epitaxial YBa$_2$Cu$_3$O$_{7-x}$ nanocomposite thin films from colloidal solutions. *Supercond. Sci. Technol.* **2015**, *28*, 124007. [CrossRef]

20. De Keukeleere, K.; Coucke, S.; De Canck, E.; Van Der Voort, P.; Delpech, F.; Coppel, Y.; Hens, Z.; Van Driessche, I.; Owen, J.S.; De Roo, J. Stabilization of Colloidal Ti, Zr, and Hf Oxide Nanocrystals by Protonated Tri-n-octylphosphine Oxide (TOPO) and Its Decomposition Products. *Chem. Mater.* **2017**, *29*, 10233–10242. [CrossRef]

21. Hammersley, A.P. *FIT2D V12. 012 Reference Manual V6.0*; ESRF Internal Report ESRF98HA01T; ESRF: Grenoble, France, 2004.

22. Juhás, P.; Davis, T.; Farrow, C.L.; Billinge, S.J. PDFgetX3: A rapid and highly automatable program for processing powder diffraction data into total scattering pair distribution functions. *J. Appl. Crystallogr.* **2013**, *46*, 560–566. [CrossRef]

23. Yang, X.; Juhás, P.; Farrow, C.L.; Billinge, S.J. xPDFsuite: An end-to-end software solution for high throughput pair distribution function transformation, visualization and analysis. *arXiv* **2014**, arXiv:1402.3163.

24. Farrow, C.; Juhas, P.; Liu, J.; Bryndin, D.; Božin, E.; Bloch, J.; Proffen, T.; Billinge, S. PDFfit2 and PDFgui: Computer programs for studying nanostructure in crystals. *J. Phys. Condens. Matter* **2007**, *19*, 335219. [CrossRef] [PubMed]

25. Smith, D.K.; Newkirk, W. The crystal structure of baddeleyite (monoclinic ZrO$_2$) and its relation to the polymorphism of ZrO$_2$. *Acta Crystallogr.* **1965**, *18*, 983–991. [CrossRef]

26. Howard, C.J.; Kisi, E.H.; Roberts, R.B.; Hill, R.J. Neutron Diffraction Studies of Phase Transformations between Tetragonal and Orthorhombic Zirconia in Magnesia-Partially-Stabilized Zirconia. *J. Am. Ceram. Soc.* **1990**, *73*, 2828–2833. [CrossRef]

27. Egami, T.; Billinge, S.J. *Underneath the Bragg Peaks: Structural Analysis of Complex Materials*; Elsevier: New York, NY, USA, 2003.

28. Golovchanskiy, I.; Pan, A.; Shcherbakova, O.; Fedoseev, S. Rectifying differences in transport, dynamic, and quasi-equilibrium measurements of critical current density. *J. Appl. Phys.* **2013**, *114*, 163910. [CrossRef]

29. Gateshki, M.; Petkov, V.; Williams, G.; Pradhan, S.; Ren, Y. Atomic-scale structure of nanocrystalline ZrO_2 prepared by high-energy ball milling. *Phys. Rev. B* **2005**, *71*, 224107. [CrossRef]

30. Gateshki, M.; Petkov, V.; Hyeon, T.; Joo, J.; Niederberger, M.; Ren, Y. Interplay between the local structural disorder and the length of structural coherence in stabilizing the cubic phase in nanocrystalline ZrO_2. *Solid State Commun.* **2006**, *138*, 279–284. [CrossRef]

31. Dippel, A.-C.; Jensen, K.; Tyrsted, C.; Bremholm, M.; Bøjesen, E.D.; Saha, D.; Birgisson, S.; Christensen, M.; Billinge, S.J.; Iversen, B.B. Towards atomistic understanding of polymorphism in the solvothermal synthesis of ZrO_2 nanoparticles. *Acta Crystallogr.* **2016**, *72*, 645–650. [CrossRef]

32. De Roo, J.; Coucke, S.; Rijckaert, H.; De Keukeleere, K.; Sinnaeve, D.; Hens, Z.; Martins, J.C.; Van Driessche, I. Amino Acid-Based Stabilization of Oxide Nanocrystals in Polar Media: From Insight in Ligand Exchange to Solution ^{1}H NMR Probing of Short-Chained Adsorbates. *Langmuir* **2016**, *32*, 1962–1970. [CrossRef] [PubMed]

33. Rijckaert, H.; De Roo, J.; Roeleveld, K.; Pollefeyt, G.; Bennewitz, J.; Bäcker, M.; Lynen, F.; De Keukeleere, K.; Van Driessche, I. Microwave-assisted $YBa_2Cu_3O_7$ precursors: A fast and reliable method towards chemical precursors for superconducting films. *J. Am. Ceram. Soc.* **2017**, *100*, 2407–2418. [CrossRef]

34. Paturi, P.; Malmivirta, M.; Palonen, H.; Huhtinen, H. Dopant Diameter Dependence of J(c)(B) in Doped YBCO Films. *IEEE Trans. Appl. Supercond.* **2016**, *26*. [CrossRef]

materials

MDPI

Article

Thickness Characterization Toolbox for Transparent Protective Coatings on Polymer Substrates

Matthias Van Zele [1], Jonathan Watté [1], Jan Hasselmeyer [1], Hannes Rijckaert [1], Yannick Vercammen [2], Steven Verstuyft [3], Davy Deduytsche [4], Damien P. Debecker [5], Claude Poleunis [5], Isabel Van Driessche [1] and Klaartje De Buysser [1,*]

[1] Sol-Gel Centre for Research on Inorganic Powders and Thin Film Synthesis (SCRiPTS),
    Department of Chemistry, Ghent University, Krijgslaan 281-S3, 9000 Ghent, Belgium;
    Matthias.VanZele@ugent.be (M.V.Z.); Jonathan.Watté@ugent.be (J.W.); Jan.Hasselmeyer@ugent.be (J.H.);
    Hannes.Rijckaert@ugent.be (H.R.); Isabel.VanDriessche@ugent.be (I.V.D.)
[2] Department of Chemistry, Universiteit Antwerpen, Campus Drie Eiken, Universiteitsplein 1,
    2610 Wilrijk, Belgium; mail@yannickv.be
[3] Photonics Research Group, Department of Information Technology, Ghent University-IMEC,
    9000 Ghent, Belgium; Steven.Verstuyft@ugent.be
[4] Conformal Coating of Nanomaterials (CoCooN), Department of Solid State Sciences, Ghent University,
    Krijgslaan 281-S1, 9000 Ghent, Belgium; davy.deduytsche@ugent.be
[5] Institute of Condensed Matter and Nanosciences, Université catholique de Louvain,
    1348 Louvain-la-Neuve, Belgium; damien.debecker@uclouvain.be (D.P.D.);
    claude.poleunis@uclouvain.be (C.P.)
*   Correspondence: Klaartje.DeBuysser@ugent.be; Tel.: +32-9-264-4441

Received: 30 May 2018; Accepted: 26 June 2018; Published: 28 June 2018

Abstract: The thickness characterization of transparent protective coatings on functional, transparent materials is often problematic. In this paper, a toolbox to determine the thicknesses of a transparent coating on functional window films is presented. The toolbox consists of a combination of secondary ion mass spectrometry and profilometry and can be transferred to other transparent polymeric materials. A coating was deposited on designed model samples, which were characterized with cross-sectional views in transmission and in scanning/transmission electron microscopy and ellipsometry. The toolbox was then used to assess the thicknesses of the protective coatings on the pilot-scale window films. This coating was synthesized using straightforward sol-gel alkoxide chemistry. The kinetics of the condensation are studied in order to obtain a precursor that allows fast drying and complete condensation after simple heat treatment. The shelf life of this precursor solution was investigated in order to verify its accordance to industrial requirements. Deposition was performed successfully at low temperatures below 100 °C, which makes deposition on polymeric foils possible. By using roll-to-roll coating, the findings of this paper are easily transferrable to industrial scale. The coating was tested for scratch resistance and adhesion. Values for the emissivity (ε) of the films were recorded to justify the use of the films obtained as infrared reflective window films. In this work, it is shown that the toolbox measures similar thicknesses to those measured by electron microscopy and can be used to set a required thickness for protective coatings.

Keywords: coating materials; inorganic materials; surfaces and interfaces; sol-gel processes; low-emissivity; SIMS

---

1. Introduction

Thickness effects of transparent thin coatings often seems to be a critical factor in the design of functional devices. These coatings should be thick enough to protect the underlying materials.

On the other hand, losses in transparency can be noticed with an increased thickness, which is often detrimental for their application.

This paper looks into the field of insulating window films. These window films are designed in such a way that heat is kept outside of buildings during summer times and vice versa. This functionality is achieved by the deposition of a far infrared reflective Fabry-Pérot stack with low emissivity (ε) on polymer films. The Fabry-Pérot stack consists of transparent layers, deposited typically by subsequent sputtering. An example is a $TiO_2/Ag/TiO_2$ multi-layer stack. This stack shows excellent low-ε properties [1].

The aforementioned stack is often used outside the field of window films and placed in between two glazing panels, where no moisture can reach the stack. This is crucial, as it is well known that the inner metallic silver layers are not stable and that they are especially sensitive to moisture. Several reports have been published on the moisture-induced degradation of sputtered low-ε coatings containing silver, concluding that high concentrations of chlorides could lead to the corrosion of the stack [2–7]. Therefore, contact with moisture and chlorides is detrimental for the low-ε properties of these stacks and should be protected with an additional coating.

When looking outside the scope of flexible polymeric window films, many coatings have been reported to effectively protect this kind of stacks on solid substrates. One way of protecting stacks is to apply ceramic materials, such as ZnO, Al-doped ZnO [8], or Al_2O_3 material, as the final hard coat [6]. These ceramic coatings are often deposited by sputtering [9], which is not very cost-effective. Another approach is to use organic coatings to protect underlying layers and materials which are prone to mechanical damage [10,11]. These organic coatings, however, absorb far infrared radiation, which renders the reflecting stacks on window films useless, as the films will heat up and loss of heat reflection will occur. For the same reason, no hybrid materials containing scratch-resistant metal oxide particles, such as silica nanoparticles, embedded in an organic matrix [12–14] could be used for this application.

To overcome the drawbacks of all known protective coatings and thus preserve the low-ε characteristics of mentioned window films, a silica thin layer was deposited through a wet chemical deposition method. This protective coating should not only meet the requirements for industrial applications, but also should be cost-effective. Chemical solution deposition (CSD) is a cost-effective deposition method for thin films. CSD starts from molecules in solution or suspension, which are deposited on a surface, which results in the growth of a new phase. The method has proven to be effective to deposit a wide range of materials [8,15–19].

Industrial requirements for the protective coating include a class 0 adhesion and 3H pencil hardness, according to ASTM D3359 [20] and ASTM D3363 [21], respectively. Because of these goals, the possibility to deposit this coating by using roll-to-roll deposition processes was further explored.

The aim of this work is to develop a thickness determination toolbox for transparent protective coatings on transparent substrates and to assess the minimal thickness to pass the predefined application tests of a protective silica coating for low-ε window films. Ellipsometry, scanning, and transmission electron microscopy (SEM and TEM) are often the characterization techniques of choice. However, ellipsometry is not suitable for fully transparent devices containing a relatively thick, transparent substrate. This is because ellipsometry is not able to distinguish thin layers from the underlying substrate in transparent devices, as the difference in refractive indices between the layers is too low. Ellipsometry fails because the signal coming from the thin layers is shadowed by the signal of the substrate. Cross-sectional view SEM and TEM measurements, on the other hand, are very good techniques to determine the thickness of top coatings, but are not suitable if polymer substrates are present, as the electron beam burns through the organic substrates [22]. To circumvent these drawbacks, we propose to combine secondary ion mass spectrometry (SIMS) and profilometry as an effective thickness determination toolbox.

The idea behind this proposed toolbox is to use the SIMS apparatus to dig a crater in the film while checking the chemical composition of the surface (by analyzing the ions that are emitted). In this way, the sudden appearance of the ions attributed to the support is a direct indication that the upmost film has been eroded. The sputtering is then interrupted and the crater is analyzed by profilometry to determine its depth, which strictly corresponds to the thickness of the film.

To develop this toolbox, model samples were designed (Figure 1a,b), comparable to window films (Figure 1c). The model samples were characterized with ellipsometry, SEM/TEM, and the proposed SIMS-profilometry toolbox. An additional gold layer was sputtered to make ellipsometry and SEM/TEM measurements possible. On the one hand, the gold layer is needed to make a distinction between the individual layers with ellipsometry. On the other hand, this conducting gold layer is necessary to make Focused ion beam (FIB)-SEM measurements possible. Otherwise, too much drifting of the sample would occur as the samples become charged during measurement. In order to make TEM measurements possible, the polymer substrate was changed to silicon.

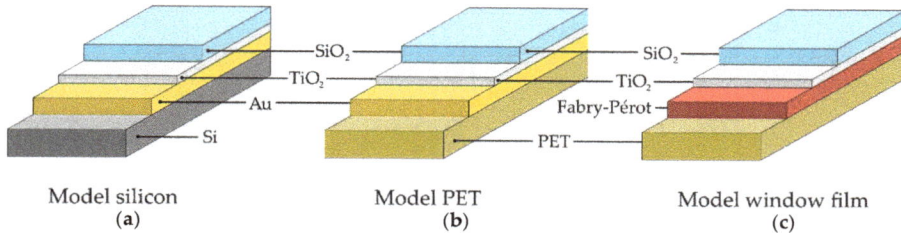

Figure 1. Composition of (**a**) the TiO_2/Au/silicon model sample; (**b**) the TiO_2/Au/polyethylene terephtalate (PET) model sample; (**c**) the window film. The model sample was built up on silicon or plain PET, on which a gold layer was deposited. The window film functionality comes forth from the functional metal/metal oxide layer. All samples were provided with a titania buffer layer, on which the protective silica coating was deposited.

In this work, it is shown that the proposed toolbox delivers similar thicknesses to those observed in cross-sectional view SEM/TEM images, while ellipsometry shows different values of thickness. Electronic microscopy (EM) methods are considered to be accurate in the determination of film thickness. In order to check this hypothesis, both results obtained by the proposed toolbox and by ellipsometry are compared to EM. The conclusions drawn from model sample characterizations are transferred to comparable functional window films, which cannot be characterized by EM. It is shown that the proposed toolbox can be used to determine the thicknesses of thin coatings on transparent polymeric substrates.

2. Materials and Methods

The chemicals were used as received. Absolute ethanol was obtained from Applichem Panreac (Darmstadt, Germany). Hydrochloric acid (36 %) and glacial acetic acid (100 m%) were obtained from Roth (Karlsruhe, Germany). The silane precursor tetraethyl orthosilicate (TEOS, 99% pure) was supplied by ABCR (Karlsruhe, Germany). Deuterated water was obtained from Sigma Aldrich (Overijse, Belgium).

Sample fabrication: A graphical representation of sample compositions can be seen in Figure 1. The substrate used as model sample was either silicon (1×1 cm^2) (Figure 1a) or plain PET (1×1 cm^2) (Figure 1b), provided with a gold layer and a titania buffer layer, both deposited by physical vapor deposition. The 100-nm thick gold layer was deposited using electron beam operation with a base pressure of 1.0×10^{-6} mbar. Also, the 5-nm thick TiO_x layer was deposited using the electron beam operation of Ti, under an oxygen partial pressure of 4.0×10^{-5} mbar.

Window films (Figure 1c) were provided by Group Michiels Advanced Materials (M.A.M.) and consisted of a 20-cm wide and 50-µm thick PET film. This PET film was functionalized with a single sputtered Fabry-Pérot stack (Zele, Belgium). On top of this stack, a titania buffer layer was sputtered.

Coating precursor synthesis: A sol-gel synthesis was adapted from previous work [15,23–25]. In a typical synthesis procedure, tetraethyl orthosilicate was added to ethanol. In a separate container, a solution of glacial acetic acid in distilled water was prepared. Acetic acid is needed in order to catalyze the condensation reaction. All reagents were mixed while stirring. An additional pre-condensation step was conducted by heating the total mixture for 3 h under refluxing conditions at 60 °C. As a final step, the reaction was quenched with ethanol. Four hundred milliliters of a transparent sol was obtained in accordance with patent EP15199592.5 [26].

Coating deposition: TiO_2/Au/silicon and TiO_2/Au/PET model substrates were coated using spin-coating by dropping 90 µL of the transparent sol on the substrate. The sample was then spun for 60 s at 2000 rpm with a spin acceleration of 40 rpm/s^2. After spin-coating, it was annealed at 80 °C for 60 s on a hot plate in an ambient atmosphere.

Window film substrates were coated via roll-to-roll coating (Figure 2), using the coating conditions summarized in Table 1. Characterized window films were coated once and coated twice using these coating settings.

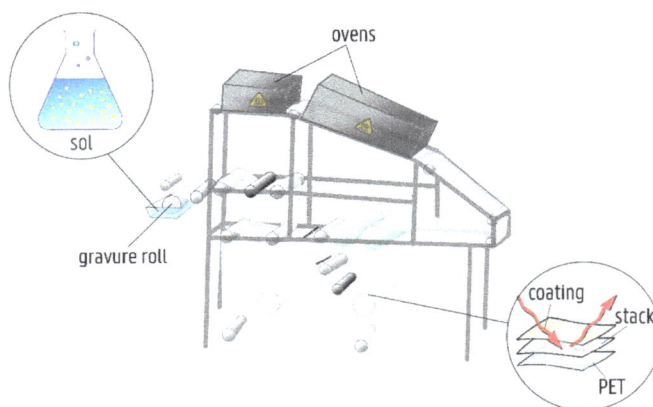

Figure 2. Schematic overview of a roll-to-roll coater. The foil is transported by multiple rolls, while deposition of the sol is achieved by a gravure roll. Before rewinding, a heat treatment takes place in the top ovens.

Table 1. Coating settings for roll-to-roll coating.

Parameter	Setting
Unwinding speed	2.4 m/min
Application speed	1 m/min
Coating method	Reverse gravure
Temperature first oven	60 °C
Temperature second oven	80 °C

Raman characterization: The kinetics of the pre-condensation step were monitored with a RamanRxn system from Kaiser optical systems (Ann Arbor, MI, USA), model Rxn1-532. The measurements were performed in situ by placing a laser probe inside the reaction mixture. The mixture was put in a three-neck flask, equipped with a reflux cooler and a dropping funnel to add reagents, together with the laser probe. The flask, the dropping funnel, and the reflux cooler were covered in aluminum foil to prevent the detection of radiation coming from the outside of the mixture. The wavelength of the laser was 532 nm with a maximum power output of 450 mW. A resampling interval of 1 cm^{-1} was used while recording 13 accumulations with an exposure time of 10 s.

Nuclear Magnetic Resonance (NMR): To evaluate the degree of condensation and the correlated shelf life of the sols, ^{29}Si-NMR measurements were performed using a Bruker (Billerica, MA, USA) Avance III spectrometer with a 1 h frequency of 500 MHz, equipped with a 5-mm BBI Z probe. Samples were prepared by adding 100 μL of D_2O to 900 μL of the sol before transferring to an NMR test tube.

Spectroscopic characterization: Optical transmission and reflection spectra were recorded with an Agilent Cary 5000 UV-Vis-NIR spectrophotometer (Santa Clara, CA, USA). The spectra were recorded from 300 to 2000 nm.

Application Tests: Crosshatch tests were performed according to ASTM D3359 standard [16]. A cutter size of 1 mm was used with six cutting edges. The adhesion test tape has an adhesive strength of 9.5 N per 25-mm width. Both were supplied by Paint Test Equipment (Congleton, UK). The sample was prepared by a first cut in the coating with the cutter. A second cut, perpendicular to the first cut was made. The adhesive tape was placed over the damaged area and, finally, the tape was removed from the surface. Using a magnifying glass, the coating was checked for damage.

The hardness of the coating was evaluated with an Elcometer (Utrecht, The Netherlands) 501 Pencil Hardness Tester. The method used was the Wolff-Wilborn method. The equipment complies with the ASTM D3363 standard [17].

The emissivity ε was measured with a TIR 100-2, supplied by Inglas (Friedrichshafen, Germany). The samples were cut to a size of 10 by 10 cm.

Layer thickness: The proposed toolbox should be compared to other layer thickness characterization techniques to verify its effectiveness. The cross-sectional view images and lamella were prepared by a FEI Nova (Hillsboro, OR, USA) 600 Nanolab Dual-Beam Focused Ion Beam system and associated SEM.

For TEM analysis, a cross-sectional lamella was obtained using ion milling techniques via the FIB in situ lift-out procedure with an Omniprobe (FEI, Hillsboro, OR, USA) extraction needle and top cleaning. High-angle annular dark-field (HAADF) and bright-field (BF) scanning TEM images were taken on a JEOL JEM-2200FS TEM (Tokyo, Japan) with a Cs corrector, operated at 200 kV. Chemical information was obtained via the phase analysis via energy dispersive X-ray spectroscopy.

Spectroscopic ellipsometry measurements were performed using a J.A. Woollam M-2000 ellipsometer (J.A. Woollam Co., Lincoln, NE, USA). The spectral range of the ellipsometer ranged from 250 to 1680 nm and the COMPLETEEASE software (version 6.34) was used for fitting and data analysis.

Ellipsometry mapping measurements were performed on a homemade mapping stage. The nominal angle of incidence for all measurements was fixed at 70°. The acquisition time for one spectrum was set at 1.5 s.

The measured data was modeled using a B-spline layer for the Au film, a Cauchy layer for the TiO_2 substrate, and another Cauchy layer for the silica film. The thicknesses of the various layers were determined using fixed parameters for the Cauchy parameters.

Depth profiles SIMS were made using a TOF.SIMS5 (IONTOF GmbH, Münster, Germany) time-of-flight secondary ion mass spectrometer [27]. This instrument is equipped with a Cs+ ion beam as a sputtering source and a liquid metal ion gun (Bi) as an analytical source, both mounted at 45° with respect to the sample surface. The time-of-flight mass analyzer is perpendicular to the sample surface. Depth profiles were carried out in the 'interlaced' mode in a cycle time of 100 μs. During this cycle time, the analytical beam (pulsed Bi_5^+ for SIMS analysis) was followed by periods of continuous Cs^+ ion sputtering. Low energy electrons were also sent to the sample during this cycle in order to recover the initial surface potential. The Cs^+ ion source was operated at 500 eV with a direct current of 40 nA. For depth profiling, the focused Cs^+ beam of primary ions was rastered over an area that typically measured 450×450 μm^2. A pulsed beam of 30 keV Bi_5^+ (alternating current of 0.11 pA) ions was employed to provide mass spectra from a 150×150 μm^2 area in the center of the sputter crater. Charge compensation was conducted using an electron flood gun ($E_k = 20$ eV). All data analyses were carried out using the software supplied by the instrument manufacturer, SurfaceLab (version 6.5). Depth profiling was stopped when the Ti^+ signal reached 50% of its maximum intensity.

Stylus profilometer was used to measure the craters obtained by SIMS measurements (DektakXT, Bruker Nano Surfaces Division, Tucson, AZ, USA). The stylus had a radius of 0.7 μm and the applied force was 0.1 mg. Four line scans were conducted over the measured crater in order to obtain average and standard deviation measurements.

3. Results and Discussion

3.1. Precursor Synthesis and Deposition

Keeping industrial upscaling in mind, one must be able to produce a coating precursor in a relatively short time-span, which can be conserved for longer times. Raman spectroscopy has proven to be a valuable technique to monitor the condensation species of silicon-based sol-gel reactions [28–30]. This technique was used to follow the condensation reaction over time.

Figure 3 shows a graphical representation of the monitored species. Table 2 describes the bands that are monitored. As can be seen, chain length is the main characteristic that is monitored with Raman spectroscopy. Figure 4 shows the Raman intensities of monomer, dimer (including end groups), trimer, and tetramer species present in the reaction mixture when using acetic acid as a catalyst. The marked drop in intensity of the monomer signal indicates the addition of catalyst and water, which are necessary for the condensation reaction. After the addition, a rise in dimer, trimer, and tetramer was recorder. This means that the sol starts condensing immediately after the addition of the diluted catalyst. What is more important is the time the reaction mixture takes to reach an equilibrium state. As can be seen in Figure 4, after 3 h almost all monomer species were depleted and converted into more condensed species.

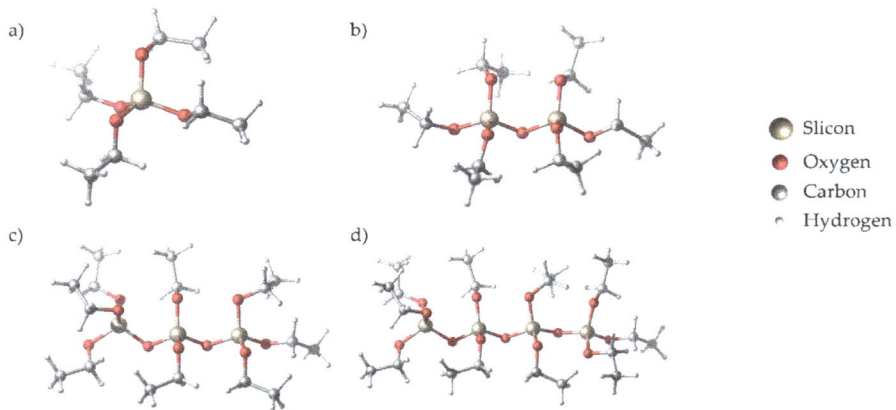

Figure 3. Graphical representation of the species monitored with Raman spectroscopy. The monomer species (**a**) reacts subsequently to dimer (**b**), trimer (**c**), and tetramer (**d**) species.

Table 2. Description of the bands monitored by Raman spectroscopy [30].

Band Profile	Description
Monomer	Height to zero, peak from 665 to 645nm
Dimer and end group	Height to zero, peak from 605 to 585 nm, referenced to monomer
Trimer	Height to zero, peak from 580 to 560 nm, referenced to monomer
Tetramer	Height to zero, peak from 560 to 540 nm, referenced to monomer

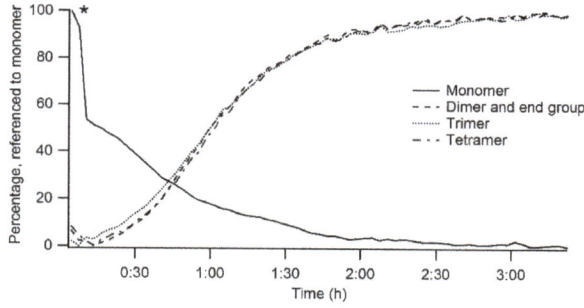

Figure 4. Raman intensities monitored during the pre-condensation step. All signals are referenced to the monomer. The marked (*) drop in intensity shows the addition of the catalyst, diluted in water.

Once the reaction mixture reaches this equilibrium state, it is important to inhibit further condensation. In order to obtain this inhibition, a quenching step is induced by the addition of ethanol. To map the stability of the sol at longer times after this quenching step, ^{29}Si-NMR was used. ^{29}Si-NMR analyses show how many times an ethoxy group is substituted by another siloxane group and thus the condensation degree. The species monitored are shown in Table 3. The spectra obtained are shown in Figure 5. Immediately after the addition of the diluted catalyst, the spectrogram labeled 'Start' was recorded. Some resonances are of particular interest. The resonance at −73 ppm corresponds to the four-times hydrolyzed species $Q^0_{hydrolized}$, with formula $Si(OH)_4$. Furthermore, −82 ppm corresponds to non-hydrolyzed TEOS monomer. The resonances between −73 and −82 ppm are due to three-, two-, and one-time hydrolyzed monomer units. The resonance of the one-time condensed species Q^1 could not be distinguished from the monomer at −82 ppm. The resonance at −92 ppm stands for the two-times condensed TEOS, labeled as Q^2. The region between −100 and −120 ppm reflects the three-times and four-times condensed species, Q^3 and Q^4, respectively [30–32]. Here, no clear distinction between three-times and four-times condensed species is possible.

Table 3. ^{29}Si chemical shifts and corresponding silicate structures [30].

Label	δ ppm	Silicate Structure
$Q^0_{hydrolized}$	−73	$Si^*(OH)_4$
Q^0	−82	$Si^*(OR)_4$
Q^2	−92	$Si^*(OR)_3(-O-Si\equiv)$
Q^3 and Q^4	−100 to −120	$Si^*(OR)_2(-O-Si\equiv)_2$ $Si^*(OR)(-O-Si\equiv)_3$

R corresponds to C_2H_5. Each peak corresponds to the marked silicon atom (*)

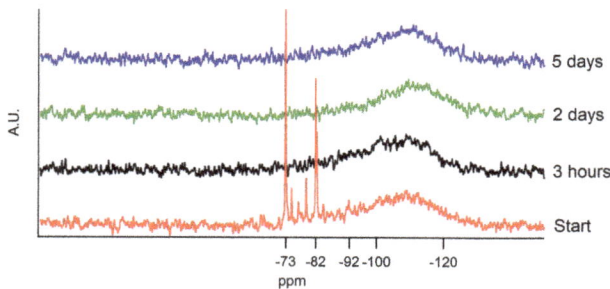

Figure 5. NMR spectrogram of the sol at different time intervals.

Figure 5 shows the ^{29}Si-NMR spectrum at the start of the condensation step. $Q^0_{\text{hydrolized}}$ species are distinguishably present at the start and are converted into other, more condensed species over time. The region between -73 and -82 ppm shows four resonances, corresponding to one-, two-, three-, and four-times hydrolyzed silicon atoms. It can be seen that already a fraction of Q^3 and Q^4 is detected.

After a condensation of 3 h and subsequent quenching, all hydrolyzed species are converted and a shift towards Q^3 and Q^4 can be noted.

^{29}Si-NMR measurements at longer gelation times were performed to monitor the shelf-life of the sol. Three possible hypotheses were set up before the experiment. The first one indicated that the sol would be stable in time. No significant change should be noticeable in the ^{29}Si-NMR spectrograms. The second hypothesis depicted the condensed material being broken up by the excess amount of ethanol present in the reaction mixture. The intensities in the region between -100 and -120 ppm should lower and shift to higher ppm values. A last hypothesis depicted the total condensation of all the species. For this hypothesis to be true, the intensities between -100 and -120 ppm should shift to lower ppm values. As can be seen in Figure 5, no significant change in the region of interest was observed. A very small shift towards -120 ppm was noticed after which an equilibrium was found. This equilibrium implies that the sol is still usable for coating processes after longer times.

The aforementioned silica precursor was deposited both on the model samples and window films.

3.2. Layer Thickness Determination

The characterization of the coating thickness was carried out by three independent techniques, namely EM, ellipsometry, and the proposed characterization toolbox, defined as a combination of SIMS with profilometry. The results are listed in Table 4. EM is reported to be the technique of choice for thickness characterization, while ellipsometry is often used when EM is not possible. When looking at the results in Table 4, one can conclude that the values for EM and ellipsometry do not correspond well to each other. This is because of the high complexity of the model sample. Ellipsometry is not able to distinguish the top thin layer from the underlying interlayers and the substrate, as it is a technique that needs a significant difference in refractive index between two layers to make a distinction. As this difference is not present prominently in window films between the top coating and the substrate, ellipsometry measurements are shadowed by the relatively thick substrate. Even the sputtering of an additional gold layer as an interface does not lead to conclusive ellipsometry measurements.

Table 4. Results overview of secondary ion mass spectrometry (SIMS), electronic microscopy (EM), and ellipsometry measurements. For each model type, two individual samples (a and b) were synthesized and characterized. A graphical representation is shown in Figure 1.

Model Substrate	SIMS (nm)	EM (nm)	Ellipsometry (nm)
TiO$_2$/Au/silicon (a)	–	63.9	88.2
TiO$_2$/Au/silicon (b)	63.6 ± 5.7	61.8	92.9
TiO$_2$/Au/PET (a)	61.0 ± 6.3	57.5	95.2
TiO$_2$/Au/PET (b)	84.7 ± 8.2	82.8	77.8
Window film (a)	126 ± 8	X	X
Window film (b)	274 ± 12	X	X

All values are noted in nm. X denotes that the measurement was not conclusive.

When comparing the proposed toolbox to EM measurements on the model substrates (Figures 6 and 7), the results match closely. Thus, the combination of SIMS and profilometry can offer a possibility to determine the thickness of the top coating, as long as the interface between the top coating and substrate is detectable. As SIMS envelops mass spectrometry, this technique can detect individual ions. Therefore, SIMS shows the possibility to distinguish different layers as soon as the atomic composition is different, meaning that other ions are present in each layer or that the relative concentration differs between layers.

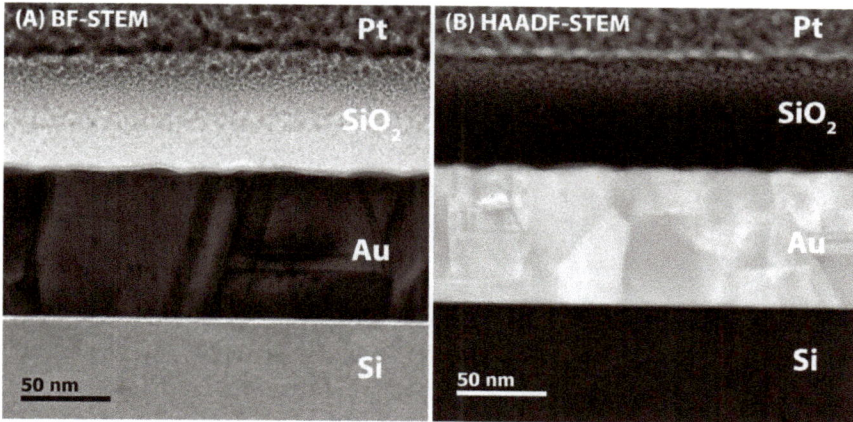

Figure 6. Cross-sectional (**A**) BF- and (**B**) HAADF-STEM image of TiO_2/Au/silicon model sample a. Individual layers can be distinguished by using Z-contrast.

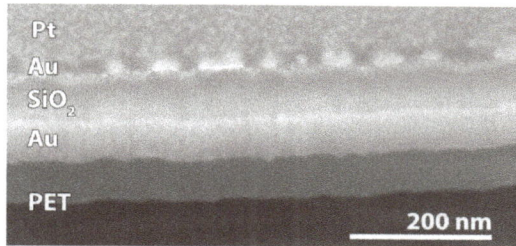

Figure 7. Cross-sectional SEM image of TiO_2/Au/PET model sample b.

Figure 8a shows a representative SIMS depth profile obtained on silicon model substrate b. To measure the crater depth, a fresh zone was selected and sputtering was stopped at the interface between the protective coating and the TiO_2 buffer layer. This was detected by a steep increase of Ti^+ ions. The depth of the sputtered crater was then determined using profilometry (Figure 8b). As can be seen in Table 4, coating thicknesses were determined to be 126 ± 8 nm for single coated window films and 274 ± 12 nm for double coated window films by the proposed toolbox.

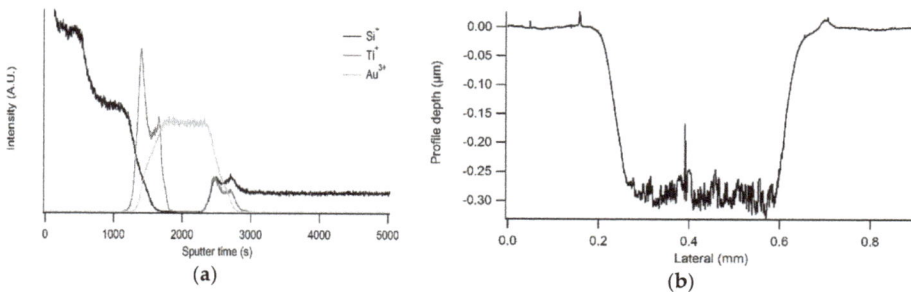

Figure 8. Complete profile of Si model sample b obtained with SIMS (**a**). Profilometry analysis (**b**) was conducted on a fresh zone and sputtering was stopped when the Ti signal reached half of the maximal intensity.

3.3. Window Film Characteristics

3.3.1. Application Tests

The coating deposited on window films was tested for adhesion, ASTM D3359 [20], and scratch resistance, ASTM D3363 [21]. To evaluate the adhesion of the layer, the standard provides a classification to compare the sample. Class 0 relates to very good adhesion, while class 5 relates to very poor adhesion. The classification is depicted in Figure 9. The scratch resistance standard used is a straightforward technique where pencils with varying hardness are used to scratch the sample.

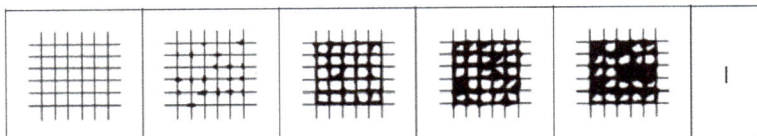

Figure 9. Crosshatch classification. On the left-hand side class 0 is depicted. On the right-hand side class 4 is depicted. Class 5 corresponds to a complete removal of the coating.

The double coated sample was able to withstand scratch attempts with a 3H pencil and showed a class 0 adhesion, using a simple crosshatch test. The single coated film, however, also showed class 0 adhesion but was not able to pass the pencil hardness test using a 3H pencil. One can say that a coating thickness of 274 ± 12 nm is needed to obtain scratch-resistant coatings with good adhesion. The minimal required thickness could be lower, but this falls out of scope of this work.

3.3.2. Visual Transmission and Infrared Reflection Properties

To be able to use a window film onto which a heat-reflecting layer is applied, good visual transparency and optimal infrared reflectivity should be achieved. UV-VIS-IR measurements were recorded in transmission and reflection modes.

As shown in Figure 10, there is a difference in visual transparency between the uncoated stack and coated samples in the region between 400 and 600 nm. Visual transparency increases when applying the silica coating, while there is almost no difference between the two coated samples. This is not the case for the reflection spectra, as shown in Figure 11; almost no difference is seen between the uncoated stack and double coated film, while a big difference is noted between the uncoated stack and single coated film on the one hand and double coated film on the other hand between 800 and 2000 nm. This drop is not present for the double coated window film.

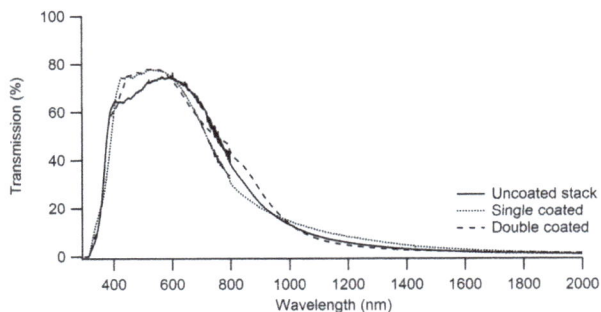

Figure 10. Transmission spectra of the uncoated stack and samples of single coated window film and double coated window film. The double coated window film is more reflective in the region between 400 nm and 600 nm than the single coated window film.

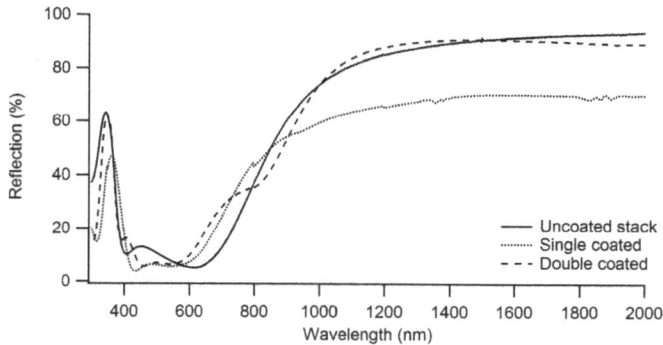

Figure 11. Reflection spectra of the uncoated stack, single coated window film, and double coated window film. The double coated is more reflective than single coated film at higher wavelengths.

It is thus shown that the silica layer interferes with the Fabry-Pérot stack in such a way that higher visual transparency is obtained, while maintaining the heat reflecting properties. The fact that this changes with coating thicknesses proves that this is due to the combination of refractive indices of the silica coating and the Fabry-Pérot stack.

These findings were confirmed by the emissivity values, which ranged from 0.022 for the uncoated stack to 0.031 for the double coated film. The last value is well below the threshold of 0.1 for heat-reflecting window films.

4. Conclusions

The main conclusion of this work is that the proposed thickness characterization toolbox for transparent protective coatings on transparent devices is proven to be a valuable method. The results obtained by this combination of SIMS and ellipsometry are in agreement with those achieved using electron microscopy for the designed model samples. The toolbox was able to determine the thicknesses of thin layers deposited on functional window films, where other techniques failed.

This research also showed synthesis and deposition methods for the large-scale application of scratch-resistant coatings on polymeric, heat-reflecting window films. Both Raman and ^{29}Si-NMR spectroscopy were used to assess precursor synthesis and the following shelf-life. The results showed that an optimal coating precursor was obtained after 3 h of condensation and that a shelf-life up to five days was guaranteed. After simple and straightforward coating deposition with roll-to-roll coating, application tests were conducted. It was shown that a coating thickness of 274 ± 12 nm leads to class 0 adhesion, 3H pencil hardness resistance, optimal visual transparency, and an even improved infrared reflectivity, compared to uncoated samples. It is possible that a lower coating thickness would be sufficient, keeping in mind the lower threshold of 126 ± 8 nm, but this was not studied further. By avoiding use of chloride containing reagents, further scaling up towards industrial settings is possible. It is proven that this chloride-free synthesis has no negative effects on the drying and curing of the coating, as the coating passed all application tests. A low temperature heat-treatment program was shown to be sufficient to obtain these scratch-resistant coatings. A maximum temperature of 80 °C was needed to fully dry and cure the coating. It is thus possible to transfer these results to other temperature-sensitive substrates.

In short, this research demonstrated an effective thickness characterization toolbox. Additionally, during this research an easily up-scalable coating was developed. This coating is intended to protect scratch-sensitive substrates, which cannot be coated with organic or hybrid state-of-the-art materials, due to their optical properties.

5. Patents

The protective coating precursor composition was patented under number EP15199592.5 [26].

Author Contributions: Conceptualization, M.V.Z., J.H., and K.D.B.; Investigation, M.V.Z., H.R., J.H., D.D., D.P.D., and C.P.; Resources, S.V., I.V.D., and K.D.B.; Writing-Original Draft Preparation, M.V.Z.

Funding: This work was supported by the Institute for Science and Technology in Flanders (IWT) through grant 140454. The authors also thank UMISOL for their partnership in this industrial grant through Flanders Innovation and Entrepeneurship grants HBC.2016.0260 and HBC.2017.0874.

Acknowledgments: J. Martins is acknowledged for NMR measurements. P. Van Der Voort of the Ceter for Ordered Materials, Organometallics & Catalysis (COMOC) is acknowledged for providing access to the Raman setup. D. Van Thourhout of the Ugent expertise center for Nano- and Microfabrication (NAMIFAB) is acknowledged for the sputtering of titania and gold on the model samples. P. Chan of the Sol-Gel Centre for Research on Inorganic Powders and Thin Film Synthesis (SCRiPTS) is acknowledged for the design of the roll-to-roll coater figure.

Conflicts of Interest: The authors declare no conflict of interest.

References

1. Fan, J.C.C.; Bachner, F.J.; Foley, G.H.; Zavracky, P.M. Transparent heat-mirror films of $TiO_2/Ag/TiO_2$ for solar energy collection and radiation insulation. *Appl. Phys. Lett.* **1974**, *25*, 693–695. [CrossRef]
2. Ando, E.; Miyazaki, M. Moisture degradation mechanism of silver-based low-emissivity coatings. *Thin Solid Films* **1999**, *351*, 308–312. [CrossRef]
3. Ando, E.; Miyazaki, M. Moisture resistance of the low-emissivity coatings with a layer structure of Al-doped ZnO/Ag/Al-doped ZnO. *Thin Solid Films* **2001**, *392*, 289–293. [CrossRef]
4. Ando, E.; Suzuki, S.; Aomine, N.; Miyazaki, M.; Tada, M. Sputtered silver-based low-emissivity coatings with high moisture durability. *Vacuum* **2000**, *59*, 792–799. [CrossRef]
5. Miyazaki, M.; Ando, E. Durability improvement of Ag-based low-emissivity coatings. *J. Non-Cryst. Solids* **1994**, *178*, 245–249. [CrossRef]
6. Ross, R. Observations on humidity-induced degradation of Ag-based low-emissivity films. *Solar Energy Mater.* **1990**, *21*, 25–42. [CrossRef]
7. Nadel, S.J. Durability of Ag based low-emissivity coatings. *J. Vac. Sci. Technol. A* **1987**, *5*, 2709–2713. [CrossRef]
8. Vernieuwe, K.; Cuypers, D.; Kirschhock, C.E.A.; Houthoofd, K.; Vrielinck, H.; Lauwaert, J.; De Roo, J.; Martins, J.C.; Van Driessche, I.; De Buysser, K. Thermal processing of aqueous AZO inks towards functional TCO thin films. *J. Alloys Compd.* **2017**, *690*, 360–368. [CrossRef]
9. Levine, J.B.; Ciraldo, J.P. Method of Growing Aluminum Oxide onto Substrates by Use of an Aluminum Source in an Environment Containing Partial Pressure of Oxygen to Create Transparent, Scratch-Resistant Windows. U.S. Patent US2016215381 (A1), 28 July 2016.
10. Sangermano, M.; Perrot, A.; Gigot, A.; Rivolo, P.; Pirri, F.; Messori, M. Hydrophobic Scratch Resistant UV-Cured Epoxy Coating. *Macromol. Mater. Eng.* **2016**, *301*, 93–98. [CrossRef]
11. Groenewolt, M.; Kues, J.B.; Schroeder, S.; Cavaleiro, P.; Karminski, H.L.; Michael, G. High-Transparency Polycarbonates with Scratch-Resistant Coating, Process for Production Thereof and Use Thereof. U.S. Patent US200913129528, 28 October 2014.
12. Amerio, E.; Fabbri, P.; Malucelli, G.; Messori, M.; Sangermano, M.; Taurino, R. Scratch resistance of nano-silica reinforced acrylic coatings. *Prog. Org. Coat.* **2008**, *62*, 129–133. [CrossRef]
13. Tahmassebi, N.; Moradian, S.; Ramezanzadeh, B.; Khosravi, A.; Behdad, S. Effect of addition of hydrophobic nano silica on viscoelastic properties and scratch resistance of an acrylic/melamine automotive clearcoat. *Tribol. Int.* **2010**, *43*, 685–693. [CrossRef]
14. Kues, J.B.; Brosseit, A.; Homann, K.; Schroeder, S. Plastic Substrates Having a Scratch-Resistant Coating, in Particular Housings of Electronic Devices, Having High Transparency, Method for the Production Thereof, and Use Thereof. U.S. Patent US201013577100, 28 April 2015.
15. Watté, J.; Van Zele, M.; De Buysser, K.; Van Driessche, I. Recent Advances in Low-Temperature Deposition Methods of Transparent, Photocatalytic TiO_2 Coatings on Polymers. *Coatings* **2018**, *8*, 131. [CrossRef]

16. Rijckaert, H.; Pollefeyt, G.; Sieger, M.; Hänisch, J.; Bennewitz, J.; De Keukeleere, K.; De Roo, J.; Hühne, R.; Bäcker, M.; Paturi, P.; et al. Optimizing Nanocomposites through Nanocrystal Surface Chemistry: Superconducting YBa$_2$Cu$_3$O$_7$ Thin Films via Low-Fluorine Metal Organic Deposition and Preformed Metal Oxide Nanocrystals. *Chem. Mater.* **2017**, *29*, 6104–6113. [CrossRef]

17. Narayanan, V.; Lommens, P.; De Buysser, K.; Vanpoucke, D.E.P.; Huehne, R.; Molina, L.; Van Tendeloo, G.; Van Der Voort, P.; Van Driessche, I. Aqueous CSD approach for the growth of novel, lattice-tuned La$_x$Ce$_{1-x}$O$_\delta$ epitaxial layers. *J. Mater. Chem.* **2012**, *22*, 8476. [CrossRef]

18. Pollefeyt, G.; Clerick, S.; Vermeir, P.; Lommens, P.; De Buysser, K.; Van Driessche, I. Influence of Aqueous Precursor Chemistry on the Growth Process of Epitaxial SrTiO$_3$ Buffer Layers. *Inorg. Chem.* **2014**, *53*, 4913–4921. [CrossRef] [PubMed]

19. Pollefeyt, G.; Rottiers, S.; Vermeir, P.; Lommens, P.; Hühne, R.; De Buysser, K.; Van Driessche, I. Feasibility study of the synthesis of YBiO$_3$ thin films by aqueous chemical solution deposition as an alternative for CeO$_2$ buffer layers in coated conductors. *J. Mater. Chem. A* **2013**, *1*, 3613–3619. [CrossRef]

20. ASTM D3359-09e2. *Standard Test Methods for Measuring Adhesion by Tape Test*; ASTM International: West Conshohocken, PA, USA, 2009.

21. ASTM D3363-05(2011)e2. *Standard Test Method for Film Hardness by Pencil Test*; ASTM International: West Conshohocken, PA, USA, 2011.

22. Yasuda, M.; Morimoto, K.; Kainuma, Y.; Kawata, H.; Hirai, Y. Analysis of Charging Phenomena of Polymer Films on Silicon Substrates under Electron Beam Irradiation. *Jpn. J. Appl. Phys.* **2008**, *47*, 4890–4892. [CrossRef]

23. Geltmeyer, J.; De Roo, J.; Van den Broeck, F.; Martins, J.C.; De Buysser, K.; De Clerck, K. The influence of tetraethoxysilane sol preparation on the electrospinning of silica nanofibers. *J. Sol-Gel Sci. Technol.* **2016**, *77*, 453–462. [CrossRef]

24. Geltmeyer, J.; Van der Schueren, L.; Goethals, F.; De Buysser, K.; De Clerck, K. Optimum sol viscosity for stable electrospinning of silica nanofibres. *J. Sol-Gel Sci. Technol.* **2013**, *67*, 188–195. [CrossRef]

25. Goethals, F.; Ciofi, I.; Madia, O.; Vanstreels, K.; Baklanov, M.R.; Detavernier, C.; Van Der Voort, P.; Van Driessche, I. Ultra-low-k cyclic carbon-bridged PMO films with a high chemical resistance. *J. Mater. Chem.* **2012**, *22*, 8281–8286. [CrossRef]

26. De Buysser, K.; Ide, M.; Van Zele, M.; Van Driessche, I. A Method of Manufacturing a Coated Polymer Substrate Having Low Emissivity. WO2016EP79943, 6 December 2016.

27. Caillot, M.; Chaumonnot, A.; Digne, M.; Poleunis, C.; Debecker, D.P.; van Bokhoven, J.A. Synthesis of amorphous aluminosilicates by grafting: Tuning the building and final structure of the deposit by selecting the appropriate synthesis conditions. *Microporous Mesoporous Mater.* **2014**, *185*, 179–189. [CrossRef]

28. De Ferri, L.; Lorenzi, A.; Lottici, P.P. OctTES/TEOS system for hybrid coatings: Real-time monitoring of the hydrolysis and condensation by Raman spectroscopy. *J. Raman Spectrosc.* **2016**, *47*, 699–705. [CrossRef]

29. Matos, M.C.; Ilharco, L.M.; Almeida, R.M. The evolution of TEOS to silica gel and glass by vibrational spectroscopy. *J. Non-Cryst. Solids* **1992**, *147–148*, 232–237. [CrossRef]

30. Depla, A.; Lesthaeghe, D.; van Erp, T.S.; Aerts, A.; Houthoofd, K.; Fan, F.; Li, C.; Van Speybroeck, V.; Waroquier, M.; Kirschhock, C.E.A.; et al. ^{29}Si NMR and UV−Raman Investigation of Initial Oligomerization Reaction Pathways in Acid-Catalyzed Silica Sol−Gel Chemistry. *J. Phys. Chem. C* **2011**, *115*, 3562–3571. [CrossRef]

31. Devreux, F.; Boilot, J.; Chaput, F.; Lecomte, A. Sol-gel condensation of rapidly hydrolyzed silicon alkoxides: A joint Si ^{29} NMR and small-angle X-ray scattering study. *Phys. Rev. A* **1990**, *41*, 6901–6909. [CrossRef] [PubMed]

32. Turner, C.W.; Franklin, K.J. Studies of the hydrolysis and condensation of tetraethylorthosilicate by multinuclear (^{1}H, ^{17}O, ^{29}Si) nmr spectroscopy. *J. Non-Cryst. Solids* **1987**, *91*, 402–415. [CrossRef]

![materials logo] *materials*

MDPI

Article

Introducing Obliquely Perforated Phononic Plates for Enhanced Bandgap Efficiency

Saeid Hedayatrasa [1,*], Mathias Kersemans [1], Kazem Abhary [2] and Wim Van Paepegem [1]

[1] Department of Materials, Textiles and Chemical Engineering, Ghent University, Technologiepark-Zwijnaarde 903, 9052 Zwijnaarde, Belgium; mathias.kersemans@ugent.be (M.K.); wim.vanpaepegem@ugent.be (W.V.P.)

[2] School of Engineering, University of South Australia, Mawson Lakes, SA 5095, Australia; kazem.abhary@unisa.edu.au

* Correspondence: saeid.hedayatrasa@ugent.be; Tel.: +32-9-331-0421

Received: 6 June 2018; Accepted: 25 July 2018; Published: 28 July 2018

Abstract: Porous phononic crystal plates (PhPs) that are produced by perpendicular perforation of a uniform plate have well-known characteristics in selective manipulation (filtration, resonation, and steering) of guided wave modes. This paper introduces novel designs of porous PhPs made by an oblique perforation angle. Such obliquely perforated PhPs (OPhPs) have a non-uniform through-the-thickness cross section, which strongly affects their interaction with various wave mode types and therefore their corresponding phononic properties. Modal band analysis is performed in unit-cell scale and variation of phononic bandgaps with respect to the perforation angle is studied within the first 10 modal branches. Unit-cells with arbitrary perforation profile as well as unit-cells with optimized topology for maximized bandgap of fundamental modes are investigated. It is observed that the oblique perforation has promising effects in enhancing the unidirectional and/or omnidirectional bandgap efficiency, depending on the topology and perforation angle of OPhP.

Keywords: Phononic Crystal; Plate; Oblique Perforation; Bandgap

1. Introduction

Phononic crystals (PhCrs) are lattice structures that can manipulate elastic waves in an extraordinarily way through their periodic microstructure [1–10]. The main characteristic of PhCrs is the existence of frequency bands (so called bandgaps), over which propagation of an incident wave is banned. Moreover, the wave may be resonated and/or guided inside an intentionally introduced defect in a PhCr, at a frequency in the bandgap frequency range [11,12]. A wider bandgap enables phononic controllability over a wider frequency range, while a lower bandgap frequency range implies that a larger incident wavelength can be manipulated by a specified phononic lattice. Therefore, relative bandgap width (RBW), which is defined as bandgap width divided by mid gap frequency, is normally used to indicate the bandgap efficiency of PhCrs. Furthermore, due to the strong anisotropy that is introduced by PhCrs, they possess flat and concave wave fronts at particular frequencies that can, respectively, be used for self-collimation and focusing of elastic waves [8,13,14].

The bandgap frequency of a PhCr depends on the dimensions, the constitutive material(s), and the topology of its irreducible periodic feature (unit-cell). Phononic crystal plates (PhPs) have promising application in manipulation of guided waves for designing low loss acoustic devices (resonators, filter, and wave guides) [3,15–17] and for structural health monitoring purposes [12,18]. PhPs may be produced by periodic placement of stiff inclusions inside a compliant base plate [3,19], by periodic through perforation of a uniform plate [5,20–24], or by attaching a periodic array of pillars on a substrate [15,25], or a combination of them [26]. Perforated PhPs are relatively easy to produce, in which porosities introduce strong reflecting boundaries (i.e., high acoustic impedance mismatch) that are free from interfacial imperfections. Moreover, finite thickness and light weight cellular design of perforated PhPs make them a promising constitutive structural material with low vibration transmission and acoustic radiation. The topology of perforation profile can be optimized such that maximized RBW of a particular mode type (symmetric or asymmetric) or a complete bandgap of mixed guided waves (combination of symmetric and asymmetric modes) is obtained [27,28] and with desired stiffness [5,22,29] or deformation induced tunability [30].

Earlier studies have shown the abnormal dispersion properties of tapered meta-surfaces with axisymmetric non-uniform through the thickness profile [31,32]. In this paper, obliquely perforated PhPs (OPhPs) are introduced for enhanced bandgap efficiency through their non-uniform through-the-thickness cross section. Two distinct designs with symmetric and asymmetric through-the-thickness constitution, with respect to the mid-plane, are proposed. Various unit-cells with arbitrary as well as optimized perforation profile are examined at perforation angles 0° to 60° with respect to the normal axis of the plate. The modal band structure of the first 10 wave modes is calculated and the variation of total RBW with respect to the perforation angle is studied. The results confirm the promising effect of oblique perforation in enhancing partial (unidirectional) and/or complete (omnidirectional) RBW of studied topologies.

The layout of the paper is, as follows. First, the two proposed designs of OPhPs, relevant unit-cells, and selected topologies to be examined are presented and constitutive equations of modal band analysis are given. Then, the calculated modal band structure and RBW of selected topologies with respect to the perforation angle are presented and discussed. Finally, an alternative topology is selected and the enhanced bandgap efficiency of its OPhP is validated through transmission spectrum of its finite phononic structure.

2. OPhP Designs and Modal Band Analysis

As schematically shown in Figure 1a,b, the proposed OPhP design can be produced by lateral perforation of a uniform background plate at an angle θ with respect to the plate's normal axis (i.e., z-axis). Figure 1b shows the irreducible unit-cell of OPhP with continuous solid boundary being chosen along the perforation path at an angle θ about y-axis. The perforation profile is assumed to have square symmetry (in xy-plane), and unit-cell with aspect ratio (width to thickness) $a/h = 2$ is considered.

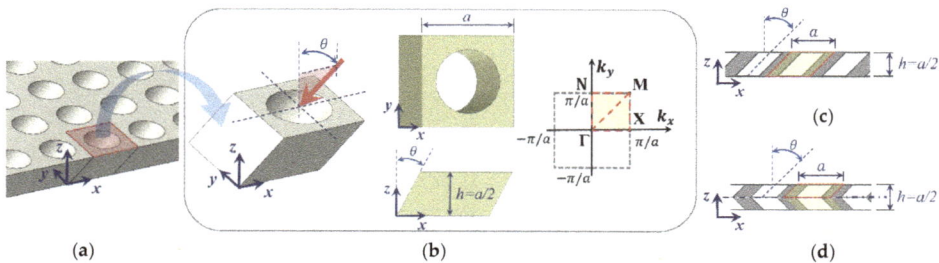

Figure 1. (**a**) Schematic of proposed obliquely perforated porous phononic crystal plates (OPhPs) produced by oblique perforation at an angle θ (about y-axis) on a uniform plate and (**b**) selected unit-cell and relevant irreducible Brillouin zone, and cross section of (**c**) an asymmetric OPhP (A-OPhP) and (**d**) a symmetric OPhP (S-PhP) lattice, respectively, with asymmetric and symmetric through-the-thickness design.

The designs shown in Figure 1c,d introduce asymmetric and symmetric OPhPs, respectively. An asymmetric OPhP (A-OPhP) may be produced by oblique thorough perforation, and a symmetric OPhP (S-OPhP) may be produced by double sided perforation to the mid-plane or by mirrored attachment of two A-OPhPs. The shaded area in Figure 1c,d indicates the border of OPhP unit-cell. In the case of S-OPhP, as shown in Figure 1d, only half of the thickness from mid-plane to the top suffices to be considered as the unit-cell due to its mirrored symmetry. However, appropriate boundary conditions have to be applied to the mid-plane to define this mirrored symmetry [29].

For modal band analysis of the unit-cell, Bloch-Floquet boundary conditions have to be applied to ensure the periodicity of opposing boundaries for the in-plane wave vectors $\mathbf{k} = \left\{ \begin{array}{cc} k_x & k_y \end{array} \right\}$ as follows [33]:

$$\mathbf{u}(\mathbf{x}, t) = \mathbf{u}(\mathbf{x} + \mathbf{A}, t)e^{i\mathbf{k} \cdot \mathbf{A}} \tag{1}$$

where $\mathbf{u} = \{u\ v\ w\}^{\mathrm{T}}$ is the displacement vector corresponding to location vector $\mathbf{x} = \{x\ y\ z\}$, $\mathbf{A} = \{a\ a\}$ is the lattice periodicity vector for unit-cell width a, and t is time and $i = \sqrt{-1}$. Contrary to perpendicular perforation, m which leads to a square symmetric unit-cell, oblique perforation reduces the symmetry while the lattice periodicity remains the same in x- and y-axis. Due to the periodicity of boundary condition (Equation (1)), $\mathbf{k}$ can be searched over the first Brillouin zone and, according to the common practice, only on its border [34]. The inset of Figure 1b shows the first Brillouin zone in which the irreducible Brillouin zone corresponding to the OPhP unit-cell (confined by the border ΓXMΓNM) is highlighted. Obviously, for perpendicular perforation (i.e., $\theta = 0°$), which leads to a square symmetric unit-cell, the irreducible Brillouin zone reduces to the triangle ΓXM. By modal analysis of the unit-cell over the n_k discrete search points i on the border of the irreducible Brillouin zone, the total RBW within the first 10 modal branches can be initially defined as:

$$RBW_i(\theta) = \sum_{j=1}^{10} \frac{max(min_{i=1}^{n_k}\omega_{j+1}^2(\mathbf{k}_i, \theta) - max_{i=1}^{n_k}\omega_j^2(\mathbf{k}_i, \theta), 0)}{0.5\,(min_{i=1}^{n_k}\omega_{j+1}^2(\mathbf{k}_i, \theta) + max_{i=1}^{n_k}\omega_j^2(\mathbf{k}_i, \theta))} \tag{2}$$

where $\omega_j(\mathbf{k}_i, \theta)$ is the modal frequency of mode j at wave vector $\mathbf{k}_i$ corresponding to point i on the border of the Brillouin zone and at perforation angle θ. RBW is calculated for 13 perforation angles between 0 to 60° with 5° increment. The upper limit of bandgap frequency within the first 10 modal branches varies with perforation angle and it reaches a maximum ω_{MB} at an angle specific to the topology. Therefore, the RBW of all the perforation angles is consistently recalculated over a constant frequency range of $0 < \omega \leq \omega_{MB}$:

$$RBW(\theta) = \sum_{j=1}^{20} \frac{max(min(min_{i=1}^{n_k}\omega_{j+1}^2(\mathbf{k}_i, \theta), \omega_{MB}) - max_{i=1}^{n_k}\omega_j^2(\mathbf{k}_i, \theta), 0)}{0.5\,(min_{i=1}^{n_k}\omega_{j+1}^2(\mathbf{k}_i, \theta) + max_{i=1}^{n_k}\omega_j^2(\mathbf{k}_i, \theta))} \tag{3}$$

Naturally, sufficiently more modal branches have to be included in Equation (3) (herein 10 more modal branches) to take into account any bandgap emerging below the upper frequency limit ω_{MB}.

The finite element method (FEM) is implemented through ANSYS APDL FEM solver (ANSYS, Inc., Canonsburg, Pennsylvania, USA, Academic Research, Release 16) for modal band analysis of OPhPs. Aluminum with elastic modulus $E_s = 70$ GPa, Poisson's ratio $\nu_s = 0.34$, and density $\rho_s = 2700$ kg/m^3 is considered as the constitutive material. However, the constitutive material properties do not significantly affect the RBW calculated for the porous design of OPhPs. Two identical unit-cell models are meshed and superimposed by constraint equations, as explained in [29], which one model accounts for the real terms and the other one accounts for the imaginary terms of the periodic boundary condition, as defined in Equation (1). Opposite faces of the unit-cell are modelled by conforming meshes (i.e., each boundary node has a mirrored counterpart on the opposite boundary) to ensure a proper definition of periodicity.

As mentioned earlier, in the case of S-OPhPs, only half of the thickness from mid-plane to the top surface is modelled as the unit-cell. Then, appropriate boundary conditions are applied to the mid-plane to calculate the modal band structure of symmetric and asymmetric guided wave modes individually [29]. However, for A-OPhPs the modal band structure of mixed guided waves is only calculated because its modes with dominant symmetric or asymmetric character cannot be easily decoupled by such boundary condition.

In order to demonstrate the bandgap efficiency of OPhPs, a set of perforation profiles with arbitrary and optimized topologies are chosen, as depicted in Figure 2.

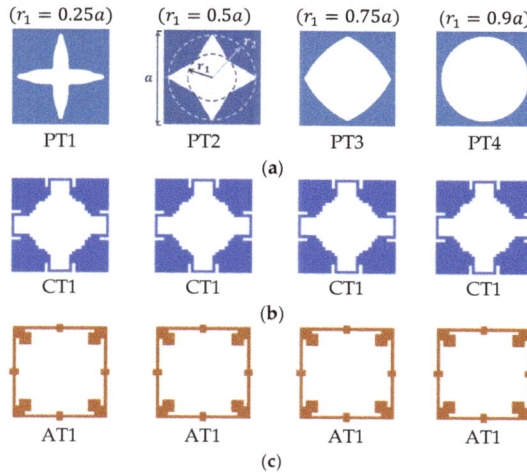

Figure 2. Perforation profiles chosen to study the bandgap efficiency of OPhPs, (**a**) prescribed topologies with arbitrary perforation profile ($r_2 = 0.9a$ and different r_1), and optimized topologies with (**b**) maximized complete bandgap of mixed guided waves and (**c**) maximized bandgap of asymmetric guided wave modes [10].

An arbitrary perforation profile with square symmetry is prescribed with smoothly varying perforation radius from an inner radius r_1 to an outer radius r_2, as shown in Figure 2a. Topologies PT1 to PT4 all have outer radius of $0.9a$ but various inner radiuses of $0.25a$, $0.5a$, $0.75a$, and $0.9a$, respectively, where a is the unit-cell width. Moreover, optimized topologies with maximized RBW of complete bandgap of mixed guided wave modes CT1–CT4 as well as optimized topologies with maximized RBW of asymmetric guided wave modes AT1–AT4 are studied [10]. The topologies were optimized through a multi-objective optimization algorithm for both maximized RBW of fundamental modes and maximized in-plane stiffness for unit-cell with aspect ratio 2. The bandgap efficiency of the selected optimized topologies reduces, while their stiffness increases from CT1 to CT4 and likewise from AT1 to AT4. When oblique perforation shows an increase in a bandgap type (symmetric, asymmetric, or mixed modes) of arbitrary profiles, it is interesting to also examine its efficiency for optimized perforation profiles of the same bandgap type. However, it is not meant that the topologies optimized for perpendicular perforation are still optimized in the case of oblique perforation.

3. Bandgap of Mixed Guided Wave Modes by A-OPhPs

In this section the variation of RBW of A-OPhPs with respect to the perforation angle is presented. The total RBW of mixed guided wave modes is calculated for prescribed topologies PT1–PT4, as well as optimized topologies CT1–CT4 for 13 perforation angles between $0°$ to $60°$ with $5°$ increments, as shown in Figure 3.

According to Figure 3a, the partial bandgap of all the prescribed topologies PT1–PT4, along Brillouin zone border ΓX, initially increases by perforation angle and declines after a particular angle depending on the topology of perforation profile.

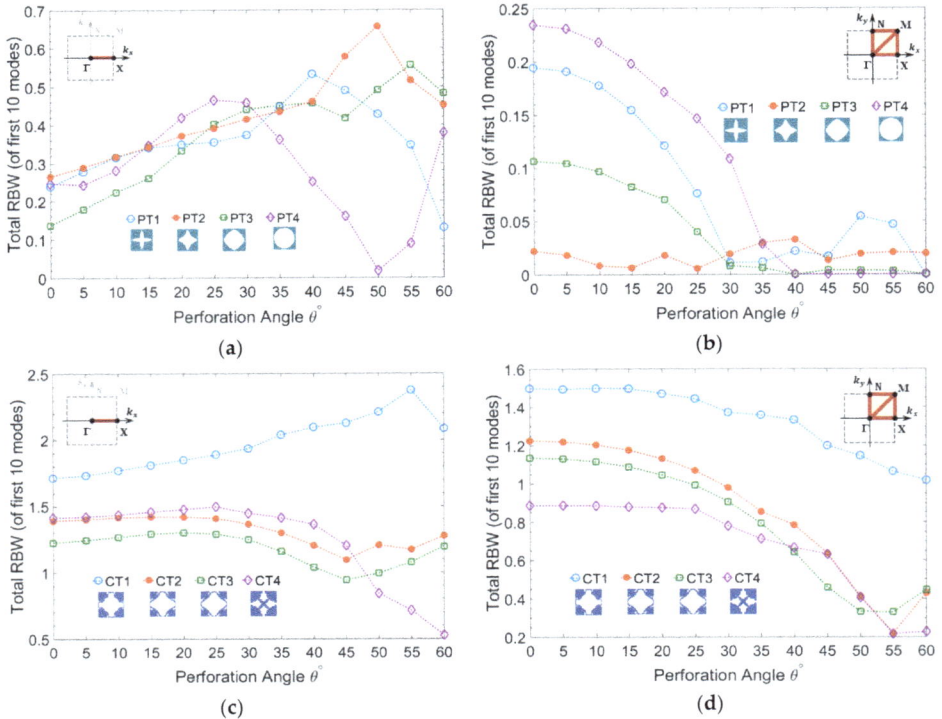

Figure 3. Total relative bandgap width (RBW) of mixed guided wave modes versus perforation angle calculated for (**a**,**b**) prescribed topologies PT1-PT4 and (**c**,**d**) optimized topologies CT1-CT4, (**a**,**c**) partial bandgap in ΓX and (**b**,**d**) complete bandgap in ΓXMΓNM.

This partial bandgap corresponds to the unidirectional wave propagation in x-axis along which oblique perforation is performed in zx-plane. Those topologies with lower initial RBW (at $\theta = 0°$) reach their first peak at a smaller angle and PT1 with highest initial RBW keeps rising up to the angle $\theta = 50°$. Likewise, as demonstrated by Figure 3c, the partial bandgap of all the optimized topologies show an initial increase by perforation angle among which CT1 with highest RBW keeps increasing up to the angle $\theta = 55°$. The topologies with lowest initial RBW e.g., PT4, CT2, and CT3, which reach their initial peak at smaller angle, start rising up again after reaching a minimum value. As expected, the RBW of optimized topologies is significantly higher than that of arbitrarily prescribed topologies.

On the contrary, the complete bandgap of all topologies PT1-PT4 and CT1-CT4 do not improve by perforation angle. Figure 3b demonstrates the steep decline of RBW for PT1-PT3 and minor fluctuations of PT4 with lowest initial RBW versus perforation angle. The rate of reduction of RBW versus perforation angle is lower in optimized topologies when compared to prescribed topologies, which are correlated to their initially much higher RBW.

The frequency ranges corresponding to the partial bandgaps of the first 10 modal branches and their variation with respect to the perforation angle are shown in Figure 4a,b for topologies PT2 and CT1, respectively. Whereas, the actual bandgap frequency of the unit-cell depends on its periodicity a and constitutive material properties, it is common practice to calculate a dimensionless frequency $f_d = \omega a / 2\pi C_p$ where $C_p = \sqrt{E_s/\rho_s}$.

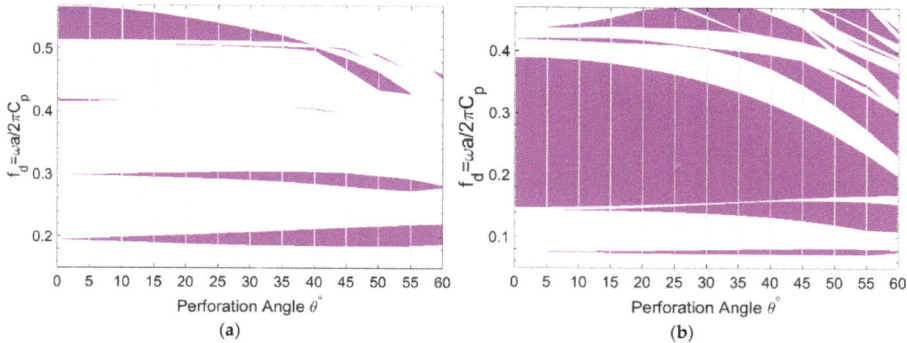

Figure 4. Variation of frequency ranges corresponding to the bandgaps of mixed guided wave modes opened within the first 10 modal branches versus perforation angle $0° \leq \theta \leq 60°$, (**a**) partial bandgap of prescribed topology PT2 and (**b**) partial bandgap of optimized topology CT1 in ΓX.

As demonstrated by Figure 4a concerning the topology PT2, two fundamental partial bandgaps, just below $f_d = 0.2$ and around $f_d = 0.3$ open, widen, and develop towards lower frequency ranges by perforation angle, which have the most contribution in increased RBW of A-OPhP. Another gap opens above $f_d = 0.5$ from perforation angle $15°$, widens towards lower frequencies, and closes at angle $55°$. However, the highest bandgap narrows down by perforation angle and closes at $40°$. Other minor gaps are also present, particularly at larger angles.

Likewise, the partial bandgaps of the topology CT1, as shown in Figure 4b are deviated towards lower frequencies and new bandgaps are introduced as the perforation angle increases. The maximized bandgap of CT1 is gradually narrowed down by perforation angle, however, two fundamental bandgaps emerge and widen at lower frequencies: one from $5°$ and the other from $10°$.

From Figure 4, it is obvious that the gradient of total RBW with respect to the perforation angle strongly depends on the number of modal branches included in calculation of RBW. However, common observations from partial bandgaps of both prescribed topology PT2 and optimized topology CT1 are:

- increasing perforation angle introduces and/or widens fundamental low frequency bandgaps;
- increasing perforation angle shifts higher order bandgaps towards lower frequency levels; and,
- narrowing of a bandgap is generally associated with development of a lower bandgap.

Obviously, all the above observed behavior contribute to the increasing of RBW with perforation angle.

4. Bandgap of Asymmetric Wave Modes by S-OPhPs

In this section, the variation of RBW of S-OPhPs with respect to the perforation angle is presented. As discussed in Section 2, half of S-OPhP's thickness from mid-plane is modelled and symmetric and asymmetric guided wave modes are decoupled by applying appropriate boundary condition to the mid-plane. Partial and complete bandgaps of both wave mode types are calculated for prescribed topologies PT1-PT4, as well as optimized topologies AT1-AT4 for the 13 perforation angles between 0 to $60°$ with $5°$ increments.

The results concerning prescribed topologies PT1-PT4 for both asymmetric and symmetric wave modes are shown in Figure 5 to compare their sensitivity to oblique perforation. According to Figure 5a, the partial RBW of asymmetric wave modes in Brillouin zone border ΓX is increased by perforation angle for all prescribed topologies, among which PT4 shows the highest increase rate.

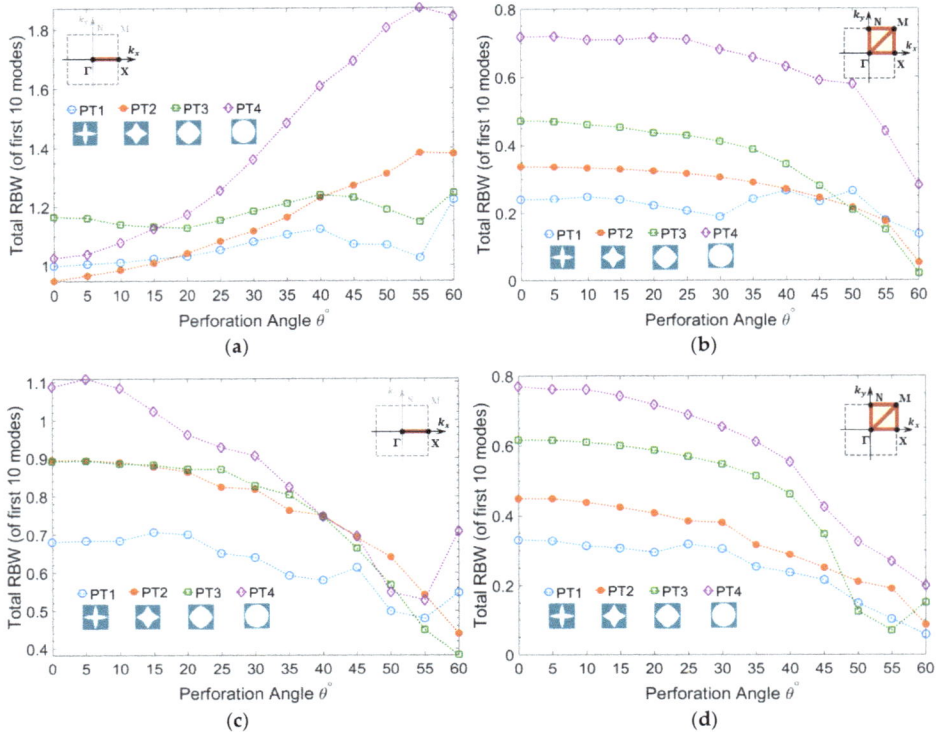

Figure 5. Total RBW of guided wave modes versus perforation angle calculated for prescribed topologies PT1-PT4, (**a**,**b**) asymmetric wave modes and (**c**,**d**) symmetric wave modes, (**a**,**c**) partial bandgap in ΓX, and (**b**,**d**) complete bandgap in ΓXMΓNM.

Regarding the complete bandgap of asymmetric wave modes, as shown in Figure 5b, the topology PT1 shows a minor increase of RBW at 40° and 50°, while the RBW of other topologies constantly decline by perforation angle. No improvement is observed in partial and complete bandgap of symmetric modes by perforation angle, as demonstrated in Figure 5c,d. It is noteworthy that RBW of mixed guided waves was also calculated (not shown) and its was observed that the bandgaps degrade by perforation angle.

Furthermore, RBW of asymmetric wave modes are calculated for optimized topologies AT1-AT4 having maximized bandgap of fundamental asymmetric wave modes, as shown in Figure 6. The results confirm that both partial and complete bandgaps of asymmetric wave modes are enhanced by perforation angle in the S-OPhP design of all optimized topologies AT1-AT4. The sensitivity of RBW to the perforation angle strongly depends on the topology. For example, both topologies AT3 and AT4 have almost the same initial complete RBW (Figure 6b). However, AT3 shows higher increase rate up to the angle 30° and AT4 shows a considerably lower increase rate up to a larger perforation angle 50°.

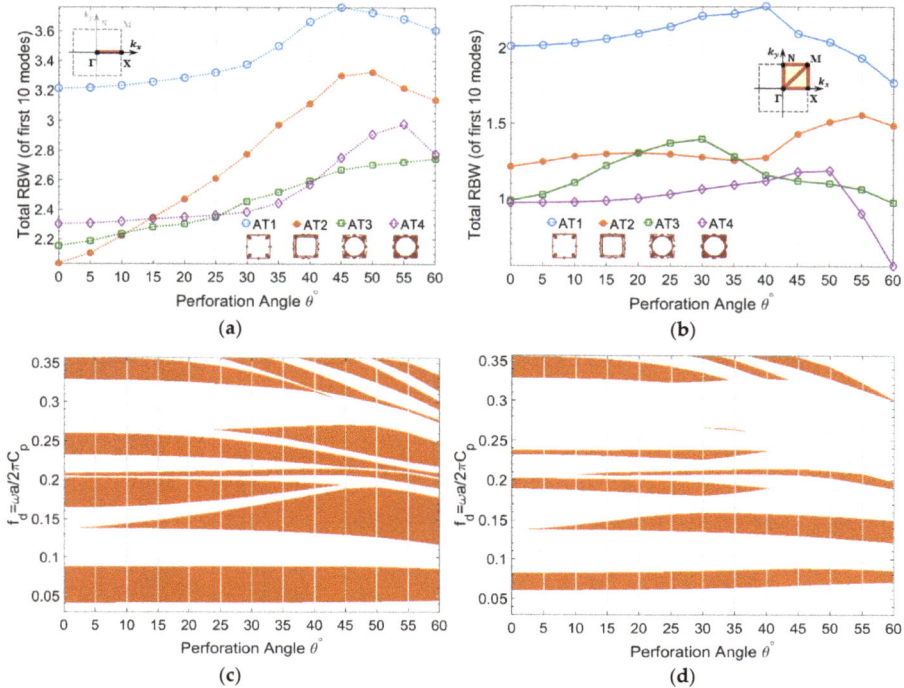

Figure 6. (**a**,**b**) Total RBW of asymmetric guided wave modes versus perforation angle calculated for optimized topologies AT1-AT4, and (**c**,**d**) corresponding bandgap frequency ranges for topology AT3, (**a**,**c**) partial bandgap in ΓX and (**b**,**d**) complete bandgap in ΓXMΓNM.

The gradient of frequency ranges corresponding to the partial and complete bandgaps of optimized topology AT3 with respect to the perforation angle is also demonstrated in Figure 6. From Figure 6c, it is evident that the lowest partial bandgap is almost insensitive to the perforation angle. A second partial bandgap emerges at around $f_d = 0.14$ below 5° and it keeps widening and inclining towards lower frequency ranges by the perforation angle, while the third bandgap above it keeps narrowing down by the perforation angle.

This trend of bandgap narrowing is generally associated with the emergence and/or widening of another bandgap between lower modes, and as such, leads to increasing RBW. The same trend is also observed for complete bandgaps of AT3, as demonstrated in Figure 6d. The lowest bandgap, which is the maximized fundamental bandgap of topology, slightly shifts towards higher frequencies by perforation angle and a second bandgap emerges at around $f_d = 0.14$ below 5° and widens and inclines towards lower frequencies.

Furthermore, the bandgap efficiency of S-OPhPs is validated by calculating the transmission spectrum of a finite phononic structure of topology AT3 for perforation angles 0° and 30° (Figure 7). The modal band structure of both perforation angles is calculated for an aluminum unit-cell of size $a = 10$ mm and thickness $h = 5$ mm, as demonstrated in Figure 7a,c. The symmetric model of a square phononic array of 7 × 7 unit-cells, as shown on top of Figure 7b is modeled and is subjected to an out-of-plane harmonic excitation $w_E = 1$ μm on the top surface at point E. This approach leads to dominant excitation of asymmetric wave modes. The transmission spectrum of the out-of-plane displacement from an arbitrary point A on the excitation side to an arbitrary point B on the other side of the phononic structure is then calculated as $20 \log \frac{w_B}{w_A}$ dB, as depicted in Figure 7b.

The dips of the transmission spectrum correspond to the phononic bandgap frequencies over which the amplitude of elastic waves is highly attenuated. The calculated modal band structures of the unit-cells are in excellent agreement with the transmission spectrum of this finite phononic structure, and thus confirm the enhancement of the complete RBW of S-OPhPs in a finite structure. The transmission spectrum of the lowest (i.e., optimized) bandgap of AT3 shows slightly higher attenuation at perforation angle 30°, which may be due to the contribution of symmetric wave modes in the transmission spectrum of perforation angle 0°. In contrast to the unchanged bandgap efficiency of the fundamental bandgap, multiple extra dips are present at higher frequency in the transmission spectrum of perforation angle 30° when compared to the transmission spectrum of perforation angle 0°.

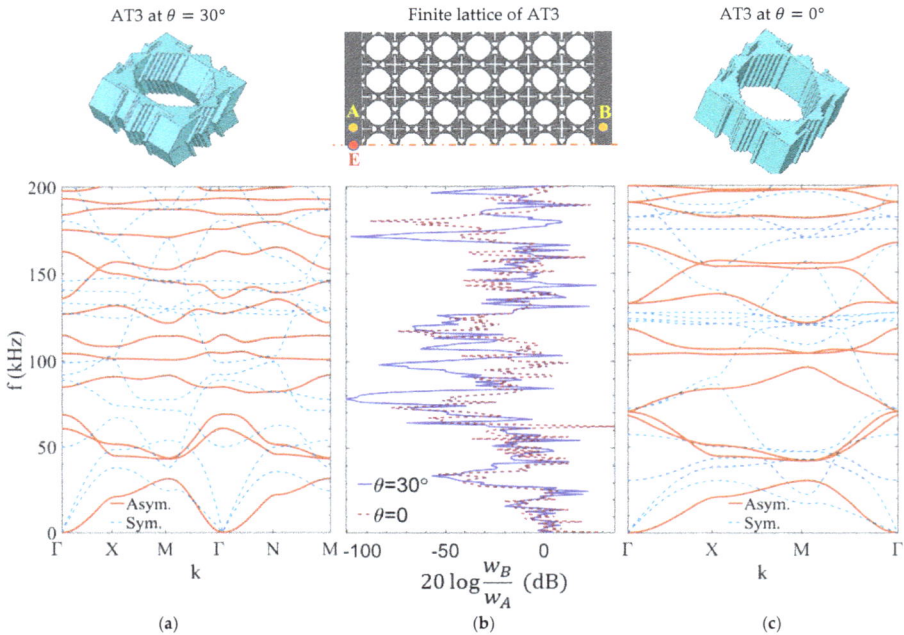

Figure 7. Modal band structure of S-OPhP of topology AT3 at perforation angles (**a**) 30° and (**c**) 0°, and (**b**) transmission spectrum in a finite phononic structure for both perforation angles.

5. Conclusions

Obliquely perforated phononic crystal plates were introduced in this paper with a symmetric (S-OPhP) and an asymmetric (A-OPhP) design. Modal band structure of various oblique perforation angles were calculated and the sensitivity of total RBW of the first 10 modal branches with respect to the perforation angle was investigated. Perforation profiles for arbitrary unit-cell topologies, as well as optimized unit-cell topologies (with maximized RBW of fundamental modes), were examined.

Partial (unidirectional) bandgaps along the plane of perforation angle as well as complete (omnidirectional) bandgaps were evaluated. It was observed that for an A-OPhP design, the partial RBW of mixed guided wave modes increases by perforation angle for all of the selected topologies, and it reaches a maximum at a perforation angle, which is specific to the topology. Moreover, by a S-OPhP design the partial bandgap of asymmetric wave modes increases for all of the selected topologies, and likewise reaches a maximum at a perforation angle specific to the topology. S-OPhP design also proved to enhance the complete bandgap of those topologies with maximized bandgap of asymmetric wave modes.

Materials **2018**, *11*, 1309

A general trend was observed that narrowing of a bandgap by perforation angle is normally associated with emergence and/or widening of another bandgap between lower modes, which leads to an increasing RBW by perforation angle.

It is concluded that introducing a constant perforation angle throughout the phononic lattice can significantly enhance its unidirectional and/or omnidirectional bandgap efficiency, depending on the topology of perforation profile. This fact inspires the idea of simultaneous optimization of perforation profile and perforation angle. Varying the azimuth angle of perforation over the perimeter of the perforation profile introduces conical like cavities, which may lead to supreme omnidirectional bandgap efficiency.

Author Contributions: S.H. and M.K. proposed the research, developed and performed the analysis and wrote the paper, K.A. and W.V.P. supervised and financed the research.

Funding: This research received no external funding.

Conflicts of Interest: The authors declare no conflict of interest.

References

1. Deymier, P. *Acoustic Metamaterials and Phononic Crystals*; Springer: Berlin, Germany, 2011.
2. Wu, T.T.; Chen, Y.T.; Sun, J.H.; Lin, S.C.S.; Huang, T.J. Focusing of the lowest antisymmetric Lamb wave in a gradient-index phononic crystal plate. *Appl. Phys. Lett.* **2011**, *98*, 171911. [CrossRef]
3. Olsson Iii, R.H.; El-Kady, I.F. Microfabricated phononic crystal devices and applications. *Meas. Sci. Technol.* **2009**, *20*, 012002. [CrossRef]
4. Celli, P.; Gonella, S. Low-frequency spatial wave manipulation via phononic crystals with relaxed cell symmetry. *J. Appl. Phys.* **2014**, *115*, 103502. [CrossRef]
5. Hedayatrasa, S.; Kersemans, M.; Abhary, K.; Uddin, M.; Van Paepegem, W. Optimization and experimental validation of stiff porous phononic plates for widest complete bandgap of mixed fundamental guided wave modes. *Mech. Syst. Signal Process.* **2018**, *98*, 786–801. [CrossRef]
6. Hussein, M.I.; Leamy, M.J.; Ruzzene, M. Dynamics of phononic materials and structures: Historical origins, recent progress, and future outlook. *Appl. Mech. Rev.* **2014**, *66*, 040802. [CrossRef]
7. Khelif, A.; Adibi, A. *Phononic Crystals: Fundamentals and Applications*; Springer: Berlin, Germany, 2015.
8. Craster, R.V.; Guenneau, S. *Acoustic Metamaterials: Negative Refraction, Imaging, Lensing and Cloaking*; Springer Science & Business Media: Berlin, Germany, 2012.
9. Liu, J.; Li, F.; Wu, Y. The slow zero order antisymmetric Lamb mode in phononic crystal plates. *Ultrasonics* **2013**, *53*, 849–852. [CrossRef] [PubMed]
10. Hedayatrasa, S. *Design Optimisation and Validation of Phononic Crystal Plates for Manipulation of Elastodynamic Guided Waves*; Springer: Berlin, Germany, 2018.
11. Shelke, A.; Banerjee, S.; Habib, A.; Rahani, E.K.; Ahmed, R.; Kundu, T. Wave guiding and wave modulation using phononic crystal defects. *J. Intell. Mater. Syst. Struct.* **2014**, *25*, 1541–1552. [CrossRef]
12. Zhu, H.; Semperlotti, F. A passively tunable acoustic metamaterial lens for damage detection applications. In *Sensors and Smart Structures Technologies for Civil, Mechanical, and Aerospace Systems*; International Society for Optics and Photonics: Bellingham, WA, USA, 2014; Volume 9061, p. 906107.
13. Semperlotti, F.; Zhu, H. Achieving selective interrogation and sub-wavelength resolution in thin plates with embedded metamaterial acoustic lenses. *J. Appl. Phys.* **2014**, *116*, 054906. [CrossRef]
14. Lin, S.-C.S. Acoustic metamaterials: Tunable gradient-index phononic crystals for acoustic wave manipulation. In *Engineering Science and Mechanics*; The Pennsylvania State University: State College, PA, USA, 2012.
15. Wu, T.-T.; Hsu, J.-C.; Sun, J.-H. Phononic plate waves. *IEEE Trans. Ultrason. Ferroelectr. Freq. Control* **2011**, *58*, 2146–2161. [PubMed]
16. Pennec, Y.; Vasseur, J.O.; Djafari-Rouhani, B.; Dobrzyński, L.; Deymier, P.A. Two-dimensional phononic crystals: Examples and applications. *Surf. Sci. Rep.* **2010**, *65*, 229–291. [CrossRef]
17. Andreassen, E.; Manktelow, K.; Ruzzene, M. Directional bending wave propagation in periodically perforated plates. *J. Sound Vib.* **2015**, *335*, 187–203. [CrossRef]

18. Miniaci, M.; Gliozzi, A.S.; Morvan, B.; Krushynska, A.; Bosia, F.; Scalerandi, M.; Pugno, N.M. Proof of concept for an ultrasensitive technique to detect and localize sources of elastic nonlinearity using phononic crystals. *Phys. Rev. Lett.* **2017**, *118*, 214301. [CrossRef] [PubMed]
19. Hedayatrasa, S.; Abhary, K.; Uddin, M. Numerical study and topology optimization of 1D periodic bimaterial phononic crystal plates for bandgaps of low order Lamb waves. *Ultrasonics* **2015**, *57*, 104–124. [CrossRef] [PubMed]
20. Hung, C.H.; Wang, W.S.; Lin, Y.C.; Liu, T.W.; Sun, J.H.; Chen, Y.Y.; Esashi, M.; Wu, T.T. Design and fabrication of an AT-cut quartz phononic Lamb wave resonator. *J. Micromech. Microeng.* **2013**, *23*, 065025. [CrossRef]
21. Foehr, A.; Bilal, O.R.; Huber, S.D.; Daraio, C. Spiral-based phononic plates: From wave beaming to topological insulators. *Phys. Rev. Lett.* **2018**, *120*, 205501. [CrossRef] [PubMed]
22. Hedayatrasa, S.; Kersemans, M.; Abhary, K.; Uddin, M.; Guest, J.K.; Van Paepegem, W. Maximizing bandgap width and in-plane stiffness of porous phononic plates for tailoring flexural guided waves: Topology optimization and experimental validation. *Mech. Mater.* **2017**, *105*, 188–203. [CrossRef]
23. Miniaci, M.; Pal, R.K.; Morvan, B.; Ruzzene, M. Observation of topologically protected helical edge modes in Kagome elastic plates. *arXiv* **2017**.
24. Miniaci, M.; Mazzotti, M.; Radzieński, M.; Kherraz, N.; Kudela, P.; Ostachowicz, W.; Morvan, B.; Bosia, F.; Pugno, N.M. Experimental observation of a large low-frequency band gap in a polymer waveguide. *Front. Mater.* **2018**, *5*, 8. [CrossRef]
25. Pourabolghasem, R.; Dehghannasiri, R.; Eftekhar, A.A.; Adibi, A. Waveguiding Effect in the Gigahertz Frequency Range in Pillar-based Phononic-Crystal Slabs. *Phys. Rev. Appl.* **2018**, *9*, 014013. [CrossRef]
26. Bilal, O.R.; Hussein, M.I. Trampoline metamaterial: Local resonance enhancement by springboards. *Appl. Phys. Lett.* **2013**, *103*, 111901. [CrossRef]
27. Bilal, O.R.; MHussein, I. Topologically evolved phononic material: Breaking the world record in band gap size. In *Photonic and Phononic Properties of Engineered Nanostructures*; International Society for Optics and Photonics: Bellingham, WA, USA, 2012; p. 826911.
28. Halkjær, S.; Sigmund, O.; Jensen, J.S. Maximizing band gaps in plate structures. *Struct. Multidiscip. Optim.* **2006**, *32*, 263–275. [CrossRef]
29. Hedayatrasa, S.; Abhary, K.; Uddin, M.; Ng, C.T. Optimum design of phononic crystal perforated plate structures for widest bandgap of fundamental guided wave modes and maximized inplane stiffness. *Mech. Phys. Solids* **2016**, *89*, 31–58. [CrossRef]
30. Hedayatrasa, S.; Abhary, K.; Uddin, M.S.; Guest, J.K. Optimal design of tunable phononic bandgap plates under equibiaxial stretch. *Smart Mater. Struct.* **2016**, *25*, 055025. [CrossRef]
31. Zhu, H.; Semperlotti, F. Phononic thin plates with embedded acoustic black holes. *Phys. Rev. B* **2015**, *91*, 104304. [CrossRef]
32. Zhu, H.; Semperlotti, F. Anomalous refraction of acoustic guided waves in solids with geometrically tapered metasurfaces. *Phys. Rev. Lett.* **2016**, *117*, 034302. [CrossRef] [PubMed]
33. Aberg, M.; Gudmundson, P. The usage of standard finite element codes for computation of dispersion relations in materials with periodic microstructure. *J. Acoust. Soc. Am.* **1997**, *102*, 2007–2013. [CrossRef]
34. Sigmund, O.; Jensen, J.S. Systematic design of phononic band-gap materials and structures by topology optimization. *Philos. Trans. R. Soc. Lond. A* **2003**, *361*, 1001–1019. [CrossRef] [PubMed]

MDPI

St. Alban-Anlage 66

4052 Basel

Switzerland

Tel. +41 61 683 77 34

Fax +41 61 302 89 18

www.mdpi.com

Materials Editorial Office

E-mail: materials@mdpi.com

www.mdpi.com/journal/materials